Plant and Nanoparticles

Jen-Tsung Chen
Editor

Plant and Nanoparticles

Springer

Editor
Jen-Tsung Chen
Department of Life Sciences
National University of Kaohsiung
Kaohsiung, Taiwan

ISBN 978-981-19-2502-3 ISBN 978-981-19-2503-0 (eBook)
https://doi.org/10.1007/978-981-19-2503-0

This Springer imprint is published by the registered company Springer Nature Singapore Pte Ltd.
The registered company address is: 152 Beach Road, #21-01/04 Gateway East, Singapore 189721, Singapore

Contents

Emerging Trends of Nanoparticles in Sustainable Agriculture: Current and Future Perspectives 1
Kanika Khanna, Nandni Sharma, Puja Ohri, and Renu Bhardwaj

Nanoparticles in Plant Disease Management 53
Masudulla Khan, Azhar U. Khan, and Aiman Parveen

Proteomics of Plant-Nanoparticle Interaction Mechanism 67
Ghazala Mustafa and Setsuko Komatsu

Importance of the Secondary Metabolites and Biological Parameter Modification by Metallic, Oxide, and Carbon-Based Nanomaterials Over Forage Plants 85
Luis Páramo, Ana A. Feregrino-Pérez, Humberto Aguirre Becerra, Ramón G. Guevara-González, and Karen Esquivel

Polymer-Based Nanoparticles (NPs): A Promising Approach for Crop Productivity 119
Fatima El Amerany, Fatima Zahra Aboudamia, Iman Janah, Moha Taourirte, and Mohammed Rhazi

Plant-Mediated Eco-Friendly Synthesis of Platinum Nanoparticles and Their Applications 155
Canh Minh Vu, Suresh Ghotekar, Thanh-Dong Pham, Nguyen Minh Viet, Rajeshwari Oza, Muhammad Bilal, and Arpita Roy

Foliar Application of Metallic Nanoparticles on Crops Under Field Conditions 171
Martin Šebesta, Ľuba Ďurišová, Dávid Ernst, Samuel Kšiňan, Ramakanth Illa, B. Ratna Sunil, Avinash P. Ingle, Yu Qian, Martin Urík, and Marek Kolenčík

Phytotoxic Effects of Nanoparticles and Defense Mechanisms in Plants 217
Muhammad Adil, Amar Nasir, Noor Muhammad Khan, and Arbab Sikandar

Plant Molecular Responses to Nanoparticle Stress 239
Ilham Khan, Murtaza Hasan, Rehana Kausar, Junaid Shehzad, and Ghazala Mustafa

Nanoelicitation: A Promising and Emerging Technology for Triggering the Sustainable In Vitro Production of Secondary Metabolites in Medicinal Plants 265
Rabia Javed, Buhara Yucesan, Muhammad Zia, and Ekrem Gurel

Nanomaterials as Unique Carriers in Agricultural Practices for Plant Growth and Development: A State of Current Knowledge 281
Amir Khan and Faheem Ahmad

Nanotechnologies and Sustainable Agriculture for Food and Nutraceutical Production: An Update 315
Pragya Tiwari

Green Synthesis of Plant-Assisted Manganese-Based Nanoparticles and Their Various Applications 339
Canh Minh Vu, Suresh Ghotekar, Nguyen Minh Viet, Harshal Dabhane, Rajeshwari Oza, and Arpita Roy

Biogenic Synthesis of Lead-Based Nanoparticles and Their Recent Applications 355
Khanderao Pagar, Suresh Ghotekar, Onkar Pardeshi, Shreyas Pansambal, Sachin Pawar, Jigna Machhi, and Balasaheb Pagar

Nanofertilizers and Nanopesticides for Crop Growth 367
Nam Nghiep Tran, Tu Nguyen Quang Le, Hue Quoc Pho, Tung Thanh Tran, and Volker Hessel

The Janus Face of Nanomaterials: Physiological Responses as Inducers of Stress or Promoters of Plant Growth? 395
Harleen Kaur, Jashanpreet Kaur, Anu Kalia, and Kamil Kuca

About the Editor

Jen-Tsung Chen is currently a professor at the National University of Kaohsiung in Taiwan. He teaches cell biology, genomics, proteomics, medicinal plant biotechnology, and plant tissue culture. His research interests include bioactive compounds, chromatography techniques, in vitro culture, medicinal plants, phytochemicals, plant physiology, and plant biotechnology. He has published over 90 scientific papers and serves as an editorial board member for Plant Methods and Plant Nano Biology.

Emerging Trends of Nanoparticles in Sustainable Agriculture: Current and Future Perspectives

Kanika Khanna, Nandni Sharma, Puja Ohri, and Renu Bhardwaj

Abstract Nanotechnology in agriculture is emerging at an escalating rate, owing to its excellent properties in plant growth and development. In the present era where climate change is most commonly observed, the global patterns of agriculture are observing unprecedented challenges. For achieving food quality and yields, nano-engineering is a novel tool that maintains sustainable crop production. Due to the disastrous effects of chemical fertilizers, there is a need to switch to safer alternatives. Nano-technology enables safe crop production by improving efficiency and reducing losses. This technology has been predominantly entered into wider areas of fertilizers and pesticides for synthesizing agrochemicals based-nanoparticles. Because of their critical and direct/indirect approach in management and regulatory inputs (herbicides, pesticides, fungicides, etc.), nanotools like nano-biosensors also support avant-garde agriculture farms. Therefore, plant biology and nanotechnology together have a great impact on the environment due to their innovative characteristics in agriculture, to meet the urgent needs of food with environmental sustainability. In this chapter, we have mainly focused on nanoparticle interactions among plants, their uptake, mobilization, and metabolic actions. Moreover, the bioactive compounds in plants possess many functions that are also modulated by nanoparticles. Therefore, nanoparticles function as elicitors in the plant's secondary metabolism. We have envisaged the multidisciplinary actions of nanoparticles with plant nanotechnology, biotechnology, genetic engineering and pushed it towards agriculture sector, as well as plant research. In particular, we have depicted the role of nanoparticles in enhancing bioactive compounds of plants, thereby improving crop productivity through boosting the nutraceutical and nutrients of plants. Here, we have also reviewed the nanoparticle abilities toward plant protection and stress management against numerous adverse conditions. This chapter will enable the researchers to understand the nanotechnology blend in agriculture, thereby

---

K. Khanna (✉) · R. Bhardwaj
Plant Stress Physiology Lab, Department of Botanical and Environmental Sciences, Guru Nanak Dev University, Amritsar, India

N. Sharma · P. Ohri
Department of Zoology, Guru Nanak Dev University, Amritsar, India

J.-T. Chen (ed.), *Plant and Nanoparticles*,
https://doi.org/10.1007/978-981-19-2503-0_1

designing the specific nanoparticles according to the agriculture needs as well as standards for promoting sustainable agriculture.

1 Introduction

Agriculture is the main source of livelihood in various developing countries, and nearly two-thirds of the world's total populace is dependent on it. The enormous agricultural demand leads to huge pressure for formulating novel agricultural techniques that enhance the yield and productivity of the crops by lessening the impact of chemical fertilizers on soils. The fertilizers are imperative for enhancing the agricultural yields, though their excessive levels alter the soil ecology and area for crop production. For meeting the rising demands of food, agriculturalists tend to use chemical-based fertilizers, and it has become an integral factor in farming systems. These are mainly classified into three types such as nitrogenous, phosphate, and potassium fertilizers. But certain complexities and other climatic or edaphic factors cause the mineral elements to immobilize within the soil, thereby hindering its availability and uptake by plants [1]. Consequently, their utilization creates significant drift on human as well as environmental health [2]. The indiscriminate usage and improper maintenance have made our focus oriented toward biofertilizers from chemical-based fertilizers. However, the use of farmyard manure, animal waste, compost, etc. has been promoted that stimulates the production in appropriate proportion. Sustainable agriculture necessitates the minimal usage of agrochemicals to protect and conserve them for posterity. To tackle the challenges for sustainable agriculture and food demand in agriculture, various technological advancements and novelties in the past few years should be explored [3]. The most important concern here is to make more efficient use of these fertilizers by substituting nanoparticles (NPs). Nanotechnology has the potential for providing an effective strategy for agricultural problems. It is a great solution to bridge the gap between bulk materials and NPs. Decades ago, noteworthy research on nanotechnology has been carried out with a special emphasis on agriculture [4]. NPs have remarkable physicochemical properties with minute size, larger surface area to mass ratio, reactivity, ionizing power, chemical stability, enhanced absorbability, pH resistance, and thermal stability. Owing to these abundant characteristics, nanomaterials combined with fertilizers enhance the potential growth and development of the plants.

Nanofertilizer using efficiency is escalated by penetrating the NPs into agrochemicals *via* different modes or methods such as encapsulation, ionic charges, hydrogen bonding, absorption, entrapment into nanomatrix, etc. [5]. Sidewise, there are numerous sectors for globally commercializing the NPs fertilizers. For example, other factors like NP absorption capacity, behavior in soil, root/shoot uptake, chemical stability, xylem/phloem loading and unloading, and interaction with other biomolecules such as protein, RNA, and DNA, respectively, also play a critical role [1]. Interestingly, the NPs improve the plant productivities by enhancing the agricultural inputs for facilitation of site-targeted regulated delivery of nutrients, to ensure the minimal usage of agri-inputs. Indeed, nanotechnology has reformed as an avant-garde technique in plant protection, and its popularity is increasing on an exponential phase with a surety for higher production levels. Besides, the important aspect in agriculture practices is to induce plant adaptation toward changing climatic conditions such as temperature, water deprivation, salinity, freezing, water logging, and heavy metal pollution without hampering the ecosystem [6, 7]. Furthermore, the formulation of nanosensors in farming for proper monitoring of crops, soil activities, pathogens, invasion of agrochemicals, and environmental pollutants ensures sustainability in terms of soil and plant health. Subsequently, it maintains quality and safety assurance for successfully mediating sustainability and environment systems [8]. Nano-engineering is a cutting-edge technology that enables the formation of high-technological aids and offers broad areas of research for agricultural systems. Henceforth, nanotechnology coordinates with sustainable agriculture practices with innovations and fixing various problems and concerns regarding modern agriculture. The present chapter summarizes the types, applications, and role of nanotechnology in agriculture for maintaining sustainability and standards.

2 Sources of Nanoparticles

2.1 Natural Sources of NPs

NPs occur naturally in all "spheres" of our planet, covering the biosphere, atmosphere, lithosphere, and hydrosphere. These particles are formed by various photochemical, chemical, thermal, mechanical, and biological processes occurring either separately or in amalgamation [9]. The natural occurrence of metallic NPs and their sulfides/oxides in hydrothermal vents, ore deposits, waters, wastewaters, and mining regions is largely controlled by the environmental conditions, such as the temperature, pH, light, oxic/anoxic conditions, and the characteristics and concentration of the naturally occurring organic matter [10]. The NPs present in the hydrosphere and the atmosphere occur at concentrations of up to 106–107 particles/mL and impose a major effect on biota because of their close association and contact with the biota.

The major processes resulting in the formation of natural NPs are entirely inorganic including nucleation, mechanical, thermal, and biological processes. Nucleation and inorganic phases in the hydrosphere, atmosphere, and the lithosphere occur purely based on inorganic reactions or may also originate from organic matter. Reactions occurring in hydrothermal vents and surface water often contribute to NP release into the environment and may advance *via* photochemical, thermal, and nonthermal processes. The presence of Fe(II) facilitates the development of ferrihydrite NPs, stabilized by silicon ions. Similarly, different NPs containing Cu, Mn, Ba, Cr, and Pb are also formed in chilly CO_2 seeps. The mechanical processes involved in NP formation include aeolian erosion resulting from desert winds, unvegetated farmlands, deforested lands, and the particles emanating during events like earthquakes. Biomass combustion such as the forest fires, mainly occurring in the equatorial regions of the earth, exemplifies the involvement of thermal processes in the generation of NPs [11]. And lastly, biomineralization, carried out by microbes, generates inorganic nanomaterials through various biological processes. Such inorganic particles include silicon and iron-based nanominerals, magnetite, calcium phosphate, and calcium carbonate [12].

2.1.1 Volcanic Eruptions and Forest Fires

The ash emitted by volcanic eruptions frequently reaches above 1400 °C and possesses a complex composition of liquid and solid particulate matter lifted in the atmosphere by hot gas currents. After ashes diffuse in the atmosphere, the temperature of the gas lowers down, and its composition changes, resulting in the accumulation of particles [13]. Volcanic eruptions result in the release of a gigantic amount of fine particles and aerosols with sizes of up to micrometers to nanometers, into the atmosphere. Ash clouds formed from volcanic eruptions contain an array of polydispersed nano- and microparticles with sizes ranging between 100 and 200 nm. About 30 million tons of NPs are emitted in the form of ash in a single volcanic eruption. Chemically these particles are composed of iron and silicate compounds which are easily suspended in the air. Volcanic eruptions also release bismuth oxide NPs in the stratosphere. These particles remain suspended in the air and may result in respiratory disorders once inhaled [14]. Volcanic ash on reaching the upper troposphere and stratosphere can readily spread worldwide affecting different regions of the globe for years. In the upper atmosphere, the nanoparticulate debris of the ash blocks and scatters sun radiation. While some effects of volcanic ash are seen throughout the world, the maximum levels of nanomaterial released are present in areas within 10 km of the volcano [15].

Lightning and anthropogenic activities are largely responsible for grass and forest across the globe. The smoke and ash released by these fires can extend over larger areas and increase the number of nanosized particulate matter, thus distressing the ambient air quality. In the inorganic sphere, fire is one such chemical process that eventually results in the formation of nanoscopic particles [14]. Numerous fires are known to occur throughout the world, in North America, Europe, Asia, Africa, Brazil, and Australia [15]. The fire smoke carries soot and black carbon in huge quantities as Asian brown clouds have been deposited over the Himalayan glaciers. The deposited particles have resulted in amplified absorption of heat from the sun, thus accelerating glacial melting. Many reported cases of forest fire reveal transportation of micro- and nanosized particulate matter through ash and smoke and are held responsible for respiratory ailments in animals and humans. Smoke containing nanosized particles can worsen cardiopulmonary problems in patients [16].

2.1.2 Water Bodies

Oceans and seas, covering the largest area on the surface of the earth, emit sea salt aerosols in huge amounts. These aerosols are created by evaporation of water and by ejection of waterdrops into the atmosphere by waves. The size of these particles ranges from 100 nm to up to several microns. Water bodies can also form NPs through precipitation resulting from evaporation and temperature changes in patients [16]. The water of Lake Michigan contains large amounts of calcium carbonate which remains dissolved in cold water, for most of the year, but when summer is about to end, the temperature of the water increases considerably, thus reducing calcium carbonate solubility in the lake water. This results in the precipitation of calcium carbonate out from the water, generating clouds of nanosized particles [14]. Inorganic sulfide, i.e., HS^- and H_2S, is an essential component of the biogeochemical sulfur cycle occurring in mining water, hydrothermal vents, and sediments under anaerobic conditions. Sulfur and metals present in the ocean are often emitted from hydrothermal vents, and they can readily react with one another, serving as a source of metal-containing sulfide NPs [11].

Also, the drinking water from underground sources and freshwater streams contains nanoscopic and microscopic materials, chemically originating from $CaSO_4$ and $CaCO_3$ along with elements like iron oxides. Nonetheless, physical and chemical processes, like weathering, precipitation, and dissolution of carbonates influenced by CO_2, intermediate hydrocarbonate (HCO^{3-}) formation, and the dawdling iron oxide precipitation, can readily generate nanosized particles [14, 15].

2.1.3 Dust Storms and Cosmic Dust

Deserts on the earth's surface are the largest permanent major source of NPs, and the dust storms originating from them are responsible for the long-distance migration of not only mineral dust but also anthropogenic pollutants, which are lifted in the atmosphere by the air currents [13]. Nearly 50% of the atmospheric aerosols in the troposphere are originated from deserts. The particle size produced by dust storms varies between 100 nm and several microns. Satellite imagery reveals the dynamics of dust migration on large scale across continents, demonstrating that the NPs generated in one region of the globe by important environmental events affect regions that are present thousand kilometers away. For instance, the dust storms stirring up during spring in the Gobi desert every year strongly influence the air quality in North America and Asia [15]. However, the composition of the dust varies with a specific place and the human actions in the adjoining areas from where the wind currents generally cross.

Throughout the universe, different kinds of nanomaterials are present which are sorted, mixed, and tailored into various forms. Electromagnetic radiations, dramatic temperatures, pressure gradients, shock waves, and physical collisions help in stimulating and generating NPs in outer space [16]. Astronomical observations along with direct analysis of stardust during space expeditions and meteorite collections affirmed the presence of a vast array of carbon, nitride, oxide, silicate, carbide, and organic nanomaterials as major stardust components [17]. The presence of nanosized diamond in Murchison meteorite exemplifies the origin of nanoparticulate matter in planetary structures other than the stars. The lunar dust when compared to terrestrial is fine grain and comprises substantial amounts of magnetic NPs that often cling to astronauts' suits which have electrostatically charged surfaces [15].

2.1.4 Biogenic Production

Nature provides an insight into the synthesis of nanomaterials. Biological systems act as "bio-laboratory" or "bio-factory" for the fabrication of metal oxide particles and pure metals at nanometer scale employing a biomimetic approach [10]. Many uni- and multicellular can generate nanoparticulate inorganic matter through intra- and extracellular processes. In microbial environments, NPs are constantly being formed indirectly *via* redox reactions related to metabolic processes [11]. On exposure to inorganic salts such as those containing Ag^+, Au^{3+}, S^{2-}, and $SeO_3{}^{2-}$, certain bacteria like *Pseudomonas aeruginosa*, *Stenotrophomonas*, *Serratia*, and *Thiobacillus* species employ an oxidizing or reductive detoxification pathway which forms nanosized elemental particles [14]. Nanobacterium synthesizes a calcium phosphate shell around itself, resembling an inorganic particle. The size of the shell ranges between 20 and 300 nm [15]. Fungi also contain enzymes that can synthesize NPs of different shapes and sizes. They have been known for generating various NPs such as silver, gold, and even alloys [16].

However various organisms can also naturally produce NPs. Plants utilize the macro- and micronutrients present in soil/water for their development and growth leading to the amassment of these minerals in nano-forms. Small insects and animals use nanostructures for protection from predators. In some insects, the lightweight wings are protected by nanowax coatings. Even humans possess organs, e.g., bones, primarily constructed by nanosized structures. Enzymes, antibodies, proteins, and DNA are also composed of nanostructures [16]. It is apparent from the aforesaid that NPs are generated in the form of aerosols, colloids, and dust, as constituents of sediments and soils, chemical or hydrothermal deposits, mineral nuclei, and lamellae. In most cases, NP formation occurs through a combination of different processes, e.g., weathering is a mechanical process that combines with precipitation or dissolution, and colloid formation in volcanic activity and rivers results from explosive eruptions expelling tephra along with rapid cooling of fumes. Based on chemical composition, naturally occurring NPs include an array of elements such as metal hydroxides/oxides, metal alloys, silicates sulfides, halides, carbonates, etc. [18].

2.2 *Anthropogenic Sources*

Anthropogenic NPs are created by humans and often lead to incidental exposure. Man-made NPs are intentionally or unintentionally released into the environment due to mechanical and industrial practices. These anthropogenic NPs are heterogeneous in nature and fall under two categories. The first category of NPs displays undefined chemistry and does not have a predetermined size. This includes combustion particulates, welding fumes, coal fly ash, and diesel exhaust. The second category includes the engineered NPs which exhibit a specific size, ranging between 1 and 100 nm, and are pure particles displaying controlled surfaces. These include carbon nanotubes, fullerenes, dendrimers, quantum dots, silver and gold NPs, etc. [9, 19]. The man-made particles are either produced by stationary or mobile sources, and their generation may or may not be deliberate.

Accurate estimation of the annual discharge of NPs in the environment is impossible. However, there is a strong increase in their production volumes every year. The exhaust gases produced from diesel engines contain huge quantities of NPs resulting from incomplete fuel combustion. Similarly industrial processes, liquid or solid waste from manufacturing units, gas boilers, oil, and coal account for the unintentional release of tons of NPs. The advancement of nanotechnology and industrial processes has led to the manufacturing of NPs on large scale, and their use in related products results in the unavoidable release of these engineered NPs into the air, soil, and water, both knowingly or unknowingly [9].

2.2.1 Mobile Sources

Due to urbanization and industrialization, automobile exhaust is the major source of atmospheric NPs. Among different kinds of automobile exhaust, diesel engines release particles in the environment with sizes range between 20 and 130 nm, whereas gasoline engines release particles of 20–60 nm size. Besides, carbon nanotubes and fibers are also liberated during the diesel combustion process as by-products. Over 90% of the carbon NPs that occur in the atmosphere are an outcome of diesel fuel combustion [16]. These particles are hydrocarbons or sulfates generated by nucleation reactions. The accumulation mode NPs from diesel exhaust are mainly sooty carbonaceous aggregates. NPs from diesel exhaust are primarily composed of lubricating oil and unburned fuel. Also, sulfuric acid which accounts for a small percentage plays a crucial role as nuclei providing for condensation of the organic particles. The nanosized diesel exhaust comprises about 95% unburned lubricating oil. In diesel engines, the lower emission standards often decrease particle mass emissions, but the concentration of NPs tends to increase due to the reduced availability of surfaces providing for condensation. Both leaded and unleaded gasoline engines emit particles with an average diameter of about 45 nm. Liquid petroleum gas is considered a cleaner fuel. However, in comparison to unleaded gasoline, it leads to more emission of NPs [20]. Thus, the pollution caused by vehicles is the main root of nanoparticulate contamination in the urban atmosphere.

2.2.2 Industrial and Stationary Sources

A spectrum of industrial settings has an immense potential to create ultrafine particles, yet to be investigated, including milling, printing, plasma cutting, powder coating, cooking, and baking. The ultrafine particles released during industrial processes include fumes produced by hot processes like welding and smelting, fumes from incomplete fuel combustion processes, and bioaerosols. Industrial setups mainly provide favorable conditions such as high temperature, the occurrence of vaporizable material, and large temperature gradient along with rapid cooling processes [20]. Ultrafine-sized particles from gas, coal, and oil-fired stationary in the size range of 15–25, 40–50, and 70–100 nm are generated on combustion of natural gas, sulfur-containing bituminous coal, and No. 6 fuel oil, respectively. The composition of used fuel largely determines the production of the end product. Industrial effluents contain carbon nanotubes, silver-, gold-, zinc oxide-, and titanium oxide-based nanoparticles normally, which are released among the wastewater effluents as well as through the chimneys. Once released into the environment, these NPs are deposited in sediments, landfills, soil, and water bodies [21].

2.2.3 Engineered NPs

Currently, many consumer products contain NP ingredients, such as paints and sunscreens (nanostructured zinc oxide and titanium dioxide), tire additives (carbon fibers, fumed silica and nanotubes), polishing slurries and lubricants (nanosized silica), and detergents and shampoos (nanosized alumina). With the advent of nanotechnology, the usage of NPs is considerably increasing [20]. These particles are often synthesized employing chemical, physical, biological, and hybrid methods. Nanomaterials such as titanium oxide NPs, carbon NPs, and hydroxyapatites occur in sporting goods, cosmetics, toothpaste, and sunscreens [16]. Through several pathways, these NPs present in consumer products enter the environmental surroundings either mechanically from the product or upon their disposal [20]. Bundschuh et al. [21] proposed that engineered NPs gain entry into our surroundings through three emission scenarios, i.e., (1) discharged during fabrication of nano-enabled products and raw material, (2) release during usage, and (3) release during waste handling of products containing NPs. These emitted nanosized particles ultimately deposit in surface water bodies and on land [18]. ZnO NP, primarily used in electronics, medicine, and cosmetics, amass in sediments, urban and natural soil, and landfills. Similarly, 90% of carbon nanotube production accumulates in landfills, soils, sediments, and air [20]. However, at times, certain kinds of NPs are intentionally released into the environment for their application. Examples include nanosized metal particles for dechlorination of groundwater, nanoenergetic explosives/materials and highly obscured nanoflakes, and application of nanopesticides and nanonutrients in agricultural fields [22].

2.2.4 Miscellaneous Sources

Besides the abovementioned sources, several human activities are responsible for the formation of NPs. Cigarette smoking releases a complex of about 100,000 compounds in the atmospheres that are nanosized, ranging between 10 and 700 nm [15]. Cooking practices like frying contribute to an indoor source of ultrafine particles. It has been reported by Wallace et al. [23] that >90% of the particulate matter produced during cooking was contributed to the ultrafine fraction exhibiting a primary and secondary peak at 60 nm and <10 nm, respectively. Heating elements of an electric stove or even a gas burner result in ultrafine particle generation at concentration levels of $1.1 \times 10^5\ cm^{-3}$. Candles, incense, mosquito coils, residential wood, coal combustion, and tobacco smoke also add to indoor NP levels [20]. Micro- and nanoparticulates of size smaller than 10 μm are often released in the environment upon demolition of large buildings. Along with the building debris, respirable asbestos fibers, glass, lead, and various toxic particles belonging to the nanometric scale are released from household materials at the demolition site [16]. NPs are employed in the fabrication of an array of products that are used on regular basis. Their production, usage, and disposal lead to continuous emission of these particles into the atmosphere.

3 Types of Nanoparticles

The NPs synthesized by a biological source such as microbes prove to be highly effective in terms of their applicability [24]. The NPs are characterized and designed in various shapes and sizes in contrast to other sources. Owing to its eco-friendly nature, low cost, and high effectiveness, the microbial sources are considered to be the best resource for manufacturing metal NPs. Various types of NPs have been discussed below.

3.1 Silver NPs

Silver NPs are frequently applied in the biomedical field (drugs/coatings, bandages, and implants), agriculture, clothing, water purification, and as antibiotics. They possess antimicrobial potential with high efficiency. Musarrat et al. [25] reported AgNPs synthesis of around 5–27 nm by *Amylomyces rouxii* strain KSU-09, screened from *Phoenix dactylifera*. They confirmed that water extracts of 72-h suspended mycelia facilitated the generation of stable, monodispersed, spherical NPs from

1 mM silver nitrate. The results were confirmed by UV-visible spectroscopy, AFM, XRD, and TEM. The fungal proteins present in the extracts revealed by infrared spectrum attributed toward NP stability. The bioreduction process of NP synthesis involves enzymes that are known to transfer electrons from donors to positively charged metal ions, thus stabilizing the end product. Certain *Bacillus* sp. synthesizing AgNPs enzymatically control the size of the NPs. α-amylase enzymes generated by these bacteria are held responsible for the same [26].

Synthesis of AgNPs extracellularly by microbes often involves a mechanism mediated by nitrate reductase [27]. Fungi like *Fusarium oxysporum*, *Verticillium*, and *Aspergillus flavus* have been employed for the synthesis of AgNPs in solutions or surfaces of fungal cells and even in the form of a film [28]. Moreover, Mishra et al. [29] reported that *Stenotrophomonas* sp. BHU-S7 could extracellularly biosynthesize spherically shaped AgNPs with an approximate diameter of 12 nm. They suggested that extracellular enzyme nitrate reductase, which appeared in the cell supernatant, facilitated the transfer of electrons to $Ag+$ ions leading to the formation of AgNPs. Besides, they also proposed the role of carbonyl groups of certain enzymes and proteins in stabilizing the NPs by attaching them to the NP surfaces.

Cell-free culture supernatants of various bacteria such as *Phaeocystis antarctica*, *Bacillus cecembensis*, *Pseudomonas meridiana*, *Pseudomonas proteolytica*, *Arthrobacter gangotriensis*, *Arthrobacter kerguelensis*, and *B. indicus* can also biosynthesize stable AgNPs in the size range of 6–13 nm. The biosynthesis and stability of AgNPs often depend upon pH, temperature, or the bacterial species from which the supernatants are collected [30].

3.2 Gold NPs

In medicine, NPs offer an alternate choice to antimicrobials, since they do not lead to multidrug resistance in bacteria. Gold NPs (AuNPs) hold promising deals in the field of antimicrobials, tumor therapy, diagnostics, and targeted gene delivery systems. In general, AuNPs exhibit negligible cell membrane permeability, except cancer/tumor cells that show enhanced uptake, permeation, and retention effect. When AuNPs are capped with proteins, their uptake is enhanced even more, since the protein cap not only stabilizes the NPs in their colloidal state but also provides a docking site for drugs'/genes' delivery [31]. Besides, gold NPs are biocompatible and nontoxic and have unique catalytic, optical, and biomedical properties [32].

In addition, cell-free extract of *Rhodopseudomonas capsulata* can synthesize gold NPs. This bacterium secretes a cofactor NADH along with NADH-dependent enzymes. The bioreduction of gold ions activates by electron transfer from NADH via NADH-dependent reductase acting as an electron carrier. After receiving the electrons, gold ions get reduced to $Au^{(0)}$ leading to the generation of gold NPs. Varying the concentration of $HAuCl_4$, different shapes of NPs can be obtained. Low concentration of Au ions in the aqueous solutions (pH of 7) generate spherical gold

NPs, whereas higher concentrations (pH of 4) generate networked gold nanowires. The diameter of these gold nanowires ranged between 50 and 60 nm [33]. Basu et al. [34] reported the extracellular synthesis of protein-coated polydispersed AuNPs by mycorrhizal, edible fungi, *Tricholoma crassum* within a size range of 5–25 nm and possessing different shapes. Spectroscopic analysis revealed that at higher pH, blue shift of absorption maxima was observed indicating inhibition of enzyme reaction involved in NP synthesis. However, substrate concentrations, temperature, and precursor concentrations largely determined the amount of AuNPs. The produced NPs even displayed antimicrobial activity against pathogenic organisms such as *E. coli*, *A. tumefaciens*, and *M. oryzae*. At higher doses, these particles induced apoptosis in eukaryotic cells, confirming its putative role in cancer therapy.

Even biomolecules obtained from different microbial species have also been employed for NP synthesis. Protein extracts of *Deinococcus radiodurans* generated AgNPs by providing functional groups like –SH, $-NH_2$, –OH, and –COOH as binding sites that facilitate the bioreduction of Au^{3+} to AuNPs. Furthermore, these proteins also encapsulated the newly formed AuNPs, as capping agents, to prevent them from aggregating, thus stabilizing the NPs [26].

3.3 *Magnetic NPs*

Magnetic NPs possess exclusive micro-configuration, superparamagnetic properties, and tremendous coercive force and, therefore, show broad application in the field of biomedicine, DNA analysis, gene therapy, and cancer treatment [28]. Gram-negative magnetotactic bacteria synthesize magnetic NPs of varied morphology and often occur in fresh and marine water sediments. These bacteria can synthesize magnetosomes, which act as magneto receptions. Magnetotactic bacteria can produce membrane-bound, intracellular magnetite, greigite, and pyrrhotite [12, 35]. They can synthesize magnetic nanoparticles made up of iron sulfides (FeS), iron oxide (FeO), or both intracellularly. Fe(III), an oxidant present in natural and contaminated areas, influences the geochemistry of aquatic sediments by increasing the concentration of dissolved iron, manganese, phosphate, and trace metals. Sulfate-reducing bacteria can readily produce magnetic iron FeSNPs that adsorb radioactive metals because of their increased surface area and thus can be used for bioremediation [28, 30]. Because of the superparamagnetic nature of Fe and FeONPs, they are extensively used in the biomedical field (tissue repair, cell labeling, magnetic resonance imaging, and drug delivery) [10].

Magnetic Fe_3O_4-NPs can be intracellularly generated by *Magnetospirillum magneticum* using $FeCl_3$ precursor. The magnetosome organelles of *M. magneticum*, present intracellularly, play a pivotal role in the creation of these NPs. Ferritin, a globular protein complex, encapsulates the magnetosomes, thus enabling the storage of Fe in bacteria in nontoxic soluble form. This protein attributes to the generation of Fe_3O_4NPs and their nucleation [26]. Bhargava et al. [36] were able to achieve FeONPs employing *Aspergillus japonicus* strain AJP01.

The fungus could hydrolyze the precursor (iron cyanide) salt solution, under suitable conditions releasing ferric and ferrous ions. These ions co-precipitated under the influence of fungal proteins and underwent controlled nucleation, ultimately leading to the formation of FeONPs. Analysis of TEM, SAED, EDS, and X-ray diffraction results verified the mycosynthesis of these cubical shaped crystalline NPs respectively (60–70 nm).

Iron(Fe)-reducing thermophilic bacteria can also substitute metals like cobalt, nickel, uranium, manganese, and chromium into biosynthesized magnetite crystals. Different bacterial species, such as *Desulfovibrio* sp., *Thermoanaerobacter ethanolicus*, *Magnetospirillum sp.*, and *Pelobacter* sp., and fungal species like *Verticillium* sp. and *Fusarium oxysporum* are reported for their high efficiency in producing magnetic NPs involving different metals [12, 30]. Certain non-magnetotactic bacteria such as *Leptospirillum ferrooxidans*, *Ferroplasma thermophilum*, and *Acidithiobacillus ferrooxidans* produce only a few magnetosomes, thus exhibiting a weak magnetic field. These are present in freshwater, seawater, trash, sulfur springs, and soil and are also employed in the production of magnetosomes owing to their easy mass cultivation [12].

3.4 Zinc NPs

Zinc oxide NPs (ZnONPs) and zinc sulfide NPs (ZnSNPs) have recently gained popularity in the scientific world because of their photocatalytic, electronic, optical, and antibacterial properties and dermatological properties and are extensively used in photocatalysis, memory resistors, chemical sensors, and photovoltaics. One-dimensional ZnONPs are recognized as major photonic materials in the UV region because of their huge exciton binding energy, broad direct band gap, and high surface-to-volume ratio. ZnO nanopowder also finds its commercial application in products including ceramics, glass, plastics, cement, lubricants, paints, rubber, pigments, foods, batteries, personal care products, fire retardants, etc. [37].

Aeromonas hydrophila, a reproducible bacterium, can synthesize ZnO NPs following a simple low-cost procedure. These particles have a size of approximately 57 nm and are spherical to oval in shape as confirmed from atomic force microscopy. The crystalline nature of these particles was revealed by X-ray diffraction. At the concentration of 25 μg/mL, these ZnONPs exhibited strong antifungal and antibacterial activity against *Aspergillus flavus* and *Pseudomonas aeruginosa*, respectively [38]. Moreover, Malarkodi et al. [39] reported the formation of ZnSNPs by *Klebsiella pneumoniae* from zinc sulfate. The spherical NPs of 65 nm size are generated extracellularly in the presence of zinc sulfate. The structural (XRD) and morphological (SAED and TEM) analysis along with spectroscopic techniques (FTIR and UV-Vis spectrophotometer) affirmed the role of bacteria in the stabilization of ZnS NPs. These NPs also exhibited strong fungicidal and bactericidal activity against *Candida albicans*, *Streptococcus* sp., and *Lactobacillus* sp.

3.5 Selenium and Tellurium NPs

Selenium because of its semiconducting and photo-optical properties finds its application in electronic circuit devices and photocopiers [30]. Also, Se compounds are used in anticancer therapy since they can lower the risk of cancers like mammary, prostate, liver, lung, and colon cancers. Researchers have suggested that the redox potential, the concentration of Se compounds, and chemical species are crucial for an anticancer response. Generally, high dosages of Se compounds exhibit substantial anticancer activity, but such high doses pose toxicity concerns. Se nanostructured particles offer an alternative for removal of Se toxicity and have been employed in cancer treatments owing to their anticancer activity along with lesser toxicity in comparison to organic and inorganic Se compounds [40]. *Stenotrophomonas maltophilia* can easily transform selenite (SeO_3^{-2}) into elemental Se and accumulate granules in cell cytoplasm or extracellular spaces. In addition, *Enterobacter cloacae*, *Desulfovibrio desulfuricans*, and *Rhodospirillum rubrum* can reduce SeO_3^{-2}, both intercellularly and extracellularly, to selenium NPs exhibiting different morphologies like fibrillar, granular, and spherical [30].

In another report by Dwivedi et al. [41], monodispersed, spherical, and stable Se NPs (average size 21 nm) were biosynthesized employing bacterial isolate of JS-11 strain of *Pseudomonas aeruginosa*. The bacteria displayed considerable tolerance to SeO_3^{-2}. The supernatant of the bacterial culture at 37 °C exhibited the potential to reduce colorless and soluble and colorless selenite into red elemental insoluble selenium nanospheres (Se^0). Phenazine-1-carboxylic acid, a metabolite secreted from strain JS-11, and NADH and NADH-dependent reductases (redox agents) were responsible for this biomimetic reduction. The authors suggested the use of red-colored Se^0 nanospheres as a biosensor for assessing nanotoxicity assessment. Similarly, Ahmad et al. [40] reported the bioreductive ability of *Streptomyces bikiniensis* Ess_amA-1 strain for the biosynthesis of Se nanorods. The strain in the presence of selenium oxide displayed a time-dependent color change from gray to red, of the liquid culture medium in which it was grown. The appearance of red-brick color after 48 h of incubation indicated the biogenic ability of the strain in reducing selenite ions into elemental Se (Se^0) insoluble form. The yield of Se nanorods was about 7.74 mg/100 mL of culture medium. The strain produced aromatic amino acids that helped in the adherence of biological macromolecules on nanorods' surfaces. These biological molecules are attributed toward reduction, nucleation growth, and stabilization of the biosynthesized Se nanorods as revealed by FTIR spectroscopy.

Different fungal isolates of *Aspergillus* were screened for their capacity to reduce potassium tellurite into elemental tellurium and generate NPs (TeNPs) in the process. *Aspergillus welwitschiae* (KY766958) was reported to be the most efficient species depending upon their enzymatic production of NPs. DLS, TEM, and FTIR techniques characterized the produced TeNPs and revealed that the spherical and oval particles that were formed had an average size of about 60.80 dnm. TeNPs showed antimicrobial activity against pathogens, *Staphylococcus aureus* and *Escherichia coli*, when applied at 25 mg/mL concentration. Exposure of the fungus in culture medium to γ-irradiation enhanced TeNP production [42]. Similarly, *Sulfurospirillum barnesii* and *Bacillus selenitireducens* can also produce NPs within size <50 nm and 10 nm diameter [27]. Generally, tellurium resistance in microbes requires either reductive precipitation or volatilization of tellurite. Generation of TeNPs involves NADH-dependent tellurite reductase which is liable for tellurite detoxification [26].

3.6 Cadmium NPs

Cadmium sulfide nanoparticles (CdSNPs) are well-known wide band gap semiconductors, and because of their optical properties, they are employed as fluorophores. Owing to their smaller size, generally 1–10 nm, they are also referred to as quantum dots. These cadmium nanocrystals are regularly used in laser technology, optoelectronics, and biomedicine [43]. Microbes like *Moorella thermoacetica*, *Klebsiella aerogenes*, *Schizosaccharomyces pombe*, *Candida glabrata*, *Coriolus versicolor*, and *Flagellospora curta* synthesize cadmium sulfide NPs [26, 30].

Klebsiella pneumoniae strain MAA in the presence of cadmium sulfate reduced sulfate into sulfide, readily synthesizing spherically shaped CdSNPs after 24-h incubation period of the bacterial biomass with $CdSO_4$, as observed from the appearance of white color in the reaction mixture [39]. The authors outlined the mechanism involved in the synthesis of these NPs and suggested that firstly the sulfate ions present in the extracellular nutrient medium are taken up by the bacteria and in the presence of ATP sulfurylase get reduced to adenosine phosphosulfate, which is then phosphorylated into 3′phosphoadenosine phosphosulfate. Next, 3′phosphoadenosine phosphosulfate gets reduced, resulting in the formation of sulfite ions in the presence of enzyme phosphoadenosine phosphosulfate reductase. The enzyme sulfite reductase reduces these sulfite ions to sulfide ions. The sulfide ions thus formed a couple with inorganic cadmium ions present in the extracellular environment and forms NPs of CdS.

3.7 Palladium NPs

Palladium (Pd) is considered a scarce natural resource that has increasing demand in industrial applications. Pd^0 NPs (PdNPs) can be biorecovered by *Enterococcus faecalis*, electron donor, and sodium formate. Successful Pd^{2+}reduction led to the generation of PdNP, 10 nm, either inside the cell or at the membrane surface, as revealed by TEM. The process occurred under optimal conditions of at 40 °C temperature, pH 3.0–3.5, with 25 mM sodium formate concentration, 1.2 g/L bacterial biomass, and 210 mg/L Pd^{2+} respectively [44].

4 Applications of Nanoparticles

The synthesis of metal-based and inorganic NPs has augmented the development of interlinkage of new fields/disciplines of science. The development and designing of affordable novel techniques for the production of NPs have provided a fascinating field of study and also address the escalating human needs in terms of health security and environmental problems. In modern times, industries utilize nanomaterial, and it is progressively being adopted anonymously and will soon replace the toxic and harmful chemicals used during traditional times. This is possibly due to NPs and their nanocomposites offering a comparatively better alternative [35]. Microbially synthesized NPs offer applications in various fields like medicines, agriculture, bioremediation, biosensing, catalysis, etc. and are briefly discussed in this section.

4.1 Biomedical Applications

Nanotechnology in medicine and healthcare has emerged as a promising deal, owing to its deployment in gene and drug delivery, biosensors, treatment of human diseases, detection of pathogens, tumor destruction, DNA analysis, and various phagokinetic examinations finds its applications in plethora of fields. The microbe-synthesized metallic NPs hold immense potential in their usage in formulating antimicrobial agents, in drug delivery, in imaging/diagnostics, and in biosensor development [45].

4.1.1 Nanomedicine

Microbial NPs exhibit strong antimicrobial activities. NPs attach to the cell membrane and penetrate the cell by communing with DNA, thus obstructing DNA replication and also even attacking pathogenic respiratory processes. In certain cases, they cause structural damage to the cell membranes resulting in pit formation accompanied by degradation and deterioration of the cellular components, ultimately leading to their death [46]. AgNPs obtained from *Bacillus cereus*, an endophytic bacteria, display bactericidal properties against certain pathogenic bacteria's such as *Klebsiella pneumonia*, *Escherichia coli*, *Salmonella typhi*, *Pseudomonas aeruginosa*, and *Staphylococcus aureus* [47]. It is suggested that silver ions ($Ag+$) of the AgNPs get discharged and adhere to the thiol (–SH) group present on the pathogen's cell membrane and interrupt its function, thus displaying antimicrobial potential [48].

Spherical-shaped AuNPs, produced from the protein extract of *Streptomyces platensis*, display inhibitory response against *S. aureus* and *B. subtilis*. Gram-positive bacteria possess a thick peptidoglycan layer to which the NPs adhere and disrupt the bonds in the cell membrane, thereby gaining entry into the microorganism [10]. Fungi-generated AgNPs show bactericidal activity against Gram-negative/-positive bacteria. They adhere and invade bacterial cell walls and modulate signaling pathways in the cells by dephosphorylating major peptide substrates present on tyrosine residues. Extracellularly synthesized AgNPs, using *Trichoderma* sp. and *Fusarium oxysporum*, can be integrated into materials like textiles. The fabrics implanted with AgNPs are antibacterial and can safely be employed in hospitals to minimize/prevent infection of certain pathogenic bacteria like *Staphylococcus aureus* [45]. Copper oxide NPs synthesized by *Bifurcaria bifurcata*, brown algae, also exhibited significant antibacterial activity against pathogenic bacteria, *Enterobacter aerogenes* and *Staphylococcus aureus* [49]. Besides, certain cobalt NPs obtained from *Bacillus thuringiensis* display larvicidal activity against dengue- and malaria-causing vectors, *Aedes aegypti* and *Anopheles subpictus*, respectively [50].

Nanomedicine has been effectively employed for the detection of tumors, site-specific drug delivery, and cancer treatment [51]. The biologically biosynthesized NPs, because of their intrinsic benefits, can readily cross biological barriers and assist molecular interactions without distressing healthy cells. Biosynthesized AgNPs affect apoptosis induction and endocytic activity of cancer cell lines. The efficacy of particles is reported to be directly proportional to the endocytic activities of the cancer cells. Silver NPs formed by *Cryptococcus laurentii* demonstrated effective antitumor activity against cancer cell lines [52]. Platinum NPs biosynthesized by *Saccharomyces boulardii* exhibit anticancer activity against MCF-7 and A431 cell lines [31]. Selenium nanorods with an average particle size of 17 nm, synthesized by *Streptomyces bikiniensis* Ess_amA-1 strain, induced cell death of MCF-7 and Hep-G2 human cancer cells at a lethal dose ($LD_{50\%}$) of 61.86 and 75.96 μg/mL, respectively [40]. Similarly, selenium nanorods and gold NPs have also been successfully employed against cancer cells, owing to their ability to stimulate mitochondrial apoptosis, DNA impairment, and cytokinesis detention in cancer cell lines [46]. PEG-coated gold NPs maximize tumor damage in comparison to TNF-α (tumor necrosis factor-alpha), a cytokine with anticancer activity [10].

4.1.2 Targeted Drug Delivery

An important application of NP is the targeted or localized delivery of biomolecules and drugs in the cells and tissues. These nanosized particles, acting as delivery vehicles, protect the biomolecule or the drug from degradation, successfully transport them to the targeted cells/tissues, and sustainably release the delivered molecules. Besides, a significantly higher cellular uptake efficiency of the bioactive molecules is observed for NPs in comparison to microparticles [53]. NP drug carriers such as AgNPs are often referred to as drug conveyors and, owing to their minute size, can easily bypass the skin's rigid epithelial junctions and blood-brain barrier that often impede drug delivery at the preferred target site. Due to the high surface/volume ratio, these nanocarriers exhibit enhanced biodistribution and pharmacokinetics of the therapeutic agents, thus minimizing toxicity at the desired site. They not only enhance the solubility of hydrophobic compounds but also make them pertinent for parenteral administration [54]. Additionally, these NPs augment the stability of therapeutic agents such as oligonucleotides and peptides. Toxicokinetics can easily be controlled in cases where the drug readily conjugates with NPs either by encapsulation or by linker molecules. Toxicity of the drug carrier systems is lowered ensuring the drug's therapeutic effects in the patients [46].

Magnetic NPs such as magnetite (Fe_3O_4) and maghemite (Fe_2O_3) are biocompatible and are extensively used in site-specific cancer treatment, guided drug delivery, stem cell manipulation and sorting, DNA analysis, and gene therapy and MRI scanning [28]. *Rhodococcus pyridinivorans*-synthesized ZnO NPs laden with anthraquinone, displayed cytotoxicity towards HT-29 colon carcinoma cells in a concentration-dependent manner, thus revealing its role during drug delivery carrier for cancer treatment [37]. NP-targeted drug delivery systems are anticipated to significantly cut down the dose of anticancer drugs with low toxicity, better specificity, and enhanced efficacy [28].

4.2 Biosensors

NPs exhibit optical and electronic properties and thus find their application in biosensing techniques. They are frequently used for sensing different biological analytes such as DNA, proteins, and small molecules like glucose. In biosensing techniques, specific receptors bound to the NP surface interact with an explicit biological analyte, and this recognition event is then translated into a detectable magnetic, optical, or electrochemical signal [53]. Ag-Au alloy NPs biosynthesized from yeast cells are used for fabricating an electrochemical sensitive vanillin sensor. This vanillin sensor can successfully determine vanillin content from vanilla tea and vanilla bean samples, suggesting its practical application in vanillin tracking systems [28]. AuNPs are utilized as biosensor labels, for curing hyperthermia, for determining glucose content in glucose injections fabricated commercially, for staining biological tissues, and for estimating biomolecules [54]. In comparison to macro-scale biosensors, the NP-based biosensors exhibit low detection limits and increased sensitivity, endorsed to NPs' high surface to volume ratio which results in greater density of specific receptors/unit volume of NP [53].

4.3 Catalytic Applications

Owing to their larger surface area and certain special characteristics, NPs have found their application in improving reaction rates, either as reductants or as catalysts. Magnetic NPs are employed for improving rates of microbiological reactions. Cells of *Pseudomonas delafieldii* coated with Fe_3O_4 magnetic NPs are used for dibenzothiophene desulfurization [28]. The PdNPs synthesized by *Enterococcus faecalis* could readily catalyze the complete reduction of chromate [44]. Certain biologically synthesized NPs can readily remove pollutants like heavy metals, pesticides, synthetic dyes, etc. from the environment by acting as catalysts. Palladium NPs obtained from bacterial biomass are used as a catalyst to produce hydrogen, using hypophosphite as substrate [26]. AuNPs biosynthesized using *Trichoderma* sp. cell-free extract (20–30 nm) along with anisotropic planar shapes are useful in optoelectronics and photonics [45]. Fe_3O_4 NPs can proficiently adsorb crystal violet dye, a model pollutant, thus offering an alternative for the removal of water pollutants. PdNPs from *Chlorella vulgaris* act as a catalyst in Mizoroki-Heck cross-coupling reaction [10].

4.4 Agriculture

Nanotechnology aims at improving agricultural practices by escalating input efficiency and reducing production losses. NPs provide a broad surface area for pesticides and fertilizers. Besides, nanomaterial-based agrochemicals facilitate targeted delivery of mineral elements along with augmented crop protection [55]. Nanonutrient/nanopesticide application to plants in the form of aerosol sprays is considered superior as compared to traditional sprays. Also, the loss of nanonutrient/nanopesticide during spray is lesser (15%) as compared to natural sprays (33%). Using particles of 20 nm or less is generally considered more beneficial [9]. Inorganic NPs like ZnO, SiO_2, TiO_2, Cu, CaO, MnO, MgO, and AgNPs play a pertinent role in plant protection against pathogens and pests [55]. Owing to distinct properties such as sensitivity and performance, NPs can also be used as biosensors for detecting crop pests and physiological stresses like drought stress and soil analysis, thereby, employing global positioning systems using field satellite images. NPs display excellent transduction properties owing to which they are explored for agricultural products. Several nanoscale carrier molecules can thus be utilized for delivery of herbicides, pesticides, fertilizers, plant growth regulators, etc. in small amounts and improved and extended management in the agriculture sector [45].

5 Nanoparticle-Plant Interactions

Plants are the fundamental and most important biotic component of the ecosystem. They play an imperative role in maintaining equilibrium *via* the transportation of nutrients across the food chain and food web of the ecosystem. These biotic components work in coordination with other abiotic components of the ecosystem like water, soil, etc. These abiotic components make a path for different components like nanoparticles (NPs) *via* specified routes [56]. So, there are different ways through which NPs interact with the plants like direct application, accidental release,

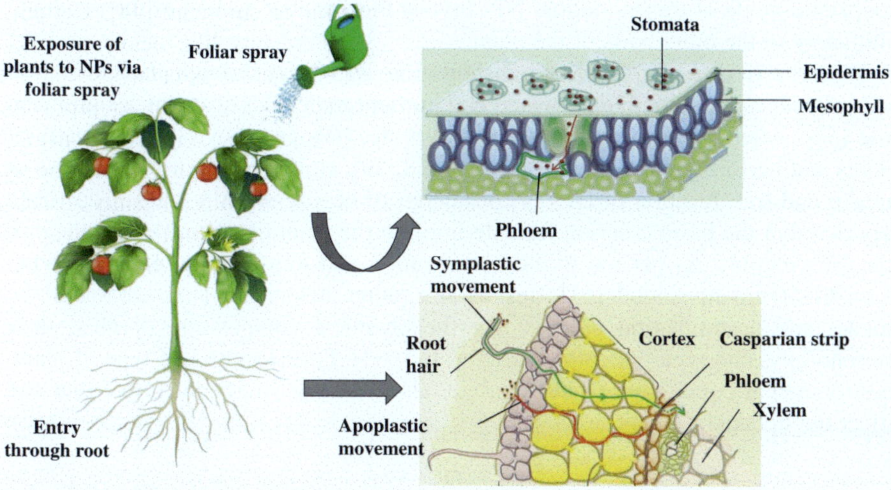

Fig. 1 Model depicting the uptake and transportation of NPs in plants

and presence in the soil as contaminants in the soil or atmosphere. A schematic overview of plant-NP interaction is given in Fig. 1. After reaching the soil-plant zone, NPs interact with plants in a non-partial manner, thus influencing the physiological processes of plants and enhancing the food security and thus ultimately the management of agronomy fields. But researchers have also documented the toxic effects of NPs on the environment and its components. The toxicity of NPs depends directly on their interaction with the specific substrate where they have been applied. In a nutshell, the phytotoxicity of NPs is based on their uptake, transport, and accumulation in plants.

5.1 Uptake and Translocation Mechanism

The researchers have reported two methods of NPs exposure to plants, i.e., root exposure and exposure to the vegetative part, especially leaves.

5.1.1 Uptake of NPs Through Root

During the encounter of NPs with plants, NPs enter the plant cell by crossing the cell wall and cell membrane of the root epidermal cells, and this penetration is followed by a cascade of events that ultimately results in the entry of NPs to the vascular tissue. Once inside the plant tissue, NPs take up either symplastic or apoplastic modes of transportation to migrate from one plant part to the other. With the help of apoplastic mode of transportation, NPs invade the vascular system of the plant and ultimately to the other parts of the plant.

Whereas, in case of symplastic transportation, NPs move through plasmodesmata and cell sieves. In both, modes, water, and nutrient molecules also play a significant role [57]. After passing *via* the symplastic route, NPs can move to other parts of plants and perform their functions (Fig. 1) [58, 59]. However, the whole process of uptake and translocation of NPs is considered to be size-specific [60]. It has been reported that the basic criterion for entering into the plant tissue and cell is the size that NPs exhibit. As per the studies, penetration and translocation of NPs having sizes 40–50 nm are smooth [61]. Besides size, other factors play a critical role in the uptake and accumulation process like the chemical composition of NPs, their morphology, and type of coating material [62, 63]. Furthermore, the type of plant, environmental conditions, microflora, etc. also serve as important factor that can affect the uptake of NPs.

5.1.2 Uptake of NPs via Foliar Spray

In the case of uptake *via* foliar spray in plants, the primary hurdle is the waxy protective layer present in the leaves, i.e., cuticle which prevents excess water loss and also controls the exchange of solutes [57]. Researchers have reported two possible pathways through which the NPs can penetrate the waxy cuticle, i.e., polar solutes can penetrate through the hydrophilic pathway, and nonpolar solutes can enter *via* lipophilic pathways that include permeation and diffusion [58, 64]. Moreover, the studies have also revealed that the uptake of hydrophilic substances can also occur through stomatal apertures. But in the case of stomatal uptake, the major influencing factors are the morphology of leaf and size and density of stomata [65]. After entering the apoplast of the leaf, the most possible route that NPs follow must be the conductive tissue or the vascular system, usually phloem, because in phloem, the flow of substances is from top to bottom, i.e., from shoot to root (Fig. 1). Thus, the NPs translocated during the foliar spray may be exudated into the phyllosphere and influence the microbial community in the phyllosphere.

Researchers have well documented the role of NPs as potential abiotic elicitors. The NPs are known to induce bioactive metabolites in plants [66]. Furthermore, researchers have constantly examined the potency of NPs in modulating the expression of genes encoding biomanufacturing of the secondary active metabolites [67]. It has been reported that apart from inducing secondary signaling cascade, NPs also induce the level of various ROS in the plant cells, which further triggers oxidative stress and thus influences the levels of primary and secondary metabolites [68]. Several studies that document the role of NPs in secondary metabolite production in plants are enlisted in Table 1.

Table 1 Role of nanoparticle-mediated secondary metabolic profiles in plants

Nanoparticles	Size	Conc.	Plant	Tissue	Effects	References
Mesoporous silica in combination with amines and TiO_2	165 nm	1, 10, 100 μg/mLand 1, 2.5 mg/mL	*Solidago nemoralis*	Hair root culture	Enhanced level of flavonoids, continued synthesis of flavonoids after harvest	[69, 70]
Ag	40 nm	30, 60, and 90 μg/L	*Caralluma tuberculata*	Callus culture	Increment in the levels of flavonoids, total phenols, phenylalanine lyase, superoxide dismutase, catalase, and ascorbate peroxidase	[71]
ZnO	–	100 and 150 mg/L	*Zataria multiflora, Thymus vulgaris, T. aenensis, T. kotschyanus*	Callus culture	Enhanced contents of thymol and carvacrol	[72]
ZnO and NaCl	10–30 nm	NaCl—0.50 and 100 mM ZnO—0, 20, 40, 80 mg/L	*Camelina sativa*	Shoot-root	Decreased antioxidant capacity, total flavonoid content. Induced total phenol, anthocyanins, carotenoid calcium, zinc, and phosphorus content	[73]
Se	50–78 nm	5 mg/L	*Apium graveolens* L.	Stems-leaves	Enhanced flavonoids, total phenols, chlorophyll, total proteins, soluble sugars, and various amino acids	[74]

(continued)

Table 1 (continued)

Nanoparticles	Size	Conc.	Plant	Tissue	Effects	References
Ag	10, 40, and 100 nm	0.5, 1.0, 5.0 mg/L	*Arabidopsis thaliana*	Seedlings	Increase in the contents of glutathione disulfide, sinapoyl malate, kaempferitrin, G(8-5)FA dihexoside, G(8-0-4)G hexoside, G(8-0-4)G(8-0-4)G hexoside, coniferyl aldehyde hexoside, 4-glucosyloxy-benzoate, 6-MeO-ICOOH, sinapic acid, 4-hydroxyglucobrassicin, and camalexin. Decreased of 4-glucosyloxy-benzoate, pinoresinol hexoside, and kaempferol-3-*O*-glucoside	[75]
$nTiO_2$ and $bTiO_2$	$nTiO_2$—less than 50 nm and $bTiO_2$—68 nm	200 and 800 mg/kg mixed in potting soil	*Abelmoschus esculentus* L.	Roots, leaves, and fruits	Enhanced seed germination, increment in the content of chlorophyll	[76]
SiO_2 and TiO_2	SiO_2: 5–15 nm TiO_2: an average of 25 nm	5, 10, and 20 ppm after 15 days of incubation	*Argania spinosa*	Callus culture (leaves)	Increased tocopherol content	[77]
SiO_2 and TiO_2 NPs and NaCl	SiO_2: 10–15 nm TiO_2: an average of 24.5 nm	SiO_2 and TiO_2 NPs: 25 and 50 mg/L NaCl: 0.3 M	*Tanacetum parthenium* L.	Leaves	Enhanced level of genes that are used in the biosynthesis of parthenolide and β-caryophyllene	[78]
Ag	–	0, 0.25, 0.5, 1, 1.5, and 2 mg/L	*Isatis constricta*	Plantlets	Increased indigo and tryptanthrin	[79]

ZnO NPs, jasmonate, and chitosan	–	ZnO; 20, 60, and 100 ppm Jasmonate: 100, 250, and 500 μM Chitosan: 10, 50, and 100 μM	*Momordica charantia* L.	Shoot	Upregulation in the activities of antioxidant enzymes, enhanced level of phenolic compounds, carotenoids, and proline content	[80]
ZnO and CuO	ZnO: 20–30 nm CuO: 25–30 nm	0, 2, 20, 200, and 2000 mg/L	*Stevia rebaudiana*	Shoot	Enhanced contents of phenolic compounds, total reducing sugars, and increased antioxidant activity	[81]
AgNPs and $AgNO_3$	–	0.5, 1.0, and 2.0 mg/L	*Cucumis anguria* L.	Hairy root	Increased levels of total phenols	[82]
Ag	30–50 nm	0, 2.5, 5, and 10 ppm	*Corylus avellana* L.	Hazel cells	Rise in taxanes content (taxol and baccatin III)	[83]
Ag, Au, and naphthalene acetic acid (NAA)	–	30 μg/L of each NPs in ratios of AgAu: 1:2; 1:3; 2:1, and 3:1 with NAA	*Prunella vulgaris* L.	Calli cultures	Enhanced biomass, total phenols and flavonoids, and DPPH-radical scavenging activity	[84]
CuO	–	1 ppm	*Withania somnifera* L.	Shoots and roots	Elevated polyphenols (total phenol content, flavonoid content) and antioxidant activities	[85]
Cerium oxide ($nCeO_2$) and citric acid-coated cerium oxide ($nCeO_2$ +CA)	$nCeO_2$: 8 ± 1 (primary size) and 231 ± 16 (in deionized water) $nCeO_2$+CA: 12.4 nm (primary size) and 189 ± 2 (in deionized water)	0, 62.5, 125, 250, and 500 mg/kg	*Solanum lycopersicum*	Fruits	Increment in the content of lycopene	[86]

(continued)

Table 1 (continued)

Nanoparticles	Size	Conc.	Plant	Tissue	Effects	References
Cu	40 nm	0, 200, 400, and 800 mg/kg	*Cucumis sativus*	Fruit	Augmentation of leucine, threonine, fructose, valine glycine, proline, benzoic acid, isoleucine, glutaric acid, tyrosine, caprylic acid, linolenic acid, xylose, and imidazole. Decrease of lysine and methionine	[87]
Cu absorbed on chitosan hydrogel	–	100 mg/kg 0.3, 0.15, 0.06, 0.03, and 0.015 g/L	*Solanum lycopersicum*	Seedlings, fruits, and leaves	Inclined lycopene, titratable acidity, and catalase activities	[88]

5.2 *Nanoparticles and Plant Genetic Engineering*

Genetic engineering in plants has enhanced the quality of crops as well as the fundamental biology of plants [89, 90]. But the presence of cell walls in plant cells acts as a hurdle in the delivery of foreign genetic material to the nucleus of the target cell. For this purpose, gene gun and *Agrobacterium*-based transfer mechanisms are widely used. But there are some disadvantages of using these methods like gene gun technology that can have destroying impact on the tissue and the *Agrobacterium*-based transformation in host-specific. So, there must be some other safe method that can be used for delivering purpose, and one such method is nanoparticle-mediated gene transformation (Fig. 2) [91]. This method has been used commonly in the case of animal cells. And now the researchers are experimenting with nanoparticle-based genetic engineering in plant cell culture too. Silicon carbide-mediated transfer of DNA in cotton plants is one such example [92]. Another successful genetic transformation using magnetic nanoparticles (MNPs) has also been reported in the case of cotton plants. During this transformation, the reported gene, GUS (β-glucosidase), was complexed with MNP, and the complex thus formed was infiltrated by the magnetic force into pollen grains of the plant, without having any effect on the pollen viability. Then by pollination of these magnetofected pollens, cotton transgenic plants with successful incorporation of exogenous DNA into the genome occur [93]. Another study documents the application of mesoporous silica nanoparticles (MSNs) in delivering Cre recombinase in immature embryos of maize plant that carries Lox sites incorporated in the chromosomal DNA. After the application of modified MSN, the lox p was found to recombine specifically, thus resulting in successful genome editing in maize plants [94].

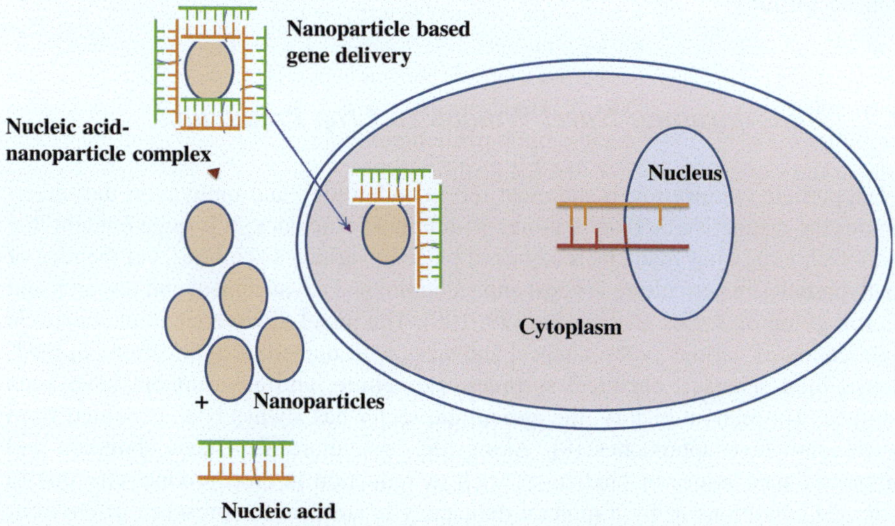

Fig. 2 Model depicting nanoparticle-mediated gene transfer in case of plants

6 Nanoparticles as Frontiers in Agriculture

Nanotechnology upholds a great position in the agriculture industry due to its extraordinary characteristics and novel strategies of nutrient acquisition, precision agricultural practices, and pathogen recognition [95]. Research is consistently being focused on discovering nano-structured catalysts that boost the competence of fertilizers, pesticides, or herbicides for their minimal use. Nonetheless, through the use of nanosensors, the farming protocols induce agricultural productivity and yield by providing accurate information about weather, soil properties, and other factors that enable the farmers in agricultural practices. For instance, the equipment could be utilized for measuring the plant health before they cause any severity. These devices may further be potent in responding to various stressful conditions and remediating them [96]. Nanotechnology paves us with novel strategies without impairing the environment through the usage of nanomaterials coated on chemical fertilizers or biofertilizers to boost their effectiveness. They possess the ability to slowly and sustainably release fertilizer that is further taken up by plants, preventing their nutritional losses along with avoidance of non-wanted nutrient associations with microbial communities [97]. Silicon nanomaterials coated on fertilizers form a film over microbial cells to prevent infections along with improvement in plant resistance mechanisms against diseases [64]. Apart from this, silicon nanoparticles improve germination rate and promote root development in plants. Moreover, few nanoparticles also enhance water retention ability in soil, for example, *Bacillus subtilis*, *Pseudomonas fluorescens*, *P. elgii*, etc. are used as biofertilizers when augmented with nanoparticles and enhance plant growth and metabolism [1]. All these positive characteristics of nanoparticles make them capable to be used as bio-nanofertilizers that are the most effective and feasible alternative of the agriculture sector [98].

6.1 Nano-farming: Novel Window in Crop Production

Nanoparticle engineering is the most recent technology and innovation that determines the distinctive targeted features with utmost efficiency. It is quite evident that nanotechnology has made its recognized place in various disciplines, yet the idea of nanoparticles in agriculture is latest and technological advancement and research and is still going on for its progression [99, 100]. The novel discoveries in nanoparticle fabrication of various types, shapes, and sizes with numerous applications in medicine, food sciences, chemical sciences, agriculture, and environment have been studied. Throughout history, the agriculture sector has always been benefited from such innovative approaches [8]. Alongside, agriculture also faces abundant and unprecedented issues or challenges such as reduction in crop productivity due to stressed conditions or by a mineral deficiency in soil and the presence of environmental pollutants. Therefore, nanotechnology has proved to offer propitious

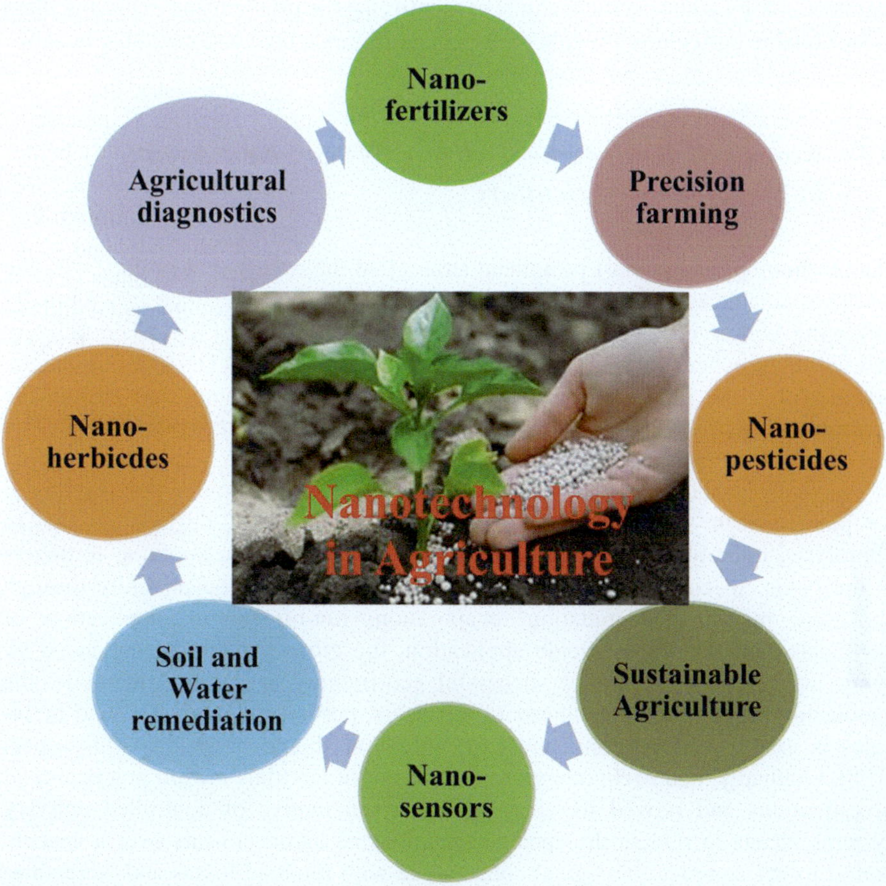

Fig. 3 Applications of nanotechnology in agriculture

applications in precision agricultural practices (Fig. 3). The term precision farming has been found to evolve in the last few years with the involvement of wireless networking and the use of miniature sensors to monitor, examine, and regulate agricultural practices. Especially, it is associated with site-specific regulation with a plethora of production strategies in agriculture starting from horticulture to field crops [55]. More recently the role of tissue engineering and engineering of nanoparticles by using a targeted delivery system based on clustered regularly interspaced short palindromic repeats (CRISPER)/Cas (CRISPER-linked proteins) mRNA and sRNA for transgenic crops is a remarkable achievement in agriculture [101, 102]. Apart from this, nanotechnology also provides an excellent approach for coping with various environmental challenges. To illustrate, nanosensors have broader prospects in evaluating environmental stresses along with the potential to protect them against diseases [103]. Thus, such consistent innovations in nanotechnology with specific features of identifying the problem and exploring them in sustainable agriculture provide them noteworthy potential and equitable benefits.

6.2 Nanoparticle-Mediated Delivery System: New Avenue in Sustainable Agriculture

Nanotechnology is the most prominent strategy of the twenty-first century with the latest sustainable agricultural practices and with the power to regulate agriculture by following conserved tactics to minimize agricultural wastes [104]. The delivery process of agrochemicals, as well as organic molecules such as DNA or oligonucleotides within the plant cells, is substantial for precise farming and sustainable agriculture [105]. While in case of traditional methods, the agrochemicals are directly sprayed onto crops that usually result in nontargets. A very limited supply of chemicals reaches the target site that is quite low in concentration that is required for plant growth. This further results in chemical leaching, degradation due to photolysis, microbial degradation, or hydrolysis [106]. While applying fertilizers in the fields, special care should be taken toward nutrient bioavailability through chelation, microbial degradation, evaporation, run-off, or overapplication of chemicals [99]. During pesticide application, the efficacy could be enhanced by spray management [107]. For successful eco-friendly agricultural methods, the nanotechnology-mediated synthesis of fertilizers, pesticides, fungicides, and herbicides is extremely important, and extra attention should be paid to this aspect [108]. Gradually, nanotechnology has moved out of lab trials toward practical field investigations and proved its efficacy. The main motive of controlled delivery systems means to release the specific quantities of agrochemicals over a specific period to get complete biological competence with minimal losses and adversities [109]. Nanoparticles provide us the effective agrochemical delivery due to their huge surface area, fast mass transfer, and ability to get easily attached to the surfaces [108]. Owing to this, micronic or sub-micronic particles are penetrated within agrochemicals by various methods such as capsulation, absorption, ionic bond attachments, or entrapment within nano-matrix [5]. The capsulation of KNO_3 by graphene oxide usually protracts from the release of fertilizer. This process is cost-effective and is used for large-scale production purposes [110]. The nanomaterial coating enhances the agrochemical stability and protects them from being degraded and released into the ecosystem that further induces its effectiveness and lowering the agrochemical quantity.

Nevertheless, the combinatorial role of nanotechnology and biotechnology also designs new kits of molecular transporters for gene modification and the production of new organisms (Fig. 4) [111]. To elucidate, nanobiotechnology offers nanoparticles, nanocapsules, and nanofibers for carrying out foreign DNA along with chemicals to facilitate the modification of target genes. At the time of genetic material delivery, viral vectors face many challenges due to limited host range, size

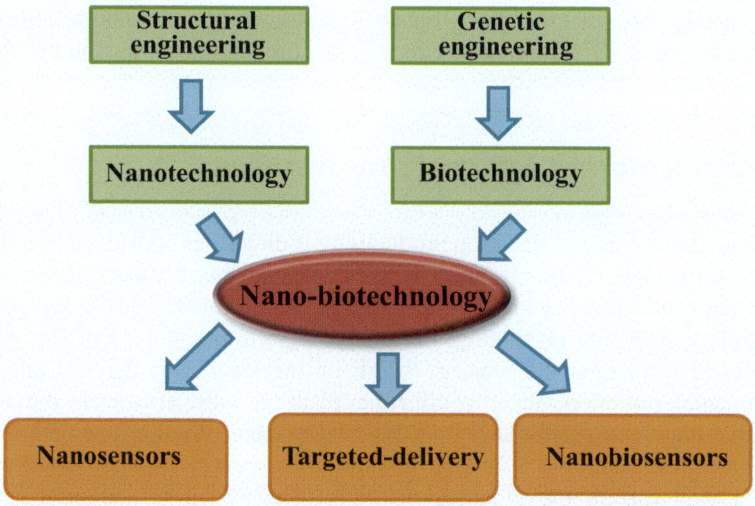

Fig. 4 Schematic representation of nano-biotechnology. The combination of nanotechnology and biotechnology gives birth to nano-biotechnology, which entails the understanding of genetic as well as structural engineering. This field has emerged as a novel for various sectors, especially agriculture with the emergence of smart approaches in farming such as nanosensors, targeted delivery systems of genetic material, and pathogen identification through nanobiosensors

of genetic material, and transport across cell membrane along with trafficking within nuclei [107]. Contrastingly, the recent quantum loop in nano-biotechnology provides many advancements to replace the genetic material completely by others [112]. Genetic engineering offers different types of nanoparticles for delivering DNA sequences in the target plant species without causing any side effects [113]. Additionally, the nanoparticle delivery system is useful in breeding programs for developing resistant varieties of crops. For example, DNA-coated nanoparticles are useful as bullets in the gene gun method for tissue and cell bombardment for transferring desirable genes at targeted locations in different plants [111]. Moreover, the recent development of the chitosan nanoparticle entrapped siRNA delivery system has attained a remarkable position in the field of crop improvement that enables target-specific control of pathogens. This is because chitosan has high binding properties with RNA along with its penetration ability across cell membranes [114]. The contemporary technological advancements in nanomaterial-mediated delivery of CRISPER/Cas9 RNA have formed a breakthrough in genetic engineering. CRISPER/Cas9 constitutes CRISPER repeat spacers along with Cas proteins, which form RNA-directed defensive strategy and are being continuously used in genome editing of plants [102]. Though, the limited delivery is still a barricade of this application and many studies are still being conducted for understanding their applicability. Strikingly, nanomaterials minimize the target changes through improvising the specificity of CRISPER/Cas systems. Like in the case of cationic arginine, AuNPs congregated Cas9En (E-tag)-RNP delivery of RNA and showed nearly 30% of efficient cytoplasmic and nuclear gene editing effectively, which would have a great impact on crop production and development [115].

6.3 Nanofertilizers: Effectual Crop Nutrition

It is unequivocally sensed that crop productivity is directly proportional to fertilizers and their efficiency. Generally, the application of mineral elements improves crop fertility and productivity; henceforth the fertilizers are considered to be the predominant prerequisites for promoting sustainable agriculture [116]. In the pool of conventional fertilizers, the majority of chemicals leach into the soil and cause environmental pollution, thereby affecting plant growth. However, the role of engineered nanoparticles for sustainable agriculture has paved a new path to overcome these uncertainties [117]. The revolution in green nano-biotechnology has played a vital role in the global agricultural canvass where nanoparticles have emerged as promising agents to meet the future projections of crop production. Nanoparticles help in the alleviation of macro- and micronutrient deficiencies by enhancing mineral use efficiencies for overcoming environmental problems of eutrophication [118]. Nanofertilizers are unique and are mainly developed as plant nutrients possessing higher absorption and utilization efficacy with the minimal losses that make them best for facilitating nutrient uptake in plants. The studies conducted by Abdel-Aziz et al. [119] observed the remarkable properties of nanochitosan-based nitrogen, potassium, and phosphorus-based fertilizer in wheat plants in terms of boosting their sugar levels and potential to grow in varied soils. Similarly, peanut crops when treated with zinc oxide nanoparticles enhanced seed germination, yield, flowering, and chlorophyll content, revealing their significance in promoting morpho-physiological characteristics [120]. Nanofertilizers are designed with a special intention to properly regulate the crop requirement along with minimizing losses, for example, nitrogen fertilizers are lost by leaching or evaporation, but their nano-formulations synchronize their release as per the crop demand [121]. They prevent the losses through direct internalization followed by avoidance of nutrients interacting with soils, microbes, water, or soil [55]. To illustrate, the supplementation of porous nanomaterials like zeolites, chitosan reduced the nitrogen loss by targeted release and enhances the uptake process of the plant [122]. Moreover, ammonium zeolites also increase phosphate nutrients and availability, while graphene oxide carbon nanomaterial enhances KNO_3 release to extend the release time and prevent leaching [123]. Furthermore, Sabir et al. [124] depicted that the nanocalcite combined with nano-silicon oxide, magnesium oxide, and iron oxide enhanced Ca, Mg, and Fe uptake along with improving Zn, Mn, and P uptake.

Based on different characteristics, nanofertilizers can be defined as slowly released fertilizers, magnetic fertilizers, controlled fertilizers, nano-composite fertilizers, and nano-device combined fertilizers, respectively [108]. They are produced through the encapsulation of minerals with nanomaterials (produced by top-down/

bottom-up approaches). This includes encapsulation with nano-porous material with a thin coating of polymer and delivered as emulsions of nanoscale or surface modification of anionic nutrients [108]. Overall agricultural yield is boosted by the use of nano-formulated fertilizers. Studies depicted that chitosan-NPK fertilizer enhances harvest index, crop index, and mobilization index along with overall crop productivity and yields. Moreover, nanomaterials also enhance other facets like root and shoot development due to their porous state that enables effective nutrient uptake [119]. Moreover, they also facilitate complex formation with molecular transporters and root exudates by generating new pores, ion channels, and endocytosis, respectively [125]. The size of nanoparticles also affects the nutrient ion adsorption/desorption process depending upon their surface charge to mass ratios; therefore they ensure a balanced amount of nutrition throughout the plant growth cycle [126]. Nanotechnology is a promising field to earn profit in agriculture by stimulating crop productivity. They also play a vital role under nutrient-deficient conditions by maintaining the nutrient balance for effective soil and plant health. Like in the case of Zn-deprived soils, zinc oxide nanoparticles at lower doses influence physiological plant processes [71, 127]. The technological and scientific advancements have benefited human welfare, where plant scientists aimed to restore the genomic diversities of various crops and reduce the chemical fertilizers without compromising the crop yields as well as sustainability. In this regard, "controlled fertilizers" are being promoted that have been engineered only to reduce pollution as well as enhance agriculture by forming a nano-networking [128]. This considerably works in a way where encapsulated fertilizer in a soil network is bound by hydrogen bonds, molecular attractions, and surface tension. As a consequence, their spatial movement allows them to be blocked by soil filtration and get attached around the plant roots in the soil for facilitating nutrient acquisition in plants throughout their growth phases [129]. However, further investigations are still carried out for exploring such promising techniques.

6.4 Nanomaterials in Modulating Crop Production, Quality, and Yield

Nanoscience is a novel field with enormous developments with broad nanotech applications in agri-tech for inducing plant processes such as seed germination, vigor index, growth, and adaptation toward varied environmental conditions. Seed germination is a very sensitive stage of plants that induces seedling formation, growth, and survival. Yet the germination is very much affected by environmental, genetic, and edaphic factors [130]. NPs have, however, played an essential role in plant growth, development, and yield. For example, the use of carbon nanotubes had a huge impact on the germination of various plant species like tomato, potato, soybean, wheat, garlic, etc. [131]. Also, silicon NPs, titanium NPs, and zeolite have a positive impact on germination [130]. Another study conducted on iron and silicon NPs showed an improved germination rate in maize and barley plants. Despite a handful of studies conducted on the positive role of NPs on plants, the exact mechanism by which they stimulate growth and yield is still unclear. A very few studies determined that NPs penetrate the seed coat to boost their absorption and water retention for the proper function of enzymes to regulate germination and growth [132].

Additionally, NPs, namely, zinc oxide NOs, titanium oxide NPs, iron oxide NPs, and zinc/iron/copper oxide NPs, are reported to mediate the crop productivity and development of plants [123, 133]. The carbon nanomaterials like fullerols show positive effects on plants in terms of stimulating hypocotyl growth, cell division, fruit number, size, yield, and bioactive ingredients (cucurbitacin B, lycopene, and inulin) [134]. Moreover, the studies conducted by Yousefzadeh and Sabaghnia [135] determined that nano-iron fertilizers induced agronomic traits of *Dracocephalum moldavica* along with their oil content. Similar to this, zinc and boron nanofertilizers enhanced fruit yield, quality, soluble sugars, and maturity index of *Punica granatum* [136]. All these investigations on NPs stimulated crop yield and quality product, but the exact underlying mechanisms are still unclear. However, it has been explained that NPs have a higher absorption capacity to absorb nutrients and water content for enhancing vigor index as well as enzymatic activities [137]. Along with this, the research carried out on the regulated release of nutrients to prevent the loss of nanofertilizers has also inferred that the availability of nutrients is essential for plants in terms of flowering, germination, fruiting, etc. [138]. Furthermore, hydroxyapatite NP coated on urea fertilizer controls the nitrogen release, thereby implicating their positive impact on crops [139]. Table 2 summarizes the reports depicting the role of NPs in mediating crop production in plants.

6.5 Nanoparticles in Stress Management and Plant Protection

The food demand is quite challenging to be fulfilled for the sky-rocketed population due to scarcity of resources and climate change all over the world. Climate change means the alteration of climate over some time such as temperature, water deprivation, low temperature, salinity, heavy metal pollution, etc. Hence, the matter of concern is to enhance the adaptation of plants without harming the ecosystem to strive to counteract environmental stressors [6]. For the accomplishment of this approach, it is imperative to strengthen the plant's enzymatic system, hormonal synthesis, stress genes and proteins, and regulation of stressed conditions through a shortening life cycle. Apart from this, NP engineering shows that nano-fertilizers are best for crop production in extreme environmental conditions (Fig. 5).

For example, silicon oxide NPs enhance seed germination, growth parameters, pigment levels, and proline synthesis in plants under saline conditions [158]. A similar study conducted by Torabian et al. [159] found that iron sulfate NPs

Table 2 Role of nanoparticles in crop production

S. No	Nanoparticles	Conc.	Plant	Response in plants	References
1.	Zero-valent iron (ZVI), Fe_3O_4, and Fe_2O_3 NPs	50, 500 mg/L	*Oryza sativa*	Induced growth parameters (biomass, root/shoot length), pigments, phytohormones along with mitigation of oxidative stress	[140]
2.	Hydroxyapatite nanoparticles (HANPs)	10,000 mg/L	*Raphanus sativus*	Stimulated shoot and root elongation, dry biomass, soluble protein, and indoleacetic acid (IAA) content	[141]
3.	ZnO NPs	100 and 500 mg/L	*Datura stramonium*	Enhanced Zn bioaccumulation and induced phenylalanine ammonia-lyase, catalase, and peroxidase activities	[54]
4.	AgNPs	25, 50, 75, 100 and 150 μM	*Nicotiana tabacum*	Improved germination rate, photosynthetic efficiency, pigments, and growth rate	[142]
5.	Fe_3O_4 NPs	2000 mg/L	*Phaseolus vulgaris*	Enhanced morphological characters, nitrogenase activity, iron content, and nitrogen fixation	[143]
6.	SiNPs	200 mg/L	*Tagetes erecta* L.	Higher plant biometrics, physiology, flowering (number, diameter, fresh, and dry masses)	[144]
7.	Fe_3O_4-NPs	50 and 100 mg iron/L	*Glycine max*	Higher root surface, shoot weight, chlorophyll content, and antioxidants	[145]
8.	AgNPs	5–100 μg/mL	*Allium cepa*	Improved plant growth and physiological responses	[146]
9.	ZnO NPs	20 and 100 mg/L	*Solanum lycopersicum*	Higher chlorophyll content, photosystem II activity with upregulated expression of genes encoding nutrient element transport, carbon/nitrogen metabolism, secondary metabolism, antioxidative enzymes, transporters, sugars, and amino acids	[147]
10.	Ceria nanoparticles (NPs)	50, 100, and 200 mg/kg	*Phaseolus vulgaris*	Modulated levels of antioxidants, nutrient quality, the flavor of pods, and mineral content	[148]
11.	CeO_2 NPs	0–2000 mg/L	*Fragaria* × *ananassa* Duch.	Improved the plant growth and colonization of plant growth-promoting rhizobacteria	[149]
12.	α- and γ-Fe_2O_3 NPs	100, 200, and 400 ppm	*Citrullus lanatus*	Promoted plant growth characteristics and nutritional qualities	[74, 150]

(continued)

Table 2 (continued)

S. No	Nanoparticles	Conc.	Plant	Response in plants	References
13.	Fe_2O_3 NPs	10–100 mg/L	*Hordeum vulgare*	Stimulated plant growth, and formulations can be best used as nanofertilizers	[151]
14.	MnO_x/FeO_x NPs	2 mM	*Lactuca sativa*	Promoted plant growth, germination rate, root growth, fresh weight, and used as micronutrient nanofertilizer	[133]
15.	Poly(vinylpyrrolidone) (PVP) stabilized platinum nanoparticles (Pt:PVP)	1 mM	*Pisum sativum*	Improved yield and modification of growth regulatory hormones or biochemicals	[152]
16.	Al_2O_3 and NiO NPs	50,100, 1000 and 2500 mg/L	*Nigella arvensis* L.	Enhanced biomass, antioxidants such as ascorbate peroxidase, catalase, superoxide dismutase, and peroxidases, total antioxidant capacity, reducing power, iridoid content, saponin content, and total phenolic content	[153]
17.	Fe_3O_4 NPs	50 and 500 mg/kg	*Zea mays*	Promoted morphological attributes and dehydrogenase enzyme activity	[154]
18.	TiO_2 NPs	50, 100, 200, and 400 mg/L	*Coriandrum sativum* L.	Higher shoot K, Ca, Mg, Fe, Mn, Zn, and B accumulation along with improved antioxidant defense system to alleviate oxidative stress	[155]
19.	Ag NPs	100–1000 mg/L	*Zea mays*	Reduced phytotoxicity and oxidative stress	[156]
20.	CuO NPs	0.2 and 300 μg/mL	*Lactuca sativa* L.	Promoted plant growth and enhanced nitrite and *S*-nitrosothiol levels in radicle with involvement in nitric oxide signaling	[157]

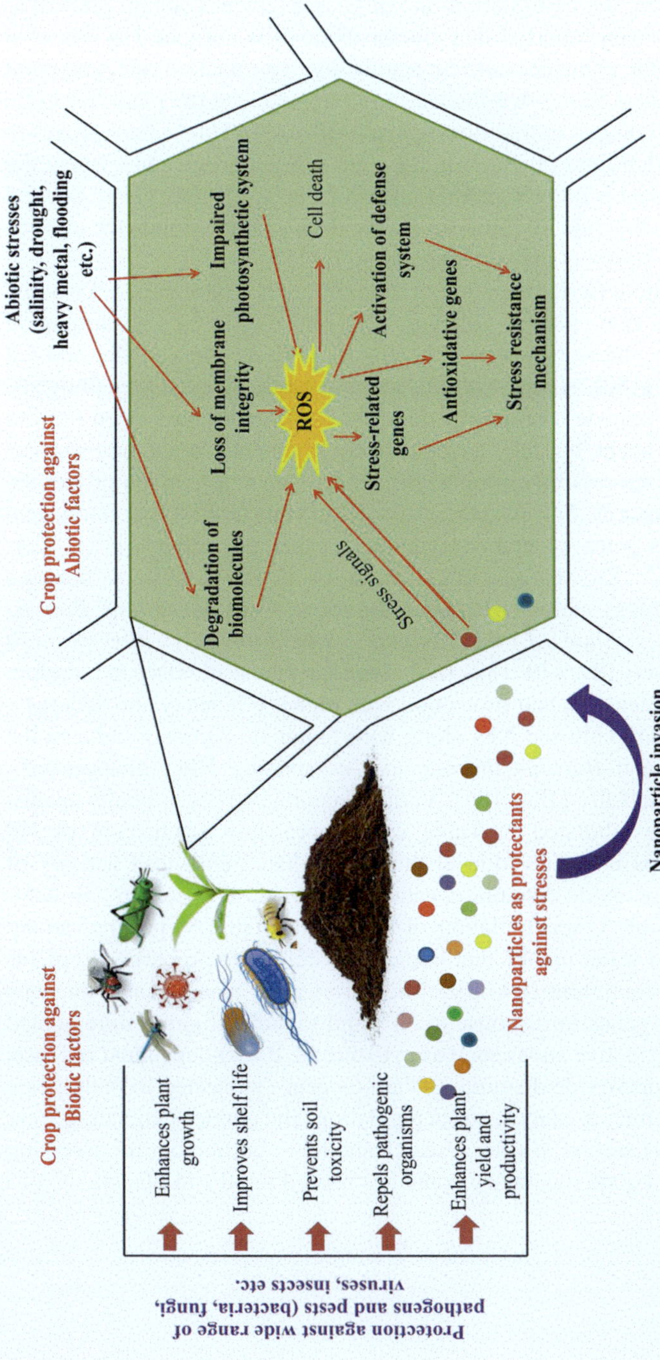

Fig. 5 Mechanistic role of nanoparticle-mediated stress tolerance in plants

mediated salinity tolerance in sunflower along with stimulating other characteristics such as leaf surface area, the assimilation rate of CO_2, pigment content, photosynthesis, etc. It has also been reported that silicon oxide NPs mitigate UV stress in wheat, while nano-zeolite enhances nutrient acquisition, germination rate, and plant growth [160]. Studies also found that the usage of NPs reduced the plant life cycle with enhanced yield in contrast to traditionally used fertilizers. This feature proves to be effective during agricultural practices in drought or flood-prone areas where the short life cycle of the plant is quintessential for sustainable agriculture [119]. Several reports encompassing the role of NPs in plants during stress tolerance against different biotic/abiotic stressors are summarized in Table 3.

NPs are also imperative for remediating or detoxifying heavy metals. Like in the case of Cd-stressed rice plants, silicon NPs enhanced Cd accumulation [89, 179]. Interestingly, these NPs were effective in curbing Pb, Cu, Zn, and Cd pollution, respectively [180]. Along with this, certain biotic agents like pathogens, pests, and insects also impede crop production [181]. Pesticides have been used to control this obnoxious agent, but they are against environmental health and sustainability. The use of NPs successfully minimizes the pathogen attack and improves the crop yield from such hazards. For instance, silver NPs synthesized from cotton stem extract showed a strong potential against bacterial diseases, respectively [182]. Certain metal oxide NPs (Cu, Zn, Mg) also inhibit various soil- and plant-borne diseases caused by pathogenic fungi such as *Alternaria alternate*, *Fusarium solani*, *Botrytis cinerea*, *Verticillium dahliae*, *Colletotrichum gloeosporioides*, *Phytophthora infestans*, etc. respectively [89, 179, 183–185]. The judicial use of NPs can therefore prove beneficial in the field of plant protection with higher efficiency and accuracy. The silver combined with chitosan NPs along with fungicide Antracol induced the antifungal properties, and *Bacillus thuringiensis* comprising NPs enhances efficiency and pesticide shelf life [184, 186]. Various reports of NPs in plants against stress management are tabulated in Table 2. The probable mechanism of NP effectively toward crops is mainly due to the NP-mediated enhanced activity of enzymatic systems (superoxide dismutase, catalase, peroxidase, ascorbate, glutathione, nitrate reductase, etc.), accumulation of metabolites (sugars, proline, amino acids), and nutrient and water uptake that imparts the defensive properties to plants for coping the stressed conditions [89, 179]. Microarray assay also revealed the gene expression of different genes in response to NPs and found that genes upregulated were concerned with oxidative stress such as cytochrome P-450-dependent oxidase, peroxidase, catalase, superoxide dismutase. Various genes subjected to pathogenic attack, hormones, systemic resistance, auxin regulation, and ethylene signaling were downregulated [132]. Therefore, further research should be carried out for assessing signaling responses and gene regulatory pathways related to NPs in plants.

Table 3 Role of nanoparticles in plants during stress resistance against various stressors

S. No	Nano-particles	Conc.	Plant	Stress	Response in plants	References
Biotic stress						
1.	CuNPs	0.1, 0.25, 0.5, 0.75, and 1.0 mg/mL	*Solanum lycopersicum*	*Fusarium oxysporum*	Induced plant growth and chlorophyll content and acted as potential fertilizer	[161]
2.	MgO NPs	25, 50, or 100 ppm	*Vigna unguiculata* L.	*Meloidogyne incognita*	Lowered nematode fecundity and galls and stimulated plant growth, chlorophyll, carotenoid, seed protein, and root/shoot nitrogen content	[162]
3.	Alginate-chitosan nanoparticles (CNPs)	–	*Capsicum annuum*	*Sclerotium rolfsii*	High IAA synthesis, nitrogen-fixing capacity, 1-aminocyclopropane-1-carboxylic acid (ACC) deaminase production, and antifungal activity	[163]
4.	SeNPs	100 ppm	*Solanum lycopersicum*	*Phytophthora infestans*	Plant growth promotion with an accumulation of lignin, callose, and hydrogen peroxide and elevated activities of lipoxygenase (LOX), phenylalanine lyase, β-1,3-glucanase, and superoxide dismutase as a cellular defense mechanism	[131]
5.	Chitosan-coupled copper nanoparticles (Ch-CuNPs)	0.2%	*Vigna unguiculata, Solanum lycopersicum*	*Rhizoctonia solani* and *Pythium aphanidermatum*	Acted as fungicide and growth promoter, and alternative to pesticides	[164]
6.	Ag NPs	50 mg/L	*Lycopersicon esculentum*	*Alternaria solani*	Higher antioxidant enzyme activity, along with enhanced plant growth attributes and lowered fungal infection	[165]

(continued)

Table 3 (continued)

S. No	Nano-particles	Conc.	Plant	Stress	Response in plants	References
7.	Fe_3O_4NPs	100 μg/mL	*Nicotiana benthamiana*	*Tobacco mosaic virus*	Induced plant biomass and antioxidants and upregulated SA synthesis along with the expression of SA-responsive PR genes (i.e., *PR1* and *PR2*), to enhance plant resistance	[166]
8.	S-NPs	0.1 mL	*Rosmarinus officinalis*	*Meloidogyne javanica*	S-NPs displayed nematicidal activity and mortality of second-stage juveniles	[167]
9.	Graphene oxide (GO), ZnO NPs	0.05 and 0.10 mg/mL	*Daucus carota*	*Pectobacterium carotovorum*, *Xanthomonas campestris* pv. *carotae*, *Meloidogyne javanica*, *Alternaria dauci*, and *Fusarium solani*	Improved plant growth, chlorophyll, carotenoid, and proline content with declined galling, nematode multiplication, and disease indices	[168]
10.	TiO_2 and ZnO NPs	0.25 and 0.50 mL/L	*Beta vulgaris*	*Pectobacterium betavasculorum*, *Xanthomonas campestris*, *Pseudomonas syringae*	Augmented plant growth, chlorophyll, carotenoid, proline, antioxidant activities, proline, and H_2O_2 contents and decreased MDA content and disease indices	[169]
Abiotic stress						
11.	Chitosan nanoparticles (CSNPs)	1%	*Catharanthus roseus*	Salinity	Enhancement in the activities of catalase, peroxidase, and glutathione reductase along with alleviation of oxidative stress	[170]
12.	Chitosan nanoparticles (CSNPs)	1%	*Catharanthus roseus*	Drought	Alkaloid and proline accumulation and higher activities of catalase and ascorbate peroxidase with induced gene expression of deacetylvindoline-4-*O*-acetyltransferase, strictosidine synthase, peroxidase 1, and	[171]

					geissoschizine synthase and reduced H_2O_2 and malondialdehyde (MDA) accumulation	
13.	ZnO NPs	25, 50, 100 mg/L	*Triticum aestivum*	Cd	Increased leaf chlorophyll content, leaf superoxide dismutase, and peroxidase and decreased oxidative stress	[172]
14.	Fe_3O_4 NP	2 mM	*Oryza sativa*	As	Significant increase in germination, root/shoot length, and As stress mitigation	[173]
15.	Zero-valent iron nanoparticles (nZVI)	5 mg/L	*Solanum lycopersicum*	Cr	Increased chlorophyll and carotenoid levels with effective Cr remediation	[174]
16.	Zn, B, Si, and zeolite NPs	20, 12, and 15 ppm	*Solanum tuberosum* L.	Salinity	Enhanced morphological characters, photosynthetic rate, stomatal conductance, chlorophyll content, tuber yield, proline, gibberellins, proteins, carbohydrates, and antioxidant enzymes	[175]
17.	ZnO-NPs	50 mg/L	*Lycopersicon esculentum*	Cd	Stimulated plant height, fresh and dry weight, leaf area, chlorophyll, photosynthetic attributes, (net photosynthetic rate, transpiration rate, internal CO_2 concentration, and stomatal conductance), protein content, and activities of nitrate reductase and carbonic anhydrase	[176]
18.	Si NPs	25, 50, and 100 mg/kg	*Triticum aestivum* L	Cd	Dwindled oxidative stress by reducing hydrogen peroxide, electrolyte leakage, and malondialdehyde, with an increase in growth,	[69, 70]

(continued)

Table 3 (continued)

S. No	Nano-particles	Conc.	Plant	Stress	Response in plants	References
					photosynthesis, superoxide dismutase, and peroxidase activities	
19.	Ag NPs	5 ppm	*Allium cepa*	Salinity	Stimulated proline, flavonoids, sugars, chlorophyll, and carotenoid content	[177]
20.	TiO_2 NPs	50, 100, and 200 mg/L	*Dracocephalum moldavica* L.	Salinity	Improved agronomic traits, highest essential oil content, and increased antioxidant enzyme activities with lowered H_2O_2 concentration	[178]

7 Conclusions and Future Perspectives

Agriculture is the only proprietor for the survival of the human race, and with the increasing population, there is a need to establish novel techniques and strategies to accelerate the production rate especially in developing countries. The green revolution has enhanced crop production worldwide but the use of massive chemical-based fertilizers dilapidates the ecosystem. The use of eco-friendly methods in plant growth promotion and maintaining the ecosystem has developed a better agriculture pattern. The latest techniques of using agro-nanofertilizers in contrast to chemical fertilizers have proved to be environment-friendly inputs to implement sustainable agriculture by the use of NPs. Nanotechnology enhances crop production along with maintaining the quality standards and proves to be an enrichment technique in farming systems. The surfacing of NP engineering and their actions in sustainable agriculture have revolutionized the global agriculture canvass with its novelty, enormous growth, and effectiveness to meet the global food demands. This also upholds the process to protect the environment from hazards, and NPs have emerged with the assurance to regulate and conserve the environment for plant production. NPs provide a new green revolution to the farmers by taking into account all the risks and mitigation practices. However, there is still an enormous information lacking about NPs in terms of their uptake, permissible limits, and ecotoxicology. Henceforth, many other investigations are on the way to untangle the behavior of NPs in the soil and its fate along with their interactions in the living systems.

References

1. Fatima, F., Hashim, A., & Anees, S. (2020). Efficacy of nanoparticles as nanofertilizer production: a review. *Environmental Science and Pollution Research, 17*, 1–2.
2. Bindraban, P. S., Dimkpa, C. O., & Pandey, R. (2020). Exploring phosphorus fertilizers and fertilization strategies for improved human and environmental health. *Biology and Fertility of Soils, 56*(3), 299–317.
3. Dwivedi, S., Saquib, Q., Al-Khedhairy, A. A., & Musarrat, J. (2016). Understanding the role of nanomaterials in agriculture. In *Microbial inoculants in sustainable agricultural productivity* (pp. 271–288). Springer.
4. Lv, M., Liu, Y., Geng, J., Kou, X., Xin, Z., & Yang, D. (2018). Engineering nanomaterials-based biosensors for food safety detection. *Biosensors & Bioelectronics, 106*, 122–128.
5. Pandey, G. (2018). Challenges and future prospects of agri-nanotechnology for sustainable agriculture in India. *Environmental Technology and Innovation, 11*, 299–307.
6. Vermeulen, S. J., Aggarwal, P. K., Ainslie, A., Angelone, C., Campbell, B. M., Challinor, A. J., Hansen, J. W., Ingram, J. S., Jarvis, A., Kristjanson, P., & Lau, C. (2012). Options for support to agriculture and food security under climate change. *Environmental Science & Policy, 15*(1), 136–144.
7. Zhao, L., Lu, L., Wang, A., Zhang, H., Huang, M., Wu, H., Xing, B., Wang, Z., & Ji, R. (2020). Nano-biotechnology in agriculture: Use of nanomaterials to promote plant growth and stress tolerance. *Journal of Agricultural and Food Chemistry, 68*(7), 1935–1947.

8. Chen, Y. W., Lee, H. V., Juan, J. C., & Phang, S. M. (2016). Production of new cellulose nanomaterial from red algae marine biomass *Gelidium elegans*. *Carbohydrate Polymers, 151*, 1210–1219.
9. Smita, S., Gupta, S. K., Bartonova, A., Dusinska, M., Gutleb, A. C., & Rahman, Q. (2012). Nanoparticles in the environment: assessment using the causal diagram approach. *Environmental Health, 11*(1), S13.
10. Sharma, V. K., Filip, J., Zboril, R., & Varma, R. S. (2015). Natural inorganic nanoparticles–formation, fate, and toxicity in the environment. *Chemical Society Reviews, 44*(23), 8410–8423.
11. Sharma, D., Kanchi, S., & Bisetty, K. (2019). Biogenic synthesis of nanoparticles: A review. *Arabian Journal of Chemistry, 12*(8), 3576–3600.
12. Ghozlan, H. A., Abouelkheir, S. S., & Sabry, S. A. (2018). Microbial fabrication of magnetic nanoparticles and their applications. In *Magnetic nanostructured materials* (pp. 117–136). Elsevier.
13. Strambeanu, N., Demetrovici, L., & Dragos, D. (2015). Natural sources of nanoparticles. In *Nanoparticles' promises and risks* (pp. 9–19). Springer.
14. Griffin, S., Masood, M. I., Nasim, M. J., Sarfraz, M., Ebokaiwe, A. P., Schäfer, K. H., Keck, C. M., & Jacob, C. (2018). Natural nanoparticles: A particular matter inspired by nature. *Antioxidants, 7*(1), 3.
15. Buzea, C., Pacheco, I. I., & Robbie, K. (2007). Nanomaterials and nanoparticles: sources and toxicity. *Biointerphases, 2*(4), MR17-MR71.
16. Jeevanandam, J., Barhoum, A., Chan, Y. S., Dufresne, A., & Danquah, M. K. (2018). Review on nanoparticles and nanostructured materials: History, sources, toxicity and regulations. *Beilstein Journal of Nanotechnology, 9*(1), 1050–1074.
17. Barnard, A. S., & Guo, H. (Eds.). (2012). *Nature's nanostructures*. CRC Press.
18. Klaine, S. J., Alvarez, P. J., Batley, G. E., Fernandes, T. F., Handy, R. D., Lyon, D. Y., Mahendra, S., McLaughlin, M. J., & Lead, J. R. (2008). Nanomaterials in the environment: Behavior, fate, bioavailability, and effects. *Environmental Toxicology and Chemistry, 27*(9), 1825–1851.
19. Sadik, O. A. (2013). Anthropogenic nanoparticles in the environment. *Environmental Science: Processes & Impacts, 15*(1), 19–20.
20. Biswas, P., & Wu, C. Y. (2005). Nanoparticles and the environment. *Journal of the Air & Waste Management Association (1995), 55*(6), 708–746.
21. Bundschuh, M., Filser, J., Lüderwald, S., McKee, M. S., Metreveli, G., Schaumann, G. E., Schulz, R., & Wagner, S. (2018). Nanoparticles in the environment: Where do we come from, where do we go to? *Environmental Sciences Europe, 30*(1), 1–17.
22. Wagner, S., Gondikas, A., Neubauer, E., Hofmann, T., & von der Kammer, F. (2014). Spot the difference: engineered and natural nanoparticles in the environment—Release, behavior, and fate. *Angewandte Chemie, International Edition, 53*(46), 12398–12419.
23. Wallace, L. A., Emmerich, S. J., & Howard-Reed, C. (2004). Source strengths of ultrafine and fine particles due to cooking with a gas stove. *Environmental Science & Technology, 38*(8), 2304–2311.
24. Zhang, X., Yan, S., Tyagi, R. D., & Surampalli, R. Y. (2011). Synthesis of nanoparticles by microorganisms and their application in enhancing microbiological reaction rates. *Chemosphere, 82*(4), 489–494.
25. Musarrat, J., Dwivedi, S., Singh, B. R., Al-Khedhairy, A. A., Azam, A., & Naqvi, A. (2010). Production of antimicrobial silver nanoparticles in water extracts of the fungus *Amylomyces rouxii* strain KSU-09. *Bioresource Technology, 101*, 8772–8776.
26. Fang, X., Wang, Y., Wang, Z., Jiang, Z., & Dong, M. (2019). Microorganism assisted synthesized nanoparticles for catalytic applications. *Energies, 12*(1), 190.
27. Hulkoti, N. I., & Taranath, T. C. (2014). Biosynthesis of nanoparticles using microbes—A review. *Colloids and Surfaces. B, Biointerfaces, 121*, 474–483.

28. Li, X., Xu, H., Chen, Z. S., & Chen, G. (2011). Biosynthesis of nanoparticles by microorganisms and their applications. *Journal of Nanomaterials*. https://doi.org/10.1155/2011/270974
29. Mishra, S., Singh, B. R., Naqvi, A. H., & Singh, H. B. (2017). Potential of biosynthesized silver nanoparticles using *Stenotrophomonas* sp. BHU-S7 (MTCC 5978) for management of soil-borne and foliar phytopathogens. *Scientific Reports, 7*, 1–15.
30. Iravani, S. (2014). Bacteria in nanoparticle synthesis: Current status and future prospects. *International Scholarly Research Notices, 2014*, 359316.
31. Borse, V., Kaler, A., & Banerjee, U. C. (2015). Microbial synthesis of platinum nanoparticles and evaluation of their anticancer activity. *International Journal of Emerging Trends in Electrical and Electronics, 11*, 26–31.
32. Ranjitha, V. R., & Rai, V. R. (2017). Actinomycetes mediated synthesis of gold nanoparticles from the culture supernatant of *Streptomyces griseoruber* with special reference to catalytic activity. *3 Biotech, 7*(5), 299.
33. He, S., Zhang, Y., Guo, Z., & Gu, N. (2008). Biological synthesis of gold nanowires using extract of *Rhodopseudomonas capsulata*. *Biotechnology Progress, 24*(2), 476–480.
34. Basu, A., Ray, S., Chowdhury, S., Sarkar, A., Mandal, D. P., Bhattacharjee, S., & Kundu, S. (2018). Evaluating the antimicrobial, apoptotic, and cancer cell gene delivery properties of protein-capped gold nanoparticles synthesized from the edible mycorrhizal fungus *Tricholoma crassum*. *Nanoscale Research Letters, 13*(1), 154.
35. Musarrat, J., Dwivedi, S., Singh, B. R., Saquib, Q., & Al-Khedhairy, A. A. (2011). Microbially synthesized nanoparticles: scope and applications. In *Microbes and microbial technology* (pp. 101–126). Springer.
36. Bhargava, A., Jain, N., Barathi, M., Akhtar, M. S., Yun, Y. S., & Panwar, J. (2013). Synthesis, characterization and mechanistic insights of mycogenic iron oxide nanoparticles. In *Nanotechnology for sustainable development* (pp. 337–348). Springer.
37. Kundu, D., Hazra, C., Chatterjee, A., Chaudhari, A., & Mishra, S. (2014). Extracellular biosynthesis of zinc oxide nanoparticles using *Rhodococcus pyridinivorans* NT2: multifunctional textile finishing, biosafety evaluation and in vitro drug delivery in colon carcinoma. *Journal of Photochemistry and Photobiology B: Biology, 140*, 194–204.
38. Jayaseelan, C., Rahuman, A. A., Kirthi, A. V., Marimuthu, S., Santhoshkumar, T., Bagavan, A., Gaurav, K., Karthik, L., & Rao, K. B. (2012). Novel microbial route to synthesize ZnO nanoparticles using *Aeromonas hydrophila* and their activity against pathogenic bacteria and fungi. *Spectrochimica Acta Part A: Molecular and Biomolecular Spectroscopy, 90*, 78–84.
39. Malarkodi, C., Rajeshkumar, S., Paulkumar, K., Vanaja, M., Gnanajobitha, G., & Annadurai, G. (2014). Biosynthesis and antimicrobial activity of semiconductor nanoparticles against oral pathogens. *Bioinorganic Chemistry and Applications, 2014*, 347167.
40. Ahmad, M. S., Yasser, M. M., Sholkamy, E. N., Ali, A. M., & Mehanni, M. M. (2015). Anticancer activity of biostabilized selenium nanorods synthesized by Streptomyces bikiniensis strain Ess_amA-1. *International Journal of Nanomedicine, 10*, 3389.
41. Dwivedi, S., AlKhedhairy, A. A., Ahamed, M., & Musarrat, J. (2013). Biomimetic synthesis of selenium nanospheres by bacterial strain JS-11 and its role as a biosensor for nanotoxicity assessment: A novel Se-bioassay. *PLoS One, 8*(3), e57404.
42. Elsoud, M. M. A., Al-Hagar, O. E., Abdelkhalek, E. S., & Sidkey, N. M. (2018). Synthesis and investigations on tellurium myconanoparticles. *Biotechnology Reports, 18*, e00247.
43. Zhuravliova, O. A., Voeikova, T. A., Khaddazh, M. K., Bulushova, N. V., Ismagulova, T. T., Bakhtina, A. V., Gusev, S. A., Gritskova, I. A., Lupanova, T. N., Shaitan, K. V., & Debabov, V. G. (2018). Bacterial synthesis of cadmium and zinc sulfide nanoparticles: Characteristics and prospects of application. *Molecular Genetics, Microbiology and Virology, 33*(4), 233–240.
44. Ha, C., Zhu, N., Shang, R., Shi, C., Cui, J., Sohoo, I., Wu, P., & Cao, Y. (2016). Biorecovery of palladium as nanoparticles by *Enterococcus faecalis* and its catalysis for chromate reduction. *Chemical Engineering Journal, 288*, 246–254.

45. Purohit, J., Chattopadhyay, A., & Singh, N. K. (2019). Green synthesis of microbial nanoparticle: Approaches to application. In *Microbial nanobionics* (pp. 35–60). Springer.
46. Fariq, A., Khan, T., & Yasmin, A. (2017). Microbial synthesis of nanoparticles and their potential applications in biomedicine. *Journal of Applied Biomedicine, 15*(4), 241–248.
47. Sunkar, S., & Nachiyar, C. V. (2012). Biogenesis of antibacterial silver nanoparticles using the endophytic bacterium *Bacillus cereus* isolated from *Garcinia xanthochymus*. *Asian Pacific Journal of Tropical Biomedicine, 2*(12), 953–959.
48. Shahzad, A., Saeed, H., Iqtedar, M., Hussain, S. Z., Kaleem, A., Abdullah, R., Sharif, S., Naz, S., Saleem, F., Aihetasham, A., & Chaudhary, A. (2019). Size-controlled production of silver nanoparticles by *Aspergillus fumigatus* BTCB10: Likely antibacterial and cytotoxic effects. *Journal of Nanomaterials, 2019*. https://doi.org/10.1155/2019/5168698
49. Abboud, Y., Saffaj, T., Chagraoui, A., El Bouari, A., Brouzi, K., Tanane, O., & Ihssane, B. (2014). Biosynthesis, characterization and antimicrobial activity of copper oxide nanoparticles (CONPs) produced using brown alga extract (*Bifurcaria bifurcata*). *Applied Nanoscience, 4*(5), 571–576.
50. Marimuthu, S., Rahuman, A. A., Kirthi, A. V., Santhoshkumar, T., Jayaseelan, C., & Rajakumar, G. (2013). Eco-friendly microbial route to synthesize cobalt nanoparticles using *Bacillus thuringiensis* against malaria and dengue vectors. *Parasitology Research, 112*(12), 4105–4112.
51. Sutradhar, K. B., & Amin, M. L. (2014). Nanotechnology in cancer drug delivery and selective targeting. ISRN Nanotechnology 2014.
52. Ortega, F. G., Fernández-Baldo, M. A., Fernández, J. G., Serrano, M. J., Sanz, M. I., Diaz-Mochón, J. J., Lorente, J. A., & Raba, J. (2015). Study of antitumor activity in breast cell lines using silver nanoparticles produced by yeast. *International Journal of Nanomedicine, 10*, 2021.
53. Moeinzadeh, S., & Jabbari, E. (2017). Nanoparticles and their applications. In *Springer handbook of nanotechnology* (pp. 335–361). Springer.
54. Moghadam, A. V., Iranbakhsh, A., Saadatmand, S., Ebadi, M., & Ardebili, Z. O. (2021). New insights into the transcriptional, epigenetic, and physiological responses to zinc oxide nanoparticles in *Datura stramonium*; potential species for phytoremediation. *Journal of Plant Growth Regulation, 29*, 1–1.
55. Shang, Y., Hasan, M., Ahammed, G. J., Li, M., Yin, H., & Zhou, J. (2019). Applications of nanotechnology in plant growth and crop protection: A review. *Molecules, 24*(14), 2558.
56. Monica, R. C., & Cremonini, R. (2009). Nanoparticles and higher plants. *Caryologia, 62*, 161–165.
57. Pollard, M., Beisson, F., Li, Y. H., & Ohlrogge, J. B. (2008). Building lipid barriers: Biosynthesis of cutin and suberin. *Trends in Plant Science, 13*, 236–246.
58. Eichert, G. H. E. (2008). Equivalent pore radii of hydrophilic foliar uptake routes in stomatous and astomatous leaf surfaces-further evidence for a stomatal pathway. *Physiologia Plantarum, 132*(491–502), 35.
59. Popp, C., Burghardt, M., Friedmann, A., & Riederer, M. (2005). Characterization of hydrophilic and lipophilic pathways of *Hedera helix* L. cuticular membranes: Permeation of water and uncharged organic compounds. *Journal of Experimental Botany, 56*, 2797–2806.
60. Rico, C. M., Majumdar, S., Duarte-Gardea, M., Peralta-Videa, J. R., & Gardea-Torresdey, J. L. (2011). Interaction of nanoparticles with edible plants and their possible implications in the food chain. *Journal of Agricultural and Food Chemistry, 59*, 3485–3498.
61. Taylor, A. F., Rylott, E. L., Anderson, C. W., & Bruce, N. C. (2014). Investigating the toxicity, uptake, nanoparticle formation and genetic response of plants to gold. *PLoS One, 9*, e93793.
62. Ma, Y. H., He, X., Zhang, P., Zhang, Z. Y., Guo, Z., Tai, R. Z., Xu, Z. J., Zhang, L. J., Ding, Y. Y., Zhao, Y. L., & Chai, Z. F. (2011). Phytotoxicity and biotransformation of La_2O_3 nanoparticles in a terrestrial plant cucumber (*Cucumis sativus*). *Nanotoxicology, 5*, 743–753.

63. Zhang, P., Ma, Y. H., Zhang, Z. Y., He, X., Zhang, J., Guo, Z., Tai, R. Z., Zhao, Y. L., & Chai, Z. F. (2012). Biotransformation of ceria nanoparticles in cucumber plants. *ACS Nano, 6*, 9943–9950.
64. Rastogi, A., Tripathi, D. K., Yadav, S., Chauhan, D. K., Živčák, M., Ghorbanpour, M., El-Sheery, N. I., & Brestic, M. (2019). Application of silicon nanoparticles in agriculture. *3 Biotech, 9*(3), 1–1.
65. Uzu, G., Sobanska, S., Sarret, G., Munoz, M., & Dumat, C. (2010). Foliar lead uptake by lettuce exposed to atmospheric fallouts. *Environmental Science & Technology, 44*, 1036–1042.
66. Hatami, M., Naghdi Badi, H., & Ghorbanpour, M. (2019). Nano-elicitation of secondary pharmaceutical metabolites in plant cells: A review. *Journal of Medicinal Plants, 18*, 6–36.
67. Yarizade, K., & Hosseini, R. (2015). Expression analysis of ADS, DBR2, ALDH1 and SQS genes in *Artemisia vulgaris* hairy root culture under nano cobalt and nano zinc elicitation. *Extension Journal of Applied Science, 3*, 69–76.
68. Marslin, G., Sheeba, C. J., & Franklin, G. (2017). Nanoparticles alter secondary metabolism in plants via ROS burst. *Frontiers in Plant Science, 8*, 832.
69. Khan, M. A., Wallace, W. T., Sambi, J., Rogers, D. T., Littleton, J. M., Rankin, S. E., & Knutson, B. L. (2020). Nanoharvesting of bioactive materials from living plant cultures using engineered silica nanoparticles. *Materials Science and Engineering C: Materials for Biological Applications, 106*, 110190.
70. Khan, Z. S., Rizwan, M., Hafeez, M., Ali, S., Adrees, M., Qayyum, M. F., Khalid, S., Ur Rehman, M. Z., & Sarwar, M. A. (2020). Effects of silicon nanoparticles on growth and physiology of wheat in cadmium contaminated soil under different soil moisture levels. *Environmental Science and Pollution Research, 27*(5), 4958–4968.
71. Ali, S., Rizwan, M., Noureen, S., Anwar, S., Ali, B., Naveed, M., Abd Allah, E. F., Alqarawi, A. A., & Ahmad, P. (2019). Combined use of biochar and zinc oxide nanoparticle foliar spray improved the plant growth and decreased the cadmium accumulation in rice (*Oryza sativa* L.) plant. *Environmental Science and Pollution Research, 26*(11), 11288–11299.
72. Mosavat, N., Golkar, P., Yousefifard, M., & Javed, R. (2019). Modulation of callus growth and secondary metabolites in different *Thymus* species and *Zataria multiflora* micropropagated under ZnO nanoparticles stress. *Biotechnology and Applied Biochemistry, 66*, 316–322.
73. Hezaveh, T. A., Rahmani, F., Alipour, H., & Pourakbar, L. (2020). Effects of foliar application of ZnO nanoparticles on secondary metabolite and micro-elements of Camelina (*Camelina sativa* L.) under salinity stress. *Journal of Stress Physiology and Biochemistry, 16*, 54–69.
74. Li, J., Wan, F., Guo, W., Huang, J., Dai, Z., Yi, L., & Wang, Y. (2020). Influence of α-and γ-Fe_2O_3 nanoparticles on watermelon (*Citrullus lanatus*) physiology and fruit quality. *Water, Air, and Soil Pollution, 231*(4), 1–2.
75. Kruszka, D., Sawikowska, A., Selvakesavan, R. K., Krajewski, P., Kachlicki, P., & Franklin, G. (2020). Silver nanoparticles affect phenolic and phytoalexin composition of *Arabidopsis thaliana*. *Science of the Total Environment, 716*, 135361.
76. Ogunkunle, C. O., Adegboye, E. F., Okoro, H. K., Vishwakarma, V., Alagarsamy, K., & Fatoba, P. O. (2020). Effect of nanosized anatase TiO_2 on germination, stress defense enzymes, and fruit nutritional quality of *Abelmoschus esculentus* (L.) Moench (okra). *Arabian Journal of Geosciences, 13*, 120.
77. Hegazi, G. A., Ibrahim, W. M., Hendawy, M. H., Salem, H. M., & Ghareb, H. E. (2020). Improving α-tocopherol accumulation in *Argania spinosa* suspension cultures by precursor and nanoparticles feeding. *Plant Archives, 20*, 2431–2437.
78. Khajavi, M., Rahaie, M., & Ebrahimi, A. (2019). The effect of TiO2 and SiO2 nanoparticles and salinity stress on expression of genes involved in parthenolide biosynthesis in Feverfew (*Tanacetum parthenium* L.). *Caryologia. International Journal of Cytology. Cytosystemics and Cytogenetics, 72*, 3–14.

79. Karakas, O. (2020). Effect of silver nanoparticles on production of indole alkaloids in *Isatis constricta*. *Iranian Journal of Sciences and Technology, Transactions A: Sciences, 44*, 621–627.
80. Sharifi-Rad, R., Bahabadi, S. E., Samzadeh-Kermani, A., & Gholami, M. (2020). The Effect of Non-biological Elicitors on Physiological and Biochemical Properties of Medicinal Plant *Momordica charantia* L. *Iranian Journal of Sciences and Technology, Transactions A: Sciences, 44*, 1315–1326.
81. Ahmad, M. A., Javed, R., Adeel, M., Rizwan, M., Ao, Q., & Yang, Y. (2020). Engineered ZnO and CuO nanoparticles ameliorate morphological and biochemical response in tissue culture regenerants of candy leaf (*Stevia rebaudiana*). *Molecules, 25*, 1356.
82. Chung, I. M., Rajakumar, G., & Thiruvengadam, M. (2018). Effect of silver nanoparticles on phenolic compounds production and biological activities in hairy root cultures of *Cucumis anguria*. *Acta Biologica Hungarica, 69*, 97–109.
83. Jamshidi, M., & Ghanati, F. (2017). Taxanes content and cytotoxicity of hazel cells extract after elicitation with silver nanoparticles. *Plant Physiology and Biochemistry, 110*, 178–184.
84. Fazal, H., Abbasi, B. H., Ahmad, N., Ali, M., Shujait Ali, S., Khan, A., & Wei, D. Q. (2019). Sustainable production of biomass and industrially important secondary metabolites in cell cultures of selfheal (*Prunella vulgaris* L.) elicited by silver and gold nanoparticles. *Artificial Cells, Nanomedicine, and Biotechnology, 47*, 2553–2561.
85. Singh, O. S., Pant, N. C., Laishram, L., Tewari, M., Dhoundiyal, R., Joshi, K., & Pandey, C. (2018). Effect of CuO nanoparticles on polyphenols content and antioxidant activity in Ashwagandha (*Withania somnifera* L. Dunal). *Journal of Pharmacognosy and Phytochemistry, 7*, 3433–3439.
86. Barrios, A. C., Medina-Velo, I. A., Zuverza-Mena, N., Dominguez, O. E., Peralta-Videa, J. R., & Gardea-Torresdey, J. L. (2017). Nutritional quality assessment of tomato fruits after exposure to uncoated and citric acid coated cerium oxide nanoparticles, bulk cerium oxide, cerium acetate and citric acid. *Plant Physiology and Biochemistry, 110*, 100–107.
87. Zhao, L., Hu, J., Huang, Y., Wang, H., Adeleye, A., Ortiz, C., & Keller, A. A. (2017). ^{1}H NMR and GC–MS based metabolomics reveal nano-Cu altered cucumber (*Cucumis sativus*) fruit nutritional supply. *Plant Physiology and Biochemistry, 110*, 138–146.
88. Juarez-Maldonado, A., Ortega-Ortíz, H., Pérez-Labrada, F., Cadenas-Pliego, G., & Benavides-Mendoza, A. (2016). Cu Nanoparticles absorbed on chitosan hydrogels positively alter morphological, production, and quality characteristics of tomato. *Journal of Applied Botany and Food Quality, 89*, 183–189.
89. Wang, S., Li, S., Liu, Q., Wu, K., Zhang, J., Wang, S., Wang, Y., Chen, X., Zhang, Y., & Gao, C. (2015). The OsSPL16-GW7 regulatory module determines grain shape and simultaneously improves rice yield and grain quality. *Nature Genetics, 47*, 949–954.
90. Wang, Y., Li, R., Li, D., Jia, X., Zhou, D., Li, J., & Lyi, S. M. (2017). NIP1;2 is a plasma membrane-localized transporter mediating aluminum uptake, translocation, and tolerance in *Arabidopsis*. *Proceedings of the National Academy of Sciences, 114*, 5047–5052.
91. Billingsley, M., Singh, N., Ravikumar, P., Zhang, R., June, C. H., & Mitchell, M. J. (2020). Ionizable lipid nanoparticle mediated mRNA delivery for human CAR t cell engineering. *Nano Letters, 20*, 1578–1589.
92. Lau, H. Y., Wu, H., Wee, E. J. H., Trau, M., Wang, Y., & Botella, J. R. (2017). Specific and sensitive isothermal electrochemical biosensor for plant pathogen DNA detection with colloidal gold nanoparticles as probes. *Scientific Reports, 7*, 1–7.
93. Demirer, G. S., Zhang, H., Matos, J., Goh, N. S., Cunningham, F. J., Sung, Y., Chang, R., Aditham, A. J., Chio, L., Che, M. J., Staskawicz, B., & Landry, M. P. (2018). High aspect ratio nanomaterials enable delivery of functional genetic material without DNA integration in mature plants. *Nature Nanotechnology, 14*(5), 456–464.
94. Martin-Ortigosa, S., Peterson, D. J., Valenstein, J. S., Lin, V. S. Y., Trewyn, B. G., Lyznik, L. A., & Wang, K. (2014). Mesoporous silica nanoparticle-mediated intracellular cre protein delivery for maize genome editing via loxP site excision. *Plant Physiology, 164*, 537–547.

95. Prasad, R., Bhattacharyya, A., & Nguyen, Q. D. (2017). Nanotechnology in sustainable agriculture: Recent developments, challenges, and perspectives. *Frontiers in Microbiology, 8*, 1014.
96. Fraceto, L. F., Grillo, R., de Medeiros, G. A., Scognamiglio, V., Rea, G., & Bartolucci, C. (2016). Nanotechnology in agriculture: Which innovation potential does it have? *Frontiers in Environmental Science, 22*(4), 20.
97. Janmohammadi, M., Navid, A., Segherloo, A. E., & Sabaghnia, N. (2016). Impact of nano-chelated micronutrients and biological fertilizers on growth performance and grain yield of maize under deficit irrigation condition. *Biologija, 62*(2), 134.
98. Karunakaran, G., Suriyaprabha, R., Rajendran, V., & Kannan, N. (2016). Influence of ZrO2, SiO_2, Al_2O_3 and TiO_2 nanoparticles on maize seed germination under different growth conditions. *IET Nanobiotechnology, 10*(4), 171–177.
99. Gogos, A., Knauer, K., & Bucheli, T. D. (2012). Nanomaterials in plant protection and fertilization: Current state, foreseen applications, and research priorities. *Journal of Agricultural and Food Chemistry, 60*(39), 9781–9792.
100. Khanna, K., Kohli, S. K., Handa, N., Kaur, H., Ohri, P., Bhardwaj, R., Yousaf, B., Rinklebe, J., & Ahmad, P. (2021). Enthralling the impact of engineered nanoparticles on soil microbiome: A concentric approach towards environmental risks and cogitation. *Ecotoxicology and Environmental Safety, 222*, 112459.
101. Kim, D. H., Gopal, J., & Sivanesan, I. (2017). Nanomaterials in plant tissue culture: the disclosed and undisclosed. *RSC Advances, 7*(58), 36492–36505.
102. Miller, J. B., Zhang, S., Kos, P., Xiong, H., Zhou, K., Perelman, S. S., Zhu, H., & Siegwart, D. J. (2007). Non-viral CRISPR/Cas gene editing in vitro and in vivo enabled by synthetic nanoparticle co-delivery of Cas9 mRNA and sgRNA. *Angewandte Chemie, International Edition, 56*(4), 1059–1063.
103. Kwak, S. Y., Wong, M. H., Lew, T. T., Bisker, G., Lee, M. A., Kaplan, A., Dong, J., Liu, A. T., Koman, V. B., Sinclair, R., & Hamann, C. (2017). Nanosensor technology applied to living plant systems. *Annual Review of Analytical Chemistry, 10*, 113–140.
104. Jampilek, J., & Kráľová, K. (2015). Application of nanotechnology in agriculture and food industry, its prospects and risks. *Ecological Chemistry and Engineering Science, 22*(3), 321–361.
105. Joga, M. R., Zotti, M. J., Smagghe, G., & Christiaens, O. (2016). RNAi efficiency, systemic properties, and novel delivery methods for pest insect control: What we know so far. *Frontiers in Physiology, 7*, 553.
106. Yang, H., Xu, M., Koide, R. T., Liu, Q., Dai, Y., Liu, L., & Bian, X. (2016). Effects of ditch-buried straw return on water percolation, nitrogen leaching and crop yields in a rice–wheat rotation system. *Journal of the Science of Food and Agriculture, 96*(4), 1141–1149.
107. Ghormade, V., Deshpande, M. V., & Paknikar, K. M. (2011). Perspectives for nano-biotechnology enabled protection and nutrition of plants. *Biotechnology Advances, 29*(6), 792–803.
108. Panpatte, D. G., Jhala, Y. K., Shelat, H. N., & Vyas, R. V. (2016). *Nanoparticles: The next generation technology for sustainable agriculture. In Microbial inoculants in sustainable agricultural productivity* (pp. 289–300). Springer.
109. Shojaei, T. R., Salleh, M. A., Tabatabaei, M., Mobli, H., Aghbashlo, M., Rashid, S. A., & Tan, T. (2019). Applications of nanotechnology and carbon nanoparticles in agriculture. In *Synthesis, technology and applications of carbon nanomaterials* (pp. 247–277). Elsevier.
110. Zhang, M., Gao, B., Chen, J., Li, Y., Creamer, A. E., & Chen, H. (2014). Slow-release fertilizer encapsulated by graphene oxide films. *Chemical Engineering Journal, 255*, 107–113.
111. Lyons, K. (2010). Nanotechnology: Transforming food and the environment. *Food First Backgrounder, 16*(1), 1–4.
112. Torney, F., Trewyn, B. G., Lin, V. S., & Wang, K. (2007). Mesoporous silica nanoparticles deliver DNA and chemicals into plants. *Nature Nanotechnology, 2*(5), 295–300.

113. Cheng, H. N., Klasson, K. T., Asakura, T., & Wu, Q. (2016). Nanotechnology in agriculture. In *Nanotechnology: Delivering on the promise* (Vol. 2, pp. 233–242). American Chemical Society.
114. Zhang, X., Zhang, J., & Zhu, K. Y. (2010). Chitosan/double-stranded RNA nanoparticle-mediated RNA interference to silence chitin synthase genes through larval feeding in the African malaria mosquito (*Anopheles gambiae*). *Insect Molecular Biology, 19*(5), 683–693.
115. Mout, R., Ray, M., Yesilbag Tonga, G., Lee, Y. W., Tay, T., Sasaki, K., & Rotello, V. M. (2017). Direct cytosolic delivery of CRISPR/Cas9-ribonucleoprotein for efficient gene editing. *ACS Nano, 11*(3), 2452–2458.
116. Li, C., Li, Y., Li, Y., & Fu, G. (2018). Cultivation techniques and nutrient management strategies to improve productivity of rain-fed maize in semi-arid regions. *Agricultural Water Management, 210*, 149–157.
117. Godfray, H. C., Beddington, J. R., Crute, I. R., Haddad, L., Lawrence, D., Muir, J. F., Pretty, J., Robinson, S., Thomas, S. M., & Toulmin, C. (2010). Food security: The challenge of feeding 9 billion people. *Science, 327*(5967), 812–818.
118. Shukla, P., Chaurasia, P., Younis, K., Qadri, O. S., Faridi, S. A., & Srivastava, G. (2019). Nanotechnology in sustainable agriculture: studies from seed priming to post-harvest management. *Nanotechnology for Environmental Engineering, 4*(1), 11.
119. Abdel-Aziz, H. M., Hasaneen, M. N., & Omer, A. M. (2016). Nano chitosan-NPK fertilizer enhances the growth and productivity of wheat plants grown in sandy soil. *Spanish Journal of Agricultural Research, 14*(1), 17.
120. Prasad, T. N., Sudhakar, P., Sreenivasulu, Y., Latha, P., Munaswamy, V., Reddy, K. R., Sreeprasad, T. S., Sajanlal, P. R., & Pradeep, T. (2012). Effect of nanoscale zinc oxide particles on the germination, growth and yield of peanut. *Journal of Plant Nutrition, 35*(6), 905–927.
121. Miao, Y. F., Wang, Z. H., & Li, S. X. (2015). Relation of nitrate n accumulation in dryland soil with wheat response to fertilizer. *Field Crops Research, 170*, 119–130.
122. Millán, G., Agosto, F., & Vázquez, M. (2008). Use of clinoptilolite as a carrier for nitrogen fertilizers in soils of the Pampean regions of Argentina. *International Journal of Agriculture and Natural Resources, 35*(3), 293–302.
123. Shalaby, T. A., Bayoumi, Y., Abdalla, N., Taha, H., Alshaal, T., Shehata, S., Amer, M., Domokos-Szabolcsy, É., & El-Ramady, H. (2016). Nanoparticles, soils, plants and sustainable agriculture. In *Nanoscience in food and agriculture* (pp. 283–312). Springer.
124. Sabir, A., Yazar, K., Sabir, F., Kara, Z., Yazici, M. A., & Goksu, N. (2014). Vine growth, yield, berry quality attributes and leaf nutrient content of grapevines as influenced by seaweed extract (*Ascophyllum nodosum*) and nanosize fertilizer pulverizations. *Scientia Horticulturae, 175*, 1–8.
125. Mastronardi, E., Tsae, P., Zhang, X., Monreal, C., & DeRosa, M. C. (2015). Strategic role of nanotechnology in fertilizers: Potential and limitations. In *Nanotechnologies in food and agriculture* (pp. 25–67). Springer.
126. Monreal, C. M., DeRosa, M., Mallubhotla, S. C., Bindraban, P. S., & Dimkpa, C. (2016). Nanotechnologies for increasing the crop use efficiency of fertilizer-micronutrients. *Biology and Fertility of Soils, 52*(3), 423–437.
127. Ali, A., Mohammad, S., Khan, M. A., Raja, N. I., Arif, M., Kamil, A., & Mashwani, Z. (2019). Silver nanoparticles elicited in vitro callus cultures for accumulation of biomass and secondary metabolites in *Caralluma tuberculata*. *Artificial Cells, Nanomedicine, and Biotechnology, 47*, 715–724.
128. Jiang, J., Cai, D. Q., Yu, Z. L., & Wu, Y. J. (2006). *Loss-control fertilizer made by active clay, flocculant and sorbent*. Chinese Patent Specification ZL200610040631. 1.
129. Cai, D., Wu, Z., Jiang, J., Wu, Y., Feng, H., Brown, I. G., Chu, P. K., & Yu, Z. (2014). Controlling nitrogen migration through micro-nano networks. *Scientific Reports, 4*(1), 1–8.

130. Manjaiah, K. M., Mukhopadhyay, R., Paul, R., Datta, S. C., Kumararaja, P., & Sarkar, B. (2019). Clay minerals and zeolites for environmentally sustainable agriculture. In *Modified clay and zeolite nanocomposite materials* (pp. 309–329). Elsevier.
131. Joshi, S. M., De Britto, S., & Jogaiah, S. (2021). Myco-engineered selenium nanoparticles elicit resistance against tomato late blight disease by regulating differential expression of cellular, biochemical and defense responsive genes. *Journal of Biotechnology, 325*, 196–206.
132. Banerjee, J., & Kole, C. (2016). Plant nanotechnology: An overview on concepts, strategies, and tools. *Plant Nanotechnology, 2016*, 1–4.
133. de França Bettencourt, G. M., Degenhardt, J., Torres, L. A., de Andrade Tanobe, V. O., & Soccol, C. R. (2020). Green biosynthesis of single and bimetallic nanoparticles of iron and manganese using bacterial auxin complex to act as plant bio-fertilizer. *Biocatalysis and Agricultural Biotechnology, 30*, 101822.
134. Kole, C., Kole, P., Randunu, K. M., Choudhary, P., Podila, R., Ke, P. C., Rao, A. M., & Marcus, R. K. (2013). Nanobiotechnology can boost crop production and quality: first evidence from increased plant biomass, fruit yield and phytomedicine content in bitter melon (*Momordica charantia*). *BMC Biotechnology, 13*(1), 1.
135. Yousefzadeh, S., & Sabaghnia, N. (2016). Nano-iron fertilizer effects on some plant traits of dragonhead (*Dracocephalum moldavica* L.) under different sowing densities. *Acta Agriculturae Slovenica, 107*(2), 429–437.
136. Davarpanah, S., Tehranifar, A., Davarynejad, G., Abadía, J., & Khorasani, R. (2016). Effects of foliar applications of zinc and boron nano-fertilizers on pomegranate (*Punica granatum* cv. Ardestani) fruit yield and quality. *Scientia Horticulturae, 210*, 57–64.
137. Dubey, A., & Mailapalli, D. R. (2016). Nanofertilisers, nanopesticides, nanosensors of pest and nanotoxicity in agriculture. In *Sustainable agriculture reviews* (pp. 307–330). Springer.
138. Lateef, A., Nazir, R., Jamil, N., Alam, S., Shah, R., Khan, M. N., & Saleem, M. (2016). Synthesis and characterization of zeolite based nano–composite: an environment friendly slow release fertilizer. *Microporous and Mesoporous Materials, 232*, 174–183.
139. Kottegoda, N., Munaweera, I., Madusanka, N., & Karunaratne, V. (2011). A green slow-release fertilizer composition based on urea-modified hydroxyapatite nanoparticles encapsulated wood. *Current Science, 10*, 73–78.
140. Li, M., Zhang, P., Adeel, M., Guo, Z., Chetwynd, A. J., Ma, C., Bai, T., Hao, Y., & Rui, Y. (2021). Physiological impacts of zero valent iron, Fe_3O_4 and Fe_2O_3 nanoparticles in rice plants and their potential as Fe fertilizers. *Environmental Pollution, 269*, 116134.
141. Madanayake, N. H., Adassooriya, N. M., & Salim, N. (2021). The effect of hydroxyapatite nanoparticles on *Raphanus sativus* with respect to seedling growth and two plant metabolites. *Environmental Nanotechnology, Monitoring and Management, 15*, 100404.
142. Biba, R., Tkalec, M., Cvjetko, P., Peharec Štefanić, P., Šikić, S., Pavoković, D., & Balen, B. (2021). Silver nanoparticles affect germination and photosynthesis in tobacco seedlings. *Acta Botanica Croatica, 80*(1), 1–1.
143. De Souza-Torres, A., Govea-Alcaide, E., Gómez-Padilla, E., Masunaga, S. H., Effenberger, F. B., Rossi, L. M., López-Sánchez, R., & Jardim, R. F. (2021). Fe_3O_4 nanoparticles and *Rhizobium* inoculation enhance nodulation, nitrogen fixation and growth of common bean plants grown in soil. *Rhizosphere, 17*, 100275.
144. Attia, E. A., & Elhawat, N. (2021). Combined foliar and soil application of silica nanoparticles enhances the growth, flowering period and flower characteristics of marigold (*Tagetes erecta* L.). *Scientia Horticulturae, 282*, 110015.
145. Iannone, M. F., Groppa, M. D., Zawoznik, M. S., Coral, D. F., van Raap, M. B., & Benavides, M. P. (2021). Magnetite nanoparticles coated with citric acid are not phytotoxic and stimulate soybean and alfalfa growth. *Ecotoxicology and Environmental Safety, 211*, 111942.
146. Casillas-Figueroa, F., Arellano-García, M. E., Leyva-Aguilera, C., Ruíz-Ruíz, B., Vázquez-Gómez, R. L., Radilla-Chávez, P., Chávez-Santoscoy, R. A., Pestryakov, A., Toledano-Magaña, Y., García-Ramos, J. C., & Bogdanchikova, N. (2020). Argovit™ silver

nanoparticles effects on *Allium cepa*: plant growth promotion without cyto genotoxic damage. *Nanomaterials, 10*(7), 1386.
147. Sun, L., Wang, Y., Wang, R., Wang, R., Zhang, P., Ju, Q., & Xu, J. (2020). Physiological, transcriptomic, and metabolomic analyses reveal zinc oxide nanoparticles modulate plant growth in tomato. *Environmental Science. Nano, 7*(11), 3587–3604.
148. Ma, Y., Xie, C., He, X., Zhang, B., Yang, J., Sun, M., Luo, W., Feng, S., Zhang, J., Wang, G., & Zhang, Z. (2020). Effects of ceria nanoparticles and CeCl3 on plant growth, biological and physiological parameters, and nutritional value of soil grown common bean (*Phaseolus vulgaris*). *Small, 16*(21), 1907435.
149. Dai, Y., Chen, F., Yue, L., Li, T., Jiang, Z., Xu, Z., Wang, Z., & Xing, B. (2020). Uptake, transport, and transformation of CeO_2 nanoparticles by strawberry and their impact on the rhizosphere bacterial community. *ACS Sustainable Chemistry & Engineering, 8*(12), 4792–4800.
150. Li, D., An, Q., Wu, Y., Li, J. Q., & Pan, C. (2020). Foliar application of selenium nanoparticles on celery stimulates several nutrient component levels by regulating the α-linolenic acid pathway. *ACS Sustainable Chemistry & Engineering, 8*, 10502–10510.
151. Rostamizadeh, E., Iranbakhsh, A., Majd, A., Arbabian, S., & Mehregan, I. (2020). Green synthesis of Fe_2O_3 nanoparticles using fruit extract of *Cornus mas* L. and its growth-promoting roles in barley. *Journal of Nanostructure in Chemistry, 17*, 1–6.
152. Rahman, M. S., Chakraborty, A., Mazumdar, S., Nandi, N. C., Bhuiyan, M. N., Alauddin, S. M., Khan, I. A., & Hossain, M. J. (2020). Effects of poly (vinylpyrrolidone) protected platinum nanoparticles on seed germination and growth performance of *Pisum sativum*. *Nano-Structures & Nano-Objects, 21*, 100408.
153. Chahardoli, A., Karimi, N., Ma, X., & Qalekhani, F. (2020). Effects of engineered aluminum and nickel oxide nanoparticles on the growth and antioxidant defense systems of *Nigella arvensis* L. *Scientific Reports, 10*(1), 1–1.
154. Yan, L., Li, P., Zhao, X., Ji, R., & Zhao, L. (2020). Physiological and metabolic responses of maize (*Zea mays*) plants to Fe3O4 nanoparticles. *Sciences of the Total Environment, 718*, 137400.
155. Hu, J., Wu, X., Wu, F., Chen, W., White, J. C., Yang, Y., Wang, B., Xing, B., Tao, S., & Wang, X. (2020). Potential application of titanium dioxide nanoparticles to improve the nutritional quality of coriander (*Coriandrum sativum* L.). *Journal of Hazardous Materials, 389*, 121837.
156. Abbas, Q., Yousaf, B., Ullah, H., Ali, M. U., Zia-ur-Rehman, M., Rizwan, M., & Rinklebe, J. (2020). Biochar-induced immobilization and transformation of silver-nanoparticles affect growth, intracellular-radicles generation and nutrients assimilation by reducing oxidative stress in maize. *Journal of Hazardous Materials, 390*, 121976.
157. Pelegrino, M. T., Kohatsu, M. Y., Seabra, A. B., Monteiro, L. R., Gomes, D. G., Oliveira, H. C., Rolim, W. R., de Jesus, T. A., Batista, B. L., & Lange, C. N. (2020). Effects of copper oxide nanoparticles on growth of lettuce (*Lactuca sativa* L.) seedlings and possible implications of nitric oxide in their antioxidative defense. *Environmental Monitoring and Assessment, 192*(4), 1–4.
158. Siddiqui, M. H., & Al-Whaibi, M. H. (2014). Role of nano-SiO2 in germination of tomato (*Lycopersicum esculentum* seeds Mill.). *Saudi Journal of Biological Sciences, 21*(1), 13–17.
159. Torabian, S., Zahedi, M., & Khoshgoftar, A. H. (2017). Effects of foliar spray of nanoparticles of $FeSO_4$ on the growth and ion content of sunflower under saline condition. *Journal of Plant Nutrition, 40*(5), 615–623.
160. Tripathi, D. K., Singh, S., Singh, V. P., Prasad, S. M., Dubey, N. K., & Chauhan, D. K. (2017). Silicon nanoparticles more effectively alleviated UV-B stress than silicon in wheat (*Triticum aestivum*) seedlings. *Plant Physiology and Biochemistry, 110*, 70–81.
161. Lopez-Lima, D., Mtz-Enriquez, A. I., Carrión, G., Basurto-Cereceda, S., & Pariona, N. (2021). The bifunctional role of copper nanoparticles in tomato: Effective treatment for Fusarium wilt and plant growth promoter. *Scientia Horticulturae, 277*, 109810.

162. Tauseef, A., Khalilullah, A., & Uddin, I. (2021). Role of MgO nanoparticles in the suppression of *Meloidogyne incognita*, infecting cowpea and improvement in plant growth and physiology. *Experimental Parasitology, 220*, 108045.
163. Panichikkal, J., Puthiyattil, N., Raveendran, A., Nair, R. A., & Krishnankutty, R. E. (2021). Application of encapsulated *Bacillus licheniformis* supplemented with chitosan nanoparticles and rice starch for the control of *Sclerotium rolfsii* in *Capsicum annuum* (L.) seedlings. *Current Microbiology, 78*(3), 911–919.
164. Vanti, G. L., Masaphy, S., Kurjogi, M., Chakrasali, S., & Nargund, V. B. (2020). Synthesis and application of chitosan-copper nanoparticles on damping off causing plant pathogenic fungi. *International Journal of Biological Macromolecules, 156*, 1387–1395.
165. Mahawar, H., Prasanna, R., Gogoi, R., Singh, S. B., Chawla, G., & Kumar, A. (2020). Synergistic effects of silver nanoparticles augmented *Calothrix elenkinii* for enhanced biocontrol efficacy against Alternaria blight challenged tomato plants. *3 Biotech, 10*(3), 1.
166. Cai, L., Cai, L., Jia, H., Liu, C., Wang, D., & Sun, X. (2020). Foliar exposure of Fe_3O_4 nanoparticles on *Nicotiana benthamiana*: Evidence for nanoparticles uptake, plant growth promoter and defense response elicitor against plant virus. *Journal of Hazardous Materials, 393*, 122415.
167. Al Banna, L. S., Salem, N. M., Jaleel, G. A., & Awwad, A. M. (2020). Green synthesis of sulfur nanoparticles using *Rosmarinus officinalis* leaves extract and nematicidal activity against *Meloidogyne javanica*. *Chemistry International, 6*(3), 137–143.
168. Siddiqui, Z. A., Khan, M. R., Abd Allah, E. F., & Parveen, A. (2019). Titanium dioxide and zinc oxide nanoparticles affect some bacterial diseases, and growth and physiological changes of beetroot. *International Journal of Vegetable Science, 25*(5), 409–430.
169. Siddiqui, Z. A., Parveen, A., Ahmad, L., & Hashem, A. (2019). Effects of graphene oxide and zinc oxide nanoparticles on growth, chlorophyll, carotenoids, proline contents and diseases of carrot. *Scientia Horticulturae, 249*, 374–382.
170. Hassan, F. A., Ali, E., Gaber, A., Fetouh, M., & Mazrou, R. (2021). Chitosan nanoparticles effectively combat salinity stress by enhancing antioxidant activity and alkaloid biosynthesis in *Catharanthus roseus* (L.) G. Don. *Plant Physiology and Biochemistry, 162*, 291–300.
171. Ali, E. F., El-Shehawi, A. M., Ibrahim, O. H., Abdul-Hafeez, E. Y., Moussa, M. M., & Hassan, F. A. (2021). A vital role of chitosan nanoparticles in improvisation the drought stress tolerance in *Catharanthus roseus* (L.) through biochemical and gene expression modulation. *Plant Physiology and Biochemistry, 161*, 166–175.
172. Adrees, M., Khan, Z. S., Hafeez, M., Rizwan, M., Hussain, K., Asrar, M., Alyemeni, M. N., Wijaya, L., & Ali, S. (2021). Foliar exposure of zinc oxide nanoparticles improved the growth of wheat (*Triticum aestivum* L.) and decreased cadmium concentration in grains under simultaneous Cd and water deficient stress. *Ecotoxicology and Environmental Safety, 208*, 111627.
173. Khan, S., Akhtar, N., Rehman, S. U., Shujah, S., Rha, E. S., & Jamil, M. (2021). Biosynthesized iron oxide nanoparticles (Fe_3O_4 NPs) mitigate arsenic toxicity in rice seedlings. *Toxics, 9*(1), 2.
174. Brasili, E., Bavasso, I., Petruccelli, V., Vilardi, G., Valletta, A., Dal Bosco, C., Gentili, A., Pasqua, G., & Di Palma, L. (2020). Remediation of hexavalent chromium contaminated water through zero-valent iron nanoparticles and effects on tomato plant growth performance. *Scientific Reports, 10*(1), 1–1.
175. Mahmoud, A. W., Abdeldaym, E. A., Abdelaziz, S. M., El-Sawy, M. B., & Mottaleb, S. A. (2020). Synergetic effects of zinc, boron, silicon, and zeolite nanoparticles on confer tolerance in potato plants subjected to salinity. *Agronomy, 10*(1), 19.
176. Faizan, M., Faraz, A., Mir, A. R., & Hayat, S. (2020). Role of zinc oxide nanoparticles in countering negative effects generated by cadmium in *Lycopersicon esculentum*. *Journal of Plant Growth Regulation, 5*, 1–5.

177. Jahangir, S., & Javed, K. (2020). Nanoparticles and plant growth promoting rhizobacteria (PGPR) modulate the physiology of onion plant under salt stress. *Pakistan Journal of Botany, 52*(4), 1473–1480.
178. Gohari, G., Mohammadi, A., Akbari, A., Panahirad, S., Dadpour, M. R., Fotopoulos, V., & Kimura, S. (2020). Titanium dioxide nanoparticles (TiO_2 NPs) promote growth and ameliorate salinity stress effects on essential oil profile and biochemical attributes of *Dracocephalum moldavica*. *Scientific Reports, 10*(1), 1–4.
179. Wang, S., Wang, F., & Gao, S. (2015). Foliar application with nano-silicon alleviates Cd toxicity in rice seedlings. *Environmental Science and Pollution Research, 22*(4), 2837–2845.
180. Wang, S., Wang, F., Gao, S., & Wang, X. (2016). Heavy metal accumulation in different rice cultivars as influenced by foliar application of nano-silicon. *Water, Air, and Soil Pollution, 227*(7), 1–3.
181. Oerke, E. C. (2006). Crop losses to pests. *The Journal of Agricultural Science, 144*(1), 31–43.
182. Vanti, G. L., Nargund, V. B., Vanarchi, R., Kurjogi, M., Mulla, S. I., Tubaki, S., & Patil, R. R. (2019). Synthesis of *Gossypium hirsutum*-derived silver nanoparticles and their antibacterial efficacy against plant pathogens. *Applied Organometallic Chemistry, 33*(1), e4630.
183. Imada, K., Sakai, S., Kajihara, H., Tanaka, S., & Ito, S. (2016). Magnesium oxide nanoparticles induce systemic resistance in tomato against bacterial wilt disease. *Plant Pathology, 65*(4), 551–560.
184. Le, V. T., Bach, L. G., Pham, T. T., Le, N. T., Ngoc, U. T., Tran, D. H., & Nguyen, D. H. (2019). Synthesis and antifungal activity of chitosan-silver nanocomposite synergize fungicide against *Phytophthora capsici*. *Journal of Macromolecular Sciences: Part A, 56*(6), 522–528.
185. Malandrakis, A. A., Kavroulakis, N., & Chrysikopoulos, C. V. (2019). Use of copper, silver and zinc nanoparticles against foliar and soil-borne plant pathogens. *Science of the Total Environment, 670*, 292–299.
186. Devi, P. V., Duraimurugan, P., & Chandrika, K. S. (2019). *Bacillus thuringiensis*-based nanopesticides for crop protection. In *Nano-biopesticides today and future perspectives* (pp. 249–260). Academic Press.

Nanoparticles in Plant Disease Management

Masudulla Khan, Azhar U. Khan, and Aiman Parveen

Abstract The agriculture system touches all aspects of our lives, and it is the cornerstone of a prosperous society. Sustainable agriculture becomes a priority to fill the increasing hunger and at the same time reduce the consumption of toxic chemicals. The agriculture sector has to deal with many issues like crop diseases, climate change, and reduction in land area. In addition, agriculture mainly relies on pesticides and fertilizers which have a toxic effect on humans and the environment. The world needs to develop more sustainable solutions for agriculture by using novel fields. Nowadays, nano-size materials are showing high applications in agriculture. To fill the food demand in the world, nano-enabled agrochemicals could play an important role.

1 Introduction

The agriculture system touches all aspects of our lives, and it is the cornerstone of a prosperous society. Sustainable agriculture becomes a priority to fill the increasing hunger and at the same time reduce the consumption of toxic chemicals. The agricultural sector has to deal with many issues like crop diseases, climate change, and reduction in land area. In addition, agriculture mainly relies on pesticides and fertilizers which have a toxic effect on humans and the environment. The world needs to develop more sustainable solutions for agriculture by using novel fields. Nowadays nano-size materials are showing high applications in agriculture. To fill the food demand in the world, nano-enabled agrochemicals could play an important role [1].

M. Khan (✉)
Botany Section, Women's College, Aligarh Muslim University, Aligarh, Uttar Pradesh, India

A. U. Khan
School of Life and Basic Sciences, SIILAS, Jaipur National University, Jaipur, Rajasthan, India

A. Parveen
Department of Botany, Aligarh Muslim University, Aligarh, Uttar Pradesh, India

J.-T. Chen (ed.), *Plant and Nanoparticles*,
https://doi.org/10.1007/978-981-19-2503-0_2

Nanotechnology is innovative science that has promising applications in many fields such as pesticides, fertilizers, plant protection, food, textiles, geosensing technology, paper, food, biofuel, agrochemical industries, biomass, biocomposites, and crop cultivation. Nanoparticles (NPs) have a specific nano-size with unique properties. Nanomaterials are stable and eco-environmental and can be used in biopesticide formulations. Nanomaterials could protect plants and are beneficial in agriculture and food [1, 2].

Nanomaterials have beneficial effects on plant metabolism, seed germination, and plant growth, and they could be used to control plant diseases [3, 4]. Recently, engineered NPs have emerged as potential candidates for improving crop yield [5]. The small size of NPs increases their chances of interaction with phytopathogens, rendering them effective against a wide range of diseases [6]. NPs could be used in plant disease control [7] because they show enhanced property at their minimum concentration. Unlike pesticides and fertilizers, the application of NPs can improve nutrient utilization and plant tolerance to biotic stresses, as well as plant growth with lesser environmental impact. The effect of NPs can also be influenced by the size, shape, concentration, chemical composition, surface structure, and solubility [8, 9]. Herbicides, fungicides, and pesticides can be used with nanoparticles to improve their impact against pests and pathogens [2].

2 Nanoparticles in Delivery of Herbicides

Herbicides' delivery by nanomaterial-based carriers is mostly focused on minimizing the toxic effect of herbicides on the environment. Nanoparticle-based herbicides include a large variety of nanoparticles. For example, iron (II, III) oxide magnetic NPs, hollow core-shell manganese carbonate, platy kaolinite, nano-sized tubular hallow site, and nano-sized rice husks [10–13]. Grillo et al. [14] found that polymeric nanoparticles consisting of ametryn, simazine, or atrazine were less toxic, compared to herbicide alone, during genotoxicity studies on *Allium cepa* cell cultures (Table 1).

Table 1 Some specific studies using nanoparticles in herbicide delivery

Nanoparticle	Herbicide	Target pest	References
Poly(lactic-*co*-glycolic acid)	Atrazine	Potato insect	[15]
Polymer	Metolachlor	Rice, *D. sanguinalis*, *A. thaliana*	[16]
Nano-sized rice husk	2,4-D	Maize	[10]
Chitosan/ tripolyphosphate/alginate	Imazapic and imazapyr combined	*B. pilosa*	[17]
Chitosan	Diuron	*E. crus-galli*	[18]
Tripolyphosphate/ chitosan	Paraquat	Mustard (*Brassica* sp.), Maize	[19]

3 Nanoparticles in Delivery of Insecticides

Nanoparticles can be used to deliver insecticides, and it was initiated in the early 2000s (Table 2). The most common target pests are *Helicoverpa armigera*, *Tetranychus urticae*, and *Spodoptera litura*. Nanoparticles can reduce toxicity and increase the solubility of insecticides. Insecticides that have low water solubility have been loaded successfully into porous silica [25] and chitosan [26, 27]. Nano-insecticides can increase the stability of the active compounds and maintain a sustained release which leads to low insecticide usage. Slow release of active compounds also reduces the toxicity of insecticides. Some studies highlighting the potential uses of nanoparticle carriers to increase the solubility of active insecticides have been shown in Table 2.

An issue with the use of insecticide is volatilization. This problem could be resolved by loading volatile active compounds inside nanoparticle carriers. For example, Oliveira et al. [28] found that toxicity was decreased in mouse fibroblast cell lines upon treatment with botanical repellents encapsulated into zinc nanoparticles, as compared to botanical repellents alone. Yang et al. [29] reported that encapsulation of garlic essential oil with polyethylene glycol (PEG) resulted in 80% mortality of red flour beetles (*Tribolium castaneum*).

4 Nanoparticles in Delivery of Fungicides

Studies on the formulation of nano-fungicides were started around 1997 on fungicidal screening into solid wood. Moreover, various research works have been conducted on conventional fungicides and biocides having antifungal properties with a wide range of nanoparticles (Table 3). Nanoparticle carriers like chitosan, polymer mixes, and silica are most commonly investigated [47, 48]. Nanoparticle-based fungicides can recover low water solubility issues and decrease volatilization while maintaining slow and sustained release. Some essential oils have fungicidal properties, but their rapid evaporation restricts large-scale commercial applications. Nanoparticles stabilize these essential oils and decrease their volatilization. The use of solid lipid nanoparticles (SL-NPs) stabilized essential oil of *Zataria multiflora*, thus protecting against six fungi [49]. Similarly, encapsulation of five essential oil

Table 2 Nanoparticle as carriers of insecticides

Nanoparticle	Insecticide	Target pest	References
Chitosan and ZnO	Azadirachtin	*Groundnut bruchid* (*C. serratus*)	[20]
Chitosan/TPP	Nicotine	Housefly (*M. domestica*)	[21]
Dendrimers	Thiamethoxam	Cotton bollworm (*H. armigera*)	[22]
Sodium alginate	Imidacloprid	Leafhopper (jassids)	[23]
Silica	Chlorfenapyr	Cotton bollworm (H. armigera) *P. xylostella*	[24]

Table 3 Nanoparticles in plant disease management

Pathogen	Concentration	Nanoparticle	Effect	References
Meloidogyne incognita, *Ralstonia solanacearum*, and *Fusarium oxysporum*	100 ppm	AgNPs	NPs found effective against test pathogens and pest	[30]
Pectobacterium betavasculorum, *Rhizoctonia solani*, and *Meloidogyne incognita*	0.20 g L^{-1}	TiO_2 NPs	Beetroot diseases reduced	[31]
Rhizoctonia solani	ZnO NPs	0.50 mL L^{-1}	NPs inhibited the growth of bacteria and fungi	[32]
Pseudomonas syringae, *Xanthomonas campestris*, *Pectobacterium carotovorum*, *Ralstonia solanacearum*, *Fusarium oxysporum*, and *Alternaria solani*	0.20 g L^{-1}	SiO_2 NPs	NPs inhibit the growth of test pathogens and also cause nematode mortality	[33]
Bipolaris sorokiniana	TiO_2 NPs	40 mg L^{-1}	NPs reduce the disease severity in wheat	[34]
Pseudomonas syringae, *Xanthomonas campestris* pv. *Vesicatoria*, *Pectobacterium carotovorum*, *and fungal diseases* (*Fusarium oxysporum f.* sp. *lycopersici and Alternaria solani*)	0.10 g L^{-1}, 0.20 g L^{-1}	ZnO NPs	NPs reduce disease in tomato plants	[35]
Pectobacterium betavasculorum, *Meloidogyne incognita*, and *Rhizoctonia solani*	0.50 mL L^{-1}	ZnO NPs	NPs reduce the growth of plant pathogenic bacteria and fungi	[32]
Ralstonia and Phomopsis	100 ppm	AgNPs	NPs inhibit the growth of bacteria and fungi	[36]
Pectobacterium carotovorum, *Fusarium solani*, *and Alternaria dauci*	100 mg L^{-1}	SiO_2 NPs	NPs inhibit the growth of bacteria and fungi	[37]
Meloidogyne incognita, *Pectobacterium betavasculorum*, and *Rhizoctonia solani*	0.2 mg L^{-1}	SiO_2 NPs	NPs reduce the disease complex of beetroot	[38]
Podosphaera xanthii	TiO_2 NPs	250 ppp L^{-1}	NPs inhibit the growth of the pathogen	[39]

(continued)

Table 3 (continued)

Pathogen	Concentration	Nanoparticle	Effect	References
Pyricularia grisea, Colletotrichum capsici and *Alternaria solani*	100 ppm	Se NPs	Inhibit the growth of tested pathogenic fungi	[40]
P. aeruginosa		ZnO NPs	ZnO NPs inhibit bacterial growth	[41]
Botrytis cinerea	1×10^{-3} to 5×10^{-3} concentration	ZnO NPs	NPs reduce the growth of *B. cinerea*	[42]
Botrytis cinerea	5×10^{-3} M	ZnO NPs	58% inactivation of bacteria found	[43]
Fusarium oxysporum and Penicillium expansum		ZnO NPs	100% inhibition of *P. expansum*	[44]
Colletotrichum sublineolum	Si	2 mmol L^{-1}	Si can reduce anthracnose by 20%	[45]
Pythium ultimum, Magnaporthe grisea, Colletotrichum gloeosporioides, Xanthomonas campestris pv. vesicatoria	100 ppm	Nano Si-Ag	The nano-sized silica-silver possesses antifungal activity	[46]

compounds individually into mesoporous silica NPs showed antifungal activity against *Aspergillus niger*, in comparison to the bulk compounds [50].

Leaching is the movement of water and chemicals into the soil that is a major issue with pesticides. The fungicide metalaxyl was loaded onto mesoporous-silica nanoparticles (MS-NPs), and leaching in the soil was observed between encapsulated metalaxyl (11.5% release) and free metalaxyl (76%) within 30 days. In water, the release rate of encapsulated metalaxyl was higher in comparison to the soil which highlights the importance of testing in farming conditions [51].

The slow release of fungicide compounds was achieved through encapsulation within nanoparticles. Kumar et al. [52] confirmed that carbendazim-loaded polymeric nanoparticles had high antifungal activity against *Aspergillus parasiticus* and *Fusarium oxysporum* and were safer for root development of *Zea mays*, *Lycopersicum esculentum*, and *Cucumis sativa*, compared to carbendazim alone. Similarly, nano-sized calcium carbonate loaded with validamycin showed the effectiveness of nano-carrier against *Rhizoctonia solani* [53]. Zhao et al. [54] observed that loading of fungicide pyrimethanil onto mesoporous silica nanoparticles reduces the risk of accumulation in edible parts of cucumber plants.

5 Nanoparticles: Carriers of Pesticides

Chitosan, silica layered double hydroxide nanoparticles (LDH-NPs), and solid lipid nanoparticles (SL-NPs) have been used as common carriers for pesticides. Silica nanoparticles (SiNPs) can be used as an influential delivery vehicle. Silica nanoparticles have the potential to develop agro-products for pest management [55]. Mesoporous-silica nanoparticles (MS-NPs) carry the pesticide into the interior to protect the active compound, thus ensuring a controlled release of the active compound.

Solid lipid NPs are solid at room temperature and could carry lipophilic active molecules [56]. Moreover, SL-NPs provide a controlled release of several lipophilic active compounds, due to a decrease in their mobility inside the matrix [57].

Chitosan has the property to adhere to the epidermis of leaves which may increase contact time and facilitate the uptake of molecules [58]. Chitosan contains highly reactive amine and hydroxyl groups which allow ionic interactions, modifications, and reactions, facilitating improvement of chitosan properties [59].

Layered double hydroxide nanoparticles (LDH-NPs) are present in the form of hexagonal sheets which contain active molecules in their interlayer space [60]. Positively charged LDH-NPs could be helpful in the transportation of biologically active molecules across the plant's cell wall [61].

6 Nano-encapsulation of Pesticides

In nano-encapsulation, pesticides are coated with nanomaterials of various shapes and sizes where inner side materials are referred to as core material and outer side materials are known as the coating nanomaterials [62]. Biodegradation, diffusion, dissolution, osmotic pressure, and a specific pH are among the action regulators of nanoencapsulation [63]. At present, nanoencapsulation is a novel and promising technology that can reduce the growth of pests. Nowadays, most pesticide companies are concentrating to develop new nanoscale pesticides by applying the nanoencapsulation technique.

The increased nanoformulations can reduce pesticidal movement, and they have beneficial properties in pesticide formulation like solubility, stability, and improved biodegradability [64]. They control the release of active ingredients and reduce the number of pesticides required for disease management. Furthermore, it ensures sustainable agriculture by minimum use of agrochemicals that may protect environmental damage and nontarget hosts and also reduces crop production costs. So, NP application is efficient in suppressing the pathogen infection and improving crop yield. For example, halloysite is a clay nanotube that serves as carriers of pesticides. These nano-sized tubes can extend the release period and also provide less environmental effect [65].

Li et al. [66] proposed that nano-silica is hydrophobic and can be absorbed into the cuticle layer of the insects upon contact which causes the death of insects. To manage the parasitic plant *Orobanche* sp., we apply herbicides sulfonylurea through the soil [67]. Encapsulated herbicide release is slow; also, we can do seed coating by applying encapsulated herbicides to avoid the multiple treatments [68]. Additionally, we can design nanocapsules and polymeric nanocapsules with different chemical compositions. Surfactants or oil drops can directly bind with the nanocapsule shell, and it improved cuticle penetration. Nanoencapsulation may improve the efficiency of herbicides that we cannot apply systemically like contact herbicides. It may result in a reduction of the delivered toxic amount. Stimulants used for better germination can be delivered within nanocapsules into the soil to protect them against degradation [69]. We can apply natural metabolites like mycotoxins inside nanocapsules [70]. Nanocapsules that improve herbicides activity penetrate through cuticles and facilitate the gradual release of the active substances. Some specific examples of nanoencapsulation:

1. Reduction in mortality among two insect pests, *Sarocladium oryzae* and *Rhyzopertha dominica*, was reported after treating with nano-alumina on wheat [71].
2. For the delivery of herbicides and pesticides in crop plants [72].
3. Halloysite nanotubes can be used for encapsulating the active agents including pesticides and herbicides [73, 74].

7 Nanoparticles: Against Plant Pest and Pathogens

Every year high yield loss in agriculture is reported due to various diseases. Generally, chemical pesticides are used on a large scale. Pesticides harm humans and the environment, so there is the need of time to search for an alternative for disease management. The major challenge that agriculture faces today is to reduce the high use of toxic agrochemicals. It is reported that these chemicals can hurt environmental factors plus human health [75]. Moreover, large quantities of pesticides may be lost during the application, due to photolysis, volatilization, and degradation, with below 0.1% of the applied pesticides acting effectively against target organisms [76]. More than 90% of the applied pesticides may either be lost in the environment or not able to reach the target sites critical for successful pest control [77]. Controlled delivery systems for pesticides proved to be a suitable alternative in combating these issues.

Nanomaterials protect the plants via two different mechanisms:

1. By direct application of nanoparticles in crop protection.
2. Nanomaterials as carriers for delivering existing pesticides.

The methods of controlled delivery ensure gradual release of agrochemicals in adequate and necessary quantities, to achieve maximum biological efficiency besides minimizing the harmful effects [78].

Nanoparticles could be delivery vehicles of pesticides. Unlike large size particles, the nanoparticles provide more effective delivery due to high surface area, fast transfer, and easy attachment. The active ingredient is encapsulated, adsorbed, or attached to the nano-matrix. The release of active ingredients may be controlled.

Zinc oxide (ZnO) NPs are the third most used NPs, 33,400 tons global annual production, and it could manage plant diseases [79, 80]. Foliar spray of ZnO NPs improved growth parameters and reduced diseases in beetroot [32]. Silicon NPs provide strength to prevent fungal and bacterial infections [81, 82]. Si was found effective in inducing host resistance against pathogens and reducing disease severity [83].

Application of TiO_2 NPs reduces disease caused by phytopathogenic bacteria [84]. *P. betavasculorum* and *P. syringae pv. aptata* pathogenic bacterial disease is reduced in beetroot. Silver NPs could be used in disease management; they show antimicrobial activity [85, 86]. Namburi et al. [86] found that biogenic Ag NPs showed antimicrobial activity against *Xanthomonas oryzae* pv. *oryzae*, the causal organism of bacterial leaf blight of rice. Ashraf et al. [87] reported antifungal activity of Ag NPs against *Fusarium oxysporum* which caused wilt in tomatoes. In vitro studies exhibited 79–98% inhibition of *F. oxysporum* as compared to the control. Silver NPs prevent the growth of fungi, rot, molds, and other plant diseases [88]. Silver NPs also showed antifungal activity against *Alternaria alternata*, *Macrophomina phaseolina*, and *Fusarium oxysporum* [89]. Zirconium dioxide NPs also show antifungal, antibacterial, and antimicrobial agents [90, 91]. ZrO NPs improved resistance against root rot disease, and application of ZrO_2 NPs suppressed the pathogen growth which could be due to its nano-size [6]. Jalill and Numan [92] found that zirconium oxide NPs could inhibit fungal growth and could decrease its pathogenicity.

8 Conclusion

This chapter gives an introduction to the role of nanoparticles in plant disease management and agricultural applications. Nanotechnology provides the development of nano-pesticides, nano-fertilizers, and nano-nutrients. In conclusion, nanoparticles may be a good alternative for plant disease management.

References

1. Khan, M., Khan, A. U., Hasan, M. A., & Yadav, K. K. (2021). Agro-nanotechnology as an emerging field: A novel sustainable approach for improving plant growth by reducing biotic stress. *Applied Sciences, 11*, 2282.
2. Lade, B. D., Gogle, D. P., Lade, D. B., Moon, G. M., Nandeshwar, S. B., & Kumbhare, S. D. (2019). Nanobiopesticide formulations: Application strategies today and future perspectives. In *Nano-biopesticides today and future perspectives* (pp. 179–206). Academic Press.

3. Khan, A. U., Khan, M., Malik, N., Cho, M. H., & Khan, M. M. (2019). Recent progress of algae and blue–green algae-assisted synthesis of gold nanoparticles for various applications. *Bioprocess and Biosystems Engineering, 42*, 1–15.
4. Gul, H. T., Saeed, S., Khan, F. Z. A., & Manzoor, S. A. (2014). Potential of nanotechnology in agriculture and crop protection: A review. *Applied Sciences and Business Economics, 1*(2), 23–28.
5. Nayantara, & Kaur, P. (2018). Biosynthesis of nanoparticles using eco-friendly factories and their role in plant pathogenicity: A review. *Biotechnology Research and Innovation, 2*(1), 63–73.
6. Derbalah, A., Elsharkawy, M. M., Hamza, A., & El-Shaer, A. (2019). Resistance induction in cucumber and direct antifungal activity of zirconium oxide nanoparticles against Rhizoctonia solani. *Pesticide Biochemistry and Physiology, 157*, 230–236.
7. Elmer, W., & White, J. C. (2018). The future of nanotechnology in plant pathology. *Annual Review of Phytopathology, 56*, 111–133.
8. Karimi, J., & Mohsenzadeh, S. (2016). Effects of silicon oxide nanoparticles on growth and physiology of wheat seedlings. *Russian Journal of Plant Physiology, 63*, 119–123.
9. Panpatte, D. G., Jhala, Y. K., Shelat, H. N., & Vyas, R. V. (2016). Nanoparticles – The next generation technology for sustainable agriculture. In D. P. Singh, H. B. Singh, & R. Prabha (Eds.), *Microbial inoculants in sustainable agricultural productivity* (Vol. 2: Functional applications, pp. 289–300). Springer.
10. Chidambaram, R. (2016). Application of rice husk nanosorbents containing 2, 4 dichlorophenoxyacetic acid herbicide to control weeds and reduce leaching from soil. *Journal of the Taiwan Institute of Chemical Engineers, 63*, 318–326.
11. Tan, D., Yuan, P., Annabi-Bergaya, F., Dong, F., Liu, D., & He, H. (2015). A comparative study of tubular halloysite and platy kaolinite as carriers for the loading and release of the herbicide amitrole. *Applied Clay Science, 114*, 190–196.
12. Kanimozhi, V., & Chinnamuthu, C. (2012). Engineering core/hallow shell nanomaterials to load herbicide active ingredient for controlled release. *Research Journal of Nanoscience and Nanotechnology, 2*, 58–69.
13. Viirlaid, E., Riiberg, R., Mäeorg, U., & Rinken, T. (2009). Glyphosate attachment on aminoactivated carriers for sample stabilization and concentration. *Agronomy Research, 13*, 1152–1159.
14. Grillo, R., dos Santos, N. Z. P., Maruyama, C. R., Rosa, A. H., de Lima, R., & Fraceto, L. F. (2012). Poly (ε-caprolactone) nanocapsules as carrier systems for herbicides: Physico-chemical characterization and genotoxicity evaluation. *Journal of Hazardous Materials, 231*, 1–9.
15. Schnoor, B., Elhendawy, A., Joseph, S., Putman, M., Chacón-Cerdas, R., Flores-Mora, D., Bravo-Moraga, F., Gonzalez-Nilo, F., & Salvador-Morales, C. (2018). Engineering atrazine loaded poly (lactic-co-glycolic acid) nanoparticles to ameliorate environmental challenges. *Journal of Agricultural and Food Chemistry, 66*, 7889–7898.
16. Tong, Y., Wu, Y., Zhao, C., Xu, Y., Lu, J., Xiang, S., Zong, F., & Wu, X. (2017). Polymeric nanoparticles as a metolachlor carrier: Water-based formulation for hydrophobic pesticides and absorption by plants. *Journal of Agricultural and Food Chemistry, 65*, 7371–7378.
17. Maruyama, C. R., Guilger, M., Pascoli, M., Bileshy-José, N., Abhilash, P. C., Fraceto, L. F., & De Lima, R. (2016). Nanoparticles based on chitosan as carriers for the combined herbicides imazapic and imazapyr. *Scientific Reports, 6*, 19768.
18. Yu, Z., Sun, X., Song, H., Wang, W., Ye, Z., Shi, L., & Ding, K. (2015). Glutathione-responsive carboxymethyl chitosan nanoparticles for controlled release of herbicides. *Materials Sciences and Applications, 6*, 591–604.
19. Grillo, R., Pereira, A. E., Nishisaka, C. S., De Lima, R., Oehlke, K., Greiner, R., & Fraceto, L. F. (2014). Chitosan/tripolyphosphate nanoparticles loaded with paraquat herbicide: An environmentally safer alternative for weed control. *Journal of Hazardous Materials, 278*, 163–171.

20. Jenne, M., Kambham, M., Tollamadugu, N. P., Karanam, H. P., Tirupati, M. K., Balam, R. R., Shameer, S., & Yagireddy, M. (2018). The use of slow releasing nanoparticle encapsulated Azadirachtin formulations for the management of Caryedon serratus O. (groundnut bruchid). *IET Nanobiotechnology, 12*, 963–967.
21. Yang, Y., Cheng, J., Garamus, V. M., Li, N., & Zou, A. (2018). Preparation of an environmentally friendly formulation of the insecticide nicotine hydrochloride through encapsulation in chitosan/tripolyphosphate nanoparticles. *Journal of Agricultural and Food Chemistry, 66*, 1067–1074.
22. Liu, X., He, B., Xu, Z., Yin, M., Yang, W., Zhang, H., Cao, J., & Shen, J. (2015). A functionalized fluorescent dendrimer as a pesticide nanocarrier: Application in pest control. *Nanoscale, 7*, 445–449.
23. Kumar, S., Bhanjana, G., Sharma, A., Sidhu, M., & Dilbaghi, N. (2014). Synthesis, characterization and on field evaluation of pesticide loaded sodium alginate nanoparticles. *Carbohydrate Polymers, 101*, 1061–1067.
24. Song, M. R., Cui, S. M., Gao, F., Liu, Y. R., Fan, C. L., Lei, T. Q., & Liu, D. C. (2012). Dispersible silica nanoparticles as carrier for enhanced bioactivity of chlorfenapyr. *Journal of Pesticide Science, 37*, 258–260.
25. Wang, Y., Cui, H., Sun, C., Zhao, X., & Cui, B. (2014). Construction and evaluation of controlled-release delivery system of Abamectin using porous silica nanoparticles as carriers. *Nanoscale Research Letters, 9*, 2490.
26. Campos, E. V. R., Proença, P. L. F., Oliveira, J. L., Melville, C. C., Della Vechia, J. F., De Andrade, D. J., & Fraceto, L. F. (2018). Chitosan nanoparticles functionalized with β-cyclodextrin: A promising carrier for botanical pesticides. *Scientific Reports, 8*, 2067.
27. Zhang, J., Li, M., Fan, T., Xu, Q., Wu, Y., Chen, C., & Huang, Q. (2013). Construction of novel amphiphilic chitosan copolymer nanoparticles for chlorpyrifos delivery. *Journal of Polymer Research, 20*, 107.
28. Oliveira, J. L. D., Campos, E. N. V., Pereira, A. E., Pasquoto, T., Lima, R., Grillo, R., Andrade, D. J. D., Santos, F. A. D., & Fraceto, L. F. (2018). Zein nanoparticles as eco-friendly carrier systems for botanical repellents aiming sustainable agriculture. *Journal of Agricultural and Food Chemistry, 66*, 1330–1340.
29. Yang, F. L., Li, X. G., Zhu, F., & Lei, C. L. (2009). Structural characterization of nanoparticles loaded with garlic essential oil and their insecticidal activity against Tribolium castaneum (Herbst) (Coleoptera: Tenebrionidae). *Journal of Agricultural and Food Chemistry, 57*, 10156–10162.
30. Khan, M., Khan, A. U., Bogdanchikova, N., & Garibo, D. (2021). Antibacterial and antifungal studies of biosynthesized silver nanoparticles against plant parasitic nematode *Meloidogyne incognita,* plant pathogens *Ralstonia solanacearum* and *Fusarium oxysporum. Molecules, 26*, 2462.
31. Khan, M. R., & Siddiqui, Z. A. (2021). Efficacy of titanium dioxide nanoparticles in the management of disease complex of beetroot (*Beta vulgaris* L.) caused by *Pectobacterium betavasculorum*, *Rhizoctonia solani,* and *Meloidogyne incognita*. *Gesunde Pflanzen, 73*, 445–464.
32. Khan, M. R., & Siddiqui, Z. A. (2021). Role of zinc oxide nanoparticles in the management of disease complex of beetroot (*Beta vulgaris* L.) caused by *Pectobacterium betavasculorum, Meloidogyne incognita* and *Rhizoctonia solani*. *Horticulture Environment and Biotechnology, 62*, 225–241.
33. Parveen, A., & Siddiqui, Z. A. (2022). Impact of silicon dioxide nanoparticles on growth, photosynthetic pigments, proline, activities of defense enzymes and some bacterial and fungal pathogens of tomato. *Vegetos, 35*, 83–93.
34. Satti, S. H., Raja, N. I., Javed, B., Akram, A., Mashwani, Z. U. R., Ahmad, M. S., & Ikram, M. (2021). Titanium dioxide nanoparticles elicited agro-morphological and physicochemical modifications in wheat plants to control *Bipolaris sorokiniana*. *PLoS One, 16*(2), e0246880.

35. Parveen, A., & Siddiqui, Z. A. (2021). Zinc oxide nanoparticles affect growth, photosynthetic pigments, proline content and bacterial and fungal diseases of tomato. *Archives of Phytopathology and Plant Protection, 54*, 1519.
36. Khan, M., Khan, A. U., Alam, M. J., Park, S., & Alam, M. (2020). Biosynthesis of silver nanoparticles and its application against phytopathogenic bacterium and fungus. *International Journal of Environmental Analytical Chemistry, 100*(12), 1390–1401.
37. Siddiqui, H., Ahmed, K. B., Sami, F., & Hayat, S. (2020). Silicon nanoparticles and plants: Current knowledge and future perspectives. *Sustainable Agriculture Reviews, 41*, 129–142.
38. Khan, M. R., & Siddiqui, Z. A. (2020). Use of silicon dioxide nanoparticles for the management of *Meloidogyne incognita, Pectobacterium betavasculorum* and *Rhizoctonia solani* disease complex of beetroot (*Beta vulgaris* L.). *Scientia Horticulturae, 265*, 109211.
39. Hafez, Y. M., Attia, K. A., Kamel, S., Alamery, S. F., El-Gendy, S., Al-Dosse, A., Mehiar, F., Ghazy, A., & Abdelaal, K. A. (2020). Bacillus subtilis as a bio-agent combined with nano molecules can control powdery mildew disease through histochemical and physiobiochemical changes in cucumber plants. *Physiological and Molecular Plant Pathology, 111*, 101489.
40. Joshi, S. M., De Britto, S., Jogaiah, S., & Ito, S. I. (2019). Mycogenic selenium nanoparticles as potential new generation broad spectrum antifungal molecules. *Biomolecules, 9*(9), 419.
41. Dobrucka, R., & Dlugaszewska, J. (2016). Biosynthesis and antibacterial activity of ZnO nanoparticles using *Trifolium pratense* flower extract. *Saudi Journal of Biological Sciences, 23*, 517–523.
42. Luksiene, Z., & Aponiene, K. (2015). First attempts to control microbial contamination of strawberries by ZnO nanoparticles. *Journal of Nanomedicine and Nanotechnology, 6*, 4.
43. Kairyte, K., Kadys, A., & Luksiene, Z. (2013). Antibacterial and antifungal activity of photoactivated ZnO nanoparticles in suspension. *Journal of Photochemistry and Photobiology, 128*, 78–84.
44. Yehia, R. S., & Ahmed, O. F. (2013). *In vitro* study of the antifungal efficacy of zinc oxide nanoparticles against *Fusarium oxysporum and Penicillium expansum. African Journal of Microbiology Research, 7*, 1917–1923.
45. Resende, R. S., Rodrigues, F. A., Gomes, R. J., & Nascimento, K. J. T. (2013). Microscopic and biochemical aspects of sorghum resistance to anthracnose mediated by silicon. *The Annals of Applied Biology, 163*, 114–123.
46. Park, H. J., Kim, S. H., Kim, H. J., & Choi, S. H. (2006). A new composition of nano sized silica- silver for control of various plant diseases. *Plant Pathology, 22*, 295–302.
47. Liu, Y., Yan, L., Heiden, P., & Laks, P. (2001). Use of nanoparticles for controlled release of biocides in solid wood. *Journal of Applied Polymer Science, 79*, 458–465.
48. Liu, Y., Laks, P., & Heiden, P. (2002). Controlled release of biocides in solid wood. I. Efficacy against brown rot wood decay fungus (Gloeophyllum trabeum). *Journal of Applied Polymer Science, 86*, 596–607.
49. Nasseri, M., Golmohammadzadeh, S., Arouiee, H., Jaafari, M. R., & Neamati, H. (2016). Antifungal activity of Zataria multiflora essential oil-loaded solid lipid nanoparticles in-vitro condition. *Iranian Journal of Basic Medical Sciences, 19*, 1231–1237.
50. Janatova, A., Bernardos, A., Smid, J., Frankova, A., Lhotka, M., Kourimská, L., Pulkrabek, J., & Kloucek, P. (2015). Long-term antifungal activity of volatile essential oil components released from mesoporous silica materials. *Industrial Crops and Products, 67*, 216–220.
51. Wanyika, H. (2013). Sustained release of fungicide metalaxyl by mesoporous silica nanospheres. *Journal of Nanoparticle Research, 15*, 1831.
52. Kumar, S., Kumar, D., & Dilbaghi, N. (2017). Preparation, characterization, and bio-efficacy evaluation of controlled release carbendazim-loaded polymeric nanoparticles. *Environmental Science and Pollution Research, 24*, 926–937.
53. Qian, K., Shi, T., Tang, T., Zhang, S., Liu, X., & Cao, Y. (2011). Preparation and characterization of nano-sized calcium carbonate as controlled release pesticide carrier for validamycin against Rhizoctonia solani. *Microchimica Acta, 173*, 51–57.

54. Zhao, P., Cao, L., Ma, D., Zhou, Z., Huang, Q., & Pan, C. (2017). Synthesis of pyrimethanil-loaded mesoporous silica nanoparticles and its distribution and dissipation in cucumber plants. *Molecules, 22*, 817.
55. Mody, V. V., Cox, A., Shah, S., Singh, A., Bevins, W., & Parihar, H. (2014). Magnetic nanoparticle drug delivery systems for targeting tumor. *Applied Nanoscience, 4*, 385–392.
56. Ekambaram, P., Sathali, A. A. H., & Priyanka, K. (2012). Solid lipid nanoparticles: A review. *Scientific Reviews & Chemical Communications, 2*, 80–102.
57. Borel, T., & Sabliov, C. (2014). Nanodelivery of bioactive components for food applications: Types of delivery systems, properties, and their effect on ADME profiles and toxicity of nanoparticles. *Annual Review of Food Science and Technology, 5*, 197–213.
58. Malerba, M., & Cerana, R. (2016). Chitosan effects on plant systems. *International Journal of Molecular Sciences, 17*, 996.
59. Li, M., Huang, Q., & Wu, Y. (2011). A novel chitosan-poly (lactide) copolymer and its submicron particles as imidacloprid carriers. *Pest Management Science, 67*, 831–836.
60. Xu, Z. P., Stevenson, G. S., Lu, C.-Q., Lu, G. Q., Bartlett, P. F., & Gray, P. P. (2006). Stable suspension of layered double hydroxide nanoparticles in aqueous solution. *Journal of the American Chemical Society, 128*, 36–37.
61. Bao, W., Wang, J., Wang, Q., O'Hare, D., & Wan, Y. (2016). Layered double hydroxide nanotransporter for molecule delivery to intact plant cells. *Scientific Reports, 6*, 26738.
62. Nguyen, T. N. Q., Le, V. A., Hua, Q. C., & Nguyen, T. T. (2015). *Enhancing insecticide activity of anacardic acid by intercalating it into MgAl layered double hydroxides nanoparticles*. Institut für Abfallwirtschaft und Altlasten, Fakultät Umweltwissenschaften, Technische Universität Dresden.
63. Vidyalakshmi, R., Bhakyaraj, R., & Subhasree, R. S. (2009). Encapsulation "the future of probiotics" – A review. *Advances in Biology Research, 3*(3–4), 96–103.
64. Gogos, A., Knauer, K., & Bucheli, T. D. (2012). Nanomaterials in plant protection and fertilization: Current state, foreseen applications, and research priorities. *Journal of Agricultural and Food Chemistry, 60*, 9781–9792.
65. Flood, J. (2010). The importance of plant health to food security. *Food Security, 2*(3), 215–231.
66. Li, Z. Z., Chen, J. F., Liu, F., Liu, A. Q., Wang, Q., Sun, H. Y., & Wen, L. X. (2007). Study of UV-shielding properties of novel porous hollow silica nanoparticle carriers for avermectin. *Pest Management Science, 63*, 241–246.
67. Joel, D. M., Hershenhorn, J., Eizenberg, H., Aly, R., Ejeta, G., & Rich, P. J. (2007). Biology and management of weedy root parasites. *Horticultural Reviews, 33*, 267–349.
68. Jurado-Exposito, M., Garcia-Torres, L., & Castejon-Munoz, M. (1997). Broad-bean and lentil seed treatments with imidazolines for the control of broomrape (Orobanche crenata). *The Journal of Agricultural Science, 129*, 307–314.
69. Lopez-Raez, J. A., Matusova, R., Cardoso, C., Jamil, M., Charnikhova, T., & Kohlen, W. (2009). Strigolactones: Ecological significance and use as a target for parasitic plant control. *Pest Management Science, 65*(5), 471–477.
70. Vurro, M., Boari, A., Evidente, A., Andolfi, A., & Zermane, N. (2009). Natural metabolites for parasitic weed management. *Pest Management Science, 65*(5), 566–571.
71. Stadler, T., Buteler, M., & Weaver, D. K. (2010). Novel use of nanostructured alumina as an insecticide. *Pest Management Science, 66*(6), 577–579.
72. Lodriche, S. S., Soltani, S., & Mirzazadeh, R. (2013). *Silicon nanocarrier for delivery of drug, pesticides and herbicides, and for waste water treatment*. US patent US 20130225412 A1, 29 August 2013.
73. Abdullayev, E., & Lvov, Y. (2011). Halloysite clay nanotubes for controlled release of protective agents. *Journal of Nanoscience and Nanotechnology, 11*(11), 10007–10026.
74. Murphy, K. (Ed.). (2008). *Nanotechnology: Agriculture's next "industrial" revolution* (pp. 3–5). Financial Partner, Yankee Farm Credit, ACA.
75. Mall, D., Larsen, A., & Martin, E. (2018). Investigating the (mis)match between natural pest control knowledge and the intensity of pesticide use. *Insects., 9*, 2.

76. Liang, Y., Guo, M., Fan, C., Dong, H., Ding, G., & Zhang, W. (2017). Development of novel urease responsive pendimethalin microcapsules using silica- IPTS-PEI as controlled release carrier materials. *ACS Sustainable Chemistry & Engineering, 5*, 4802–4810.
77. Nguyen, H., Hwang, I., Park, J., & Park, H. (2012). Enhanced payload and photo-protection for pesticides using nanostructured lipid carriers with corn oil as liquid lipid. *Journal of Microencapsulation, 29*, 596–604.
78. Tsuji, K. (2001). Microencapsulation of pesticides and their improved handling safety. *Journal of Microencapsulation, 18*, 137–147.
79. Sabir, S., Arshad, M., & Chaudhari, S. K. (2014). Zinc oxide nanoparticles for revolutionizing agriculture: Synthesis and applications. *The Scientific World Journal, 2014*, 1–8.
80. Bondarenko, O., Juganson, K., Ivask, A., Kasemets, K., Mortimer, M., & Kahru, A. (2013). Toxicity of Ag, CuO and ZnO nanoparticles to selected environmentally relevant test organisms and mammalian cells in vitro: A critical review. *Archives of Toxicology, 87*(7), 1181–1200.
81. Strout, G., Russell, S. D., Pulsifer, D. P., Erten, S., Lakhtakia, A., & Lee, D. W. (2013). Silica nanoparticles aid in structural leaf coloration in the Malaysian tropical rainforest understorey herb *Mapania caudata. Annals of Botany, 112*(6), 1141–1148.
82. Rastogi, A., Tripathi, D. K., Yadav, S., Chauhan, D. K., Zivcak, M., Ghorbanpour, M., El-Sheery, N. I., & Brestic, M. (2019). Application of silicon nanoparticles in agriculture. *3 Biotech, 9*, 90.
83. Ye, M., Song, Y. Y., Long, J., Wang, R. L., Baerson, S. R., Pan, Z., Zhu-Salzman, K., Xie, J., Cai, K., Luo, S., & Zeng, R. (2013). Priming of jasmonate-mediated antiherbivore defense responses in rice by silicon. *Proceedings of the National Academy of Sciences, 110*(38), E3631–E3639.
84. Siddiqui, Z. A., Khan, M. R., Abd Allah, E. F., & Parveen, A. (2018a). Titanium dioxide and zinc oxide nanoparticles affect some bacterial diseases, and growth and physiological changes of beetroot. *International Journal of Vegetable Science, 25*(5), 409–430.
85. Kaur, J., Singh, J., & Rawat, M. (2019). An efficient and blistering reduction of 4-nitrophenol by green synthesized silver nanoparticles. *SN Applied Sciences, 1*(9), 1–6.
86. Namburi, K. R., Kora, A. J., Chetukuri, A., & Kota, V. S. M. K. (2021). Biogenic silver nanoparticles as an antibacterial agent against bacterial leaf blight causing rice phytopathogen *Xanthomonas oryzae* pv. *oryzae. Bioprocess and Biosystems Engineering, 44*, 1975–1988.
87. Ashraf, H., Anjum, T., Riaz, S., & Naseem, S. (2020). Microwave-assisted green synthesis and characterization of silver nanoparticles using Melia azedarach for the management of *Fusarium* wilt in tomato. *Frontiers in Microbiology, 11*, 238.
88. Ramezani, M., Ramezani, F., & Gerami, M. (2019). Nanoparticles in pest incidences and plant disease control. In *Nanotechnology for agriculture: Crop production and protection* (pp. 233–272). Springer.
89. Bahrami-Teimoori, B., Nikparast, Y., Hojatianfar, M., Akhlaghi, M., Ghorbani, R., & Pourianfar, H. R. (2017). Characterisation and antifungal activity of silver nanoparticles biologically synthesised by *Amaranthus retroflexus* leaf extract. *Journal of Experimental Nanoscience, 12*(1), 129–139.
90. Renuka, L., Anantharaju, K. S., Sharma, S. C., Nagaswarupa, H. P., Prashantha, S. C., Nagabhushana, H., & Vidya, Y. S. (2016). Hollow microspheres Mg-doped ZrO2 nanoparticles: Green assisted synthesis and applications in photocatalysis and photoluminescence. *Journal of Alloys and Compounds, 672*, 609–622.
91. Gowri, S., Gandhi, R. R., & Sundrarajan, M. (2014). Structural, optical, antibacterial and antifungal properties of zirconia nanoparticles by biobased protocol. *Journal of Materials Science and Technology, 30*(8), 782–790.
92. Jalill, R. D. A., & Numan, R. S. (2016). Silver nitrate and zirconium oxide nanoparticles as management of wheat damping-off caused by *Fusarium graminearum. Journal of Genetic and Environmental Resources Conservation, 4*(2), 85–93.

Proteomics of Plant-Nanoparticle Interaction Mechanism

Ghazala Mustafa and Setsuko Komatsu

Abstract Plants are continuously exposed to a broad array of hostile environmental conditions that confine plant growth and restrict the yield of crop plants. Knowledge of the underlying molecular mechanisms of plant responses to nanoparticles stress is especially relevant for environmental assessments. Silver nanoparticles mediated the metabolic change from the anaerobic conditions toward the normal cellular process in soybean under flooding stress. Aluminum oxide nanoparticles regulated the ascorbate/glutathione pathway, thereby elevating reactive oxygen species scavenging. In the late and recovery stages, the differentially changed proteins were mainly linked to stress, cell wall, protein synthesis, and signaling. In wheat, copper nanoparticles improved salt stress tolerance by facilitating the method of glycolysis and tricarboxylic acid cycle. Iron nanoparticles enhanced the growth of wheat seedlings by regulating photosynthesis-related proteins in wheat. Though many studies are performed to identify the impacts of nanoparticles on plant development and growth, however, the molecular basis of nanoparticles' effect on plants is still unclear. This chapter will emphasize proteomics analysis of the nanoparticle's interaction mechanism with crop plants, particularly soybean and wheat.

1 Introduction

Nanotechnology is an emerging field dealing with the synthesis and utilization of nanomaterials. Latest advancements in nanotechnology have transformed various areas of science including life sciences and many others [1]. Increasing innovations in the field of nanoscience have led to the documentation of novel approaches for the synthesis of different types of nanoparticles that are separated into four main groups:

G. Mustafa
Department of Plant Sciences, Quaid-i-Azam University, Islamabad, Pakistan

S. Komatsu (✉)
Faculty of Environment and Information Sciences, Fukui University of Technology, Fukui, Japan
e-mail: skomatsu@fukui-ut.ac.jp

J.-T. Chen (ed.), *Plant and Nanoparticles*,
https://doi.org/10.1007/978-981-19-2503-0_3

metal or metal oxides, carbon-based nanoparticles, dendrimers, and bioinorganic complexes [2]. Along with the chemical synthesis, biological methods of nanoparticles formulations are also of concern. Nanoparticle synthesis by biological methods is gaining attention where nanoparticles are synthesized through biological ways with the involvement of plants, microorganisms, or their byproducts [3]. Nanoparticles prepared through biological routes are less toxic than chemically synthesized nanoparticles.

Plants are at continuous exposure to nanoparticles which often pose biological risks to them. Nanoparticles are the materials having special nano-size that lies between 1 and 100 nm and identified for a long time due to their important part in sustainable agriculture and effectively covering pathways allowing widespread advances in plant science [4–6]. Therefore, the toxicity caused by these nanoparticles attracted attention for its possible evaluation [7]. Nanoparticles are synthesized by biological methods. Along with the conventional methods, the use of biological materials acts as stabilizing agents in synthesis methods that led to the production of eco-friendly nanoparticles [8]. Nanoparticles are effectively used for crop improvements and tolerance against stress conditions.

The high reactivity and variety of biochemical actions of nanoparticles are particularly dependent on their unique chemical properties, making them suitable objects for altering different biological functions [9]. Various types of nanoparticles are synthesized and used for growth improvement and stress tolerance [8]. Nanoparticles improved plant growth and metabolism in a species-specific manner even under stress conditions [10]. Based on different synthesis processes, nanoparticles are ultimately released into the environment which causes toxicity [11, 12]. Regardless of these worries, various studies identified the mechanisms of their interactions with plants and the environment. Different nanoparticles (metal or metal oxides) are known for their important roles including silver [13], aluminum oxide [14], cerium oxide [15], titanium oxide [16, 17], zinc oxide [18, 19], and iron oxide [20, 21]. All these studies reported the growth-enhancing effects of nanoparticles along with protection from different abiotic stresses through the regulation of metabolism.

2 Proteomics to Understand the Interaction Between Plant and Nanoparticles

Nanoparticles are extensively used in the agriculture sector as favorable agents in fertilizers, pesticides, and plant growth for sustainable crop production [22].

Nanoparticles are considered to cause genotoxic effects on plants, thereby limiting their process of development and growth [23]. Along with toxic impacts, engineered nanoparticles are key players in plant growth regulation [24]. Nanoparticles equally caused both constructive and destructive effects on different crop plants [7]. Growing data indicates that nanoparticles can improve the damaging impacts of salt stress in crops [25]. Thus, the molecular basis for the interaction mechanism of the nanoparticles with plants needs to be explored.

2.1 Silver Nanoparticles

Silver is present as a rare element in the earth's crust. Silver is used as a probable antimicrobial agent because it activates reactive oxygen species generation in the bacterial cells leading to disabled microbial enzymes [26]. As a result, silver ions are produced which are toxic and function as bioactive molecules [27]. Silver ion initiated oxidative injury to DNA through the stimulation of antioxidant enzymes and reduction of the antioxidants [27]. Silver nanoparticles are utilized in different ways in the agriculture field. Silver nanoparticles promoted the growth of *Crocus sativus* through ethylene signaling blockage under flooding stress [28]. Silver nanoparticles encouraged *Arabidopsis* growth because of the inhibition of ethylene perception and buildup of reactive oxygen species [29]. Molecular investigations revealed that silver nanoparticles also function as gene regulators in plants. Silver nanoparticles enhanced the chlorophyll and protein amounts in *Phaseolus vulgaris* and maize [30]. In rice, silver nanoparticles alter amounts of proteins correlated with the oxidative stress tolerance along with differential regulations of cell division, apoptosis, and transcription/degradation-related proteins [31]. In *Eruca sativa*, sulfur metabolism and redox regulation-related proteins were changed on exposure to silver nanoparticles [32]. Reactive oxygen species are aggregated under silver nanoparticles leading to reduced plant growth.

Silver nanoparticles lowered the growth of soybean contrasted to aluminum and zinc oxide nanoparticles. In leaves of soybean, subtilase family protein and DAHP synthase levels were decreased which led to reduced growth under silver nanoparticles [33]. In soybean, silver nanoparticles regulated growth under flooding via regulation of amino acid metabolism and wax formation-related proteins [13]. Silver nanoparticles mixed with nicotinic acid and potassium nitrate promoted soybean growth through the control of protein quality of misfolded protein [34]. In wheat, silver nanoparticles promoted growth by changing the plant metabolism [35]. Silver nanoparticles interact with amino acid metabolism and metabolism which led to enhanced growth.

2.2 Aluminum Oxide Nanoparticles

Aluminum is present as the most abundant metal on earth. Because of its toxic features, it is considered as a main limiting factor for crop growth. Aluminum restricts plant growth in low pH acidic soils [36]. Aluminum ions, formed from aluminosilicates, are the cause of its toxicity [37]. Physiological investigations have highlighted that plants can procure aluminum tolerance by two different strategies.

One strategy to combat aluminum toxicity is blocking the ion uptake of aluminum. The second strategy is to detoxify cellular aluminum via the creation of unhazardous complexes by binding with organic ligands leading to their sequestration in subcellular organelles in plant cells [38, 39]. In aluminum susceptible plants, higher amounts of aluminum limit plant growth that ultimately causes crop yield reductions [40]. Aluminum tends to bind with different functional groups inside cells, thus damaging the cellular components [41]. Regardless of causing different toxic consequences, various studies reported positive impacts of aluminum on the plants.

Aluminum oxide nanoparticle application on ryegrass, rape, and radish improved the root elongation, while growth was reduced in cucumber, lettuce, and ryegrass [42, 43]. miRNA expression levels were significantly altered by aluminum oxide nanoparticles in tobacco which perform a critical function in stress response [44]. Antioxidant enzymes remove free radicals to eliminate the deleterious effects of aluminum oxide nanoparticles in wheat [45]. Poborilova et al. [46] suggested that aluminum oxide nanoparticle application on BY-2 cell suspension significantly reduces mitochondrial activity causing programmed cell death.

Aluminum oxide nanoparticles enhanced proteins synthesis, development, and transport-related proteins in soybean [47]. Moreover, aluminum oxide nanoparticles differentially altered proteins related to the ascorbate glutathione pathway and ribosomal proteins [14]. In another study, proteins related to oxidation-reduction, hormonal pathways, and stress signaling were distinctively changed in soybean treated with aluminum oxide nanoparticles [33]. Aluminum oxide nanoparticles enhanced growth under flooding stress by controlling mitochondrial proteins via tricarboxylic acid cycle and membrane permeability in soybean [48]. These studies suggest that aluminum oxide nanoparticles also have growth-promoting effects through the regulation of oxidative and hormonal pathways.

2.3 Iron Nanoparticles

Iron is significant micronutrient for the plant, and its absence can cause prominent growth obstruction. Iron nanoparticles are among metal-based nanoparticles which are frequently used for biomedical and commercial purposes [49]. In agriculture, iron nanoparticles are useful for crop betterment as an important fortifying agent. Iron nanoparticles are reported to lessen the stress in plants via the regulation of different metabolic reactions [50]. In *Arabidopsis*, iron oxide nanoparticles activate the proton ATPase leading to increased stomatal opening [51]. Iron nanoparticles have an important role in environmental bioremediation [52]. Iron nanoparticle at higher concentrations is toxic while beneficial at low concentrations [53]. In wheat varieties, iron nanoparticles differentially changed photosynthesis proteins [49]. In wheat grain, glycolysis and tricarboxylic acid cycle-related proteins were increased [50]. Iron nanoparticles are reported to cause mitigative properties of plants under drought stress [54]. Manzoor et al. [21] reported that iron oxide nanoparticles exhibit co-amelioration properties under heavy metal stress. These studies suggest that iron oxide nanoparticles could be effectively used against stress mitigation in different crop plants.

2.4 Zinc Nanoparticles

Zinc regulates various cellular processes as a vital micronutrient in plants. Understanding the strategy of zinc uptake and transfer is important to find out the interaction mechanisms of plants with zinc nanoparticles. Nano-zinc oxide is widely applied in different industries for many decades. Phytotoxicity studies suggest that the interaction mechanism of zinc in plants is highly specific [55]. Zinc oxide nanoparticles caused cytotoxicity in seedlings of soybean at a greater concentration [56]. Zinc nanoparticles have the ability to perform a function in the food production sector and are further working to discover it at the molecular level. In soybean, zinc oxide nanoparticles are reported to impart stress mitigation [33]. Proteins associated with redox, cell organization, lipid metabolism, and stress were identified in this study. These results suggest that zinc oxide nanoparticles can endorse plant defense through the regulation of different mechanisms for understanding plant nanoparticle relations and proteomic responses.

2.5 Other Nanoparticles

Recent studies highlighted the importance of nanotechnology in today's research. In the agriculture sector, researchers are expanding their knowledge to use these nanosystems for the betterment of crops. Various nanoparticles are prepared through different routes and are successfully applied on crops for improving their growth and yield (Table 1). Metabolomic and proteomic analyses are performed on different plants under nanoparticles stress. Salehi et al. [15] studied changes in *Phaseolus vulgaris* on exposure to cerium oxide nanoparticles. Cerium oxide nanoparticles are important metal oxide nanoparticles that are produced at larger scales due to their abundant functions in industrial products [62–64]. Cerium oxide nanoparticles' introduction to *Phaseolus vulgaris* induced differential expression of protein biosynthesis-related proteins. Foliar spray of cerium nanoparticles modulates protein folding and turnover in plants [15]. Titanium dioxide nanoparticles altered different metabolic reactions including photosynthesis [24]. In spinach leaves, titanium oxide nanoparticles reduced oxidative stress on exposure to ultraviolet light along with enhanced enzyme activities. These enzymes potentially reduce the reactive oxygen species [65]. Jha and Pudake [2] described titanium oxide nanoparticles as enhanced photosynthesis which ultimately increased plant growth. Titanium dioxide nanoparticles also reduced plant growth with varying concentrations and induced genotoxicity [66]. These findings suggest the interaction of different nanoparticles

Table 1 Nano-proteomics of crops reported in the last 10 years

Plant species	Exposed plant organ	Nanoparticles (size/dose)	Proteomic technique	Major proteins found/ associated pathways	References
Tobacco	Tissue Callus	Carbon nanoparticles 28–77 nm/ 0–100 mg/L	Gel based (2DE), LC MS/MS, iTRAQ	The majority of proteins are involved in mitochondrial functions and Ca signaling. Calmodulin (CaM) and protein expression level significantly enhanced. Furthermore, among the top 20 upregulated proteins, cytochrome c oxidase and cytochrome c are two other increased expressed proteins	Zhenjie et al. [57]
	Roots/ water supplement	Silver oxide 50 nm/100 μM	Gel-based (2-DE), MALDI/ TOF MS	CAP, cap-binding protein 20 (CBP20), β-1,3-glucanase, and Mn-SOD in roots. Glutathione S-transferase, glycine-rich RNA-binding protein, an mRNA binding protein, nucleotide metabolism in leaf	Peharec Štefanić et al. [58]
Soybean	2-day-old seedlings	Biologically synthesized AgNP (16 nm)	Immunoblot	Differentially regulated proteins were mainly involved in protein degradation and stress	Mustafa et al. [59]
		Chemically synthesized AgNP (15 nm) 10 ppm	Label-free nanoLC/ MS		
	21-day-old plants	AgNPs 60 nm/ 50 mg/kg Ag	Gel based (2D-DIGE), nanoEI-LC-MS/MS	Nucleoside diphosphate kinase (NDK), chlorophyll a-b binding protein 6A, gamma-glutamyl hydrolase	Galazzi et al. [60]
	Root	Al_2O_3 NPs 30–60 nm/5, 50, and 500 ppm (50 ppm for proteomics)	Gel-free	Proteins related to protein synthesis, stress, cell wall, and signaling changed in abundance. S-adenosyl-L-methionine-dependent methyltransferases and enolase might be involved in mediating	Yasmeen et al. [49]
			Nanospray LTQ XL Orbitrap MS		

(continued)

Table 1 (continued)

Plant species	Exposed plant organ	Nanoparticles (size/dose)	Proteomic technique	Major proteins found/ associated pathways	References
				recovery responses by Al_2O_3 NPs	
	Root/leaf	Al_2O_3 NPs 30–60 nm/ 500 ppm, 30–60 nm; ZnO <50 nm/ 500 ppm; Ag 15 nm/50 ppm	Gel-free Nanospray LTQ XL Orbitrap MS	GDSL motif lipase 5, galactose oxidase, and quinone reductase were upregulated	Hossain et al. [33]
Wheat	Seeds and 5-day-old seedlings	AgNPs (1, 10 mg/L)	2-DE IEF/ SDS-PAGE, LC–ESI-MS/MS	Proteins related to cell defense and primary metabolism and were altered	Vannini et al. [61]
	Root exposure	AgNP 15–20 nm/ 5 ppm	Gel free	Redox and mitochondrial electron transport chain proteins decreased. Glyceraldehyde-3-phosphate dehydrogenase, and glucose-6-phosphate-1-epimerase increased/decreased, while phosphoenolpyruvate carboxylase was decreased	Jhanzab et al. [35]
Bean	Foliar and root	Cerium oxide 10–30 nm/ 0–2000 mg/L	Gel-free Q-TOF tandem MS/MS	Enzymes involved in protein biosynthesis or proteases, lysine biosynthetic intermediates, and glutamine were altered	Salehi et al. [15]
Rice	Root/soil irrigation	AgNPs 18.34 nm/30 and 60 μg/mL	Gel-based (2-DE), nano LC/ FT-ICR MS	Proton motive force, oxidative stress tolerance, Ca^{2+} regulation and signaling, cell wall and DNA/RNA/ protein direct damage, cell division, and apoptosis	Mirzajani et al. [31]
Eruca sativa	Seedlings	AgNPs (10 nm/ 10 mg/L)	2DE, nanoLC-ESI-MS/MS	Proteins involved in sulfur metabolism significantly changed and V-ATPase subunits downregulated	Vannini et al. [32]

with the plant proteins and to regulate the enzyme activities which might help the plant to combat stress conditions.

3 Molecular Mechanisms Altered by Nanoparticles

Nanoparticles enter plants through the root system. Cell wall, as the first defense barrier, hinders the nanoparticles' entry allowing the particles which are tinier than cell wall pore size [67]. Tinier nanoparticles can cross cell walls and enter the cell. On the other hand, larger nanoparticles did not influence the metabolic pathways directly because they are blocked by the pore size of the cell wall [68]. Nanoparticles interfere with different cellular components which lead to alterations in metabolic/ signaling pathways. The initial response of nanoparticle interaction with plants affects the production of reactive oxygen species, energy changes, and different signaling mechanisms [69]. Thus, it would be interesting to explore the interaction system of nanoparticles at the molecular levels.

3.1 Energy Regulation in Plants on Exposure to Nanoparticles

Nanoparticles are engaged in the management of different metabolic pathways including energy metabolism [70]. In soybean, aluminum oxide nanoparticles structured the energy metabolism that ultimately led to improved growth [33]. Aluminum oxide nanoparticles controlled proteins linked with energy metabolism in soybean. Proteins related to fermentation and glycolysis were altered to adjust the energy requirements of the cell under stress conditions [14]. Different sizes of aluminum oxide nanoparticles differentially changed energy metabolism in soybean [48]. In soybean, one of the main proteins of the glycolysis pathway, glyceraldehyde-3-phosphate dehydrogenase, enhanced under stress while decreased through the use of nanoparticles [14]. Silver nanoparticles regulated the proteins related to the fermentation pathway. Pyruvate decarboxylase and alcohol dehydrogenase were enhanced under flooding while declined with nanoparticle application [71]. In this way, nanoparticles helped the plant to combat the stress situations through the control of energy metabolism including glycolysis and alcoholic fermentation.

3.2 Oxidative Stress

Nanoparticle-mediated manufacture of reactive oxygen species disturbs the cellular redox metabolism causing oxidative stress which damages internal cellular elements.

Different nanoparticles are reported to impose oxidative stress. Reactive oxygen species interfere with different cellular components which lead to mitochondrial dysfunction [72]. In soybean, silver and zinc oxide nanoparticles caused oxidative burst through the accumulation of superoxides inside the cell [33]. Thioredoxin family proteins managed the thiol-disulfide exchange reactions that could lead to the redox status of target proteins. Plant quinone reductases help in the detoxification of free radicals. The abundance of quinone reductase was reduced in soybean treated with silver and zinc oxide nanoparticles [33]. Aluminum oxide nanoparticles raised the accumulation of thioredoxin family protein that protected the cell against oxidative stress [33]. Aluminum oxide nanoparticles lowered the buildup of ascorbate/glutathione pathway protein that led to increased oxidative stress in soybean [48]. In another comparative study, the chemically synthesized silver nanoparticles decreased oxidative stress in soybean under stress [59]. Silver nanoparticles boosted the accumulation of peroxidases in soybean leading to increased reactive oxygen species scavenging activity [59]. These outcomes imply that nanoparticles regulate the abundance of proteins related to the oxidation-reduction process which led to the oxidative burst inside cells.

4 Nanoparticle Perception and Method of Action Under Stress Conditions

Plants are susceptible to changing ecological conditions including biotic and abiotic which results in a reduction of different biological functions. Despite the extensive information available to understand the consequences of stresses on plants, differences in the stress response strategies at the proteomic level are still required to be identified. Response mechanisms of the plant to stress condition are accelerated by redox reaction cascades including various biological mechanisms that are controlled by enzymes and proteins. The inequality in the quality and quantity of protein variations under stress conditions is critical for genuine diversity of protein richness.

4.1 Nanoparticles' Interaction with Soybean

In soybean, different nanoparticles are used for the mitigation strategies against flooding stress. Silver nanoparticles reduced the harmful effects of stress by regulating the cleansing of cytotoxic products. Moreover, silver nanoparticle-treated soybean faced reduced oxygen deprivation stress and helped metabolic transfer from anaerobic to the regular aerobic pathway [71]. Among different sizes of silver nanoparticles, 15 nm boosted the soybean seedling development under flooding through wax formation and amino acid synthesis proteins regulation [13]. Positive alleviation of flooding stress effects is further enhanced by the supplementation of nicotinic acid and potassium nitrate via control of protein quality of poorly folded protein in the endoplasmic reticulum [34]. In a comparative analysis, biosynthesized silver nanoparticles promoted soybean growth, through protein degradation-related proteins and ATP contents, compared to the chemically synthesized silver nanoparticles [59]. Silver nanoparticles regulated energy pathways under flooding stress and help soybean to mitigate the harmful impacts of stress (Table 2).

Aluminum oxide nanoparticles are progressively employed in different agricultural products [73, 74]. Aluminum oxide nanoparticles controlled the metabolism for energy production and cell death to enhance the growth of flooded soybean. Protein synthesis/degradation, glycolysis, and lipid metabolism-related proteins predominantly responded to aluminum oxide nanoparticles [71]. Size-dependent effects of aluminum oxide nanoparticles depicted that 30–60 nm nanoparticles boosted the growth of soybean under flooding stress through energy metabolism. On the other hand, 135 nm aluminum oxide nanoparticles increased the mitochondrial membrane permeability indicating the membrane leakage that ultimately led to cell death [48]. During soybean recovery from flooding stress, aluminum oxide nanoparticles regulated the protein synthesis, development, and transport-related proteins that helped the plant to recover from drastic impacts of flooding stress [47]. Bringing together these findings, the results indicate that nanoparticle application enhances soybean survival under stress mainly through the adjustment of energy metabolism, mitochondrial function, and lipid oxidation (Fig. 1).

4.2 Nanoparticles' Interaction with Wheat

In wheat, silver nanoparticles altered the accumulation of primary metabolism- and cell defense-related proteins in shoot and root [61]. Silver nanoparticles strengthened protein synthesis- and photosynthesis-related proteins while lessened proteins related to signaling, glycolysis, and cell wall [35]. Iron nanoparticles regulated growth in the shoot of drought- and salinity-tolerant wheat varieties through the alteration of photosynthesis-related proteins [49]. Photosynthesis- and the protein metabolism-associated proteins were lessened in wheat varieties treated with iron nanoparticles. Iron nanoparticles enhanced the growth of wheat through enhancement in photosynthesis-related protein [49]. Copper nanoparticles regulated glycolysis and tricarboxylic acid cycle and thus enhanced tolerance of wheat toward stress. Iron nanoparticles increased the proteins related to glycolysis, starch degradation, and tricarboxylic acid cycle and thus improved the stress tolerance [50]. Exposure of copper nanoparticles recovered growth of wheat seedling that could be associated with improvement of energy-related protein abundance in wheat. These results suggest that nanoparticles' interaction with wheat intervened in the photosynthesis and the tricarboxylic acid cycle.

Table 2 Proteomic technologies adapted by Plant Sciences for abiotic stress tolerance using nanoparticles

Plant species	Growth stage	Nanoparticles (size/dose)	Abiotic stress	Proteomic technique	Major proteins found/associated pathways	References
Soybean	Seedlings	Ag NP (15 nm/5 ppm) + organic/inorganic chemicals	Flooding	Gel free	Protein degradation/synthesis related proteins and calnexin/calreticulin related glycoproteins were significantly increased	Hashimoto et al. [34]
				NanoLC/MS		
				Immunoblot		
	2-day-old seedlings	Al_2O_3 (30–60 nm/50 ppm)	Flooding	Gel-free nanoLTQ XL Orbitrap MS/MS	The abundance of cell wall-related proteins was significantly increased. The most interactive protein was glyceraldehyde-3-phosphate dehydrogenase C subunit 1	Mustafa et al. [14]
	Seedlings	AgNPs 15 nm/2 ppm	Flooding	Gel-free	Commonly regulated protein glyoxalase II 3. Aspartyl protease family protein and expansin-like B1 increased in response to AgNPs in roots, while beta-ketoacyl reductase increased in cotyledons	Mustafa et al. [71]
				NanoLC-ESI-MS/MS		
	2-day-old seedlings	Al_2O_3 (30–60 nm/50 ppm)	Flooding	Immunoblot	Proteins of the ascorbate glutathione pathway and ribosomal proteins were significantly enhanced	Mustafa and Komatsu [48]
				NanoLC/MS		
	Root	Ag-NPs (2, 15, and 50–80 nm/5 ppm)	Flooding	Gel-free	Most interactive proteins	Mustafa et al. [13]
				nanoLC/MS	(1) 2 and 15 nm NP exposure: cytochrome family, beta-ketoacyl reductase 1	
					(2) 50–80 nm: enolase	

(continued)

Table 2 (continued)

Plant species	Growth stage	Nanoparticles (size/dose)	Abiotic stress	Proteomic technique	Major proteins found/associated pathways	References
Wheat	Shoot	Fe-NPs (20 nm/1, 5, 10, 50 ppm)	Salt stress	Gel-free NanoLC/MS Immunoblot	Proteins related to photosynthesis, cell, and protein metabolism; ribulose bisphosphate carboxylase/oxygenase small chain abundance increased	Yasmeen et al. [47]
	Seeds	Cu (15–30 nm) and Fe (20–30 nm)/20–40 ppm	Salt and drought	Gel-free Nanospray LTQ XL Orbitrap MS/MS	Proteins involved in starch degradation, glycolysis, and tricarboxylic acid cycle enhanced	Yasmeen et al. [50]
	Seedlings	Cu (1, 5, 10, 50 ppm/<50 nm)	Salt and drought	Gel-free NanoLC/MS	The protein of photosynthesis and tetra-pyrrole synthesis decreased, while those of glycolysis and TCA cycle enhanced	Yasmeen et al. [50]

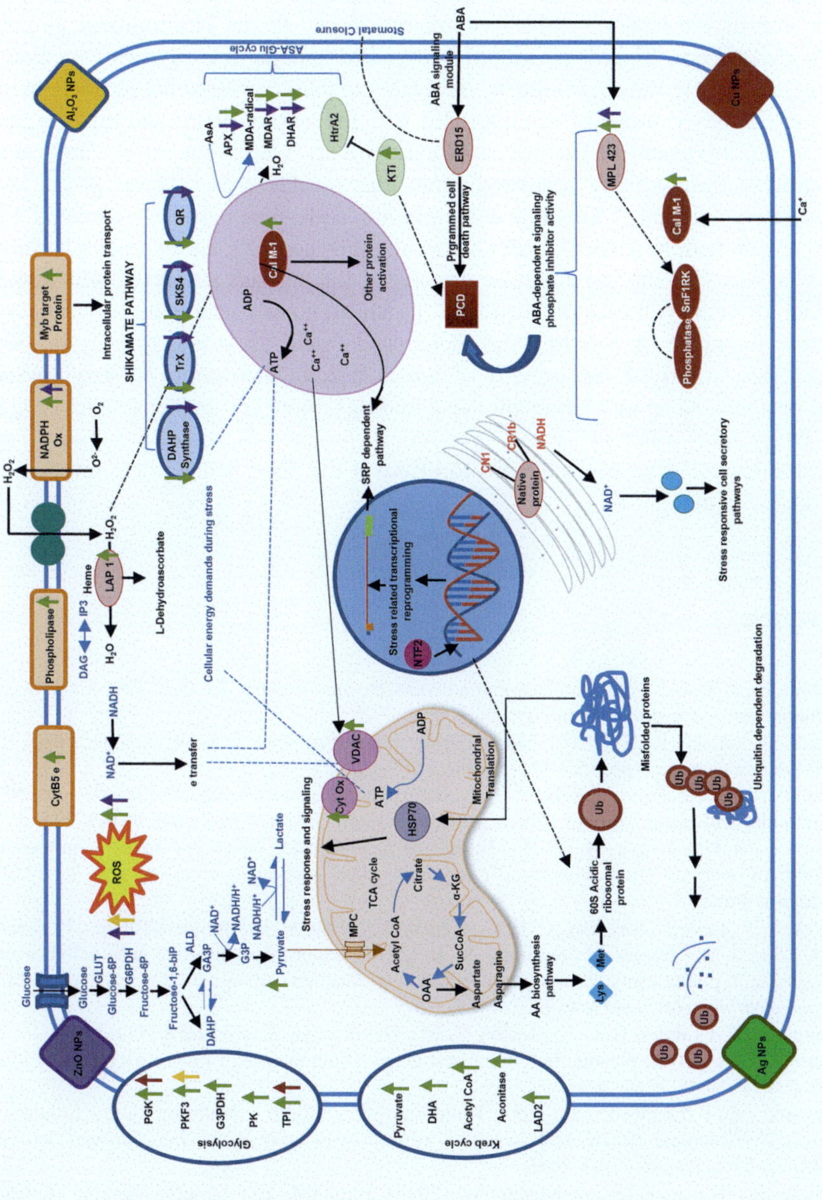

Fig. 1 Possible mechanisms on stress tolerance of soybean that are enhanced by nanoparticles

5 Conclusion and Future Perspective

Nanotechnology has gained a very important place in today's world due to its unique applications in different sectors including agriculture. However, the inadvertent deliverance of artificially synthesized nanomaterials to the environment elicited worldwide concern. Significant regard is given to nanoparticle synthesis methods, delivery, and fate in the environment. As related to their conventional physicochemical methods, biosynthesis of nanoparticles with the help of plants and microorganisms is an environment-friendly method. Silver, aluminum, iron, and zinc nanoparticles differentially regulated molecular mechanisms of crop plants and help in stress tolerance. Silver and aluminum oxide nanoparticles regulated the energy metabolism in soybean under stress and thus helped plant to relieve harmful impacts of stress. In the late and recovery stages, the proteins are linked with stress, signaling, protein synthesis, and cell wall. In wheat, copper nanoparticles improved salt stress tolerance by assisting the tricarboxylic acid cycle and glycolysis. Iron nanoparticles improved the growth of wheat seedlings through photosynthesis-related proteins. Keeping in view all these findings, more comprehensive investigations integrating different omics, genomics, proteomics, transcriptomics, and metabolomics could be advantageous to discover the plant-nanoparticle interaction mechanisms in detail.

References

1. Biswas, P., & Wu, C. Y. (2005). Nanoparticles and the environment. *Journal of the Air & Waste Management Association, 55*(6), 708–746.
2. Jha, S., & Pudake, R. N. (2016). Molecular mechanism of plant–nanoparticle interactions. In *Plant nanotechnology* (pp. 155–181). Springer.
3. Shah, V., & Belozerova, I. (2009). Influence of metal nanoparticles on the soil microbial community and germination of lettuce seeds. *Water, Air, and Soil Pollution, 9*, 143–148.
4. Khan, Z., & Upadhyaya, H. (2019). Impact of nanoparticles on abiotic stress responses in plants: An overview. In *Nanomaterials in plants, algae and microorganisms* (pp. 305–322). Academic Press.
5. Kwak, S. Y., Lew, T., Sweeney, C. J., Koman, V. B., Wong, M. H., Bohmert-Tatarev, K., Snell, K. D., Seo, J. S., Chua, N. H., & Strano, M. S. (2019). Chloroplast-selective gene delivery and expression in planta using chitosan-complexed single-walled carbon nanotube carriers. *Nature Nanotechnology, 14*(5), 447–455.
6. Wang, J. W., Grandio, E. G., Newkirk, G. M., Demirer, G. S., Butrus, S., Giraldo, J. P., & Landry, M. P. (2019). Nanoparticle-mediated genetic engineering of plants. *Molecular Plant, 12*(8), 1037–1040.
7. Hossain, Z., Yasmeen, F., & Komatsu, S. (2020). Nanoparticles: Synthesis, morphophysiological effects, and proteomic responses of crop plants. *International Journal of Molecular Sciences, 21*(9), 3056.
8. Saxena, R., Tomar, R. S., & Kumar, M. (2016). Exploring nanobiotechnology to mitigate abiotic stress in crop plants. *Journal of Pharmaceutical Sciences and Research, 8*(9), 974.

9. Dubchak, S., Ogar, A., Mietelski, J. W., & Turnau, K. (2010). Influence of silver and titanium nanoparticles on arbuscular mycorrhiza colonization and accumulation of radiocaesium in *Helianthus annuus*. *Spanish Journal of Agricultural Research, 1*, 103–108.
10. Giraldo, J. P., Landry, M. P., Faltermeier, S. M., McNicholas, T. P., Iverson, N. M., Boghossian, A. A., Reuel, N. F., Hilmer, A. J., Sen, F., Brew, J. A., & Strano, M. S. (2014). Plant nanobionics approach to augment photosynthesis and biochemical sensing. *Nature Materials, 13*(4), 400–408.
11. Nair, R., Varghese, S. H., Nair, B. G., Maekawa, T., Yoshida, Y., & Kumar, D. S. (2010). Nanoparticulate material delivery to plants. *Plant Science, 179*(3), 154–163.
12. Singh, A., Singh, N. B., Hussain, I., Singh, H., & Singh, S. C. (2015). Plant-nanoparticle interaction: An approach to improve agricultural practices and plant productivity. *International Journal of Pharmaceutical Science Invention, 4*(8), 25–40.
13. Mustafa, G., Sakata, K., & Komatsu, S. (2016). Proteomic analysis of soybean root exposed to varying sizes of silver nanoparticles under flooding stress. *Journal of Proteomics, 148*, 113–125.
14. Mustafa, G., Sakata, K., & Komatsu, S. (2015a). Proteomic analysis of flooded soybean root exposed to aluminum oxide nanoparticles. *Journal of Proteomics, 128*, 280–297.
15. Salehi, H., Chehregani, A., Lucini, L., Majd, A., & Gholami, M. (2018). Morphological, proteomic and metabolomic insight into the effect of cerium dioxide nanoparticles to *Phaseolus vulgaris* L. under soil or foliar application. *Science of the Total Environment, 616*, 1540–1551.
16. Hu, J., Wu, X., Wu, F., Chen, W., White, J. C., Yang, Y., Wang, B., Xing, B., Tao, S., & Wang, X. (2020). Potential application of titanium dioxide nanoparticles to improve the nutritional quality of coriander (*Coriandrum sativum* L.). *Journal of Hazardous Materials, 389*, 121837.
17. Poščić, F., Mattiello, A., Fellet, G., Miceli, F., & Marchiol, L. (2016). Effects of cerium and titanium oxide nanoparticles in soil on the nutrient composition of barley (*Hordeum vulgare* L.) Kernels. *International Journal of Environmental Research and Public Health, 13*(6), 577.
18. Faizan, M., Hayat, S., & Pichtel, J. (2020). Effects of zinc oxide nanoparticles on crop plants: A perspective analysis. *Sustainable Agriculture Reviews, 41*, 83–99.
19. Taran, N., Storozhenko, V., Svietlova, N., Batsmanova, L., Shvartau, V., & Kovalenko, M. (2017). Effect of zinc and copper nanoparticles on drought resistance of wheat seedlings. *Nanoscale Research Letters, 12*(1), 1–6.
20. Adrees, M., Khan, Z. S., Ali, S., Hafeez, M., Khalid, S., Ur Rehman, M. Z., Hussain, A., Hussain, K., Shahid Chatha, S. A., & Rizwan, M. (2020). Simultaneous mitigation of cadmium and drought stress in wheat by soil application of iron nanoparticles. *Chemosphere, 238*, 124681.
21. Manzoor, N., Ahmed, T., Noman, M., Shahid, M., Nazir, M. M., Ali, L., Alnusaire, T. S., Li, B., Schulin, R., & Wang, G. (2021). Iron oxide nanoparticles ameliorated the cadmium and salinity stresses in wheat plants, facilitating photosynthetic pigments and restricting cadmium uptake. *Science of the Total Environment, 769*, 145221.
22. Khot, L. R., Sankaran, S., Maja, J. M., Ehsani, R., & Schuster, E. W. (2012). Applications of nanomaterials in agricultural production and crop protection: A review. *Crop Protection, 35*, 64–70.
23. Ghosh, M., Ghosh, I., Godderis, L., Hoet, P., & Mukherjee, A. (2019). Genotoxicity of engineered nanoparticles in higher plants. *Mutation Research, 842*, 132–145.
24. Landa, P. (2021). Positive effects of metallic nanoparticles on plants: Overview of involved mechanisms. *Plant Physiology and Biochemistry, 161*, 12–24.
25. Zulfiqar, F., & Ashraf, M. (2021). Nanoparticles potentially mediate salt stress tolerance in plants. *Plant Physiology and Biochemistry, 160*, 257–268.
26. Matsumura, Y., Yoshikata, K., Kunisaki, S., & Tsuchido, T. (2003). Mode of bactericidal action of silver zeolite and its comparison with that of silver nitrate. *Applied and Environmental Microbiology, 69*, 4278–4281.
27. McShan, D., Ray, P. C., & Yu, H. (2014). Molecular toxicity mechanism of nanosilver. *Journal of Food and Drug Analysis, 22*, 116–127.

28. Rezvani, N., Sorooshzadeh, A., & Farhadi, N. (2012). Effect of nano-silver on growth of saffron in flooding stress. *World Academy of Science, Engineering and Technology, 6*, 519–524.
29. Syu, Y. Y., Hung, J. H., Chen, J. C., & Chuang, H. W. (2014). Impact of size and shape of silver nanoparticles on *Arabidopsis* plant growth and gene expression. *Plant Physiology and Biochemistry, 83*, 57–64.
30. Salama, H. M. H. (2012). Effects of silver nanoparticles in some crop plants, common bean (*Phaseolus vulgaris* L.) and corn (*Zea mays* L.). *International Research Journal of Biotechnology, 3*, 190–197.
31. Mirzajani, F., Askari, H., Hamzelou, S., Schober, Y., Römpp, A., Ghassempour, A., & Spengler, B. (2014). Proteomics study of silver nanoparticles toxicity on *Oryza sativa* L. *Ecotoxicology and Environmental Safety, 108*, 335–339.
32. Vannini, C., Domingo, G., Onelli, E., Prinsi, B., Marsoni, M., Espen, L., & Bracale, M. (2013). Morphological and proteomic responses of *Eruca sativa* exposed to silver nanoparticles or silver nitrate. *PLoS One, 8*(7), e68752.
33. Hossain, Z., Mustafa, G., Sakata, K., & Komatsu, S. (2016). Insights into the proteomic response of soybean towards Al_2O_3, ZnO, and Ag nanoparticles stress. *Journal of Hazardous Materials, 304*, 291–305.
34. Hashimoto, T., Mustafa, G., Nishiuchi, T., & Komatsu, S. (2020). Comparative analysis of the effect of inorganic and organic chemicals with silver nanoparticles on soybean under flooding stress. *International Journal of Molecular Sciences, 21*(4), 1300.
35. Jhanzab, H. M., Razzaq, A., Bibi, Y., Yasmeen, F., Yamaguchi, H., Hitachi, K., Tsuchida, K., & Komatsu, S. (2019). Proteomic analysis of the effect of inorganic and organic chemicals on silver nanoparticles in wheat. *International Journal of Molecular Sciences, 20*(4), 825.
36. von Uexküll, H. R., & Mutert, E. (1995). Global extent, development and economic impact of acid soils. *Plant and Soil, 171*, 1–15.
37. Kochian, L. V., Hoekenga, O. A., & Pineros, M. A. (2004). How do crop plants tolerate acid soils? Mechanisms of aluminum tolerance and phosphorous efficiency. *Annual Review of Plant Biology, 55*, 459–493.
38. Ma, J. F. (2000). Role of organic acids in detoxification of aluminum in higher plants. *Plant and Cell Physiology, 41*(4), 383–390.
39. Ma, J. F., & Furukawa, J. (2003). Recent progress in the research of external Al detoxification in higher plants: A minireview. *Journal of Inorganic Biochemistry, 97*(1), 46–51.
40. Valle, S. R., Carrasco, J., Pinochet, D., & Calderini, D. F. (2009). Grain yield, above-ground and root biomass of Al-tolerant and Al-sensitive wheat cultivars under different soil aluminum concentrations at field conditions. *Plant and Soil, 318*, 299–310.
41. Poschenrieder, C., Gunsé, B., Corrales, I., & Barceló, J. (2008). A glance into aluminum toxicity and resistance in plants. *Science of the Total Environment, 400*, 356–368.
42. Lin, D., & Xing, B. (2007). Phytotoxicity of nanoparticles: Inhibition of seed germination and root growth. *Environmental Pollution (Barking, Essex: 1987), 150*(2), 243–250.
43. Yang, L., & Watts, D. J. (2005). Particle surface characteristics may play an important role in phytotoxicity of alumina nanoparticles. *Toxicology Letters, 158*, 122–132.
44. Burklew, C. E., Ashlock, J., Winfrey, W. B., & Zhang, B. (2012). Effects of aluminum oxide nanoparticles on the growth, development, and microRNA expression of tobacco (*Nicotiana tabacum*). *PLoS One, 7*, e34783.
45. Riahi-Madvar, A., Rezaee, F., & Jalili, V. (2012). Effects of alumina nanoparticles on morphological properties and antioxidant system of *Triticum aestivum*. *Iranian Journal of Plant Physiology, 3*, 595–603.
46. Poborilova, Z., Opatrilova, R., & Babula, P. (2013). Toxicity of aluminium oxide nanoparticles demonstrated using a BY-2 plant cell suspension culture model. *Environmental and Experimental Botany, 91*, 1–11.
47. Yasmeen, F., Raja, N. I., Mustafa, G., Sakata, K., & Komatsu, S. (2016a). Quantitative proteomic analysis of post-flooding recovery in soybean root exposed to aluminum oxide nanoparticles. *Journal of Proteomics, 143*, 136–150.

48. Mustafa, G., & Komatsu, S. (2016). Insights into the response of soybean mitochondrial proteins to various sizes of aluminum oxide nanoparticles under flooding stress. *Journal of Proteome Research, 15*(12), 4464–4475.
49. Yasmeen, F., Raja, N. I., Razzaq, A., & Komatsu, S. (2016b). Gel-free/label-free proteomic analysis of wheat shoot in stress tolerant varieties under iron nanoparticles exposure. *Biochimica et Biophysica Acta. Proteins and Proteomics, 1864*(11), 1586–1598.
50. Yasmeen, F., Raja, N. I., Razzaq, A., & Komatsu, S. (2017). Proteomic and physiological analyses of wheat seeds exposed to copper and iron nanoparticles. *Biochimica et Biophysica Acta. Proteins and Proteomics, 1865*(1), 28–42.
51. Kim, J. H., Oh, Y., Yoon, H., Hwang, I., & Chang, Y. S. (2015). Iron nanoparticle-induced activation of plasma membrane H(+)-ATPase promotes stomatal opening in *Arabidopsis thaliana*. *Environmental Science and Technology, 49*(2), 1113–1119.
52. Yan, W., Lien, H. L., Koel, B. E., & Zhang, W. X. (2013). Iron nanoparticles for environmental clean-up: Recent developments and future outlook. *Environmental Science. Processes & Impacts, 15*(1), 63–77.
53. Mushtaq, Y. K. (2011). Effect of nanoscale $Fe_{(3)}O_{(4)}$, $TiO_{(2)}$ and carbon particles on cucumber seed germination. *Journal of Environmental Science and Health, Part A, Toxic/Hazardous Substances & Environmental Engineering, 46*(14), 1732–1735.
54. Jalil, S. U., Zahera, M., Khan, M. S., & Ansari, M. I. (2019). Biochemical synthesis of gold nanoparticles from leaf protein of *Nicotiana tabacum* L. *cv.* xanthi and their physiological, developmental, and ROS scavenging responses on tobacco plant under stress conditions. *IET Nanobiotechnology, 13*(1), 23–29.
55. Mahajan, P., Dhoke, S. K., & Khanna, A. S. (2011). Effect of nano-ZnO particle suspension on growth of mung (*Vigna radiata*) and gram (*Cicer arietinum*) seedlings using plant agar method. *Journal of Nanotechnology, 2011*, 696535.
56. López-Moreno, M. L., de la Rosa, G., Hernández-Viezcas, J. A., Castillo-Michel, H., Botez, C. E., Peralta-Videa, J. R., & Gardea-Torresdey, J. L. (2010). Evidence of the differential biotransformation and genotoxicity of ZnO and CeO_2 nanoparticles on soybean (*Glycine max*) plants. *Environmental Science and Technology, 44*(19), 7315–7320.
57. Zhenjie, Z., Hu, L., Chen, Q., Dai, H., Meng, X., Yin, Q., & Liang, T. (2020). iTRAQ-based comparative proteomic analysis provides insights into tobacco callus response to carbon nanoparticles. *Plant Molecular Biology Reporter*. https://doi.org/10.21203/rs.3.rs-109134/v1
58. Peharec Štefanić, P., Jarnević, M., Cvjetko, P., Biba, R., Šikić, S., Tkalec, M., Cindrić, M., Letofsky-Papst, I., & Balen, B. (2019). Comparative proteomic study of phytotoxic effects of silver nanoparticles and silver ions on tobacco plants. *Environmental Science and Pollution Research, 26*(22), 22529–22550.
59. Mustafa, G., Hasan, M., Yamaguchi, H., Hitachi, K., Tsuchida, K., & Komatsu, S. (2020). A comparative proteomic analysis of engineered and bio synthesized silver nanoparticles on soybean seedlings. *Journal of Proteomics, 224*, 103833.
60. Galazzi, R. M., Lopes Júnior, C. A., de Lima, T. B., Gozzo, F. C., & Arruda, M. A. Z. (2019). Evaluation of some effects on plant metabolism through proteins and enzymes in transgenic and non-transgenic soybeans after cultivation with silver nanoparticles. *Journal of Proteomics, 191*, 88–106.
61. Vannini, C., Domingo, G., Onelli, E., De Mattia, F., Bruni, I., Marsoni, M., & Bracale, M. (2014). Phytotoxic and genotoxic effects of silver nanoparticles exposure on germinating wheat seedlings. *Journal of Plant Physiology, 171*(13), 1142–1148.
62. Abbas, Q., Liu, G., Yousaf, B., Ali, M. U., Ullah, H., Mujtaba Munir, M. A., Ahmed, R., & Rehman, A. (2020). Biochar-assisted transformation of engineered-cerium oxide nanoparticles: Effect on wheat growth, photosynthetic traits and cerium accumulation. *Ecotoxicology and Environmental Safety, 187*, 109845.
63. Keller, A. A., McFerran, S., Lazareva, A., & Suh, S. (2013). Global life cycle releases of engineered nanomaterials. *Journal of Nanoparticle Research, 15*(6), 1–17.

64. Majumdar, S., Peralta-Videa, J. R., Trujillo-Reyes, J., Sun, Y., Barrios, A. C., Niu, G., Margez, J., & Gardea-Torresdey, J. L. (2016). Soil organic matter influences cerium translocation and physiological processes in kidney bean plants exposed to cerium oxide nanoparticles. *Science of the Total Environment, 569-570*, 201–211.
65. Lei, Z., Mingyu, S., Xiao, W., Chao, L., Chunxiang, Q., Liang, C., Hao, H., Xiaoqing, L., & Fashui, H. (2008). Antioxidant stress is promoted by nano-anatase in spinach chloroplasts under UV-B radiation. *Biological Trace Element Research, 121*(1), 69–79.
66. Castiglione, M. R., Giorgetti, L., Geri, C., & Cremonini, R. (2011). The effects of nano-TiO 2 on seed germination, development and mitosis of root tip cells of *Vicia narbonensis* L. and *Zea mays* L. *Journal of Nanoparticle Research, 13*(6), 2443–2449.
67. Fleischer, A., O'Neill, M. A., & Ehwald, R. (1999). The pore size of non-graminaceous plant cell walls is rapidly decreased by borate ester cross linking of the pectic polysaccharide rhamnogalacturonan. *Plant Physiology, 121*, 829–838.
68. Verano-Braga, T., Miethling-Graff, R., Wojdyla, K., Rogowska- Wrzesinska, A., Brewer, J. R., Erdmann, H., & Kjeldsen, F. I. (2014). Insights into the cellular response triggered by silver nanoparticles using quantitative proteomics. *ACS Nano, 8*, 2161–2175.
69. Marslin, G., Sheeba, C. J., & Franklin, G. (2017). Nanoparticles alter secondary metabolism in plants via ROS burst. *Frontiers in Plant Science, 8*, 832.
70. Gomes, A., Sengupta, J., Datta, P., Ghosh, S., & Gomes, A. (2016). Physiological interactions of nanoparticles in energy metabolism, immune function and their biosafety: A review. *Journal of Nanoscience and Nanotechnology, 16*(1), 92–116.
71. Mustafa, G., Sakata, K., Hossain, Z., & Komatsu, S. (2015b). Proteomic study on the effects of silver nanoparticles on soybean under flooding stress. *Journal of Proteomics, 122*, 100–118.
72. Horie, M., & Tabei, Y. (2020). Role of oxidative stress in nanoparticles toxicity. *Free Radical Research, 18*, 1–12.
73. Navrotsky, A. (2004). Energetic clues to pathways to biomineralization: Precursors, clusters, and nanoparticles. *Proceedings of the National Academy of Sciences of the United States of America, 101*(33), 12096–12101.
74. Stadler, T., Buteler, M., & Weaver, D. K. (2010). Novel use of nanostructured alumina as an insecticide. *Pest Management Science, 66*(6), 577–579.

Importance of the Secondary Metabolites and Biological Parameter Modification by Metallic, Oxide, and Carbon-Based Nanomaterials Over Forage Plants

Luis Páramo, Ana A. Feregrino-Pérez, Humberto Aguirre Becerra, Ramón G. Guevara-González, and Karen Esquivel

Abstract Nowadays, agricultural changes have been proposed due to more common oxidative stress and nutrient imbalance presented in plants. Also, a new challenge must be overcome by the production areas that are expanding into inadequate crop growth regions to fulfill food security targets to meet the needs of the continuously increasing human population. Evidence showed along the past decade that the use of nanostructured materials (NMs) and especially metallic nanoparticles (NPs) and carbon-based nanomaterials could work as nanofertilizers showing promising results by measuring the morphological affectation in stem and leaves' sizes as well as chlorophyll content, secondary metabolite content (total phenolic, flavonoids), and antioxidant capacity by DPPH radical assay.

In this chapter, it has recapitulated recent results and progress made toward the effect of diverse nanomaterials on the secondary metabolism of forage plants and future research directions to explain the roles of the morphology, chemical species, surface area, crystallinity, and concentration of the NMs to support public acceptance and safe use in the agricultural area, to be able to ensure sustainable growth in crop production in various fields.

Keywords Secondary metabolites · Nanomaterials · Nanostructure materials · Forage plants · Abiotic stress

---

L. Páramo · A. A. Feregrino-Pérez · H. Aguirre Becerra · R. G. Guevara-González · K. Esquivel (✉)
Graduate and Research Division, Engineering Faculty, Universidad Autónoma de Querétaro, Santiago de Querétaro, Qro, México
e-mail: karen.esquivel@uaq.mx

J.-T. Chen (ed.), *Plant and Nanoparticles*,
https://doi.org/10.1007/978-981-19-2503-0_4

1 Introduction

1.1 Forage Plants

Forage plants refer to the plants eaten by livestock, poultry, and wild animals; these include grasses, legumes, and browse [1]. Grasslands used mainly for grazing livestock cover more than 30% of the land area [2]. Forage plants are considered a natural strategic axis in maintaining global ecosystems and fundamental support in raising livestock worldwide [3]. The nutritional value of fodder is affected by the amount and accessibility of metabolic and anabolic products, including cell components and cell wall minerals, as well as the phenological state of the plant, climatic factors, overexploitation, variety or cultivars, etc. [4–6]. In this sense, in various regions of the world, the value of forages is recognized not only as food for livestock, and measures have been established for their protection.

However, in other regions, these measures do not exist, which causes an overexploitation of grazing, excessive logging, generating soil erosion, and a decrease in productive capacity, reflected in a decrease in arable area and supply of forage for several years [7, 8]. At the same time, fodder in the past was considered an essential element of the animal feed ration in all production areas. The adoption of industrial indoor grain-fed production systems generated a scarcity in developing countries on the nutritional value of forage plants [5]. However, various studies indicate that feeding supplemented with forages provides higher nutritional quality to livestock and therefore higher protein quality and amino acid profiles [9, 10] for the consumer. For this reason, the feed was reused as an additional dietary supplement to provide fiber, proteins, amino acids, and minerals to create and produce domestic herbivores [5, 11].

On the other hand, the demand for food has increased with population growth [12], and the area of agricultural production is limited [13] coupled with global changes such as climate change and management of the earth; in that sense, the generation of sustainable agriculture is the key to guarantee the supply of food and food security in future climatic conditions for future generations [14]. In this sense, forages are necessary to balance necessary compounds for the agricultural and livestock sectors. The use of technology or nanotechnology that provides tools to improve the quality and quantity of forage is of utmost importance.

1.2 Nanomaterials and Their Uses in the Agriculture

Nanomaterials are objects with at least one of their dimensions (length, width, depth) in the nanometric scale, which for the area of nanoscience is defined as 1–100 nm. These structures can be classified into multiple structures depending on the dimensions present in the nanoscale [15]. Among these classifications, we find the nanomaterials 0D, 1D, 2D, and 3D. The nanometric objects in 0D are those in which none of their dimensions is outside the scale already mentioned. Within this classification, we find materials such as quantum dots, nanospheres, and nanoparticles. The 1D classification encompasses materials whose one dimension exceeds the nanoscale, such as nanotubes, nanorods, and nanowires. Materials with two larger dimensions (2D) encompass structures such as sheets, thin films, and coatings, and in the 3D classification, the tree dimensions exceed 100 nm. Structures found in this classification are nanocomposites, powders, sets of nanotubes or multilayers, and MEMS [15–17]. These classifications can be observed in Fig. 1.

Nanomaterials can also be classified by their chemical composition, the most used categories being organic materials, metallic materials, metallic oxides, semiconductors, and carbon-based materials [15], as presented in Fig. 2.

Each of these materials finds interesting applications in an infinity of fields, due to that the nanoscience has given the possibility of finding new properties in materials. These discoveries are within the area of agriculture. The different types of nanomaterials already mentioned above find multiple applications that allow the plants to be elicited, which is the process in which a material can modify the secondary metabolism of a plant for promoting the synthesis of molecules of interest [18–22]. Other applications are as pesticides, fertilizers, and sensors.

1.2.1 Nanofertilizers

The use of standard fertilizer can amend deficient soils. However, the chemicals can alter soil quality over time since they are persistent on soils causing environmental pollution and affecting post-crop productions and microbial communities. Nanofertilizers can solve the problems caused by the traditional chemicals where nutrients can be delivered by slow release to the soil, helping to prevent diseases [18], regulating growth, and water release and retention. Most of the

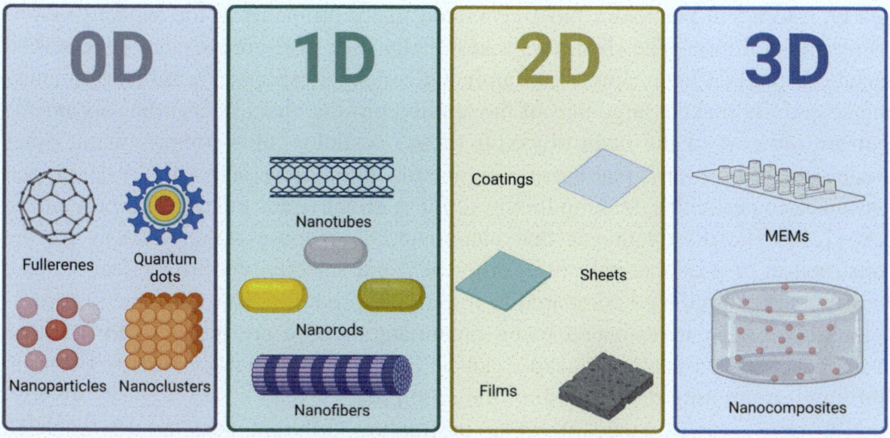

Fig. 1 Nanomaterial classification by dimension

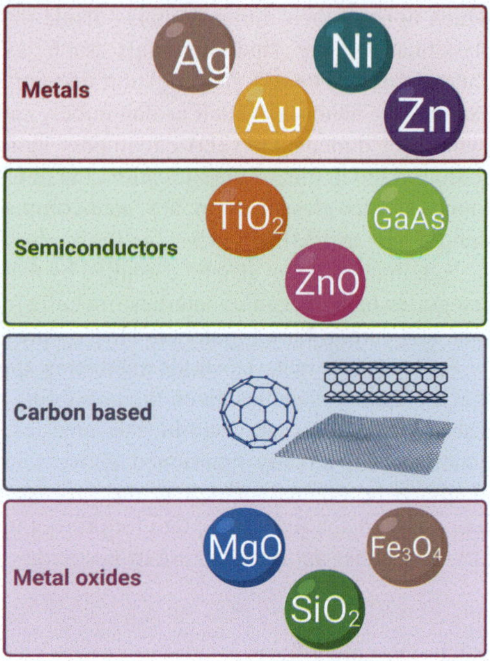

Fig. 2 Nanoparticle classification by chemical composition

nanoformulations used consists of emulsions or encapsulations. However, some nanoparticles of metal oxides have been found effective in micronutrient delivery due to the release of metal ions to the soil media [23].

1.2.2 Nanopesticides

Lately, traditional pesticides have generated many problems in the aspect of environmental pollution. The chemicals traditionally used to eliminate pests have served as substances that have slowly contaminated soils and waters. Their low efficiency and selectivity make a large part of the applied product that affects other organisms that are not part of the main objective of the pesticide, also compromising other organisms [24]. Many pathogenic organisms that affect crops have developed resistance to pesticides, making them even less effective for treating and preventing pests [25]. The disadvantages that nanomaterials propose to solve focus on the construction of products with high efficiency and selectivity whose effects on the environment are reduced compared to traditional products. That said, many formulations have been investigated using nanomaterials that are effective for treating insect pests and preventing diseases caused by pathogens such as viruses, bacteria, and fungi, promoting the survival of the crop [26].

NPs have proven to be effective in treating different pest species such as *Callosobruchus maculatus* or commonly called pulse beetle which affects cowpea grain storage. Zinc oxide nanoparticles (NPs) coated with *Pongamia pinnata* leaf extract, which has pesticide activity, delayed the development of *C. maculatus* at larval and pupal stages at a concentration of 25 µg/mL. Furthermore, 100% mortality was achieved at the same concentration. Also, it was found that several enzymes in the beetle were decreased, affecting its development [27].

1.2.3 Nanofungicides and Nanobactericides

Some bacterial strains and fungal communities can decrease crop production by producing various diseases which are detrimental to plant health. These organisms are commonly treated with chemical products that can affect non-targeted organisms due to lower selectivity, so modern pesticides are being substituted with more selective and secure products to protect humans and organisms present in crops [25]. Some of the mechanisms involved in pathogen elimination rely on membrane disruption, DNA damage, and production of reactive oxygen species ROS [28]. More research on nano-fungi and bactericides is shown in Table 1.

1.2.4 Protection Against Environmental Stresses

Some of the actual losses in forage crop productions are due to environmental factors called stress factors. They can be divided into two groups: abiotic and biotic stress. The first classification involves environmental factors such as salinity, light exposure, wind, extreme temperatures, drought, and flooding, for mentioning some examples. Living organisms generate biotic stresses, for example, pests, competitor species, bacteria, and fungi [34].

Table 1 NP applications for treating pathogenic diseases

NPs	Plant	Disease	Effect	References
TiO_2	Wheat rust	Wheat rust	Antifungal response against *Ustilago tritici* (wheat rust)	Irshad et al. [29]
ZnO MnO_2 MgO	Rice	Leaf blight	NPs suppressed leaf blight disease expression	Ogunyemi et al. [30]
Ag	-	Wheat Fusarium head blight	NPs caused DNA and protein leakage due to morphological deformities	Ibrahim et al. [31]
ZnO	Pearl millet	Downy mildew	NPs caused inhibition of spore germination and reduction of 35% of the disease after foliar treatment. NPs treated seedlings showed enhanced content of defense enzymes	Nandhini et al. [32]
Ag	Wheat	Yellow rust	NPs help to reduce yellow rust in wheat	Bano and Ummat-ul-Habib [33]

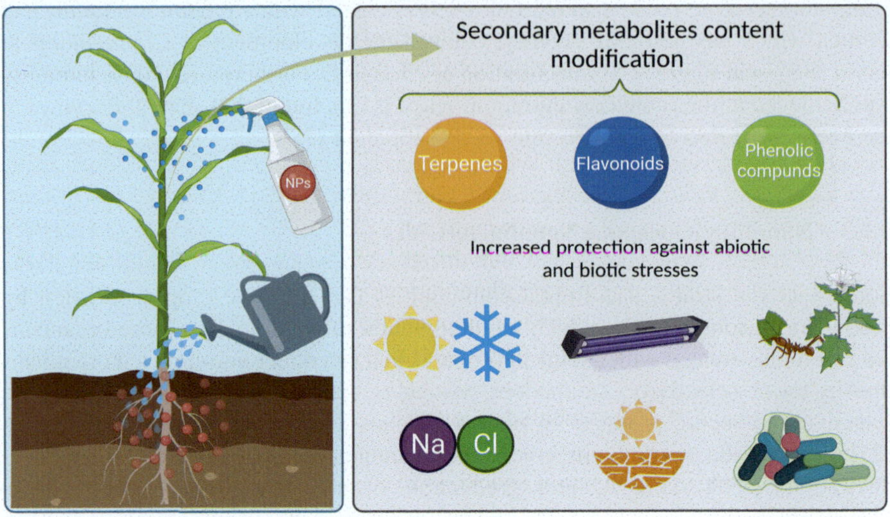

Fig. 3 Induction of secondary metabolites by NPs and their application in protection against environmental stresses

For fighting against stresses, the plants modify their primary metabolic routes to create compounds proceeding from the secondary metabolic routes. These compounds have unique features that can protect crops and improve survival [35]. Also, these compounds have special features for plants. Many of these are used as pharmacologic products, colorants, or pigments, among others [36]. Nowadays, NPs are used for enhancing the production of several metabolic compounds (Fig. 3), but the NP exposure to living organisms carries several nanotoxicological criteria. When NPs are used for secondary metabolic production, work needs to be done to find the adequate conditions of exposure and physicochemical characteristics to obtain the best result without affecting plant health [34].

1.2.5 Seed Priming

Seed priming involves deregulation of the water uptake during the seed development; researchers have found that priming techniques allow improving the development of plants, improving their grow quality, and reducing the time taken for germination, making the crop more adaptable to different environments compared to untreated seeds; this can be done with multiple approaches like wetting, humidification, or presoaking [37].

The priming process can be done in solutions containing specific solutes like KH_2PO_4, KCl, NaCl, and $CaCl_2$, micronutrients, or bacteria, that seed development needs, thus improving certain plant qualities like defense against abiotic and biotic stresses [38]. New priming techniques involve the use of nanomaterials where metals and metal oxides have been demonstrated to improve germination, growth,

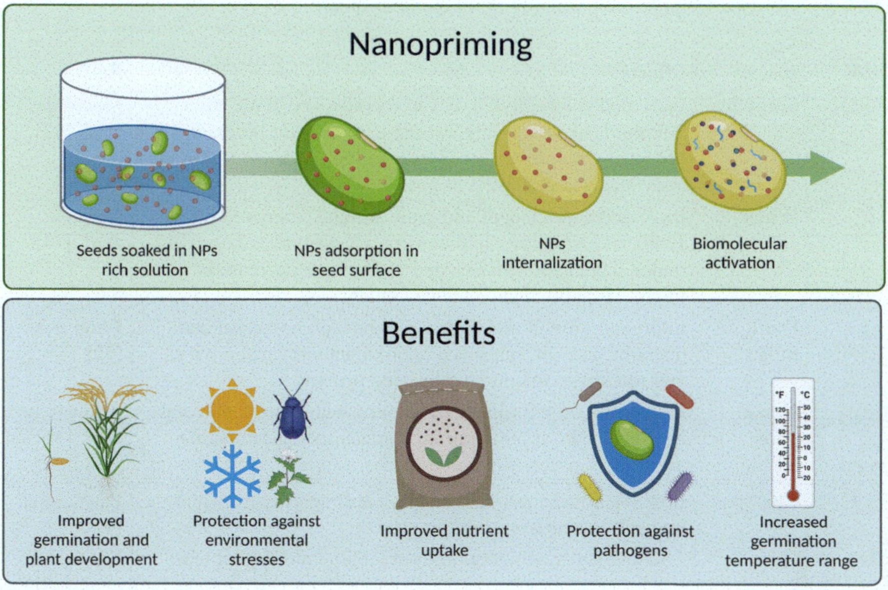

Fig. 4 Nano-seed priming and its benefits

yield, and quality [39]. Nanomaterials for seed priming can deliver nutrients at specific sites for their more efficient utilization in seed development processes [40]. As said before, there are several methods for priming seeds. The simplest method involves soaking the seed in water which is called hydropriming. A solution containing inorganic salt is called halopriming. Solutions containing nanomaterials are called nanopriming. Hormopriming involves using growth regulators, matrix priming involves using low and high temperatures, and bio-priming requires organic materials [41]. Nanomaterials are also capable of activating genes improving plant stress tolerance. In nanopriming, most of the nanomaterials (NMs) dispersed in water are retained on the seed wall with the capability of a certain fraction of NMs to be translocated into the seed interior (Fig. 4). The NMs' coating helps seeds against pathogens that affect seed health [42]; more research on nano-seed priming is compiled in Table 2.

Although several NMs are beneficial for plant development, enhancement, and protection, the toxicological aspect of the NMs and their interactions with the living organisms must be taken care of. The detrimental effect has been observed when nanoparticles are applied into crops environment, not only to the plant but also beneficial bacteria and organisms that are involved in crop development [24].

NMs are new materials whose toxicology is not yet fully understood, since their effects depend on their physicochemical characteristics like size, surface charge, and morphology [53]. The lack of safety concerns toward the use of NMs makes them a new type of environmental contamination that, in the future, will affect several living

Table 2 NPs used for nano-seed priming in forage crop species

NMs	Seed species	Effect	Ref
Fe_3O_4	Sorghum	Increased chlorophyll a, b, carotenoids, and RWC after 500 mg/L treatment, highest increase in plant biomass obtained with primed seeds over salinity treatment, priming alleviated salinity stress in plants.	Maswada et al. [43]
Si	Wheat	Increased shoot length and grain weight, improved photosynthesis and chlorophyll content; electrolyte leakage (EL), malondialdehyde (MDA), and H_2O_2 were decreased, improved antioxidant enzyme content	Hussain et al. [44]
Ag	Pearl millet	Improved growth and biomass, enhanced flavonoid and phenolic content, improved antioxidant enzyme activity, and salt stress ameliorated by seed priming	Khan et al. [45]
CeO_2 Se-CeO_2	Horse gram	Effective seed germination achieved with nano-CeO_2 and Se-CeO_2, NPs at 300 ppm accelerates seed germination	Antony et al. [46]
Fe_2O_3	Chickpea	Low NP concentration enhances seedling growth, while higher concentrations inhibit growth	Pawar et al. [47]
ZnO	Corn	ZnO increases germination, root length, and dry biomass	Esper Neto et al. [48]
Fe_3O_4	Corn	NPs promoted beneficial responses for seedling growth compared to other Fe sources	Esper Neto et al. [49]
Ag	Pearl millet	20 mM concentration improved chlorophyll, carotenoid, photosynthetic rate, stomatal conductance, and transpiration rate. Ag NPs positively influence antioxidant enzymes under salt stress, increase Cl uptake in roots	Khan et al. [50]
ZnO	Wheat	Increased root and shoot length, improved photosynthetic efficiency	Kalal and Jajoo [51]
Mn_3O_4	Corn	Improved seedling growth, higher germination vigor, dry biomass, and length	Esper Neto et al. [52]

organisms [54]. To understand NMs' interaction with plants and other types of organisms, a new area of nanotechnology has emerged, which is nanotoxicology. Although some NM treatments have improved several aspects in plant development, managing higher concentrations, modifying NM characteristics, and the exposure method could negatively affect the crops [55].

Some of the adverse effects observed with NMs' application in crops are the penetration of the cell walls, alteration of plant metabolisms, generation of reactive oxygen species (ROS), changes in plant morphology, and, finally, plant death [56]. The nanotoxicology research allows us to understand the mechanism of the NMs, know how to reduce or diminish their ecological impact, and create safety protocols to assure correct manipulation and use of these materials to protect humans and other living organisms.

2 Nanomaterials and the Secondary Metabolism in Plants

The term metabolite refers to intermediate and end products of enzymatic reactions of biological pathways of the primary or secondary metabolism in living organisms. Primary metabolites are compounds associated with essential cellular functions and cellular functionality, e.g., basic sugars, carbohydrates, amino acids, lipids, nucleotides, and small organic acids. They are considered indispensable for the viability of an organism under any condition as they are directly involved in growth, development, reproduction, respiration, and photosynthesis [57–59].

Secondary metabolites are naturally occurring bioactive compounds that are not considered part of the primary plant life processes. However, they play a fundamental role in protection against insects, parasites, herbivores, phytopathogens, and other adverse environmental variables [35, 60, 61]. They have a short-term existence, and their absence does not cause direct death but threatens the plant's defense system against biotic and abiotic menaces [59, 62, 63]. Plants generate a wide variety of secondary metabolites such as alkaloids, anthocyanins, flavonoids, quinones, lignans, steroids, and terpenoids, commonly used as medicines, agrochemicals, flavors, fragrances, dyes, biopesticides, and food additives to promote health due to their properties for disease prevention and treatment [35, 57, 61]. The production of secondary metabolites is insufficient since it composes less than 1% of the dry weight of the plant [64, 65]; therefore, several tactics have been applied to increase the productivity of these compounds by exposing the plant to stressful conditions [61].

According to the hormonal curve of each plant model, stress can be divided into discomfort (bad stress, which leads to developmental deficiencies and ultimately to plant death) or eustress (good stress that leads to activation of secondary metabolism) [35, 57]. Hormesis is a term used in medicine to apply toxins in low doses [66] but has been applied in horticultural practices. Components of the environment that induce stress in plants can be divided into biotic and abiotic, which have the potential to be supplied artificially to activate plant conservation responses that lead to increased production of secondary metabolites. In recent decades, various approaches have been proposed for this process called elicitation with various types of metabolism-inducing factors or elicitor compounds [35, 67].

The main intention of elicitation is to enhance the production of secondary metabolites due to the relationship between the elicitor and biochemical pathways, mainly with adverse outcomes on biomass development such as morphology, although augmenting some plant quality traits, aroma, taste, or color. However, responses to secondary metabolite induction rely upon the plant model, elicitor concentration, and stimulation time [35, 68]. Therefore, the impact of elicitation methods cannot be generalized since results have indicated that the dose response is not mainly linear [57, 67, 69]. Novel elicitation using abiotic factors includes electromagnetic radiation, sound, volatile compounds, nutrient privation, metals, salt soil pollutants, and nanomaterials.

The concept of nanotechnology, consolidated in the 1980s, deals with the creation, use, manipulation, control, and characterization of structures, devices, or materials with at least one quantity in the range from 1 to 100 nm [70, 71]. Nanoparticles have unique physicochemical characteristics such as large surface area, high reactivity, controlled pore size, and particle morphology, suitable for new horticultural applications [72].

The use of nanomaterials in horticulture has been a controversial issue. The adverse effect of these components has been studied on several organisms since the liberation of wastes carrying nanomaterials has become a menace to the environment since they cause contamination to the air, water, and soil [73]. However, using this type of material as an elicitor for increasing the secondary metabolites in plants has become an essential issue in the scientific community. Nanoparticles can be focused on specific cellular organelles in plants to release herbicides and nanopesticide fertilizers or allow gene manipulation and expression for enhancing plant metabolism [72, 74].

Fertilizing is a fundamental activity for crop production involved directly in plant development. Unavailability to plants of the applied fertilizers is a common problem caused by many variables, such as leaching, degradation by photolysis, hydrolysis, and decomposition [72]. In this way, nanotechnology and nanomaterials have lined new methods for minimizing nutrient losses in fertilization through nanofertilizers or nano-encapsulated nutrients, which make possible an enhanced target activity by releasing nutrients on-demand as plants grow [72, 75, 76].

The size of structural components of plants allows this technology to allocate through the organs of plants efficiently. In the same line, the size of cellular components is equivalent to nanoparticles, allowing them to effortlessly permeate cells, which unfavorable biological effects have been widely studied. However, depending on the material and dose, the exposure might cause beneficial effects on the plant performance [35, 77, 78]. For example, stomata are turgor-operated valves that govern water loss and CO_2 uptake during photosynthesis and which guard cells are relatively small and considerably variated in size between species [79].

The dimensions of the outer limits of their cell walls may vary according to the stomatal aperture, from under 10 to almost 80 μm in length and from a few micrometers to about 50 μm in width. Also, considering that the total percentage of possible pore area (with apertures of 6 μm) for a leaf is similar for most species which can reach as much as 5% when the stomata are very wide open, but usually, the value remains below 2% [79, 80]. That is why the NMs can easily pass through the different plant systems.

Likewise, bordered pits, the cavities in the lignified cell walls of the xylem ducts, are fundamental elements in higher plants' water transport systems. The pit membrane allows water to pass between the xylem ducts. However, it limits the spread of embolism and vascular pathogens in the xylem, acting as a security valve in the plant's hydraulic system [81]. Observations using particle perfusion techniques provide maximum membrane porosities of 5–420 nm for different species of angiosperms, although it should be noted that these values are generally <100 nm [81–83]. Moreover, at a cellular level, pores of the primary cell wall are pathways for penetration of molecules into the cell, consisting of a polysaccharide-protein structure with a size ranging from 3.5 to 20 nm [84–86].

In the same way, the transport of molecules from cell to cell is possible through plasmodesmata, cytoplasmic channels with 20–50 nm in diameter, which usually let the pass of particles, around 3 nm [55, 85]. In chloroplasts, the import of proteins is a prerequisite for photosynthesis and, consequently, plants' growth and development, where this process is feasible through the translocation of proteins from the outer and inner membranes of the chloroplast, which are responsible for importing about 95% of each of the chloroplast proteins from the cytoplasm and are highly conserved among all land plants [87], and they are considered to have a pore size greater than 2.56 nm [87, 88].

Despite its toxic potential, which has been thoroughly studied, new results indicate positive effects on plant development and physiology, depending on the nature of the nanomaterial, dose and time of exposure, the plant species, and growing conditions. In this chapter, the application of metallic and oxide nanoparticles and carbon-based nanomaterials in forage plants is presented as an elicitation method to increase secondary metabolites in forage plants with an essential role in the production of horticultural crops.

3 Metallic Nanoparticle Effects Over Forage Plants

Pearl millet seed priming with silver NPs showed a beneficial effect on salt stress tolerance. In another study, pearl millet was exposed to Ag NPs during their growth in pots with different salinity concentrations at different doses of Ag NPs. The study found that the presence of NPs helped reduce the salt ion content in plants and primed seeds. The ionic balance was maintained between sodium and potassium ions. Antioxidant enzyme activity was improved by reducing the adverse effects caused by the salinity stress. Pearl millet crops increased yield, height, and photosynthetic activity [50]. Another study with pearl millet and Ag NPs in concentrations of 2, 4, and 6 mM increased ROS production reducing growth, root, shoot length, and fresh and dry biomass. Ag NPs' toxicity was compared with $AgNO_3$ treatment, showing that NPs have a lower phytotoxic level. However, due to a blockage in the electron transport chain, NPs and silver nitrate exposure caused changes in antioxidant enzyme activities and reduced photosynthetic pigment contents [89].

NPs' accumulation in the soil due to contamination is also a big concern. Li et al. [90] studied the effects of Ag NP exposure in soybean and rice where the phytotoxic levels of the metal NPs were assessed; results showed that a foliar exposure caused an increase of Ag content up to 17–200 times compared to root exposure.

Even though the root exposure led to lower Ag accumulation levels than those observed in leaves, the root exposure caused the higher biomass reductions, increasing malondialdehyde and H_2O_2 content. Storage of NPs in cell walls was also observed after leaf exposure, where cell wall-trapped NPs showed changes in size distribution compared to the initial sizes.

These results demonstrate the negative side of the exposure of NMs to plants where risk protocols need to be improved for healthy protection when NPs are used in agronomical fields [90]. Translocation of root to shoot was also observed for alfalfa crops treated with zero-valent iron; interaction with OH radicals caused cell wall loosening, which was reflected in an increase in root length; chlorophyll content was also increased by the NP treatments [91].

Not only can nanomaterials act against pests that affect crop development, but also the NMs work as a protection against diseases caused by pathogens such as bacteria or fungi. It can be alleviated using materials that have antimicrobial properties, such as silver nanoparticles or metal oxides like the semiconductors such as titanium dioxide (TiO_2) and zinc oxide (ZnO) [92].

The materials with antimicrobial properties are producers of reactive oxygen species (ROS), molecules capable of affecting the development of bacteria, disrupting cell walls leading to cell death [93]. The application of these nanomaterials for protection against diseases can be used from the seed stage, which allows increasing the life of the seeds and protecting them during their development. Maity et al. [94] investigated forage sorghum exposed with multiple metallic oxide nanoparticles and silver nanoparticles. It was possible to observe that silver nanoparticles, including metallic oxides, influenced specific development characteristics such as germination, stem, and root size. Apart from these observed effects, the nature of the silver nanoparticles added an antimicrobial factor. The effects of improvement in germination were observed at low concentrations, where, at the same time, this was lost at higher concentrations.

The use of nanoparticles can improve abiotic stress defense; as shown with Ag NPs at 5, 10, and 15 ppm, salinity stress can be alleviated in grass pea crops. The interaction between plant and nanoparticle leads to improved germination, shoot, and root length, which was initially reduced by the presence of different levels of salinity [95]. Recent metallic NP effects on forage species are compiled in Table 3.

4 Metal Oxides' Effects Over Forage Plants

Seed priming has shown positive effects when using metal oxide NPs, as seen in fodder maize where the lowest concentration (20 and 40 mg/L) showed promising results increasing several parameters like the number of plants, height, biomass, and yield when increasing to 40 mg/L. NPs increased chlorophyll content and nitrogen and phosphorus availability [103]. ZnO NPs can also help plants against abiotic stress such as salinity which affects soil, nutrient availability, and plant development. When cowpea and okra crops were exposed to several seawater concentrations, plants showed a reduction of root and shoot length. After the exposure with ZnO as a foliar spray, the measured parameters improved compared to NPs' untreated plants. As told by the author, the positive effects observed in crops could be due to the release of zinc, a vital micronutrient involved in several metabolic processes in

Table 3 Effects of metallic NPs over forage crop species

Nanomaterial	Concentration	Plant	Effect	Ref.
Copper nanowires	80 and 280 mg Cu/kg	Alfalfa	Increased iron and zinc content in roots, increased number of microorganisms involved in nutrient uptake, copper nanowires can be considered as a potential nanofertilizer	Cota-Ruiz et al. [96]
Cu-chitosan		Maize	Antioxidant enzyme activities were increased by the presence of Cu-chitosan NPs, increasing plant defenses, plant height, stem diameters, and root length, and chlorophyll content was also increased when maize was cultivated in pot; plant growth and protection against disease were also promoted due to the release of copper	Choudhary et al. [97]
Ag NPs and antimony (III) exposure		Soybean	Antimony uptake improved by the presence of silver NPs and pigment content, and enzyme activity was altered	Weicheng et al. [98]
Mn, Cu, Zn, Ag, and Fe colloidal solution	120 and 240 mg/L	Soybean	Seed and foliar application improved yield, and NPK fertilizer utility increases using a colloidal suspension of nanoparticles	Batsmanova et al. [99]
Ag	20, 200, and 2000 mg/kg	Wheat	Phytotoxicity upon NP treatment decreased biomass, plant height, and grain weight, reduction in micronutrient content and arginine and histidine amino acids	Yang et al. [100]
Fe, Cu, and Co	Fe and Cu (25–250 mg/L)	Soybean	Enhanced germination yield was obtained with Fe and Cu NPs at 50 mg/L, leaves' growth was faster after 39 h of germination in Co NPs treatment, all NPs treatments enhanced cell division under optimal concentrations	Hoe et al. [101]
	Co (0.05–2.5 mg/L)			
Ag	0, 25, 50, and 100 ppm	Soybean	NPs caused increased seed germination, root and shoot length at 50 ppm, while at 25 ppm, chlorophyll carotenoids and proteins increased while all other concentrations decreased those compounds	Sharif et al. [102]

plants, such as enzyme activity, cell division and development, nutrient metabolisms, and photosynthesis [104].

Chickpea, which can be used as a forage cultivar for feeding ruminants, showed that when exposed to iron oxide NPs capped with leaf extract at 1, 5, 10, and 15 mg/

L helped different types of chickpeas for generation in callus and root induction as well as shoot.

This effect observed helped crops to be protected against infections. Iron oxide NPs were also translocated into the regenerated roots increasing the iron content. The smaller nanoparticles can dissolve iron easier into roots serving as an essential micronutrient involved in several crop development processes [105].

NPs can also be coated for improving the material interactions with increasing crop biocompatibility, thus obtaining more beneficial results. Iannone et al. [106] exposed alfalfa and soybean crops with magnetite NPs coated with citric acid. The NPs were added in solution to the crop pots. The results obtained show absorption and translocation in several organs of both crops due to paramagnetic signals measured in crop tissues. The presence of NPs also increased shoot weight and chlorophyll content. As presented by the author, magnetite nanoparticles coated with citric acid have no phytotoxic effect on both crops improving their status after exposure [106]. Red clover crops exposed to TiO_2 and cerium oxide (CeO_2) NPs showed no affectations in plan biomass. Shoot/root ratio, nitrogen fixation levels, and the number of flowers were also not affected [107].

The plant can exudate compounds through the root to modulate specific interactions with the media surrounding the roots. These exudates can have the ability to interact with NMs when they are present in the soil, as seen with CuO NPs interacting with maize roots. The exudates inhibited NP aggregation and promoted dissolution. Copper accumulation in root tissue was also observed when 20 mg/L was added into the media [108]. Recent studies relating the effect of metal oxides on forage crops are detailed in Table 4.

5 Carbon-Based Nanomaterials' (CBNs) Effects Over Forage Plants

All nanomaterials made of carbon atoms are called carbon-based or carbon nanomaterials. In addition, its unique characteristics have generated significant interest, contributing to the development of procedures for large-scale industrial, technical, medical, environmental, and agricultural applications, constantly increasing the commercial use of this technology [129]. Carbon is an interesting ingredient in a wide variety of designs, often with different characteristics. The list of known carbon allotropes has been expanded, primarily due to the discovery of many new forms of low-dimensional carbon, which comprise novel materials frequently classified as carbon nanotubes (CNTs), fullerenes, graphene, and carbon dots [129–132]. Several works have been written describing the classification and characteristics of carbon-based nanomaterials ([129] broadly describe this technology). In this chapter, the effect of this technology on important horticultural crops is the main objective.

Table 4 Effects of metal oxide NPs on forage crop species

Nanomaterial	Concentration	Plant	Effect	Ref.
Cr_2O_3	0.01, 0.05, 0.1, and 0.5 g/L	Soybean	Shoot and root biomass were reduced, NPs caused damage to the photosynthetic system, reducing the photochemical quenching, and NPs caused chloroplast thylakoid structure damage, reducing electron acceptors' activity	Li et al. [109]
ZnO	0, 10, 100, and 500 mg/L	Wild oat and wheat	Plumule length was increased by 100 ppm ZnO treatment in wheat, NPs caused, and increase in germination rate and percentage in wild oat	Zeidali et al. [110]
AgO, CeO, CuO, MoO_3, SiO, TiO, and ZnO	500 or 1000 μg/mL	Soybean exposed to *Fusarium virguliforme*	NP treatment restored nutrient content at the same level as healthy control increasing defense mechanisms against pathogen although NPs did not inhibit pathogen presence as viewed in the in vitro assays	Peréz et al. [111]
Fe_3O_4	0, 50 and 500 mg/kg	Maize	No effect in plant biomass and chlorophyll content was observed during NPs' exposure, Fe accumulation was higher in roots compared to leaves, the highest concentration caused an increase in dehydrogenase enzyme activity	Yan et al. [112]
TiO_2	0, 100, and 250 mg/L	Maize	Hydroponic exposure of NPs and cadmium showed	Lian et al. [113]

(continued)

Table 4 (continued)

Nanomaterial	Concentration	Plant	Effect	Ref.
			that TiO_2 through root exposure increased cadmium uptake, increasing the phytotoxicity compared to foliar exposure; chlorophyll content and dry plant weight were reduced, compared to root exposure; foliar application of NPs helped to reduce shoot cadmium content reducing the phytotoxicity through enzymatic and metabolic alterations	
CeO_2, polyvinylpyrrolidone (PVP)-coated CeO_2	100 mg/kg	Soybean cultivated at the different soil moisture content (55, 70, 85, or 100%)	Biomass was not affected by NPs at 55%, while the rest of the treatments increased in fresh weight, and all of the treatments increased dry weight. No total chlorophyll content differences were determined for all treatments, although the effect of NPs on photosynthesis was dependent on moisture content	Cao et al. [114]
TiO_2	(0, 5, 10, and 20 mg/L)	Maize	Increasing NP concentration reduced chlorophyll, shoot, and root biomass while positively affecting macro- and micronutrient content except for iron increasing nutrient concentration with increasing NP dose	Daghan [115]

(continued)

Table 4 (continued)

Nanomaterial	Concentration	Plant	Effect	Ref.
TiO_2	250–1000 mg/L	Soybean	Soybean growth was inhibited due to the adsorption to root surface causing physical damage, reducing cell viability, root hair number, and fresh and dry weight	de Melo et al. [116]
CuO		Soybean	Antioxidant biomarkers were altered by the presence of CuO at different sizes (25, 50, and 250 nm), showing dependence on NP size, the smaller NPs highly inhibited seed yield	Yusefi-Tanha et al. [117]
ZnO		Maize	Maize seeds primed with ZnO NPs showed improved germination and growth parameters compared to normal priming	Itroutwar et al. [118]
NiO	0, 200, and 400 mg/L	Soybean	An increase of NP concentration caused augmentation of enzyme activity in catalase and ascorbate peroxidase	Tohidiyan et al. [119]
Fe_2O_3	15, 30, and 60 mg/pot	Soybean	Both types of NPs incremented the chlorophyll content and plant biomass, higher iron content in the shot was obtained with foliar exposure compared to the soil, foliar exposure also stimulated nitrogen fixation	Yang et al. [120]
Fulvic acid-coated Fe_2O_3		Foliar or soil exposure		
Y_2O_3	0–500 mg/L	Maize	No effect on germination was observed, root	Gong et al. [121]

(continued)

Table 4 (continued)

Nanomaterial	Concentration	Plant	Effect	Ref.
			elongation was inhibited, increasing the concentration of NPs increased the concentration of peroxidase and catalase in shoots, NPs also altered carbohydrate metabolic pathway and amino acid synthesis	
La_2O_3	5 mg/L	Maize	NPs had phytotoxic effects on maize, affecting and decreasing shoot, root biomass, and chlorophyll content	Liu et al. [122]
ZnO		Sorghum	Grain yields increase at a low NPK level	Dimkpa et al. [123]
$MnFe_2O_4$	62.5, 125, 250, 500, and 1000 mg/L	Barley	NPs promoted fresh plant weight at low concentrations, NPs were translocated from root to leaves	Tombuloglu et al. [124]
CuO	CuO (5, 10, 15, and 20 mg/L)	Berseem, Cowpea, Sorghum, and Oats	Antifungal activity against seed microflora, enhanced seed germination, root length, shoot length, and dry weight	Manjunatha et al. [125]
ZnO	ZnO			
ZnO	750, 1000, and 1250 mg/kg of seed	Oat and berseem	ZnO NPs enhanced seed germination at low concentrations and a reduction in root and shoot length was caused at higher doses. All NPs enhanced germination at the lowest concentration	Maity et al. [126]
TiO_2				
CuO				
Hematite	500–8000 mg/kg	Maize	Enhanced maize growth at 500 mg/kg, increased lipid peroxidation and superoxide	Youssef et al. [127]

(continued)

Table 4 (continued)

Nanomaterial	Concentration	Plant	Effect	Ref.
			dismutase activity 4000 and 8000 mg/kg, NP aggregates were found inside vacuoles	
Fe_3O_4	125, 250, 500, and 1000 mg/L	Barley	Increased chlorophyll, soluble protein, and dry weight	Tombuloglu et al. [128]

CBNs cover around 40% of all engineered nanomaterials used for agricultural applications [133]. The interaction between plant organelles and nonbiological nanostructures is a novel topic studied since it brings the potential to deliver organelles with new and improved functionality [74]. Several experiments using CBNs have been performed in horticultural crops where the secondary metabolism and other critical biological parameters have been evaluated.

López-Vargas et al. [134] studied the effects on growth and biochemical compounds of tomato (*Lycopersicon esculentum*) seedlings in an experiment with five concentrations of CNTs and graphene.

The López-Vargas et al. experiments results in adverse effects in vigor variables, with a decrease in root length (39.2%), hypocotyl biomass (33%), plant height (29%), stem diameter (20%), fresh shoot biomass (63%), fresh root biomass (63%), and dry shoot biomass (71%) with CNTs (1000 mg/L), however, the root biomass was increased up to 127%. With graphene, the content of chlorophyll-a and chlorophyll-b (111%), vitamin C (78%), β-carotene (up to 11-fold), phenols (85%), and flavonoids (45%), as well as the H_2O_2 content (215%), increased. In this context, carbon-based nanomaterials can interact with the primary and accessory pigments in the photosynthesis process and the allocation of nutrients within the plant. CNTs present in the soil can move into the plant to access the cells of the leaves and finally reach the chloroplasts [74].

The ability of plants to harvest light energy has been improved by placing CNTs in the chloroplasts. This technology could work as artificial antennae allowing chloroplasts to take advantage of green and near-infrared wavelengths, considering that red, blue, far-red, and UV-A are the wavelengths that most plants use for photosynthesis [68, 72, 74, 135]. An increase of the chlorophyll content in plants is directly related to the ability to capture light energy and, therefore, the efficiency in producing carbohydrates needed to create and accumulate biomass [134]. Moreover, Liu et al. [136] established that CNTs act as a molecular carrier in the cell walls, stimulating metabolism and crop growth.

Contrary, other investigations have established that they accumulate inside the cell, causing adverse consequences by blocking nutrient transport, especially in the roots, producing retardation in the plant growth [136, 137]. As for stress induction, this technology has been found to generate oxidative stress, modifying plants' physiological and biochemical responses as they increase the production of reactive oxygen species [138]. CBNs can work as an elicitor, causing the upregulation of various genes and changing the signal transduction patterns involved in secondary metabolite biosynthesis, such as phenols and flavonoids [139].

Continuing with tomato, González-García et al. [140] evaluated the effect of CNTs and graphene in different doses applied via foliar or drench, resulting in beneficial effects in all the measured variables with both materials and both application methods, where the best results presented an increase of 87% in root biomass, 120% in ascorbic acid, 65.5% in glutathione, and 56.8% in the antioxidant capacity by the DPPH (2,2-diphenyl-1-picrylhydrazyl) radical with CNTs via drench.

An increase of 196% in the activity of the APX enzyme and 281% in the activity of the phenylalanine ammonia-lyase (PAL) enzyme was observed with CNTs application via foliar. An increment of 72% in chlorophyll-a, 39% in chlorophyll-b, 74% of the phenols content, 28.6% in flavonoids, 127% in the GPX activity was observed with the application of graphene via drench, and 135% in the catalase activity with graphene via drench. And finally, an increase of 19.4% in the protein content for both CNTs and graphene materials was observed when the treatments were applied via foliar.

Furthermore, González-García et al. [141] evaluated the effect of CNTs in tomato crops infected with *Alternaria solani*, a fungal disease that causes severe damage and substantial yield losses, reporting a reduction of 44% in the severity and an increase of 5% of ascorbic acid, 11% in the net photosynthesis rate, and 20% of the flavonoid content. Moreover, in the uninfected crop treated with CNTs, there was an increase of 14% in plant biomass, 11% in the GPX activity, and 18% in fruit production, leading to a 21% increase in fruit yield. CBTs can act as elicitors, playing an essential role in plant growth regulation because they trigger the biosynthesis of indoleacetic acid and abscisic acid and the expression of marker genes for cell division and cell wall elongation [142, 143]. In addition, this technology has been shown to increase the activity SOD, POD, CAT, and APX enzymes, which results in the buildup of proteins in the roots, increasing resistance to pathologies [144]. Furthermore, ascorbic acid acts as a critical substrate for detoxification and stable maintenance of ROS in chloroplasts, peroxisomes, mitochondria, and cell cytosol [145].

In addition, antioxidant enzymes defend cells from modifications caused by oxidative reactive species since they neutralize those radicals [146]. Among these enzymes, APX helps to diminish the oxidative state of chloroplasts [145]. GPX and CAT are responsible for eliminating ROS and catalyzing the reduction of H_2O_2 to prevent cell damage [147]. Finally, PAL participates in the phenylpropanoid pathway, necessary in the antioxidant defense system [148]. These enzymes are precursors of the biosynthesis of phenolic compounds, which increases the antioxidant capacity of plants and enhances protection against pathogens, improving tolerance against biotic and abiotic stress [35, 57, 149].

However, it should be emphasized that the responses displayed depend on the concentration of CNMs used. A hormetic response is seen in most experiments, where low concentrations can cause eustress (beneficial effects) but high concentrations cause the opposite effect, called distress. This hormetic reaction is constantly observed in plants as it is a dynamic adaptive response of complex biological systems to various stress inducers [67, 150].

Younes et al. [151] studied the effect of graphene nanosheets on pepper (*Capsicum annuum* L.) and eggplant (*Solanum melongena* L.) in 2 years of crop (Year 1 and Year 2), finding an increase of 21.8% and 23.1% in plant length, 41.6% and 41.4% in number of fruits per plant, 80.4% and 81.2% in fruit yield, 144% in chlorophyll-a, 132% in chlorophyll-b, 143% in carotenoids, 124% in phenols, and 100% in flavonoids for eggplant and an increase of 21.43% and 22.8% in plant length, 87.78% and 79% in number of fruits per plant, 121% and 119% in fruit yield, 126% in chlorophyll-a, 134% in chlorophyll-b, 42% in carotenoids, 53% in phenols, and 95% in flavonoids for pepper.

However, a contrary effect was found by Hao et al. [152] with mesoporous carbon in rice (*Oryza sativa* L.). Where produced a negative effect in all treatments with a decrease up to 70% in the root length, 57.1% in shoot length, 33% in fresh root weight, and 45% in fresh shoot weight and suppression of the brassinolide (BR), indole propionic acid (IPA), and dihydrozeatin riboside (DHZR) phytohormone concentrations. Phytohormones regulate various cellular processes in plants. They function as chemical messengers to communicate cellular occupations and coordinate specific signal transduction pathways throughout the abiotic stress response [153, 154].

BR, IPA, and DHZ are essential to plant hormones that mediate plant growth, metabolic processes, stimulation of cellular division, and generation of branches and stem and promote leaf elongation and stress tolerance [155–157]. CBNs can induce phytotoxicity, affecting the balance of phytohormones, including BR, JA, and ABA [158, 159]. Nanomaterials can induce excessive amounts of ROS and can manipulate the expression of genes, resulting in changes in levels of phytohormones [160, 161].

Carbon dots are new fluorescent materials characterized by a diameter below 10 nm, becoming a promising alternative because of their composition and biocompatibility [162].

Wang et al. [163] described the effect of carbon dots in mung bean sprouts (*Vigna radiata*), reporting positive effects with an increase in the carbohydrate content (up to 21.9%), total fresh biomass (up to 14.9%), root vigor (36.1%), stem length (up to 18.3%), electron transportation rate of the photosystem 1 (PS1) of leaves (8.8%), chlorophyll content (14.8%), and general photosystem activity (10.4%), compared to control.

In the same line, Park and Ahn [164] assessed the effect of multi-walled CNTs (up to 2000 mg/L final concentration) in carrot (*Daucus carota* L.) seedlings, reporting a significant enhancement in the total chlorophyll concentration in the leaf tissue by 25–30%, but the seedling growth decreased by 10%. And in the same specie, Siddiqui et al. [165] evaluated the effect of foliar spray of graphene oxide solution reporting a significant increase in root dry weight (36%), shoot dry weight (50.7%), plant fresh weight (15.8%), total chlorophyll (27%), carotenoid (18%), and proline (7.6%).

In contrast, Zhang et al. [166] evaluated different doses of graphene in roots and shoots of wheat (*Triticum aestivum* L.) plants, reporting an inhibition in the formation of root hairs, inhibition of long-term biomass production, and as the chlorophyll content. The PSII (photosystem II) activity was reduced but found a hormetic response with a rise in the activities of the enzymes SOD and POD and promoted root elongation up to 287%, which they explained might be attributable to the increased levels of ROS and a change in the hormone-mediated pathways of auxin and abscisic acid (ABA) that might also play a vital role in graphene-induced root cell elongation as described by Cheng et al. [142]. Moreover, Vochita et al. [146] studied the effects of graphene oxide in seedlings of the same species, reporting an increase of 7.11% in root length, 11.21% in the SOD activity, 4.65% in the CAT activity, and 6.50% in the POD activity but a decrease up to 22.48% in chlorophyll-a.

Likewise, Li et al. [167] evaluated graphene oxide (GO), GO quantum dots (GOQDs), and reduced GO (rGO) in wheat grains resulting in a decrease of 14–22%, 8–14%, and 22–28% in globulin; 11–25%, 16–17%, and 22–24% in prolamin; 15.1%, 8.7%, and 24.1% in the ratios of the amide II to I; and 22–34%, 24–31%, and 25–36% in the total starch, composed primarily of amylopectin and amylose, but a significant increase in soluble sugar accumulation, by 23–36%, 19–25%, and 19–35% for GO, GOQD, and rGO, respectively. Starch is the main carbohydrate stored in crops. It is crucial to crop quality and nutrition [168, 169].

A decrease in the total starch content of plants is estimated to be a mechanism used by plants to withstand stress. It involves the remobilization of contained starch to release energy, sugars, and derived metabolites [169]. The accumulation of soluble sugar (e.g., sucrose, glucose, and fructose) is an indispensable adaptive mechanism for responding to abiotic environmental stresses [170].

Even though carbon-based nanomaterials boost industrial advancement, there are concerns regarding possible environmental emissions. Interactions between released nanomaterials and live species, as well as their incorporation into food chains, have yet to be determined with yet unknown consequences [129]. It is critical to know with certainty the toxicological effects that nanostructured systems can have on plant organisms, regardless of their origin, whether as a product with a particular purpose for agriculture or their entrance into the environment owing to contamination [34].

The research mentioned in this chapter showed that both positive and negative effects could be observed with the application of CNMs, following a hormetic curve where small doses may cause eustress and high doses may cause distress. It was clear that this depended on the application route, dose, type of material, and exposure times [134].

To react to the continuously changing environment, plants have evolved complex internal and interplant signaling channels with different architectures. Plant nanobionics describes the interface between living plants and nanotechnology, where structural merits of plant organs and organelles have inspired the creation of plant-derived structures through bio-interfacing with nanoparticles [171].

Plant nanobionics engineering can help to develop biomimetic materials for light-harvesting and biochemical detection with regenerative properties and enhanced efficiency [74]. This new scientific field can contribute to the specific study of the effect of nanomaterials on plants, investigating how to produce beneficial effects on the primary and secondary metabolism and assuring that adverse effects will not be produced in the health of people that consumes those crops.

6 Conclusion

The application of nanotechnology in the agricultural sector has generated positive and promising prospects, which can be used to comply with food safety, producing safe and adequate food for consumption. Nanostructured materials have revolutionized the agricultural sector, as documented in forage plants. The use of metallic nanoparticles (NPs) and carbon-based nanomaterials increased the concentration of secondary metabolites, helping to counteract various types of stress (biotic and abiotic). Also, it has generated modification of the morphology of the plant, photosynthesis, and plant-substrate-microorganism interaction, as well as an increase in the concentration of nutrients used by various types of animals, including those for human consumption. However, although there is strong evidence of positive results from the use of nanostructured materials in forage plants, more studies should be generated in this regard since the effect of the nanoparticles and/or nanomaterials used is variable depending on the type of nanostructured material used, the concentration, type, and time of application, as well as the variety of forage plant to be used. These studies will allow us to know with greater certainty the type of nanostructured material that provides ecological, toxicological, and food safety in favor of general well-being.

Acknowledgments L. Páramo thanks CONACyT for the scholarship granted. The authors thank the Engineering Faculty-UAQ for the financial support granted through the Attention to national problems fund FI-UAQ-2021 (FIN202106) and to the Universidad Autónoma de Querétaro through the FONDEC-UAQ-2021 fund (FIN202116 and FIN202115).

References

1. Khasbagan, & Shengji, P. (2000). Ethnobotany of forage plants: A case study in Arhorchin Banner of inner Mongolia. *Acta Prataculturał Science, 9*, 74–81.
2. Del Grosso, S. J., Parton, W. J., Derner, J. D., Chen, M., & Tucker, C. J. (2018). Simple models to predict grassland ecosystem C exchange and actual evapotranspiration using NDVI and environmental variables. *Agricultural and Forest Meteorology, 249*, 1–10. https://doi.org/10.1016/j.agrformet.2017.11.007
3. Kemp, D., Han, G., Hou, X., Michalk, D., Fujiang, H., Jianping, W., & Zhang, Y. (2013). Innovative grassland management systems for environmental and livelihood benefits. *Proceedings of the National Academy of Sciences of the United States of America, 110*, 8369. https://doi.org/10.1073/pnas.1208063110

4. Alldredge, M., Peek, J., & Wall, W. (2002). Nutritional quality of forages used by elk in northern Idaho. *Journal of Range Management, 55*, 253–259. https://doi.org/10.2307/4003131
5. Kambashi, B., Boudry, C., Picron, P., & Bindelle, J. (2014). Forage plants as an alternative feed resource for sustainable pig production in the tropics: A review. *Animal, 8*(8), 1298–1311. https://doi.org/10.1017/S1751731114000561
6. Lin, C.-H., McGraw, M., George, M., & Garrett, H. (2001). Nutritive quality and morphological development under partial shade of some forage species with agroforestry potential. *Agroforestry Systems, 53*, 269–281. https://doi.org/10.1023/A:1013323409839
7. Gang, C., Zhou, W., Chen, Y., Zhaoqi, W., Sun, Z., Li, J., Qi, J., & Odeh, I. (2014). Quantitative assessment of the contributions of climate change and human activities on global grassland degradation. *Environment and Earth Science, 72*, 1–10. https://doi.org/10.1007/s12665-014-3322-6
8. Ramírez Lozano, R. G. (2009). *Nutrición de rumiantes: Sistemas extensivos*. Trillas.
9. Miranda, C., Prasad, R., & Jayalaxmi, P. (2013). A study on economics of inclusion of azolla pinnata in swine rations. *International Journal of Agricultural Sciences and Veterinary Medicine, 1*, 50.
10. Phengsavanh, P., & Lindberg, J. E. (2013). Effect of replacing soybean protein with protein from porcupine joint vetch (Aeschynomene histrix BRA 9690) and stylo (Stylosanthes guianensis Composite) leaf meal on growth performance of native (Moo Lath) Lao pigs. *Tropical Animal Health and Production, 45*(8), 1795–1802.
11. Lemaire, G., Tang, L., Bélanger, G., Zhu, Y., & Jeuffroy, M.-H. (2021). Forward new paradigms for crop mineral nutrition and fertilization towards sustainable agriculture. *European Journal of Agronomy, 125*, 126248.
12. Palombi, L., & Sessa, R. (2013). *Climate-smart agriculture: Sourcebook*. FAO.
13. Stoll-Kleemann, S., & O'Riordan, T. (2015). The sustainability challenges of our meat and dairy diets. *Environment: Science and Policy for Sustainable Development, 57*, 34–48. https://doi.org/10.1080/00139157.2015.1025644
14. Berauer, B. J., Wilfahrt, P. A., Reu, B., Schuchardt, M. A., Garcia-Franco, N., Zistl-Schlingmann, M., Dannenmann, M., Kiese, R., Kühnel, A., & Jentsch, A. (2020). Predicting forage quality of species-rich pasture grasslands using vis-NIRS to reveal effects of management intensity and climate change. *Agriculture, Ecosystems & Environment, 296*, 106929. https://doi.org/10.1016/j.agee.2020.106929
15. Saleh, T. A. (2020). Nanomaterials: Classification, properties, and environmental toxicities. *Environmental Technology and Innovation, 20*, 101067. https://doi.org/10.1016/j.eti.2020.101067
16. Dolez, P. I. (2015). Chapter 1.1 - Nanomaterials definitions, classifications, and applications. In P. I. Dolez (Ed.), *Nanoengineering* (pp. 3–40). Elsevier.
17. Sudha, P. N., Sangeetha, K., Vijayalakshmi, K., & Barhoum, A. (2018). Chapter 12 - Nanomaterials history, classification, unique properties, production and market. In A. Barhoum & A. S. H. Makhlouf (Eds.), *Emerging applications of nanoparticles and architecture nanostructures* (pp. 341–384). Elsevier.
18. Butt, B. Z., & Naseer, I. (2020). Nanofertilizers. In S. Javad (Ed.), *Nanoagronomy* (pp. 125–152). Springer International Publishing.
19. Fraceto, L., Castro, V., Grillo, R., Avila, D., Oliveira, H., & Lima, R. (2020). *Nanopesticides - From research and development to mechanisms of action and sustainable use in agriculture*. Springer.
20. Javed, R., Yucesan, B., Zia, M., & Gurel, E. (2018). Elicitation of secondary metabolites in callus cultures of Stevia rebaudiana Bertoni grown under ZnO and CuO nanoparticles stress. *Sugar Tech, 20*(2), 194–201. https://doi.org/10.1007/s12355-017-0539-1
21. Siddiqui, M. H., Al-Whaibi, M. H., & Mohammad, F. (2015b). *Nanotechnology and plant sciences*. Springer International Publishing.

22. Wang, P., Lombi, E., Zhao, F.-J., & Kopittke, P. M. (2016). Nanotechnology: A new opportunity in plant sciences. *Trends in Plant Science, 21*(8), 699–712. https://doi.org/10.1016/j.tplants.2016.04.005
23. Sun, H., Du, W., Peng, Q., Lv, Z., Mao, H., & Kopittke, P. M. (2020). Development of ZnO nanoparticles as an efficient Zn fertilizer: Using synchrotron-based techniques and laser ablation to examine elemental distribution in wheat grain. *Journal of Agricultural and Food Chemistry, 68*(18), 5068–5075. https://doi.org/10.1021/acs.jafc.0c00084
24. Côa, F., Bortolozzo, L. S., Petry, R., Da Silva, G. H., Martins, C. H. Z., de Medeiros, A. M. Z., Sabino, C. M. S., Costa, R. S., Khan, L. U., Delite, F. S., & Martinez, D. S. T. (2020). Environmental toxicity of nanopesticides against non-target organisms: The state of the art. In L. F. Fraceto, V. L. S. S. de Castro, R. Grillo, D. Ávila, H. Caixeta Oliveira, & R. Lima (Eds.), *Nanopesticides: From research and development to mechanisms of action and sustainable use in agriculture* (pp. 227–279). Springer International Publishing.
25. Chand Mali, S., Raj, S., & Trivedi, R. (2020). Nanotechnology a novel approach to enhance crop productivity. *Biochemistry and Biophysics Reports, 24*, 100821. https://doi.org/10.1016/j.bbrep.2020.100821
26. Pradhan, S., & Mailapalli, D. R. (2020). Nanopesticides for pest control. In E. Lichtfouse (Ed.), *Sustainable agriculture reviews* (Vol. 40, pp. 43–74). Springer International Publishing.
27. Malaikozhundan, B., & Vinodhini, J. (2018). Nanopesticidal effects of Pongamia pinnata leaf extract coated zinc oxide nanoparticle against the Pulse beetle, Callosobruchus maculatus. *Materials Today Communications, 14*, 106–115. https://doi.org/10.1016/j.mtcomm.2017.12.015
28. Ul Haq, I., Ijaz, S., & Khan, N. A. (2020). Application of nanotechnology for integrated plant disease management. In I. Ul Haq & S. Ijaz (Eds.), *Plant disease management strategies for sustainable agriculture through traditional and modern approaches* (pp. 173–185). Springer International Publishing.
29. Irshad, M. A., Nawaz, R., ur Rehman, M. Z., Imran, M., Ahmad, J., Ahmad, S., Inam, A., Razzaq, A., Rizwan, M., & Ali, S. (2020). Synthesis and characterization of titanium dioxide nanoparticles by chemical and green methods and their antifungal activities against wheat rust. *Chemosphere, 258*, 127352. https://doi.org/10.1016/j.chemosphere.2020.127352
30. Ogunyemi, S. O., Zhang, M., Abdallah, Y., Ahmed, T., Qiu, W., Ali, M. A., Yan, C., Yang, Y., Chen, J., & Li, B. (2020). The bio-synthesis of three metal oxide nanoparticles (ZnO, MnO2, and MgO) and their antibacterial activity against the bacterial leaf blight pathogen. *Frontiers in Microbiology, 11*, 3099. https://doi.org/10.3389/fmicb.2020.588326
31. Ibrahim, E., Zhang, M., Zhang, Y., Hossain, A., Qiu, W., Chen, Y., Wang, Y., Wu, W., Sun, G., & Li, B. (2020). Green-synthesization of silver nanoparticles using endophytic bacteria isolated from garlic and its antifungal activity against wheat fusarium head blight pathogen fusarium graminearum. *Nanomaterials, 10*(2), 219. https://doi.org/10.3390/nano10020219
32. Nandhini, M., Rajini, S. B., Udayashankar, A. C., Niranjana, S. R., Lund, O. S., Shetty, H. S., & Prakash, H. S. (2019). Biofabricated zinc oxide nanoparticles as an eco-friendly alternative for growth promotion and management of downy mildew of pearl millet. *Crop Protection, 121*, 103–112. https://doi.org/10.1016/j.cropro.2019.03.015
33. Bano, A., & Ummat-ul-Habib. (2020). Interactive effects of Ag-nanoparticles, salicylic acid, and plant growth promoting rhizobacteria on the physiology of wheat infected with yellow rust. *Journal of Plant Pathology, 102*(4), 1215–1225. https://doi.org/10.1007/s42161-020-00626-y
34. Paramo, L. A., Feregrino-Pérez, A. A., Guevara, R., Mendoza, S., & Esquivel, K. (2020). Nanoparticles in agroindustry: Applications, toxicity, challenges, and trends. *Nanomaterials, 10*(9), 1654.
35. Aguirre-Becerra, H., Vazquez-Hernandez, M. C., Alvarado-Mariana, A., Guevara-Gonzalez, R. G., Garcia-Trejo, J. F., & Feregrino-Perez, A. A. (2021). Role of stress and defense in plant secondary metabolites production. In *Bioactive natural products for pharmaceutical applications* (pp. 151–195). Springer.

36. Verpoorte, R., & Alfermann, A. W. (2000). *Metabolic engineering of plant secondary metabolism*. Springer Science & Business Media.
37. Waqas, M., Korres, N. E., Khan, M. D., Nizami, A.-S., Deeba, F., Ali, I., & Hussain, H. (2019). Advances in the concept and methods of seed priming. In M. Hasanuzzaman & V. Fotopoulos (Eds.), *Priming and pretreatment of seeds and seedlings: Implication in plant stress tolerance and enhancing productivity in crop plants* (pp. 11–41). Springer.
38. Farooq, M., Usman, M., Nadeem, F., ur Rehman, H., Wahid, A., Basra, S. M. A., & Siddique, K. H. M. (2019). Seed priming in field crops: Potential benefits, adoption and challenges. *Crop & Pasture Science, 70*(9), 731–771.
39. Acharya, P., Jayaprakasha, G. K., Crosby, K. M., Jifon, J. L., & Patil, B. S. (2020). Nanoparticle-mediated seed priming improves germination, growth, yield, and quality of watermelons (Citrullus lanatus) at multi-locations in Texas. *Scientific Reports, 10*(1), 5037. https://doi.org/10.1038/s41598-020-61696-7
40. Pawar, V., & Laware, S. (2018). Seed priming: A critical review. *International Journal of Scientific Research in Biological Sciences, 5*, 94–101. https://doi.org/10.26438/ijsrbs/v5i5.94101
41. Abbasi Khalaki, M., Moameri, M., Asgari Lajayer, B., & Astatkie, T. (2021). Influence of nano-priming on seed germination and plant growth of forage and medicinal plants. *Plant Growth Regulation, 93*(1), 13–28. https://doi.org/10.1007/s10725-020-00670-9
42. do Espirito Santo Pereira, A., Caixeta Oliveira, H., Fernandes Fraceto, L., & Santaella, C. (2021). Nanotechnology potential in seed priming for sustainable agriculture. *Nanomaterials, 11*(2), 267. https://doi.org/10.3390/nano11020267
43. Maswada, H. F., Djanaguiraman, M., & Prasad, P. V. V. (2018). Seed treatment with nano-iron (III) oxide enhances germination, seeding growth and salinity tolerance of sorghum. *Journal of Agronomy and Crop Science, 204*(6), 577–587. https://doi.org/10.1111/jac.12280
44. Hussain, A., Rizwan, M., Ali, Q., & Ali, S. (2019). Seed priming with silicon nanoparticles improved the biomass and yield while reduced the oxidative stress and cadmium concentration in wheat grains. *Environmental Science and Pollution Research, 26*(8), 7579–7588. https://doi.org/10.1007/s11356-019-04210-5
45. Khan, I., Raza, M. A., Awan, S. A., Shah, G. A., Rizwan, M., Ali, B., Tariq, R., Hassan, M. J., Alyemeni, M. N., Brestic, M., Zhang, X., Ali, S., & Huang, L. (2020). Amelioration of salt induced toxicity in pearl millet by seed priming with silver nanoparticles (AgNPs): The oxidative damage, antioxidant enzymes and ions uptake are major determinants of salt tolerant capacity. *Plant Physiology and Biochemistry, 156*, 221–232. https://doi.org/10.1016/j.plaphy.2020.09.018
46. Antony, D., Yadav, R., & Kalimuthu, R. (2021). Accumulation of Phyto-mediated nano-CeO2 and selenium doped CeO2 on Macrotyloma uniflorum (horse gram) seed by nano-priming to enhance seedling vigor. *Biocatalysis and Agricultural Biotechnology, 31*, 101923. https://doi.org/10.1016/j.bcab.2021.101923
47. Pawar, V., Ambekar, J., Kale, B., Apte, S., & Laware, S. (2019). Response in chickpea (Cicer arietinum L.) seedling growth to seed priming with iron oxide nanoparticles. *International Journal of Biosciences and the Law, 14*, 82. https://doi.org/10.12692/ijb/14.3.82-91
48. Esper Neto, M., Britt, D. W., Lara, L. M., Cartwright, A., dos Santos, R. F., Inoue, T. T., & Batista, M. A. (2020). Initial development of corn seedlings after seed priming with nanoscale synthetic zinc oxide. *Agronomy, 10*(2), 307. https://doi.org/10.3390/agronomy10020307
49. Esper Neto, M., Britt, D. W., Jackson, K. A., Coneglian, C. F., Inoue, T. T., & Batista, M. A. (2021b). Early growth of corn seedlings after seed priming with magnetite nanoparticles synthetised in easy way. *Acta Agriculturae Scandinavica, 71*(2), 91–97. https://doi.org/10.1080/09064710.2020.1852304
50. Khan, I., Awan, S. A., Raza, M. A., Rizwan, M., Tariq, R., Ali, S., & Huang, L. (2021a). Silver nanoparticles improved the plant growth and reduced the sodium and chlorine accumulation in pearl millet: A life cycle study. *Environmental Science and Pollution Research, 28*(11), 13712–13724. https://doi.org/10.1007/s11356-020-11612-3

51. Kalal, P., & Jajoo, A. (2021). Priming with Zinc oxide nanoparticles improve germination and photosynthetic performance in wheat. *Plant Physiology and Biochemistry, 160*, 341. https://doi.org/10.1016/j.plaphy.2021.01.032
52. Esper Neto, M., Britt, D. W., Jackson, K. A., Coneglian, C. F., Cordioli, V. R., Braccini, A. L., Inoue, T. T., & Batista, M. A. (2021a). Assessments in early growth of corn seedlings after hausmanite (Mn3O4) nanoscale seed priming. *Journal of Plant Nutrition, 44*(11), 1611–1620. https://doi.org/10.1080/01904167.2021.1871745
53. Khan, S. A. (2020). Chapter 1 - Metal nanoparticles toxicity: Role of physicochemical aspects. In M. R. Shah, M. Imran, & S. Ullah (Eds.), *Metal nanoparticles for drug delivery and diagnostic applications* (pp. 1–11). Elsevier.
54. Bundschuh, M., Filser, J., Lüderwald, S., McKee, M. S., Metreveli, G., Schaumann, G. E., Schulz, R., & Wagner, S. (2018). Nanoparticles in the environment: Where do we come from, where do we go to? *Environmental Sciences Europe, 30*(1), 6. https://doi.org/10.1186/s12302-018-0132-6
55. Dietz, K.-J., & Herth, S. (2011). Plant nanotoxicology. *Trends in Plant Science, 16*(11), 582–589.
56. Singh, R. P., Handa, R., & Manchanda, G. (2021). Nanoparticles in sustainable agriculture: An emerging opportunity. *Journal of Controlled Release, 329*, 1234–1248. https://doi.org/10.1016/j.jconrel.2020.10.051
57. Alvarado, A. M., Aguirre-Becerra, H., Vázquez-Hernández, M. C., Magaña-Lopez, E., Parola-Contreras, I., Caicedo-Lopez, L. H., Contreras-Medina, L. M., Garcia-Trejo, J. F., Guevara-Gonzalez, R. G., & Feregrino-Perez, A. A. (2019). Influence of elicitors and eustressors on the production of plant secondary metabolites. In *Natural bio-active compounds* (pp. 333–388). Springer.
58. Böttger, A., Vothknecht, U., Bolle, C., & Wolf, A. (2018). Plant secondary metabolites and their general function in plants. In *Lessons on caffeine, cannabis & co* (pp. 3–17). Springer.
59. Saddique, M., Kamran, M., & Shahbaz, M. (2018). Differential responses of plants to biotic stress and the role of metabolites. In *Plant metabolites and regulation under environmental stress* (pp. 69–87). Elsevier.
60. Akula, R., & Ravishankar, G. A. (2011). Influence of abiotic stress signals on secondary metabolites in plants. *Plant Signaling & Behavior, 6*(11), 1720–1731.
61. Thakur, M., Bhattacharya, S., Khosla, P. K., & Puri, S. (2019). Improving production of plant secondary metabolites through biotic and abiotic elicitation. *Journal of Applied Research on Medicinal and Aromatic Plants, 12*, 1–12.
62. Firn, R. D., & Jones, C. G. (2009). A Darwinian view of metabolism: Molecular properties determine fitness. *Journal of Experimental Botany, 60*(3), 719–726.
63. Kumar, S., Bhanjana, G., Sharma, A., Sidhu, M. C., & Dilbaghi, N. (2014). Synthesis, characterization and on field evaluation of pesticide loaded sodium alginate nanoparticles. *Carbohydrate Polymers, 101*, 1061–1067.
64. Dixon, R. A. (2001). Natural products and plant disease resistance. *Nature, 411*(6839), 843–847.
65. Oksman-Caldentey, K.-M., & Inzé, D. (2004). Plant cell factories in the post-genomic era: New ways to produce designer secondary metabolites. *Trends in Plant Science, 9*(9), 433–440.
66. Calabrese, E. J. (2004). Hormesis: A revolution in toxicology, risk assessment and medicine: Re-framing the dose–response relationship. *EMBO Reports, 5*(S1), S37–S40.
67. Vázquez-Hernández, M. C., Parola-Contreras, I., Montoya-Gómez, L. M., Torres-Pacheco, I., Schwarz, D., & Guevara-González, R. G. (2019). Eustressors: Chemical and physical stress factors used to enhance vegetables production. *Scientia Horticulturae, 250*, 223–229.
68. Aguirre-Becerra, H., García-Trejo, J. F., Vázquez-Hernández, C., Alvarado, A. M., Feregrino-Pérez, A. A., Contreras-Medina, L. M., & Guevara-Gonzalez, R. G. (2020). Effect of extended photoperiod with a fixed mixture of light wavelengths on tomato seedlings. *HortScience, 55*(11), 1832–1839.

69. Agathokleous, E., Kitao, M., & Calabrese, E. J. (2019). Hormesis: A compelling platform for sophisticated plant science. *Trends in Plant Science, 24*(4), 318–327.
70. Contado, C. (2015). Nanomaterials in consumer products: A challenging analytical problem. *Frontiers in Chemistry, 3*, 48.
71. Kreyling, W. G., Semmler-Behnke, M., & Chaudhry, Q. (2010). A complementary definition of nanomaterial. *Nano Today, 5*(3), 165–168.
72. Siddiqui, M. H., Al-Whaibi, M. H., Firoz, M., & Al-Khaishany, M. Y. (2015a). Role of nanoparticles in plants. In *Nanotechnology and Plant Sciences* (pp. 19–35). Springer.
73. Oberdörster, G., Oberdörster, E., & Oberdörster, J. (2005). Nanotoxicology: An emerging discipline evolving from studies of ultrafine particles. *Environmental Health Perspectives, 113*(7), 823–839.
74. Giraldo, J. P., Landry, M. P., Faltermeier, S. M., McNicholas, T. P., Iverson, N. M., Boghossian, A. A., Reuel, N. F., Hilmer, A. J., Sen, F., & Brew, J. A. (2014). Plant nanobionics approach to augment photosynthesis and biochemical sensing. *Nature Materials, 13*(4), 400–408.
75. DeRosa, M. C., Monreal, C., Schnitzer, M., Walsh, R., & Sultan, Y. (2010). Nanotechnology in fertilizers. *Nature Nanotechnology, 5*(2), 91.
76. Mastronardi, E., Tsae, P., Zhang, X., Monreal, C., & DeRosa, M. C. (2015). Strategic role of nanotechnology in fertilizers: Potential and limitations. In *Nanotechnologies in food and agriculture* (pp. 25–67). Springer.
77. Shang, L., Nienhaus, K., & Nienhaus, G. U. (2014). Engineered nanoparticles interacting with cells: Size matters. *Journal of Nanobiotechnology, 12*(1), 1–11.
78. Vecchio, G., Galeone, A., Brunetti, V., Maiorano, G., Rizzello, L., Sabella, S., Cingolani, R., & Pompa, P. P. (2012). Mutagenic effects of gold nanoparticles induce aberrant phenotypes in Drosophila melanogaster. *Nanomedicine: Nanotechnology, Biology and Medicine, 8*(1), 1–7.
79. Camargo, M. A. B., & Marenco, R. A. (2011). Density, size and distribution of stomata in 35 rainforest tree species in Central Amazonia. *Acta Amazonica, 41*(2), 205–212.
80. Willmer, C., & Fricker, M. (1996). *Stomata*. Springer Science & Business Media.
81. Choat, B., Cobb, A. R., & Jansen, S. (2008). Structure and function of bordered pits: New discoveries and impacts on whole-plant hydraulic function. *New Phytologist, 177*(3), 608–626.
82. Choat, B., Brodie, T. W., Cobb, A. R., Zwieniecki, M. A., & Holbrook, N. M. (2006). Direct measurements of intervessel pit membrane hydraulic resistance in two angiosperm tree species. *American Journal of Botany, 93*(7), 993–1000.
83. Shane, M. W., Cully, M. E. M. C., & Canny, M. J. (2000). Architecture of branch-root junctions in maize: Structure of the connecting xylem and the porosity of pit membranes. *Annals of Botany, 85*(5), 613–624.
84. Carpita, N., Sabularse, D., Montezinos, D., & Delmer, D. P. (1979). Determination of the pore size of cell walls of living plant cells. *Science, 205*(4411), 1144–1147.
85. Chichiriccò, G., & Poma, A. (2015). Penetration and toxicity of nanomaterials in higher plants. *Nanomaterials, 5*(2), 851–873.
86. Fleischer, A., O'Neill, M. A., & Ehwald, R. (1999). The pore size of non-graminaceous plant cell walls is rapidly decreased by borate ester cross-linking of the pectic polysaccharide rhamnogalacturonan II. *Plant Physiology, 121*(3), 829–838.
87. Shi, L.-X., & Theg, S. M. (2013). The chloroplast protein import system: From algae to trees. *Biochimica et Biophysica Acta (BBA) - Molecular Cell Research, 1833*, 314–331.
88. Ganesan, I., Shi, L.-X., Labs, M., & Theg, S. M. (2018). Evaluating the functional pore size of chloroplast TOC and TIC protein translocons: Import of folded proteins. *The Plant Cell, 30*(9), 2161–2173.
89. Khan, I., Raza, M. A., Khalid, M. H., Awan, S. A., Raja, N. I., Zhang, X., Min, S., Wu, B. C., Hassan, M. J., & Huang, L. (2019). Physiological and biochemical responses of pearl millet (Pennisetum glaucum L.) seedlings exposed to silver nitrate (AgNO3) and silver nanoparticles (AgNPs). *International Journal of Environmental Research and Public Health, 16*(13), 2261. https://doi.org/10.3390/ijerph16132261

90. Li, C.-C., Dang, F., Li, M., Zhu, M., Zhong, H., Hintelmann, H., & Zhou, D.-M. (2017). Effects of exposure pathways on the accumulation and phytotoxicity of silver nanoparticles in soybean and rice. *Nanotoxicology, 11*(5), 699–709. https://doi.org/10.1080/17435390.2017.1344740
91. Kim, J.-H., Kim, D., Seo, S. M., & Kim, D. (2019). Physiological effects of zero-valent iron nanoparticles in rhizosphere on edible crop, Medicago sativa (Alfalfa), grown in soil. *Ecotoxicology, 28*(8), 869–877. https://doi.org/10.1007/s10646-019-02083-5
92. Pandey, S., Giri, K., Kumar, R., Mishra, G., & Raja Rishi, R. (2018). Nanopesticides: Opportunities in crop protection and associated environmental risks. *Proceedings of the National Academy of Sciences, India Section B: Biological Sciences, 88*(4), 1287–1308. https://doi.org/10.1007/s40011-016-0791-2
93. Wang, L., Hu, C., & Shao, L. (2017). The antimicrobial activity of nanoparticles: Present situation and prospects for the future. *International Journal of Nanomedicine, 12*, 1227–1249. https://doi.org/10.2147/IJN.S121956
94. Maity, A., Natarajan, N., Pastor, M., Dunna, V., Gupta, C., & Wasnik, V. (2018a). Nanoparticles influence seed germination traits and seed pathogen infection rate in forage sorghum (Sorghum bicolour) and cowpea (Vigna unguiculata). *Indian Journal of Experimental Biology, 56*, 363.
95. Hojjat, S. (2019). Effect of interaction between ag nanoparticles and salinity on germination stages of Lathyrus sativus L. *Open Access Journal of Environmental and Soil Sciences, 2*, 193. https://doi.org/10.32474/OAJESS.2019.02.000132
96. Cota-Ruiz, K., Ye, Y., Valdes, C., Deng, C., Wang, Y., Hernández-Viezcas, J. A., Duarte-Gardea, M., & Gardea-Torresdey, J. L. (2020). Copper nanowires as nanofertilizers for alfalfa plants: Understanding nano-bio systems interactions from microbial genomics, plant molecular responses and spectroscopic studies. *Science of the Total Environment, 742*, 140572. https://doi.org/10.1016/j.scitotenv.2020.140572
97. Choudhary, R. C., Kumaraswamy, R. V., Kumari, S., Sharma, S. S., Pal, A., Raliya, R., Biswas, P., & Saharan, V. (2017). Cu-chitosan nanoparticle boost defense responses and plant growth in maize (Zea mays L.). *Scientific Reports, 7*(1), 9754. https://doi.org/10.1038/s41598-017-08571-0
98. Weicheng, C., Gong, J.-L., Zeng, G., Song, B., Zhang, P., Li, J., Fang, S., Qin, L., Jun, Y., & Zhe, C. (2020). Mutual effects of silver nanoparticles and antimony (III)/(V) co-exposed to Glycine max (L.) Merr. in hydroponic system: Uptake, translocation, physiochemical responses and potential mechanisms. *Environmental Science: Nano, 7*, 2691. https://doi.org/10.1039/D0EN00519C
99. Batsmanova, L., Taran, N., Konotop, Y., Kalenska, S., & Novytska, N. (2020). Use of a colloidal solution of metal and metal oxide-containing nanoparticles as fertilizer for increasing soybean productivity. *Journal of Central European Agriculture, 21*, 311–319. https://doi.org/10.5513/JCEA01/21.2.2414
100. Yang, J., Jiang, F., Ma, C., Rui, Y., Rui, M., Adeel, M., Cao, W., & Xing, B. (2018). Alteration of crop yield and quality of wheat upon exposure to silver nanoparticles in a life cycle study. *Journal of Agricultural and Food Chemistry, 66*(11), 2589–2597. https://doi.org/10.1021/acs.jafc.7b04904
101. Hoe, P. T., Mai, N. C., Lien, L., Ban, N., Minh, C., Chau, N. H., Ngo, Q., Hien, D. T., Van, N. T., Hien, L. T. T., & Linh, T. (2018). Germination responses of soybean seeds under Fe, ZnO, Cu and Co nanoparticle treatments. *International Journal of Agriculture and Biology, 20*, 1562–1568. https://doi.org/10.17957/IJAB/15.0670
102. Sharif, H., Mehmood, A., Ulfat, A., Ahmad, K. S., Hussain, I., & Khan, R. T. (2021). Environmentally sustainable production of silver nanoparticles and their effect on glycine max L. seedlings. *Gesunde Pflanzen, 73*(1), 95–103. https://doi.org/10.1007/s10343-020-00532-4
103. Tondey, M., Kalia, A., Singh, A., Dheri, G. S., Taggar, M. S., Nepovimova, E., Krejcar, O., & Kuca, K. (2021). Seed priming and coating by nanoscale zinc oxide particles improved

vegetative growth, yield and quality of fodder maize (Zea mays). *Agronomy, 11*(4), 729. https://doi.org/10.3390/agronomy11040729

104. Alabdallah, N. M., & Alzahrani, H. S. (2020). Impact of ZnO nanoparticles on growth of cowpea and okra plants under salt stress conditions. *Biosciences, Biotechnology Research Asia, 17*(2), 329–340. https://doi.org/10.13005/bbra
105. Irum, S., Jabeen, N., Ahmad, K., Shafique, S., Khan, T., Gul, H., Altaf, S., Shafi, N., Mehmood, A., & Hussain, S. (2020). Biogenic iron oxide nanoparticles enhance callogenesis and regeneration pattern of recalcitrant Cicer arietinum L. *PLoS One, 15*, e0242829. https://doi.org/10.1371/journal.pone.0242829
106. Iannone, M. F., Groppa, M. D., Zawoznik, M. S., Coral, D. F., Fernández van Raap, M. B., & Benavides, M. P. (2021). Magnetite nanoparticles coated with citric acid are not phytotoxic and stimulate soybean and alfalfa growth. *Ecotoxicology and Environmental Safety, 211*, 111942. https://doi.org/10.1016/j.ecoenv.2021.111942
107. Moll, J., Gogos, A., Bucheli, T. D., Widmer, F., & van der Heijden, M. G. A. (2016). Effect of nanoparticles on red clover and its symbiotic microorganisms. *Journal of Nanobiotechnology, 14*(1), 36. https://doi.org/10.1186/s12951-016-0188-7
108. Shang, H., Guo, H., Ma, C., Li, C., Chefetz, B., Polubesova, T., & Xing, B. (2019). Maize (Zea mays L.) root exudates modify the surface chemistry of CuO nanoparticles: Altered aggregation, dissolution and toxicity. *Science of the Total Environment, 690*, 502–510. https://doi.org/10.1016/j.scitotenv.2019.07.017
109. Li, J., Song, Y., Wu, K., Tao, Q., Liang, Y., & Li, T. (2018a). Effects of Cr2O3 nanoparticles on the chlorophyll fluorescence and chloroplast ultrastructure of soybean (Glycine max). *Environmental Science and Pollution Research, 25*(20), 19446–19457. https://doi.org/10.1007/s11356-018-2132-x
110. Zeidali, E., Moradi, R., Darabi, F., & Rostami, Z. (2017). Assessing germination response of wheat and wild oat to different levels of ZnO nanoparticles. *Journal of Plant Protection, 31*, 617–627. https://doi.org/10.22067/jpp.v31i4.60735
111. Peréz, C. D. P., De La Torre Roche, R., Zuverza-Mena, N., Ma, C., Shen, Y., White, J. C., Pozza, E. A., Pozza, A. A. A., & Elmer, W. H. (2020). Metalloid and metal oxide nanoparticles suppress sudden death syndrome of soybean. *Journal of Agricultural and Food Chemistry, 68*(1), 77–87. https://doi.org/10.1021/acs.jafc.9b06082
112. Yan, L., Li, P., Zhao, X., Ji, R., & Zhao, L. (2020). Physiological and metabolic responses of maize (Zea mays) plants to Fe3O4 nanoparticles. *Science of the Total Environment, 718*, 137400. https://doi.org/10.1016/j.scitotenv.2020.137400
113. Lian, J., Zhao, L., Wu, J., Xiong, H., Bao, Y., Zeb, A., Tang, J., & Liu, W. (2020). Foliar spray of TiO2 nanoparticles prevails over root application in reducing Cd accumulation and mitigating Cd-induced phytotoxicity in maize (Zea mays L.). *Chemosphere, 239*, 124794. https://doi.org/10.1016/j.chemosphere.2019.124794
114. Cao, Z., Rossi, L., Stowers, C., Zhang, W., Lombardini, L., & Ma, X. (2018). The impact of cerium oxide nanoparticles on the physiology of soybean (Glycine max (L.) Merr.) under different soil moisture conditions. *Environmental Science and Pollution Research, 25*(1), 930–939. https://doi.org/10.1007/s11356-017-0501-5
115. Daghan, H. (2018). Effects of tio2 nanoparticles on maize (Zea mays L.) growth, chlorophyll content and nutrient uptake. *Applied Ecology and Environmental Research, 16*, 6873–6883. https://doi.org/10.15666/aeer/1605_68736883
116. de Melo, G. S. R., Constantin, R. P., Abrahão, J., de Paiva Foletto-Felipe, M., Constantin, R. P., dos Santos, W. D., Ferrarese-Filho, O., & Marchiosi, R. (2021). Titanium dioxide nanoparticles induce root growth inhibition in soybean due to physical damages. *Water, Air, & Soil Pollution, 232*(1), 25. https://doi.org/10.1007/s11270-020-04955-7
117. Yusefi-Tanha, E., Fallah, S., Rostamnejadi, A., & Pokhrel, L. R. (2020). Particle size and concentration dependent toxicity of copper oxide nanoparticles (CuONPs) on seed yield and antioxidant defense system in soil grown soybean (Glycine max cv. Kowsar). *Science of the Total Environment, 715*, 136994. https://doi.org/10.1016/j.scitotenv.2020.136994

118. Itroutwar, P. D., Kasivelu, G., Raguraman, V., Malaichamy, K., & Sevathapandian, S. K. (2020). Effects of biogenic zinc oxide nanoparticles on seed germination and seedling vigor of maize (Zea mays). *Biocatalysis and Agricultural Biotechnology, 29*, 101778. https://doi.org/10.1016/j.bcab.2020.101778
119. Tohidiyan, Z., Hashemi, S., & Boroujeni, K. P. (2019). Facile microwave-assisted synthesis of NiO nanoparticles and its effect on soybean (Glycine max). *IET Nanobiotechnology, 13*(2), 101–106. https://doi.org/10.1049/iet-nbt.2018.5003
120. Yang, X., Alidoust, D., & Wang, C. (2020). Effects of iron oxide nanoparticles on the mineral composition and growth of soybean (Glycine max L.) plants. *Acta Physiologiae Plantarum, 42*(8), 128. https://doi.org/10.1007/s11738-020-03104-1
121. Gong, C., Wang, L., Li, X., Wang, H., Jiang, Y., & Wang, W. (2019). Responses of seed germination and shoot metabolic profiles of maize (Zea mays L.) to Y 2 O 3 nanoparticle stress. *RSC Advances, 9*(47), 27720–27731.
122. Liu, Y., Xu, L., & Dai, Y. (2018). Phytotoxic effects of lanthanum oxide nanoparticles on maize (Zea mays L.). *IOP Conference Series: Earth and Environmental Science, 113*, 012020. https://doi.org/10.1088/1755-1315/113/1/012020
123. Dimkpa, C. O., White, J. C., Elmer, W. H., & Gardea-Torresdey, J. (2017). Nanoparticle and ionic Zn promote nutrient loading of sorghum grain under low NPK fertilization. *Journal of Agricultural and Food Chemistry, 65*(39), 8552–8559. https://doi.org/10.1021/acs.jafc.7b02961
124. Tombuloglu, H., Tombuloglu, G., Slimani, Y., Ercan, I., Sozeri, H., & Baykal, A. (2018). Impact of manganese ferrite (MnFe2O4) nanoparticles on growth and magnetic character of barley (Hordeum vulgare L.). *Environmental Pollution, 243*, 872–881. https://doi.org/10.1016/j.envpol.2018.08.096
125. Manjunatha, N., Prajapati, M., Dunna, V., Maity, A., Wasnik, V. K., Gupta, C. K., & Parmar, S. S. (2018). Effect of cupric oxide and zinc oxide nanoparticles on seed mycoflora and seed quality of fodder crops. *Journal of Environmental Biology, 39*(6), 973–979. https://doi.org/10.22438/jeb/39/6/MRN-667
126. Maity, A., Natarajan, N., Vijay, D., Srinivasan, R., Pastor, M., & Malaviya, D. R. (2018b). Influence of metal nanoparticles (NPs) on germination and yield of oat (Avena sativa) and Berseem (Trifolium alexandrinum). *Proceedings of the National Academy of Sciences, India Section B: Biological Sciences, 88*(2), 595–607. https://doi.org/10.1007/s40011-016-0796-x
127. Youssef, O. A., Tammam, A. A., El-Bakatoushi, R. F., Alframawy, A. M., Emara, M. M., & El-Sadek, L. M. (2020). Hematite nanoparticles influence ultrastructure, antioxidant defenses, gene expression, and alleviate cadmium toxicity in Zea mays. *Journal of Plant Interactions, 15*(1), 54–74. https://doi.org/10.1080/17429145.2020.1745307
128. Tombuloglu, H., Slimani, Y., Tombuloglu, G., Almessiere, M., & Baykal, A. (2019). Uptake and translocation of magnetite (Fe3O4) nanoparticles and its impact on photosynthetic genes in barley (Hordeum vulgare L.). *Chemosphere, 226*, 110–122. https://doi.org/10.1016/j.chemosphere.2019.03.075
129. Zaytseva, O., & Neumann, G. (2016). Carbon nanomaterials: Production, impact on plant development, agricultural and environmental applications. *Chemical and Biological Technologies in Agriculture, 3*(1), 1–26.
130. Adams, M. P., Koh, E. J. Y., Vilas, M. P., Collier, C. J., Lambert, V. M., Sisson, S. A., Quiroz, M., McDonald-Madden, E., McKenzie, L. J., & O'Brien, K. R. (2020). Predicting seagrass decline due to cumulative stressors. *Environmental Modelling & Software, 130*, 104717.
131. In, J. B., & Noy, A. (2012). Nanotechnology's wonder material: Synthesis of carbon nanotubes. In *Hierarchical nanostructures for energy devices* (p. 26). RSC.
132. Zhu, D., Zhang, Z., Cui, P., & Zhu, W. (2019). Robust graph convolutional networks against adversarial attacks. In *Proceedings of the 25th ACM SIGKDD International Conference on Knowledge Discovery & Data Mining* (pp. 1399–1407).

133. Gogos, A., Knauer, K., & Bucheli, T. D. (2012). Nanomaterials in plant protection and fertilization: Current state, foreseen applications, and research priorities. *Journal of Agricultural and Food Chemistry, 60*(39), 9781–9792.
134. López-Vargas, E. R., González-García, Y., Pérez-Álvarez, M., Cadenas-Pliego, G., González-Morales, S., Benavides-Mendoza, A., Cabrera, R. I., & Juárez-Maldonado, A. (2020). Seed priming with carbon nanomaterials to modify the germination, growth, and antioxidant status of tomato seedlings. *Agronomy, 10*(5), 639.
135. Cossins, D. (2014). Next generation: Nanoparticles augment plant functions. The incorporation of synthetic nanoparticles into plants can enhance photosynthesis and transform leaves into biochemical sensors. *The Scientist*. News & Opinion.
136. Liu, Q., Chen, B., Wang, Q., Shi, X., Xiao, Z., Lin, J., & Fang, X. (2009). Carbon nanotubes as molecular transporters for walled plant cells. *Nano Letters, 9*(3), 1007–1010.
137. Chen, M., Zhou, S., Zhu, Y., Sun, Y., Zeng, G., Yang, C., Xu, P., Yan, M., Liu, Z., & Zhang, W. (2018). Toxicity of carbon nanomaterials to plants, animals and microbes: Recent progress from 2015-present. *Chemosphere, 206*, 255–264.
138. Burman, U., & Kumar, P. (2018). Plant response to engineered nanoparticles. In *Nanomaterials in plants, algae, and microorganisms* (pp. 103–118). Elsevier.
139. Samadi, S., Saharkhiz, M. J., Azizi, M., Samiei, L., & Ghorbanpour, M. (2020). Multi-walled carbon nanotubes stimulate growth, redox reactions and biosynthesis of antioxidant metabolites in Thymus daenensis Celak. in vitro. *Chemosphere, 249*, 126069.
140. González-García, Y., López-Vargas, E. R., Cadenas-Pliego, G., Benavides-Mendoza, A., González-Morales, S., Robledo-Olivo, A., Alpuche-Solís, Á. G., & Juárez-Maldonado, A. (2019). Impact of carbon nanomaterials on the antioxidant system of tomato seedlings. *International Journal of Molecular Sciences, 20*(23), 5858.
141. González-García, Y., Cadenas-Pliego, G., Alpuche-Solís, Á. G., Cabrera, R. I., & Juárez-Maldonado, A. (2021). Carbon nanotubes decrease the negative impact of Alternaria solani in tomato crop. *Nanomaterials, 11*(5), 1080.
142. Cheng, F., Liu, Y.-F., Lu, G.-Y., Zhang, X.-K., Xie, L.-L., Yuan, C.-F., & Xu, B.-B. (2016). Graphene oxide modulates root growth of Brassica napus L. and regulates ABA and IAA concentration. *Journal of Plant Physiology, 193*, 57–63.
143. Khodakovskaya, M. V., De Silva, K., Biris, A. S., Dervishi, E., & Villagarcia, H. (2012). Carbon nanotubes induce growth enhancement of tobacco cells. *ACS Nano, 6*(3), 2128–2135.
144. Patel, A., Tiwari, S., Parihar, P., Singh, R., & Prasad, S. M. (2019). Carbon nanotubes as plant growth regulators: Impacts on growth, reproductive system, and soil microbial community. In *Nanomaterials in plants, algae and microorganisms* (pp. 23–42). Elsevier.
145. Akram, N. A., Shafiq, F., & Ashraf, M. (2017). Ascorbic acid-a potential oxidant scavenger and its role in plant development and abiotic stress tolerance. *Frontiers in Plant Science, 8*, 613.
146. Vochita, G., Oprica, L., Gherghel, D., Mihai, C.-T., Boukherroub, R., & Lobiuc, A. (2019). Graphene oxide effects in early ontogenetic stages of Triticum aestivum L. seedlings. *Ecotoxicology and Environmental Safety, 181*, 345–352.
147. Ozyigit, I. I., Filiz, E., Vatansever, R., Kurtoglu, K. Y., Koc, I., Öztürk, M. X., & Anjum, N. A. (2016). Identification and comparative analysis of H2O2-scavenging enzymes (ascorbate peroxidase and glutathione peroxidase) in selected plants employing bioinformatics approaches. *Frontiers in Plant Science, 7*, 301.
148. Huang, J., Gu, M., Lai, Z., Fan, B., Shi, K., Zhou, Y.-H., Yu, J.-Q., & Chen, Z. (2010). Functional analysis of the Arabidopsis PAL gene family in plant growth, development, and response to environmental stress. *Plant Physiology, 153*(4), 1526–1538.
149. Astaneh, R. K., Bolandnazar, S., Nahandi, F. Z., & Oustan, S. (2018). Effect of selenium application on phenylalanine ammonia-lyase (PAL) activity, phenol leakage and total phenolic content in garlic (Allium sativum L.) under NaCl stress. *Information Processing in Agriculture, 5*(3), 339–344.

150. Bell, I. R., Ives, J. A., & Wayne, B. J. (2014). Nonlinear effects of nanoparticles: Biological variability from hormetic doses, small particle sizes, and dynamic adaptive interactions. *Dose-Response, 12*(2), 202.
151. Younes, N. A., Dawood, M. F. A., & Wardany, A. A. (2019). Biosafety assessment of graphene nanosheets on leaf ultrastructure, physiological and yield traits of Capsicum annuum L. and Solanum melongena L. *Chemosphere, 228*, 318–327.
152. Hao, Y., Xu, B., Ma, C., Shang, J., Gu, W., Li, W., Hou, T., Xiang, Y., Cao, W., & Xing, B. (2019). Synthesis of novel mesoporous carbon nanoparticles and their phytotoxicity to rice (Oryza sativa L.). *Journal of Saudi Chemical Society, 23*(1), 75–82.
153. Voß, U., Bishopp, A., Farcot, E., & Bennett, M. J. (2014). Modelling hormonal response and development. *Trends in Plant Science, 19*(5), 311–319.
154. Wani, S. H., Kumar, V., Shriram, V., & Sah, S. K. (2016). Phytohormones and their metabolic engineering for abiotic stress tolerance in crop plants. *The Crop Journal, 4*(3), 162–176.
155. Blanco, M. H., del Carmen Quintana, M., & Hernández, L. (1999). Determination of dihydrozeatin and dihydrozeatin riboside in apples by anodic stripping voltammetry with a carbon fiber microelectrode. *Fresenius' Journal of Analytical Chemistry, 364*(3), 254–260.
156. Haubrick, L. L., Torsethaugen, G., & Assmann, S. M. (2006). Effect of brassinolide, alone and in concert with abscisic acid, on control of stomatal aperture and potassium currents of Vicia faba guard cell protoplasts. *Physiologia Plantarum, 128*(1), 134–143.
157. Ma, Q.-H. (2008). Genetic engineering of cytokinins and their application to agriculture. *Critical Reviews in Biotechnology, 28*(3), 213–232.
158. Hao, Y., Ma, C., Zhang, Z., Song, Y., Cao, W., Guo, J., Zhou, G., Rui, Y., Liu, L., & Xing, B. (2018). Carbon nanomaterials alter plant physiology and soil bacterial community composition in a rice-soil-bacterial ecosystem. *Environmental Pollution, 232*, 123–136.
159. Hao, Y., Yu, F., Lv, R., Ma, C., Zhang, Z., Rui, Y., Liu, L., Cao, W., & Xing, B. (2016). Carbon nanotubes filled with different ferromagnetic alloys affect the growth and development of rice seedlings by changing the C:N ratio and plant hormones concentrations. *PLoS One, 11*(6), e0157264.
160. Syu, Y., Hung, J.-H., Chen, J.-C., & Chuang, H. (2014). Impacts of size and shape of silver nanoparticles on Arabidopsis plant growth and gene expression. *Plant Physiology and Biochemistry, 83*, 57–64.
161. Zhang, H., Yue, M., Zheng, X., Xie, C., Zhou, H., & Li, L. (2017). Physiological effects of single-and multi-walled carbon nanotubes on rice seedlings. *IEEE Transactions on Nanobioscience, 16*(7), 563–570.
162. Tuerhong, M., Yang, X. U., & Xue-Bo, Y. I. N. (2017). Review on carbon dots and their applications. *Chinese Journal of Analytical Chemistry, 45*(1), 139–150.
163. Wang, H., Zhang, M., Song, Y., Li, H., Huang, H., Shao, M., Liu, Y., & Kang, Z. (2018). Carbon dots promote the growth and photosynthesis of mung bean sprouts. *Carbon, 136*, 94–102.
164. Park, S., & Ahn, Y.-J. (2016). Multi-walled carbon nanotubes and silver nanoparticles differentially affect seed germination, chlorophyll content, and hydrogen peroxide accumulation in carrot (Daucus carota L.). *Biocatalysis and Agricultural Biotechnology, 8*, 257–262.
165. Siddiqui, Z. A., Parveen, A., Ahmad, L., & Hashem, A. (2019). Effects of graphene oxide and zinc oxide nanoparticles on growth, chlorophyll, carotenoids, proline contents and diseases of carrot. *Scientia Horticulturae, 249*, 374–382.
166. Zhang, P., Zhang, R., Fang, X., Song, T., Cai, X., Liu, H., & Du, S. (2016). Toxic effects of graphene on the growth and nutritional levels of wheat (Triticum aestivum L.): Short-and long-term exposure studies. *Journal of Hazardous Materials, 317*, 543–551.
167. Li, X., Mu, L., Li, D., Ouyang, S., He, C., & Hu, X. (2018b). Effects of the size and oxidation of graphene oxide on crop quality and specific molecular pathways. *Carbon, 140*, 352–361.

168. Dong, X., Zhang, D., Liu, J., Liu, Q. Q., Liu, H., Tian, L., Jiang, L., & Qu, L. Q. (2015). Plastidial disproportionating enzyme participates in starch synthesis in rice endosperm by transferring maltooligosyl groups from amylose and amylopectin to amylopectin. *Plant Physiology, 169*(4), 2496–2512.
169. Thalmann, M., & Santelia, D. (2017). Starch as a determinant of plant fitness under abiotic stress. *New Phytologist, 214*(3), 943–951.
170. Shi, Y., Wang, Z., Meng, P., Tian, S., Zhang, X., & Yang, S. (2013). The glutamate carboxypeptidase AMP 1 mediates abscisic acid and abiotic stress responses in Arabidopsis. *New Phytologist, 199*(1), 135–150.
171. Lew, T. T. S., Koman, V. B., Gordiichuk, P., Park, M., & Strano, M. S. (2020). The emergence of plant nanobionics and living plants as technology. *Advanced Materials Technologies, 5*(3), 1900657.

Polymer-Based Nanoparticles (NPs): A Promising Approach for Crop Productivity

Fatima El Amerany, Fatima Zahra Aboudamia, Iman Janah, Moha Taourirte, and Mohammed Rhazi

Abstract Recently, the production of nanoparticles (NPs) has become a new and pioneering approach that can be exploited in several areas such as medicine, agriculture, cosmetics, engineering, and environmental fields. The wide applicability is due to their particular physicochemical properties, like small size with the big superficial area, and high electronic properties. Regarding agricultural fields, different kinds of NPs were applied. Among the successful nanomaterials that were used to promote plant growth and productivity of crops, there are polymer-based NPs (i.e., polysaccharide NPs, protein NPs, and lipid NPs). These nanomaterials were used as a plant growth promoter and nanostructured carriers of antioxidants, phytohormones, fertilizers (NPK), volatile organic compounds, and agrochemicals (herbicides, fungicides, and so on). Although there has been remarkable progress in the development of polymer-based NPs to induce vegetative growth and the quality of fruits and vegetables as well as to cope with environmental stresses, their mechanisms remain unclear. Therefore, the present chapter will address three important

F. El Amerany (✉)
Laboratory of Sustainable Development and Health Research, Faculty of Science and Technology of Marrakech, Cadi Ayyad University, Marrakesh, Morocco

Interdisciplinary Laboratory in Bio-Resources, Environment and Materials, Higher Normal School, Cadi Ayyad University, Marrakesh, Morocco
e-mail: fatima.elamerany@ced.uca.ma

F. Z. Aboudamia
Research Team of Innovation and Sustainable Development & Expertise in Green Chemistry, "ERIDDECV", Faculty of Science Semlalia, Cadi Ayyad University, Marrakesh, Morocco

I. Janah
Laboratory of Agro-Food, Biotechnologies and Valorization of Plant Bioresources, Faculty of Science Semlalia, Cadi Ayyad University, Marrakesh, Morocco

M. Taourirte
Laboratory of Sustainable Development and Health Research, Faculty of Science and Technology of Marrakech, Cadi Ayyad University, Marrakesh, Morocco

M. Rhazi
Interdisciplinary Laboratory in Bio-Resources, Environment and Materials, Higher Normal School, Cadi Ayyad University, Marrakesh, Morocco

J.-T. Chen (ed.), *Plant and Nanoparticles*,
https://doi.org/10.1007/978-981-19-2503-0_5

points: (1) the effects of polymer-based NPs on plant physiology and fruits, (2) their mechanisms in protecting plants from biotic and abiotic stress, and (3) how they affect beneficial microorganisms (arbuscular mycorrhizal fungi and rhizobacteria). The application of NPs in sustainable agriculture is a promising solution instead of agrichemicals, which are unsafe to the ecosystem. The investigation of this novel approach could be an efficient application in plant pathology and crop productivity.

Keywords Sustainable agriculture · Nanoparticles (NPs) · Crop production · Plant protection

1 Introduction

Nanotechnology has been defined for the first time by the US Environmental Protection Agency [1] as the discipline of studying and managing either substances, atom arrangement, or procedures that function at a scale less than 10^{-7} m, where unique physical properties make novel applications possible [2]. This science is a pioneering field of interdisciplinary research that includes several fields such as physics, electronics, chemistry, biology, and medicine. In the nonagricultural field, this term is related to materials science and biomass conversion technologies, by applying colloidal particles between 10 and 1000 nm in size [3]. In addition, it has found several uses in the agricultural fields, such as nanofertilizers, nanopesticides, nanobiosensors, or as environmental remediation agents [4]. The application of nanomaterials (NMs) in agronomy wants to substitute the use of chemical substances by smart delivery of compounds, to reduce the depletion and the loss of fertilizers' nutrients, and to raise yields by providing an adequate amount of water and nutrients [5]. Recently, with the circumstances of climate change such as high temperature, drought, lack of arable land, salinity, water scarcity, frequent flash floods, urbanization, and high demand for food, the application of nanotechnology for the agricultural countries could be a good solution to increase the farm productivity, to control the nutritional needs of the plant, and to minimize the use of fertilizers with the costs of agricultural production and environmental pollution [6]. In the NMs, particle dimension and their form, pore volume, and raggedness are physical characteristics that precise the surface-to-volume ratio, which in turn influences other features especially the charge or the fee energy per unit area [7]. Recently, the manufacture and creation of nanoparticles (NPs) have grabbed considerable attention because of their faster impact with a lower dose, their faster penetration through membranes, and their great potential in the different application fields, especially agriculture [8]. Nowadays, enormous innovative NP products were produced. Some of them were synthesized naturally by plants or microbial species (i.e., *Limnothrix* sp. and *Lyngbya majuscula*), such as gold and silver NPs, while others were developed by a human for multiple uses [9–11]. Three kinds of NPs (i.e., metal, polymeric, and semiconductor NPs) have been engineered to ensure a slow release of fertilizers, phytohormones, and active compounds, to encapsulate microorganisms, to trigger signals, as well as to increase plant growth, its development, its resistance against phytopathogens, and the yield [12–15]. Polymeric NPs are particles made from a class of multitalented materials such as poly-sugars (i.e., alginate, starch), polypeptides (i.e., gelatin), or lipids (i.e., beeswax) [16]. These materials have been extensively used in agriculture because of their diverse features like biodegradability, not a hazard to the environment, biocompatibility, as well as adsorbing ions [17]. Moreover, polymer-based NPs have additional features, essentially large surface area, high reaction activity, and surface modifiability, which allow them to be used for several purposes than raw material or metallic and semiconductor NPs [18, 19]. Recently, different methods were developed for the production of polymeric NPs depending on the type of polymer used, its molecular weight, its dose, and the goal of their application [20]. Ultimately, they have been appraised as a crucial element for the development of sustainable agricultural practices due to their positive impact on the environment as well as their beneficial outcome on many fruits and vegetables; however, in some cases, their application shows neutral or negative effects depending on plant species, application mode, and soil composition. The purpose of this chapter is to present the latest advances in the production of polymer-based NPs as promising approaches of nanotechnology, as well as the major results regarding their application on plants' and fruits' production.

2 Polymer NPs: Types and Preparation Methods

The polymeric NPs could be nanocapsules or nanospheres in shape; this difference was explained in the study of Yoo and Park [21] where they mentioned that the nanocapsules are characterized with a polymeric cavity structure and inner hydrophobic phase (Fig. 11.1a), while nanospheres are described as an organized solid matrix with the polymeric chains (Fig. 11.1b). Several kinds of natural or synthetic

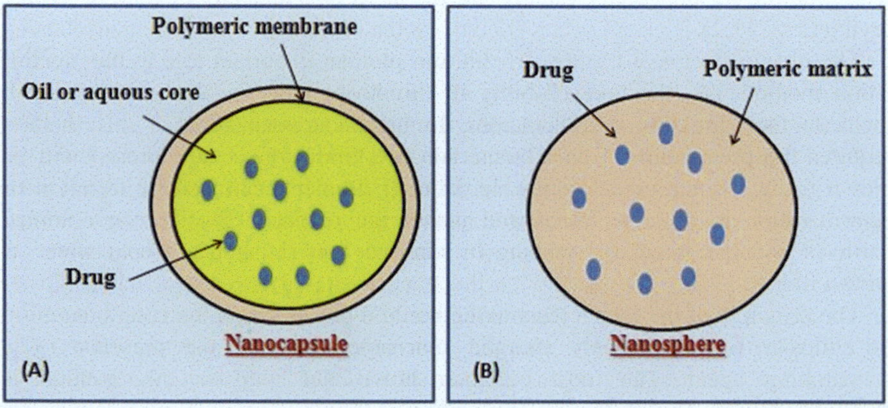

Fig. 11.1 Schematic illustration of (**a**) nanospheres and (**b**) nanocarpsules applied in agriculture as polymeric NPs

polymers could be a source of polymer-based nanoparticles (NPs) by top-down or bottom-up methods.

The most functional natural nanopolymers that are widely used in agriculture are chitosan, alginate, pectin, cellulose, starch, lignin, polyaspartic acid, and beeswax [22–24]. Their use is related to their low cost, simple and mild preparation methods, stiffness, eco-friendliness, and low toxicity, and they can be applied by spray or as a soil drench.

2.1 Chitosan NPs

The *N*-deacetylation of chitin leads to obtaining chitosan, a natural and plenty polysaccharide worldwide. This biopolymer can be derived from different sources like the exoskeleton of crustaceans [25], fish scales [26], insects [27], and fungi [28]. The deacetylation of chitin could be chemically or biologically. The study of Ohya et al. [29] gave rise to chitosan NPs for the delivery of antineoplastic agents like 5-fluorouracil. Over the years, researchers have developed new methods based on varied parameters such as stability retention time, size, and drug-loading capacity [30]. Chitosan NPs could be obtained using various methods such as emulsification; precipitation, in addition to ionic or covalent crosslinking; or amalgamation between the two last methods. The primary method reported for the formulation of chitosan NPs is emulsification and crosslinking based on the interaction between an aldehyde group of the crosslinker and the amino group of chitosan. To carry out this method, it is essential to prepare an emulsion formed from a mixture between an aqueous solution and an oily phase. In the study of Jameela et al. [31], glutaraldehyde crosslinker agent is mixed with toluene, and Span 80 is used as a stabilizer. Briefly, after homogenization of the phases intensively, NP beads were formed after crosslinking. Centrifugation could be used to separate easily NPs from the emulsion with multiple washing steps. Finally, the obtained NPs undergo vacuum or freeze-drying (Fig. 11.2).

The physicochemical features of chitosan play an important role in the precipitation method, i.e., the non-solubility of this biopolymer in alkaline media will facilitate the formation of precipitates. In brief, the realization of this method requires the preparation of an alkaline solution from which the chitosan will be blown using a compressed air nozzle forming droplets, followed by filtration or centrifugation to separate, filter, and purify the particles. Finally, the obtained particles undergo intensive washing by altering between hot and cold water as shown in Fig. 11.3.

The principle of the ionic crosslinking method depended on the conglomeration of chitosan with negatively charged macromolecules or the presence of a crosslinking agent. The most common crosslinker used in this method is tripolyphosphate. This method is recognized as the ionic gelation process because there is a formation of gels due to ionic linkage as highlighted in Fig. 11.4.

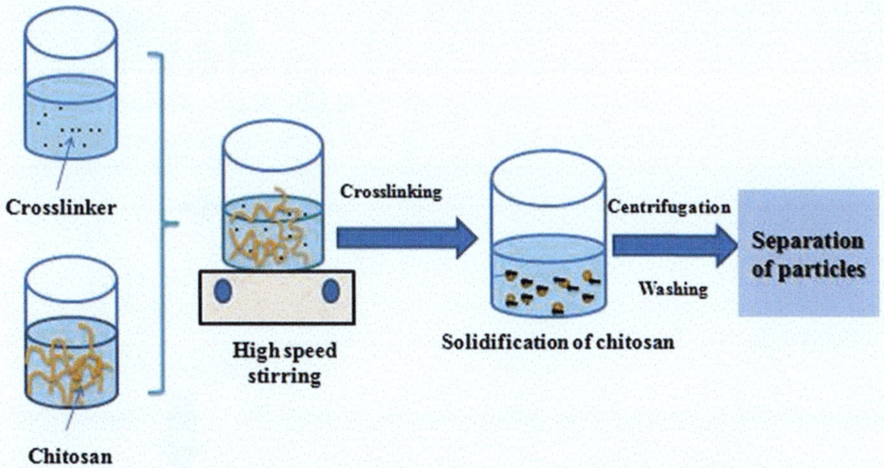

Fig. 11.2 Preparation of chitosan NPs by emulsification and crosslinking method

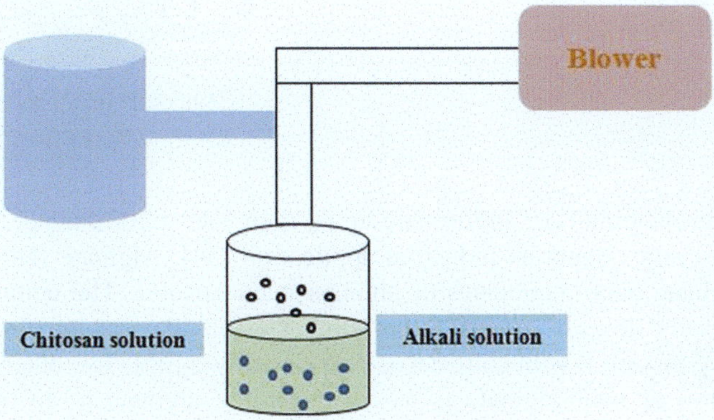

Fig. 11.3 Chitosan NPs' preparation by precipitation method

Chitosan NPs play an important role as nanocarriers that ameliorate the stability and the delivery of active compounds [32]. The application of chitosan in agriculture as NPs is advantageous because a small dose has a great effect. In addition, its biological properties give them the possibility of being applied without any risk to humans or the environment [33].

2.2 *Alginate NPs*

Alginate is a natural hydrophilic polysaccharide extracted from the rigid layer of various species of brown algae. Alginate is characterized by a linear block polymer

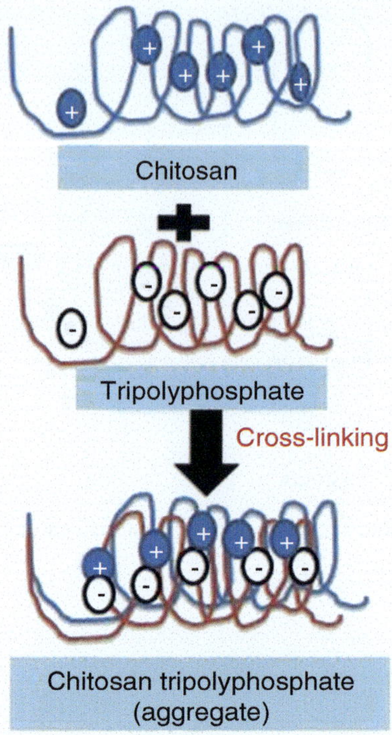

Fig. 11.4 Chitosan NPs' preparation by crosslinking method

of β-D-mannuronate and α-L-guluronate units; their rate will influence its physical characteristics and orients its industrial application [34]. Alginate (NPs) could be found as alginate nano-aggregates or alginate nanocapsules. The complexation of alginate is easily done by adding a crosslinker agent like calcium from calcium chloride or by mixing it with chitosan as polyelectrolyte [35]. The other method for the preparation of nano-alginate is based on adding alginate to an oil emulsion solution [36].

2.3 *Pectin NPs*

Pectin is a biocompatible and negatively charged polysaccharide that consists of α-D-galactopyranosyluronate units [37]. Several studies have shown the possibility of the formation of NPs based on pectin. Nanospheres of pectin were formed by adding Ca^{2+} and CO_3^{2-} ions [38] or glutaraldehyde as crosslinking agents [39]. In the agricultural field, calcium pectinate NPs were formulated as water reservoir, and their application could resolve the problem of scarcity of water in drought-stricken countries [40].

2.4 Cellulose and Starch NPs

Cellulose is the largest ubiquitous polysaccharide in the world. This linear polysaccharide is a repetition of glucose units linked with β-D-(1 → 4) linkages. Starch is widely spread in plant tissues, as glucose units linked with α-D-(1 → 4) and/or α-D-(1 → 6) linkages [37]. According to the literature, these two famous polysaccharides have shown their feasibility of being in the form of NPs in the agricultural field. The study of Zhang et al. [41] highlighted a facile synthesis of cellulose NPs with high yield. In addition, many studies reported the preparation of cellulose NPs as nanowhiskers [42–44], i.e., the study of Gu et al. [45] mentioned a simple method of preparation of cellulose NPs via an ultrasound-assisted etherification and a subsequent sonication process. On the other side, the study of Hasanin [46] reported the possibility of the extraction of starch NPs from potato peel waste via simple alkali extraction followed by ultrasonic treatment.

2.5 Lignin NPs

After cellulose, lignin is the most spread and renewable organic material on the earth extracted from biomass [47]. This biopolymer is characterized by its irregularity and complexity, containing three phenylpropanoid compounds (sinapyl, coniferyl, and coumaryl) [48]. The insolubility property of lignin in water restricts its applications. However, recently some papers have been shown the possibility of preparing aqueous lignin NP dispersions [49, 50], and it has also been reported to have the ability to improve the water absorption of materials [22]. Spherical lignin NPs were prepared using a simple method by dissolving the non-hardwood in tetrahydrofuran [51]. In addition, Yearla and Padmasree [52] have fabricated dioxane lignin NPs by nanoprecipitation method from the solid wood dioxane lignin and nonsolid wood alkali lignin.

2.6 Polyaspartate NPs

Polyaspartate (PASP) is a biodegradable functional and hydrophilic material. This biopolymer is often used in wastewater treatment due to its high potential of metal chelating via the carboxyl and amino groups [53]. PASP is developed as a nanoprobe for the detection of iron ions [54]. In addition, another study highlighted a new nonstoichiometric polyelectrolyte complex NPs based on chitosan and PASP sodium salt [55].

2.7 Beeswax NPs

In 1960, gas-liquid chromatography determined the components of beeswax [56]; this material as other lipids formed from a mixture of several classes of components composed from a series of substances with a long chain varying by two carbon atoms. Beeswax NPs are considered solid lipid NPs (SLNs); these kinds of NPs were introduced in 1991. Their good tolerability, biodegradability, physical stability, macroscale production efficiency, and feasibility to include in their lipid core the lipophilic drugs have attracted big attention in different fields. In the same sense, a study focused on the feasibility of preparing SLN by employing beeswax and carnauba wax to incorporate ketoprofen [57].

3 Effects of Polymer-Based NPs on Plants

3.1 Application of NPs as a Growth Promoter

Plant growth promoters are all substances produced by plants to regulate their biological processes and to ensure their safety. Even though a little amount of these substances trigger major plant physiological and biochemical changes, in some cases, they can have a neutral or an inhibitory effect, depending on the target part of the plant and the moment of their biosynthesis. Therefore, an exogenous application of natural or synthetic plant growth promoters at the most suitable growth stages will ensure rapid growth [58].

Few reports showed that polymer-based NPs can act as a plant growth promoter and induce seed germination and vegetative growth [59]. For instance, chitosan NPs promote the growth of biophysical properties of coffee plants in greenhouse conditions by increasing the pigment content, mineral and water assimilation, and respiration rate [60]. Also, a low dose of chitosan NPs increases the germination and the growth of wheat compared to bulk chitosan [61]. Due to the different properties of NPs, their interaction with the plant allows morphological, genetic, and physiological changes. Several factors can affect the NPs' effect on the plant such as surface covering, chemical composition, and reactivity. Regarding alginate NPs, the study of Sharma et al. [62] showed that these NPs play an important role as water reservoirs, which mixing them with plant's soil exhibits better growth compared with the control. This study proved that alginate NPs could be a good material to ensure agricultural sustainability due to their high water retention capacity. For pectin NPs, the study of Abdelrahman et al. [63] depicted large information about the stimuli-responsive formulation made from mesoporous silica NPs-NH_2-Pectin, a delivery system for agrochemicals into plant cells. In addition, nanocellulose is known also to be applied in agriculture due to its surface dimension, high width-to-length rate, great transparency, and high toughness [64]. In the same sense, cellulose NPs can also be used as a nanocarrier system of macronutrients (i.e., nitrogen) and to release slowly for 2 months [65]. Moreover, lignin NPs (LNPs) have been successfully applied in the agricultural field as biostimulants of maize [66]. It has been found that the maize seeds treated with LNP revealed important physiological and biochemical responses such as an increase in seed germination rate, radical length, and the content of photosynthetic pigments and flavonoids. The benefic effect of nano-biopolymers in the agricultural field is related to the high potential of those biomaterial properties including mechanical, thermal, barrier, and physicochemical properties [67]. As reported in the literature, few experiments have studied polymeric NPs' effects on plants grown in normal conditions (without stress or biostimulants presence). Therefore, further studies should shed light on the beneficial and adverse effects of NPs on cell growth, metabolic responses, and gene expression.

3.2 *Application of NPs for Controlling Environmental Stresses*

Plants are sessile living organisms, which makes them very sensitive to sudden and slow changes in their living environment. These adverse environmental conditions are coming from living and nonliving factors of an ecosystem such as fungi, herbivores, bacteria, insects, salinity, drought, heavy metal, as well as low and high temperatures. These stresses could affect the different aspects of the plant through disruption of metabolic pathways and physiological and biochemical activities as well as the induction of cell dysfunction [68]. These changes vary depending not only on the type and period of stress but also on the stage of the plant [69]. To sustain life and to overcome the adverse effects of stress, plants must evolve a wide range of defense mechanisms such as reducing the stomatal aperture, producing phytohormones, changing gene expression, accumulating osmolytes, and scavenging the ROS system [70–73]. These mechanisms might improve plant tolerance against environmental constraints and protect cells from oxidative damages [74]. Nevertheless, when stress is severe and unmanageable by plants, an effective approach must be taken into consideration to control stress. The application of nanotechnology techniques has already been considered, in regard to improving the growth and protection of plants. It has been proven that NPs acting as biofertilizers, biopesticides, or microbicides could help plants deal with different environmental stresses and improve their productivity [75–77]. Moreover, the application of NPs under environmental stresses has a regulatory effect on plants [78–80]. The positive effect of NPs has been observed on several plants such as broad bean [81], rice [82], tomato [83], wheat [84], cowpea [85], and cumin [86]. Numerous investigations exhibited that the impact of NPs on plants could be phytotoxic [87, 88] depending on the nature and dose of NPs [89]. Nevertheless, polymer-based NPs which are known for their low toxicity would be the promising approach that could maintain green agriculture [90]. Chitosan-based NPs are bio-elicitors and nontoxic with high permeability and high film-forming capacity [91], which possess various physicochemical characteristics and biological activities. To date, there are only a few reports that showed the effect of polysaccharide NPs on plant defense and protection. Thus, in this section, we will review the effects of chitosan NPs as an eco-friendly mechanism which enhance plant tolerance to some environmental stresses such as salt, drought, heavy metal, and biotic stress.

3.2.1 Role of Chitosan-Based NPs in Plants Exposed to Salt Stress

Salinity is regarded as one of the major nonliving factors hindering plant growth and productivity [92]. The excess of Na^+ and Cl^- ions in irrigation water or soil could cause many metabolic disorders such as oxidative stress, nutrient imbalance, and ionic toxicity [93]. Thus, chitosan NPs' application has shown great potential in mitigating adverse effects of salt stress (Table 11.1).

For example, Sheikhalipour et al. [95] suggested that the treatment with 20 mg L^{-1} of selenium functionalized using chitosan NPs enhances the shoot height (19%), root length (13%), leaves' fresh weight (16%), root fresh weight (12%), leaves' dry weight (24%), and root dry weight (13%) under 100 mM of salt stress, compared to control. The growth-promoting effect of chitosan-based NPs might be due to the stimulation of the biosynthesis of some phytohormones such as auxin and gibberellins which are involved in plant growth [110]. Also, Sen et al. [96] showed that proline, sugar, and protein content were substantially increased by the application of nano-chitosan under salt stress. The accumulation of osmoprotectant compounds in the cytoplasm plays a critical role in osmotic adjustment in plants (Fig. 11.5). This accumulation is maybe due to the increase in the synthesis of these compounds or reduction of their degradation in response to saline conditions. Analogous responses have been denoted by Hassan et al. [98], which added that the increase in the number of osmoprotectant compounds is mainly due to the increase in the level of both enzymatic antioxidants, such as glutathione S-transferase (GST), and nonenzymatic antioxidants (phenolic compounds, ascorbate (ASA), carotenoids, flavonoids, and glutathione (GSH)) (Fig. 11.5). Furthermore, the use of chitosan NPs induced the expression of superoxide dismutase (SOD) and jasmonic acid genes, which activate the antioxidant system [111]. In this way, the unpleasant effects of saline stress could be decreased or avoided.

3.2.2 Effect of Chitosan-Based NPs in Plants Exposed to Drought Stress

Drought stress is the major environmental problem that humanity is facing today. During water deficiency, closing stomata is the first reaction done by the plant to limit water loss. But, if it lasts for a long time, it could drastically reduce agricultural yield by more than 50% [112]. Overall, drought stress could affect the morphological, physiological, and nutritional characteristics of plants [113]. The use of chitosan-based NPs can mitigate its adverse effects (Table 11.1 and Fig. 11.5). Leung and Giraudat [114] reported that chitosan application could induce hydrogen

Table 11.1 Different results regarding the nano-chitosan impact on plant growth and its metabolism under environmental stresses

Plant species	Polymer-based NP treatments	Type of stress	Impacts	References
Maize (*Zea mays* L.)	Nitrogen monoxide delivering chitosan NPs	Salt stress	–Amelioration of Photosystem II activity, chlorophyll content, and plant growth	[94]
Stevia (*Stevia rebaudiana* Bertoni)	Selenium-chitosan NPs	Salt stress	–Increase of plant growth, photosynthetic performance, and antioxidant enzyme activity –Reduction in electrolyte leakage and the content of H_2O_2 and MDA	[95]
Mung bean (*Vigna radiate* L.)	Nano-sized chitosan	Salt stress	–Improvement of plant development –Increase in photosynthetic pigment –Reduction in H_2O_2 and MDA content	[96]
Bean (Phaseolus vulgaris L.)	Chitosan NPs	Salt stress	–Promotion of seed germination –Increase in the level of chlorophyll a and b, CAT, proline, RWC, carotenoids, and antioxidant enzymes	[97]
Periwinkle (Catharanthus roseus L.)	Chitosan NPs	Salt stress	–Accumulation of CAT, APX, and glutathione reductase	[98]
Periwinkle (Catharanthus roseus L.)	Chitosan NPs	Drought stress	–Enhancement of plant growth –Induction of CAT and APX enzyme activity –Raise in the content of alkaloid and the expression of *deacetylvindoline-4-O-acetyltransferase*, *geissoschizine synthase*, *peroxidase 1*, and *strictosidine synthase* genes	[99]
Sugarcane (*Saccharum* spp.)	*S*-Nitroso-glutathione chitosan NPs	Drought stress	–Increase in photosynthetic rates and root/shoot ratio	[100]
Cotton (*Gossypium* L.)	Nano-chitosan-NPK	Drought stress	–Increase in stem length and number of fruiting nodes per seedling –Improvement in the number and weight of open bolls, seed ratio, and lint percentage	[101]

(continued)

Table 11.1 (continued)

Plant species	Polymer-based NP treatments	Type of stress	Impacts	References
Wheat (*Triticum aestivum* cv. Pishtaz)	Chitosan NPs	Drought Stress	–Increase in leaf area, the content of relative water and photosynthetic pigments, activity of catalase and superoxide dismutase, and rate of photosynthesis –Improvement of yield and shoot and root biomass	[84]
Tomato (*Solanum lycopersicum* L.)	Chitosan NPs	Cadmium stress	–Reduction in MDA content –Improvement in the content of enzymatic and nonenzymatic antioxidants and osmoprotectants	[102]
Moldavian balm (*Dracocephalum moldavica* L.)	Chitosan-selenium NPs	Cadmium stress	–Reduction in MDA and H_2O_2 levels –Improvement of agronomic traits, photosynthetic pigments, and chlorophyll fluorescence parameters –Increase in the level of proline, phenols, essential oils, and antioxidant enzyme activities	[103]
Thorn apple (*Datura stramonium* L.)	Chitosan NPs	Cadmium stress	–Increase in peroxidase and polyphenol oxidase activities	[104]
Tomato (*Solanum lycopersicum*)	Copper-chitosan NPs	Pathogenic fungi	–Inhibition of mycelial growth –Delay in spore germination	[105]
Tomato (*Solanum lycopersicum* L.)	Chitosan NPs	Wilt disease (*Fusarium andiyazi*)	–High antifungal activity –Upregulation of PR proteins and antioxidant genes	[106]
Chick pea (*Cicer arietinum* L.)	Chitosan NPs	–*Fusarium oxysporum* –*Pyricularia grisea* –*Alternaria solani*	–Promotion of the growth of chickpea seedlings –Inhibition of the growth of phytopathogens	[107]
Maize (*Zea mays* L.)	Copper-chitosan NPs	Curvularia leaf spot	–High activity of plant defense enzymes such as phenylalanine ammonia-lyase (PAL) and polyphenol oxidase (PPO)	[108]
Wheat (*Triticum turgidum* L.)	Chitosan NPs	Fusarium head blight disease	–Reduction in the density of hyphal branches and number of colonies formed	[109]

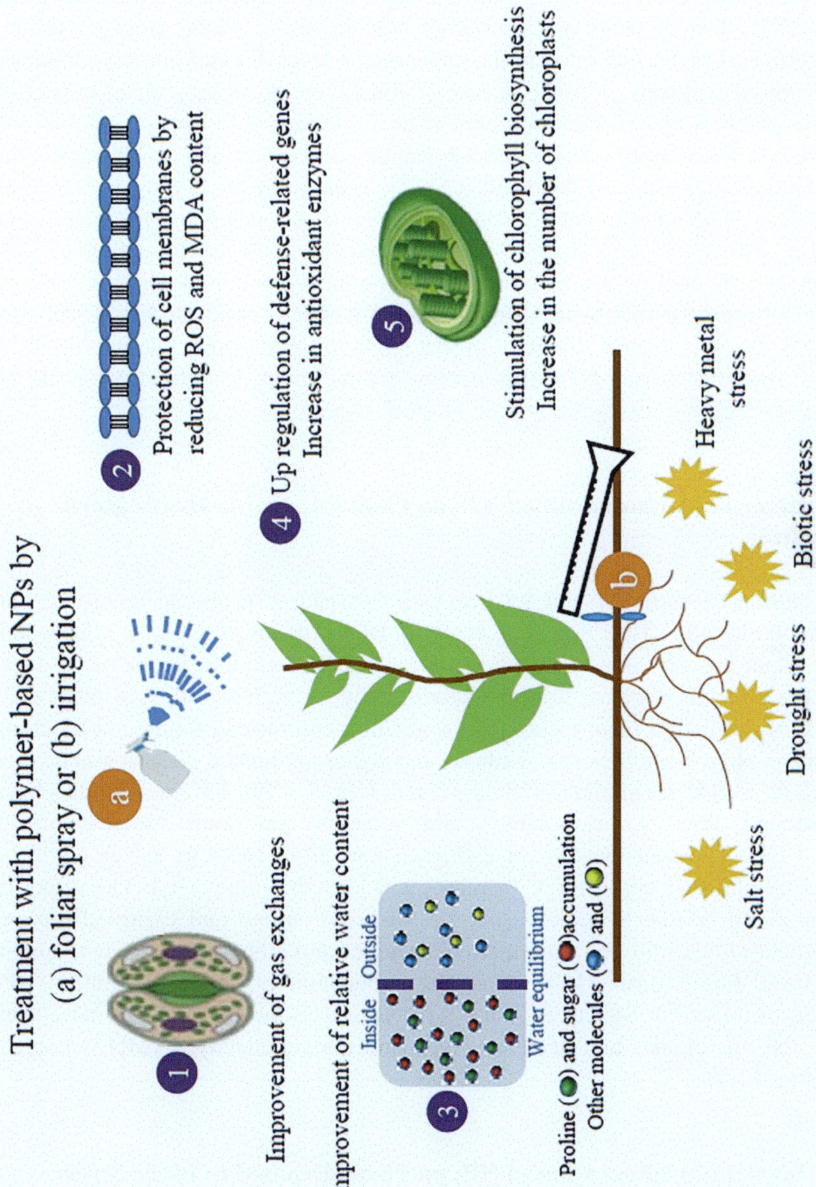

Fig. 11.5 Schematic diagram shows the effect of chitosan NPs on plants exposed to abiotic/biotic stresses

peroxide (H_2O_2) production in the guard cells which therefore increased the levels of abscisic acid leading to inhibiting the opening of stomata. Furthermore, a recent study has shown that the application of chitosan NPs at 90 ppm on wheat significantly increased leaf area, SOD and catalase (CAT) activities, and chlorophyll content [84]. The increase in chlorophyll content under water stress conditions could be linked to the induction in the endogenous level of cytokinins, a stimulator of chlorophyll synthesis, or to the release of mineral elements from chitosan such as nitrogen which is an essential component in the tetrapyrrole ring of chlorophyll. Furthermore, foliar spray of 0.1% chitosan nanoemulsion has induced changes in the photosynthetic apparatus of pearl millet [115]. These changes might be referred to the increase in the number of chloroplasts; the improvement of the activity of RuBisCo, which is a key enzyme of photosynthesis; and the limitation of CO_2 diffusion into the leaf, which therefore caused stomatal closure. In addition, Silveira et al. [100] reported that S-nitrosoglutathione filled into chitosan NPs effectively alleviated the detrimental effect of water stress on sugarcane plants. Also, the potential role of chitosan NPs in protecting plants under drought stress could be due to their ability to delay the release of nitric oxide.

3.2.3 Impact of Chitosan-Based NPs on Plant Exposed to Heavy Metal Stress

The pollution of water or soil by heavy metals causes a crucial environmental concern worldwide. The increased level of heavy metal negatively affects soil quality, plant growth, as well as human living [116]. Cadmium is one of the most toxic heavy metals because it is characterized by its high mobility in plant cells [117]. It could disrupt plant metabolism and cell membrane structure and function, dysregulate phytohormones' biosynthesis and signaling, reduce photosynthetic efficiency, and lead to membrane lipid peroxidation [118, 119]. Different studies have demonstrated that chitosan NPs could alleviate cadmium toxicity effects (Table 11.1). The translocation of cadmium ions from roots to leaves could be reduced by chitosan application; thereby plant growth is enhanced. The effect of chitosan could be due to its large number of active amino and hydroxyl groups, which chelate Cd^{2+} ions and store them in vacuoles and other organelles to maintain heavy metal homeostasis [120]. Also, the application of chitosan-selenium NPs could be an effective tool against the adverse effects of cadmium because they protect cell membranes and decrease H_2O_2 and malondialdehyde (MDA) content [103] (Fig. 11.5).

3.2.4 Impact of Chitosan-Based NPs on Plants Exposed to Biotic Stress

Regarding biotic stresses, plants can defend themselves and minimize damages occurred from herbivores' attacks and pathogen disease through multiple defense mechanisms. Among them, there is the building of a rigid cell wall structure to limit the invasion of microorganisms inside the plant cells and the production of lignin, toxic chemicals such as α-tomatine, and pathogen-degrading enzymes or making symbiotic relationship [121–125].

To prevent the loss of agricultural products, farmers are focusing now on the cultivation of plants with genetic disease resistance or applying the plant resistance inducer like NPs (Table 11.1 and Fig. 11.5). In tomato, Chun and Chandrasekaran [106] depicted that chitosan NPs have antifungal activity against phytopathogenic fungi, and this is probably due to the activation of plant defense genes such as chitinases and glucanases. Likewise, Cu-chitosan NPs were successfully controlled by *Alternaria alternate*, *Macrophomina phaseolina*, and *Rhizoctonia solani* which their growth was reduced by about 89.5, 63.0, and 60.1%, respectively, in comparison to the control [106]. Also, another study showed that the application of a very low dose of chitosan NPs (0.1%) could inhibit spore germination (up to 87.6%) [106]. Furthermore, their application on finger millet leaves has delayed the blast symptom expression by about 64% [107]. This protection could be due to either the activation of the defense response in plants or the inhibition of fungal RNA synthesis. Although chitosan is effective against bacteria [126]; however, to our knowledge, there are no data about the antibacterial potential of nano-chitosan. Thus, it might be interesting to study their ability to protect plants from harmful bacteria (i.e., *Pseudomonas syringae* pv) and their mechanisms on defense phytohormones, plant metabolome, and defense enzymes.

3.3 Application of NPs with Biostimulants

Over the past three decades, several innovative agricultural methods and techniques have been proposed to improve agricultural production and to reduce the use of agrochemicals which have adverse effects on the ecosystem. Among the promising innovation approach that respects the environment, there is the use of biostimulants [127]. They are generally substances and microorganisms which when either primed seedlings and seeds or applied to a growing substrate could modify all plant processes (physiology, biochemistry, and molecular responses) [127]. Indeed, the application of biostimulants offers potential benefits, such as the improvement of seedlings' development, yield, and nutrient assimilation, the acceleration of flowering and fruit set, as well as the increase of plant tolerance to environmental stresses [128, 129]. Biostimulants could be divided into two main groups: i) beneficial microorganisms, such as plant growth-promoting (PGP) fungi, endomycorrhizal fungi (AMF), and PGP rhizobacteria (PGPR), and ii) substances from plants and animal materials, such as chitosan, alginate, and humic acid or organic materials like vermicompost, compost, and seaweed extracts that are rich in organic and amino acids as well as phytohormones [130]. Recently, several scientific researchers have tested whether the effect of biostimulants on plants could be promoted by adding either chitosan NPs, alginate NPs, or fullerene NPs.

3.3.1 Beneficial Microorganisms

In the earth, various fungal and bacterial strains could colonize the rhizosphere and plant tissues for a mutually beneficial exchange. For example, some of them could suppress pathogenic microorganisms through competition for food, while others could fix the atmospheric nitrogen via root nodules; increase the uptake of nutrients that are limited in the soil and those that are inaccessible by plants, such as polyphosphates, via mycelium; as well as produce secondary metabolites (volatile compounds, phytohormones, polymers) that could either control cell functions and metabolisms or prime plants for defense responses [131–133].

Recently, there has been a major increase in the synthesis of metal NPs [134]. Most of these synthesized NPs were used to fertilize soil without altering the growth of the microbial community. Additionally, some antagonistic microorganisms that belong to the PGP fungi class, such as *Trichoderma viride and T. harzianum*, can naturally synthesize chitosan NPs and silver NPs, respectively, with a very low average size, around 25–89 nm [135, 136]. However, the mechanisms leading to the biosynthesis of polymer-based NPs by *T. viride* remain unknown. Some studies have demonstrated that the biosynthesis of these NPs would require the presence of pathogens, such as *Sclerotinia sclerotiorum* [137]. Upon sensing the cell wall of pathogens, *Trichoderma* strains could secrete NADA co-enzyme and NADA-dependent enzymes (nitrate reductase) which are important for NP biosynthesis [137, 138].

Moreover, the application of chitosan NPs, with a diameter size between 50.748 and 141.772 nm, in combination with *T. viride* has increased plant disease resistance and reduced maize late wilt caused by *Cephalosporium maydis* [139]. This reduction was related to the increase in the activity of acid phosphatase enzyme.

As far as we know, no study had studied yet the impact of chitosan NPs produced by *Trichoderma* spp. on their growth as well as whether the exogenous application of these NPs affects the action mechanisms of these fungi. Moreover, some studies reported that chitosan could interact with *Trichoderma* spp. in different ways, depending on the dose applied. Zavala-González et al. [140] have reported that the growth rate of *T. pseudokoningii*, *T. harzianum*, and *T. koningiopsis* was around 22–53%, 68–81%, and 100%, respectively, under the application of 0.5 mg mL^{-1} of chitosan to different growing media. Additionally, when a high dose of chitosan (1 to 2 mg mL^{-1}) was applied, the growth of *T. koningiopsis* strain was not affected because its cell wall had a lower level of linolenate (C18:3) and a high level of stearate (C18:0), in contrast to those chitosan sensitive, like *T. harzianum*, which had a high level of polyunsaturated free fatty acids [140–142].

Chitosan NPs may have positive or negative effects on AMF growth as well as their impacts on plants. The effect of chitosan NPs on mycorrhizal symbiosis could be related to various factors including size, concentration, and crosslinker used for the preparation of NPs, physicochemical properties of soil, and fungal species [143]. El-Gazzar et al. [139] reported that chitosan NPs (0.1 g mL^{-1}) applied to seeds increased the frequency and intensity of *Glomus mosseae* colonization as well as the number of arbuscules, therefore controlling late wilt. However, El Amerany et al. [144] reported that foliar spraying of leaves with chitosan NPs (250 mg L^{-1} to 1000 mg L^{-1}) reduced mycorrhization rate as well as the transcript level of fungal biomass (*RiBtub*) and AM-specific phosphate transporter (*SlPT4*) genes, and therefore, tomato plant biomass and flowering of mycorrhizal plants were as well affected. The failure of the mycorrhizal network was supposed to be related to the down expression of an acidic isoform of the chitinase (*Chi3*) gene that is important for fungal growth.

Regarding PGPR, some studies demonstrated that chitosan NP application has strengthened the activity of *Bacillus* spp. and *Pseudomonas aeruginosa* due to the increase in the level of enzymatic indicators of soil health (dehydrogenase, fluorescein diacetate, alkaline phosphatase, nitrogenase reductase, nitrous oxide reductase, nitrite reductase, nitrate reductase), the growth (seed germination, leaf area, stem height, root length) of *Zea mays* and *Bidens pilosa*, as well as the accumulation of plant metabolism (alcohol, add ester, hydrocarbons) [145, 146]. Additionally, Atalla et al. [147] reported that the use of Cu-chitosan NPs, with a diameter size of 220 nm, in combination with fermented maize wastes by *Pseudomonas fluorescens* and *T. harzianum* reduced maize diseases. The production of antimicrobial protein, such as α-amylase and β-amylase, by beneficial microorganisms under Cu-chitosan NPs application, was demonstrated to be the most important metabolites involved in reducing disease severity. Moreover, chitosan NPs could be used as herbicide (imazapic and imazapyr) nanocarrier to reduce their toxicity on plant development; however, the encapsulated herbicides could alter the growth of beneficial bacteria [146].

Polymer-based NPs are used not only for the encapsulation of substances but also nano-enclosing of beneficial microorganisms. For example, alginate-silica NPs were used for enclosing either *Pseudomonas fluorescens* VUPF5, *Bacillus subtilis* VRU1, or their metabolites (i.e., auxin), and these nanoformulations had a positive impact on shoot number and length, root architecture, and fresh weight of pistachio [148, 149].

A summary that shows polymer-based NP effect on different beneficial microorganisms was illustrated in Fig. 11.6.

3.3.2 Substances and Organic Materials

Despite the outstanding results that have been achieved through the application of beneficial microorganisms, in terms of improving yield and soil fertility, however, in some cases, their application is not feasible in the presence of harmful or competing microorganisms and because of the inappropriate soil conditions, such as high level of phosphate, as well as a slow population development [150, 151]. Thus, emerging interest has been focused recently on providing an organic amendment like compost which could be produced in a huge quantity through the controlled fermentation (or composting) of different kinds of organic wastes (plant residues, animal manure, and urban, municipal, and industrial waste) and for the shortest period [152].

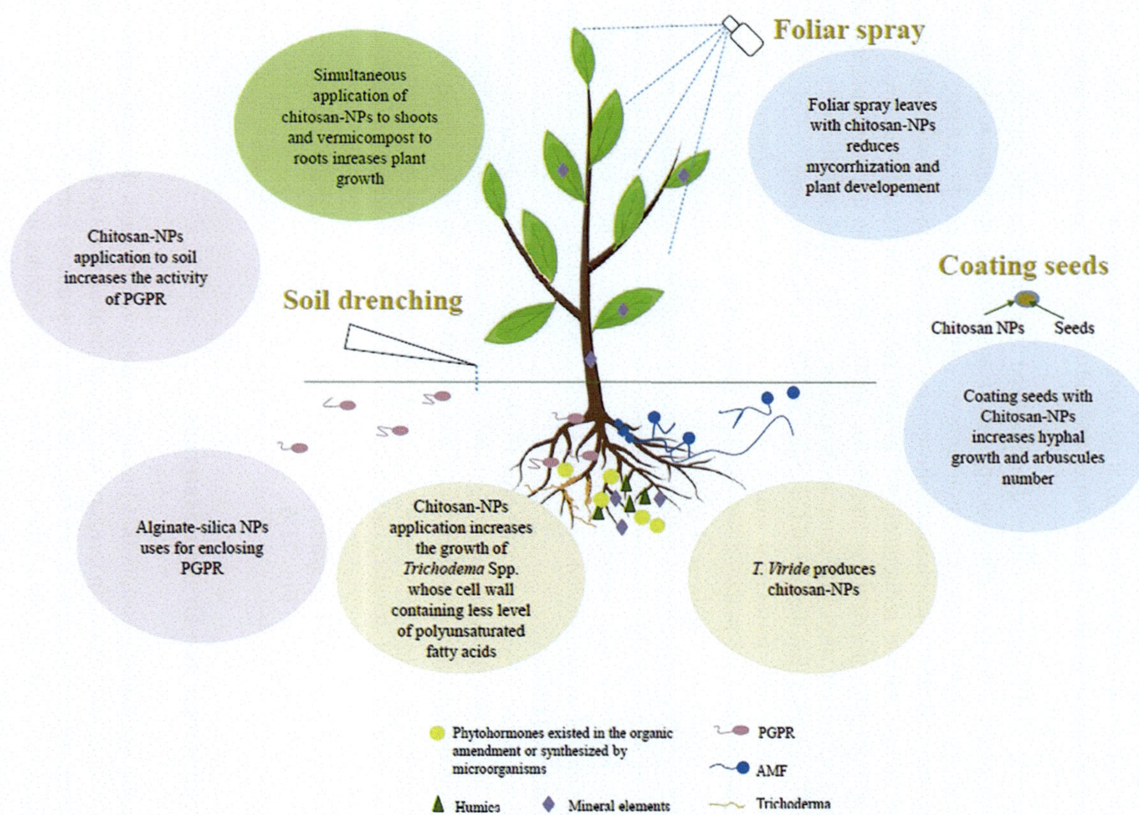

Fig. 11.6 Illustrations of chitosan and alginate NPs' impacts on soil beneficial microorganisms

The production of compost has dual benefits. It could solve the problem of the deposition of a large number of unwanted residues as well as be an amendment for improving crops' traits. The agronomic effects of compost are due to its richness in nutrients; phytohormones (i.e., indole acetic acid); simple compounds of low structural complexity, such as phenols, quinones, carboxylic acids, and ethers; and the presence of bacterial and fungal groups [153, 154]. The richness of compost makes them able to improve water and nutrient retention capacity, plant growth and its development, and disease suppression [155]. Also, its components can interact with other biostimulants, therefore affecting their impacts on plants. For instance, some studies showed that fullerene (C60) NPs could absorb humic acids in the presence of Ca^{2+} [156], while El Amerany et al. [151] showed that compost components could bind the amino groups of chitosan and this combination has increased shoot and root biomass, leaf area, stem development, chlorophyll fluorescence, and nutrient uptake (i.e., N, P, Mn, Mg, Na, Ca, Si, Fe, and Zn) of tomato plants. In regard to chitosan nanoscopic morphology, Ibrahim et al. [157] showed that chitosan NP (40 ppm) application in combination with vermicompost (6 tons/feddan) has increased plant height, branch and capsule number, herb fresh and dry weights, volatile oil level, and mineral nutrient (N, P, and K) content of black cumin (*Nigella sativa* L).

3.4 *Effects of Polymer-Based NPs on Fruit Development and Quality*

Fruit production is a critical phase for any fruiting crop species. It depends on several stages including flower formation, anthesis, fertilization, fruit set initiation, growth, maturation, and ripening. To get high fruit quality with freshness and long shelf life, scientists are trying to find methods or chemical additives that could trigger positive effects on the plant system [158–160]. Among these methods, they are the use of polymer-based NPs that have proven their effectiveness on plant growth performance. Chandra et al. [161] showed that chitosan NPs' application to shoots has induced the transcript level of β-1,3-glucanase gene that involves in increasing plant protection against harmful microorganisms and ameliorates plant development and physiology which include cellular division, flower production, seed maturation, and pedicel abscission. This is in concurrence with the acquired results by Behboudi et al. [84]. Recently, new formulations based on chitosan NPs containing phytohormones or inorganic elements were tested on shoots, roots, and seeds, to see whether they could impact fruit growth and its quality. For example, Kumaraswamy et al. [162] reported that applying SA-chitosan NPs to either seeds or leaves of maize was an effective approach that reduced flower alteration and yield loss caused by *Fusarium verticillioides*, while other studies showed that priming seeds and leaves of maize with either Cu-chitosan NPs, Zn-chitosan NPs, or chitosan-silicon NPs has increased grain yield by about 29%, 51%, and 187%, respectively, in comparison to non-treated plants [108, 163, 164]. Promotion in the number of spikelets and grain yield was also observed after foliar spray of chitosan NPK NPs and chitosan NPs containing N-acetyl cysteine on wheat (Table 11.2).

Also, the study carried out on tomatoes showed that priming seeds with chitosan (or chitosan-alginate) NPs containing GA_3 has a positive effect on fruit productivity [171]. Fruit number was induced by about 60% under the application of chitosan NPs containing GA_3, and fruits' fresh weight was increased by about 28% after the application of chitosan-alginate NPs containing GA_3, in comparison to control. The difference in effect could depend on the divergence in physicochemical features of polymers, chitosan, and alginate, which influence the duration of GA_3 release. Santo Pereira et al. [171] demonstrated that the release of GA_3 from chitosan NPs was faster than from the chitosan-alginate NP system. During the bud formation phase, the exogenous application of GA_3 could stimulate flower bud development and flowering-associated gene (squamosa promoter-binding protein-like (*SPL*) and suppressor of overexpression of CO1 (*SOC*)) expression as well as increase flowering quality and the level of endogenous hormones such as GA and IAA [180, 181]. Also, its application during flower opening and fruit production could increase the development of fruit (size, growth, and setting) and its nutritional composition (sugars, flavonoids, anthocyanins, phenolic compounds, and antioxidants) [182]. Thus, the initial release of GA_3 from chitosan stimulated flower formation, and therefore fruit number is induced; however, the induction in fruit biomass under the application of chitosan-alginate NPs could be explained by their release of GA_3 during fruit growth and ripening. Even many reports showed to what extent can the application of polymer-based NPs, during seed germination to fruit maturation, be effective on agricultural productivity, but there is no study regarding their effect on fruit quality.

The totality of agricultural products intended for human consumption does not reach its recipients due to losses during the so-called "postharvest" which is a phase that includes stages between harvest and processing of products for food. This phase can affect the quality of climacteric fruits as well as the ripening of those non-climacteric fruits. To decrease the unfavorable effects of chemicals and environmental stresses on the quality of fruits during storage, the coating was applied. It is used as a passive or as an inactivate barrier to regulate transpiration and volatile compound loss [183]. During the twelfth and thirteenth centuries, a monomer such as wax was the first film used on fruits [184]. Then, new and transparent materials have been developed to be used for short-term storage. Nowadays, different kinds of polymer-based NPs are used to maintain the quality of bananas, grapes, raspberries, persimmon, mango, sweet cherries, strawberry, and guava (Table 11.2). Coating fruits with chitosan NPs could maintain the color and the firmness due to delaying the maturation process and the microbial growth (Fig. 11.7 and Table 11.2). Also, their incorporation in hydroxypropyl methylcellulose films could prevent water and aroma loss and protect fruits from oxidative reactions and mechanical damage [185].

Moreover, the incorporation of various compounds such as amino acids (i.e., phenylalanine), unsaturated fatty acids, polysaccharides (i.e., xanthan gum), synthetic additive (i.e., propylene glycol and polyethylene glycol), and mineral ions (i.e., Ag^+) into polymer-based NPs improves their efficiency and functionality (Table 11.2). Releasing these compounds (or ions) from the coating materials onto

Table 11.2 Various impacts of polymer-based NPs on fruit attributes and composition

Polymer-based NPs	Doses	Application modes	Species	Effects	References
Polysaccharide NPs: –Chitosan NPs	0.2%	Coating fruit	Cavendish bananas (*Musa acuminate* AAA group)	–Increased smoothness of the skin and slower discoloration	[165]
	Stock solution			–Inhibited the activities of hydrolytic enzymes that involved in starch degradation –Reduced water loss and the pulp to peel ratio and, therefore, prolong the shelf life of fruits	[166]
	1–12 g L^{-1}	Coating fruit	Grapes (*Vitis labrusca* L.)	–Delayed the ripening process of fruits with no alteration of sensory characteristics	[167]
	90 ppm	Foliar/soil application	Wheat (*Triticumaestivum* cv. pishtaz)	–Increased spike weight and length, number of grain per spike, and the level of proteins	[84]
	5 g L^{-1}	Coating fruit	Raspberry (*Rubus sanctus Schreber*)	–Decreased decay extension rate –Increased the amounts of flavonoids, anthocyanins, and phenolic compounds and the activity of α-diphenyl-β-picrylhydrazyl (DPPH), phenylalanine ammonia-lyase, and guaiacol peroxidase	[168]
–Chitosan NPK NPs	10%	Foliar spray	Wheat (*Triticum Aestivum*)	–Increased spikelet number, kernel weight, grain number, and weight	[169, 170]
–Zn-chitosan NPs	0.04–0.08%	Foliar spray and coating seeds	Maize (*Zea mays* L.)	–Increased grain zinc content and yield	[163]

(continued)

Table 11.2 (continued)

Polymer-based NPs	Doses	Application modes	Species	Effects	References
–Chitosan-silicon NPs	0.8%	Foliar spray and coating seeds	Maize (*Zea mays* L.)	–Increased yield/plot	[164]
–Cu-chitosan NPs	0.16%	Foliar spray and coating seeds	Maize (*Zea mays* L.)	–Increased ear length and weight as well as grain yield	[108]
–Chitosan/TPP NPs containing gibberellic acid (GA_3)	0.5 mg mL^{-1}	Seed priming	Tomato (*Solanum lycopersicum*)	–Increased the number of fruits and productivity	[171]
–Chitosan/alginate NPs containing GA_3					
–Chitosan NPs containing *N*-acetyl cysteine (0.2 mg mL^{-1})	Stock solution	Foliar spray	Durum wheat (*Triticum durum* cv. Fabulis)	–Increased grain yield	[172]
–Chitosan NPs loaded with phenylalanine	5 mM	Coating fruit	Persimmon (*Diospyros kaki*)	–Reduced H_2O_2 –Increased the level of MDA, soluble tannin, and total carotenoid as well as SOD, CAT, and APX activities and therefore chilling injury decreased	[173]
–Guar gum/AgNPs	–	Coating fruit	Mango fruit (cv. Seddik)	–Maintained the texture, titratable acidity, and the level of carotenoids –Delayed ripening process	[174]
–Alginate NPs containing soybean oil with or without a $CaCl_2$ crosslinker	–	Coating fruit	Sweet cherries (cv. Bing)	–Increased titratable acidity and the level of soluble solid content, total soluble phenolics, anthocyanins, and antioxidant Capacity (DPPH)	[175]
Protein NPs: Proteinoid polymer NPs containing auxin	–	Dispersion in the soil media	Lettuce (*Lactuca sativa*)	–Increased root length	[176]

Lipid NPs: –Beeswax containing polysaccharides and propylene glycol	10 g L^{-1}	Coating fruit	Strawberry (*Fragaria ananassa*)	–Reduced weight and firmness loss, decay index, and color changes	[177]
–Candeuba® wax containing polysaccharide and polyethylene glycol	65 g L^{-1}	Coating fruit	Guava (*Psidium guajava* L.)	–Reduced weight loss –Increased total solid soluble –Maintained acidity value	[178]
–Candeuba® wax containing xanthan gum	65 g L^{-1}	Coating fruit	Guava (*Psidium guajava* L.)	–Reduced total color difference, respiration rates, and pectin methylesterase activity –Increased ascorbic acid and phenolic compounds levels	[179]

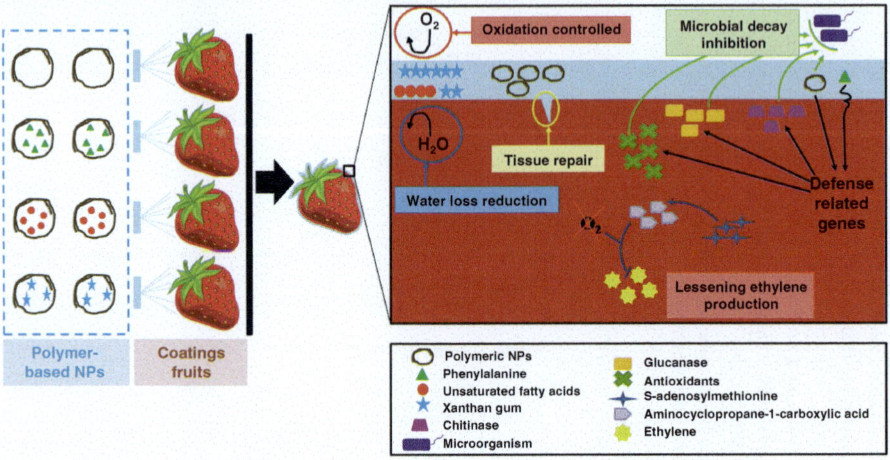

Fig. 11.7 Proposed model illustrating the impact of coating fruits with polymer-based NPs

the surface of fruits increases the level of antioxidant (i.e., ascorbic acid, sugars, phenolic compounds) and antioxidant enzymes (SOD, CAT, and peroxidase ascorbate (APX)) and stops unwanted reactions, especially discoloration and enzymatic browning (Table 11.2 and Fig. 11.7).

3.5 The Fate of Polymeric NPs

Polymer-based NPs are complex molecules that are composed of three layers: (1) the outer layer or crust could interact and make complexes with materials, molecules, and ions; (2) the mantle is made of a chemically different material from the core, and this characteristic allows them to protect the encapsulated stuff; (3) the core is the inner space responsible for establishing the encapsulation property of NPs [186].

Polymer-based NPs are unstable particles due to their ability to change rapidly or to react with the active agent to reach a state considered relatively stable [187]. However, this interaction could lead to oxidation, ion exchange, deformation, assembly, and aggregation of polymer-based NPs and affect their sizes, structures, chemical composition, and their impacts on plant tissues as well soil structure and microbiota [187]. The literature shows that the nanometer size of NPs allows them to go through cells and react with molecules; but, as far as we know, no report has shown yet the fate of NPs after their application either on plants or soil microbia.

The cell wall of plants is a selective and protective barrier that protects cells from environmental factors and restricts the entry and the egress of substances. But, it is well known that only substances of small diameter size can enter plant cells through leaf's stomata which makes a hypothesis that leaf guard cells could be the target of polymer-based NPs [102]. Avellan et al. [188] hypothesized that NP uptake could be done via a plant cuticle that contains small hydrophilic pores or a stomatal pathway. Aziz et al. [189] showed that chitosan application could enter the stomata and translocate inside wheat plants through phloem cells. Also, Abdel-Aziz et al. [170] proved that chitosan NPs could be taken up by plants and move between leaves and roots through the phloem route.

Moreover, applying NPs, with a diameter size less than 50 nm, to shoots has been shown to accumulate in younger leaves, roots, and the rhizosphere soil; but their uptake and translocation are depended on their size and structure [188]. The accumulation of NPs in leaf surface increases for smaller sizes NPs. Regarding larger NPs, the particles efficiently cross the plant cuticle layer and then accumulate in the mesophyll cells or plant vasculature.

Even though various studies examined the foliar uptake of NPs as well as their mobility outside or inside plant cells, it remains unclear whether they were degraded when they passed through cells and what form of NPs was taken up by the plant.

3.6 Conclusion and Future Perspective

It was concluded that the nature and the chemical properties of polymer NPs play an important role regarding their impacts on the plant system. They could increase plant growth and development, ameliorate fruit attributes and quality, stimulate fungal and bacterial growth, and enhance plant tolerance to water, salt, toxic elements, and pathogens. Despite many reports demonstrating their benefits on crops, there is no certain information regarding their action mechanisms. This could be attributed to the structure of polymer used, size of particles, or coated substances. But a clear insight to understand their impact on plant physiology, metabolic, and gene machinery are still missing. Also, given the importance of chitosan NPs on plants and microbiota, it remains unknown how another type of polysaccharide, such as lignin, starch, and cellulose as well as those from proteins and lipids, could affect the agricultural products.

References

1. Morris, J., & Willis, J. (2007). *US Environmental Protection Agency nanotechnology white paper*. US Environ. Prot. Agency.
2. Mukhopadhyay, S. S. (2014). Nanotechnology in agriculture: Prospects and constraints. *Nanotechnology, Science and Applications, 7*, 63.
3. Mousavi, S. R., & Rezaei, M. (2011). Nanotechnology in agriculture and food production. *Journal of Applied Environmental and Biological Sciences, 1*, 414–419.
4. Singh, H., Sharma, A., Bhardwaj, S. K., Arya, S. K., Bhardwaj, N., & Khatri, M. (2021). Recent advances in the applications of nano-agrochemicals for sustainable agricultural development. *Environmental Science: Processes & Impacts, 23*(2), 213–239.

5. Shafi, A., Qadir, J., Sabir, S., Zain Khan, M., & Rahman, M. M. (2020). Nanoagriculture: A holistic approach for sustainable development of agriculture. *Handbook of Nanomaterials and Nanocomposites for Energy and Environmental Applications*, 1–16.
6. DeRosa, M. C., Monreal, C., Schnitzer, M., Walsh, R., & Sultan, Y. (2010). Nanotechnology in fertilizers. *Nature Nanotechnology, 5*, 91.
7. Haryadi, B. M., Hafner, D., Amin, I., Schubel, R., Jordan, R., Winter, G., & Engert, J. (2019). Nonspherical nanoparticle shape stability is affected by complex manufacturing aspects: Its implications for drug delivery and targeting. *Advanced Healthcare Materials, 8*, 1900352.
8. Jampílek, J., & Kráľová, K. (2015). Application of nanotechnology in agriculture and food industry, its prospects and risks. *Ecological Chemistry and Engineering S, 22*, 321–361.
9. Murthy, S. K. (2007). Nanoparticles in modern medicine: State of the art and future challenges. *International Journal of Nanomedicine, 2*, 129.
10. Panpatte, D. G., Jhala, Y. K., Shelat, H. N., & Vyas, R. V. (2016). Nanoparticles: The next generation technology for sustainable agriculture. In D. Singh, H. Singh, & R. Prabha (Eds.), *Microbial inoculants in sustainable agricultural productivity* (pp. 289–300). Springer.
11. Singh, D. (2020). Cyanobacteria as a source of nanoparticle: Application and future projections. In P. K. Singh, A. Kumar, V. K. Singh, & A. K. Shrivastava (Eds.), *Advances in cyanobacterial biology* (pp. 319–331). Academic Press.
12. Duhan, J. S., Kumar, R., Kumar, N., Kaur, P., Nehra, K., & Duhan, S. (2017). Nanotechnology: The new perspective in precision agriculture. *Biotechnology Reports, 15*, 11–23.
13. Wang, Y., Sun, C., Zhao, X., Cui, B., Zeng, Z., Wang, A., Liu, G., & Cui, H. (2016). The application of nano-TiO_2 photo semiconductors in agriculture. *Nanoscale Research Letters, 11*, 1–7.
14. Kong, S.-Z., Li, D.-D., Luo, H., Li, W.-J., Huang, Y.-M., Li, J.-C., Hu, Z., Huang, N., Guo, M.-H., & Chen, Y. (2018). Anti-photoaging effects of chitosan oligosaccharide in ultraviolet-irradiated hairless mouse skin. *Experimental Gerontology, 103*, 27–34.
15. Xin, X., Zhao, F., Rho, J. Y., Goodrich, S. L., Sumerlin, B. S., & He, Z. (2020). Use of polymeric nanoparticles to improve seed germination and plant growth under copper stress. *Science of The Total Environment, 745*, 141055.
16. Zielińska, A., Carreiró, F., Oliveira, A. M., Neves, A., Pires, B., Venkatesh, D. N., Durazzo, A., Lucarini, M., Eder, P., & Silva, A. M. (2020). Polymeric nanoparticles: Production, characterization, toxicology and ecotoxicology. *Molecules, 25*, 3731.
17. Kurita, K. (2006). Chitin and chitosan: Functional biopolymers from marine crustaceans. *Marine Biotechnology, 8*, 203.
18. Moreno-Vega, A.-I., Gomez-Quintero, T., Nunez-Anita, R.-E., Acosta-Torres, L.-S., & Castaño, V. (2012). Polymeric and ceramic nanoparticles in biomedical applications. *Journal of Nanotechnology, 2012*, 1–10.
19. Mohammed, M. A., Syeda, J., Wasan, K. M., & Wasan, E. K. (2017). An overview of chitosan nanoparticles and its application in non-parenteral drug delivery. *Pharmaceutics, 9*, 53.
20. George, A., Shah, P. A., & Shrivastav, P. S. (2019). Natural biodegradable polymers based nano-formulations for drug delivery: A review. *International Journal of Pharmaceutics, 561*, 244–264.
21. Yoo, H. S., & Park, T. G. (2001). Biodegradable polymeric micelles composed of doxorubicin conjugated PLGA–PEG block copolymer. *Journal of Controlled Release, 70*(1–2), 63–70.
22. Chang, L., Xu, L., Liu, Y., & Qiu, D. (2021). Superabsorbent polymers used for agricultural water retention. *Polymer Testing, 94*, 107021.
23. Moradi, R., Pourghasemian, N., & Naghizadeh, M. (2019). Effect of beeswax waste biochar on growth, physiology and cadmium uptake in saffron. *Journal of Cleaner Production, 229*, 1251–1261.
24. Muñoz-Almagro, N., Herrero-Herranz, M., Guri, S., Corzo, N., Montilla, A., & Villamiel, M. (2021). Application of sunflower pectin gels with low glycemic index in the coating of fresh strawberries stored in modified atmospheres. *Journal of the Science of Food and Agriculture, 101*(14), 5775–5783.

25. Limam, Z., Selmi, S., Sadok, S., & El Abed, A. (2011). Extraction and characterization of chitin and chitosan from crustacean by-products: Biological and physicochemical properties. *African Journal of Biotechnology, 10*(4), 640–647.
26. Aboudamia, F. Z., Kharroubi, M., Neffa, M., Aatab, F., Hanoune, S., Bouchdoug, M., & Jaouad, A. (2020). Potential of discarded sardine scales (Sardina pilchardus) as chitosan sources. *Journal of the Air & Waste Management Association, 70*(11), 1186–1197.
27. da Silva Lucas, A. J., Oreste, E. Q., Costa, H. L. G., López, H. M., Saad, C. D. M., & Prentice, C. (2021). Extraction, physicochemical characterization, and morphological properties of chitin and chitosan from cuticles of edible insects. *Food Chemistry, 343*, 128550.
28. George, T. S., Guru, K. S. S., Vasanthi, N. S., & Kannan, K. P. (2011). Extraction, purification and characterization of chitosan from endophytic fungi isolated from medicinal plants. *World Journal of Science and Technology, 1*(4), 43–48.
29. Ohya, Y., Shiratani, M., Kobayashi, H., & Ouchi, T. (1994). Release behavior of 5-fluorouracil from chitosan-gel nanospheres immobilizing 5-fluorouracil coated with polysaccharides and their cell specific cytotoxicity. *Journal of Macromolecular Science, Part A: Pure and Applied Chemistry, 31*(5), 629–642.
30. Rampino, A., Borgogna, M., Blasi, P., Bellich, B., & Cesàro, A. (2013). Chitosan nanoparticles: Preparation, size evolution and stability. *International Journal of Pharmaceutics, 455*(1–2), 219–228.
31. Jameela, S. R., Kumary, T. V., Lal, A. V., & Jayakrishnan, A. (1998). Progesterone-loaded chitosan microspheres: A long acting biodegradable controlled delivery system. *Journal of Controlled Release, 52*(1–2), 17–24.
32. Perez, J. J., & Francois, N. J. (2016). Chitosan-starch beads prepared by ionotropic gelation as potential matrices for controlled release of fertilizers. *Carbohydrate Polymers, 148*, 134–142.
33. Yanat, M., & Schroën, K. (2021). Preparation methods and applications of chitosan nanoparticles; with an outlook toward reinforcement of biodegradable packaging. *Reactive and Functional Polymers, 161*, 104849.
34. Jahandideh, A., Ashkani, M., & Moini, N. (2021). Biopolymers in textile industries. In S. Thomas, S. Gopi, & A. Amalraj (Eds.), *Biopolymers and their Industrial Applications* (pp. 193–218). Elsevier.
35. Rajaonarivony, M., Vauthier, C., Couarraze, G., Puisieux, F., & Couvreur, P. (1993). Development of a new drug carrier made from alginate. *Journal of Pharmaceutical Sciences, 82*, 912–917.
36. Fessi, H., Puisieux, F., Devissaguet, J. P., Ammoury, N., & Benita, S. (1989). Nanocapsule formation by interfacial polymer deposition following solvent displacement. *International Journal of Pharmaceutics, 55*(1), R1–R4.
37. Fathi, M., Martín, Á., & McClements, D. J. (2014). Nanoencapsulation of food ingredients using carbohydrate based delivery systems. *Trends in Food Science and Technology, 39*(1), 18–39.
38. Yu, C.-Y., Cao, H., Zhang, X.-C., Zhou, F.-Z., Cheng, S.-X., Zhang, X.-Z., & Zhuo, R.-X. (2009). Hybrid nanospheres and vesicles based on pectin as drug carriers. *Langmuir, 25*, 11720–11726.
39. Chang, C., Wang, Z.-C., Quan, C.-Y., Cheng, H., Cheng, S.-X., Zhang, X.-Z., & Zhuo, R.-X. (2007). Fabrication of a novel pH-sensitive glutaraldehyde cross-linked pectin nanogel for drug delivery. *Journal of Biomaterials Science. Polymer Edition, 18*, 1591–1599.
40. Sharma, R., Bajpai, J., Bajpai, A. K., Acharya, S., Kumar, B., & Singh, R. K. (2017). Assessment of water retention performance of pectin-based nanocarriers for controlled irrigation in agriculture. *Agriculture Research, 6*, 139–149.
41. Zhang, J., Elder, T. J., Pu, Y., & Ragauskas, A. J. (2007). Facile synthesis of spherical cellulose nanoparticles. *Carbohydrate Polymers, 69*, 607–611.
42. Pandey, J. K., Lee, J.-W., Chu, W.-S., Kim, C.-S., Ahn, S.-H., & Lee, C. S.-Y. (2008). Cellulose nano whiskers from grass of Korea. *Macromolecular Research, 16*, 396–398.

43. Thou, C. Z., Khan, F. S. A., Mubarak, N. M., Ahmad, A., Khalid, M., Jagadish, P., Walvekar, R., Abdullah, E. C., Khan, S., & Khan, M. (2021). Surface charge on chitosan/cellulose nanowhiskers composite via functionalized and untreated carbon nanotube. *Arabian Journal of Chemistry, 14*, 103022.
44. Thomas, S. K., Begum, P. M. S., Midhun Dominic, C. D., Salim, N. V., Hameed, N., Rangappa, S. M., Siengchin, S., & Parameswaranpillai, J. (2021). Isolation and characterization of cellulose nanowhiskers from Acacia caesia plant. *Journal of Applied Polymer Science, 138*, 50213.
45. Gu, H., Gao, X., Zhang, H., Chen, K., & Peng, L. (2020). Fabrication and characterization of cellulose nanoparticles from maize stalk pith via ultrasonic-mediated cationic etherification. *Ultrasonics Sonochemistry, 66*, 104932.
46. Hasanin, M. S. (2021). Simple, economic, ecofriendly method to extract starch nanoparticles from potato peel waste for biological applications. *Starch, 73*, 2100055.
47. Hofrichter, M. (2001). *Biopolymers, volume 1, lignin, humic substances and coal.* Wiley-Blackwell.
48. Iravani, S., & Varma, R. S. (2020). Greener synthesis of lignin nanoparticles and their applications. *Green Chemistry, 22*, 612–636.
49. Nasrollahzadeh, M., Bidgoli, N. S. S., Issaabadi, Z., Ghavamifar, Z., Baran, T., & Luque, R. (2020). Hibiscus Rosasinensis L.: Aqueous extract-assisted valorization of lignin: Preparation of magnetically reusable pd NPs@ Fe3O4-lignin for Cr (VI) reduction and Suzuki-Miyaura reaction in eco-friendly media. *International Journal of Biological Macromolecules, 148*, 265–275.
50. Gual, A., Delgado, J. A., Godard, C., Castillón, S., Curulla-Ferré, D., & Claver, C. (2013). Novel Polymer Stabilized Water Soluble Ru-Nanoparticles as Aqueous Colloidal Fischer–Tropsch Catalysts. *Topics in Catalysis, 56*, 1208–1219.
51. Lievonen, M., Valle-Delgado, J. J., Mattinen, M. L., Hult, E. L., Lintinen, K., Kostiainen, M. A., Paananen, A., Szilvay, G. R., Setälä, H., & Österberg, M. (2016). A simple process for lignin nanoparticle preparation. *Green Chemistry, 18*, 1416–1422.
52. Yearla, S. R., & Padmasree, K. (2016). Preparation and characterisation of lignin nanoparticles: Evaluation of their potential as antioxidants and UV protectants. *Journal of Experimental Nanoscience, 11*, 289–302.
53. Ding, X., Liu, Y., Chen, X., Liu, W., & Li, J. (2021). Simultaneous removal of antibiotics and heavy metals with poly (aspartic acid)-based Fenton micromotors. *Chemistry, an Asian Journal, 16*(14), 1930–1936.
54. Li, H., Wang, H., & Wang, L. (2013). Synthesis and sensing application of highly luminescent and water stable polyaspartate functionalized LaF 3 nanocrystals. *Journal of Materials Chemistry C, 1*, 1105–1110.
55. Zheng, Y., Yang, W., Wang, C., Hu, J., Fu, S., Dong, L., Wu, L., & Shen, X. (2007). Nanoparticles based on the complex of chitosan and polyaspartic acid sodium salt: Preparation, characterization and the use for 5-fluorouracil delivery. *European Journal of Pharmaceutics and Biopharmaceutics, 67*, 621–631.
56. Tulloch, A. P. (1980). Beeswax—Composition and analysis. *Bee World, 61*, 47–62.
57. Kheradmandnia, S., Vasheghani-Farahani, E., Nosrati, M., & Atyabi, F. (2010). Preparation and characterization of ketoprofen-loaded solid lipid nanoparticles made from beeswax and carnauba wax. *Nanomedicine: Nanotechnology Biology Medicine, 6*, 753–759.
58. Waqas, M. A., Khan, I., Akhter, M. J., Noor, M. A., & Ashraf, U. (2017). Exogenous application of plant growth regulators (PGRs) induces chilling tolerance in short-duration hybrid maize. *Environmental Science and Pollution Research International, 24*(12), 11459.
59. Choudhary, R. C., Kumaraswamy, R. V., Kumari, S., Pal, A., Raliya, R., Biswas, P., & Saharan, V. (2017). Synthesis, characterization, and application of chitosan nanomaterials loaded with zinc and copper for plant growth and protection. In R. Prasad, M. Kumar, & V. Kumar (Eds.), *Nanotechnology* (Vol. 2, pp. 227–247). Springer.

60. Nguyen Van, S., Dinh Minh, H., & Nguyen Anh, D. (2013). Study on chitosan nanoparticles on biophysical characteristics and growth of Robusta coffee in green house. *Biocatalysis and Agricultural Biotechnology, 2*, 289–294.
61. Li, R., He, J., Xie, H., Wang, W., Bose, S. K., Sun, Y., Hu, J., & Yin, H. (2019). Effects of chitosan nanoparticles on seed germination and seedling growth of wheat (*Triticum aestivum* L.). *International Journal of Biological Macromolecules, 126*, 91–100.
62. Sharma, R., Bajpai, J., Bajpai, A. K., Acharya, S., Shrivastava, R. B., & Shukla, S. K. (2014). Designing slow water-releasing alginate nanoreserviors for sustained irrigation in scanty rainfall areas. *Carbohydrate Polymers, 102*, 513–520.
63. Abdelrahman, T. M., Qin, X., Li, D., Senosy, I. A., Mmby, M., Wan, H., Li, J., & He, S. (2021). Pectinase-responsive carriers based on mesoporous silica nanoparticles for improving the translocation and fungicidal activity of prochloraz in rice plants. *Chemical Engineering Journal, 404*, 126440.
64. Kargarzadeh, H., Mariano, M., Huang, J., Lin, N., Ahmad, I., Dufresne, A., & Thomas, S. (2017). Recent developments on nanocellulose reinforced polymer nanocomposites: A review. *Polymer (Guildf)., 132*, 368–393.
65. Kottegoda, N., Munaweera, I., Madusanka, N., Sandaruwan, C., Sirisena, D., Disanayake, N., Ismail, M., De Alwis, A., Karunaratne, V., & Malwana, W. (2011). Plant nutrient nanoparticles encapsulated cellulose matrix for slow and sustained release of nitrogen. *Current Science, 101*, 73–78.
66. Del Buono, D., Luzi, F., & Puglia, D. (2021). Lignin nanoparticles: A promising tool to improve maize physiological, biochemical, and chemical traits. *Nanomaterials, 11*, 846.
67. Sorrentino, A., Gorrasi, G., & Vittoria, V. (2007). Potential perspectives of bio-nanocomposites for food packaging applications. *Trends in Food Science and Technology, 18*, 84–95.
68. Pathak, M. R., Silva-Teixeira, J. A., & Wani, S. H. (2014). Polyamines in response to abiotic stress tolerance through transgenic approaches. *GM Crops Food, 5*, 87–96.
69. Feller, U., & Vaseva, I. I. (2014). Extreme climate events: Impacts of drought and high temperature on physiological processes in agronomically important plants. *Frontiers in Environmental Science, 39*, 1–17.
70. Fahad, S., Hussain, S., Matloob, A., Khan, F. A., Khaliq, A., Saud, S., & Huang, J. (2015). Phytohormones and plant responses to salinity stress. *Plant Growth Regulation, 75*, 391–404.
71. Chaves, M. M., Costa, J. M., Zarrouk, O., Pinheiro, C., Lopes, C. M., & Pereira, J. S. (2016). Controlling stomatal aperture in semi-arid regions—The dilemma of saving water or being cool? *Plant Science, 251*, 54–64.
72. Zhao, P., Wang, D., Wang, R., Kong, N., Zhang, C., Yang, C., & Chen, Q. (2018). Genome-wide analysis of the potato Hsp20 gene family: Identification, genomic organization and expression profiles in response to heat stress. *BMC Genomics, 19*(1), 1–13.
73. Khan, M. N., AlSolami, M. A., Basahi, R. A., Siddiqui, M. H., Al-Huqail, A. A., Abbas, Z. K., & Khan, F. (2020). Nitric oxide is involved in nano-titanium dioxide-induced activation of antioxidant defense system and accumulation of osmolytes under water-deficit stress in *Viciafaba* L. *Ecotoxicology and Environmental Safety, 190*, 110152.
74. Abdel Latef, A. A. H., Omer, A. M., Badawy, A. A., Osman, M. S., & Ragaey, M. M. (2021). Strategy of salt tolerance and interactive impact of *Azotobacter chroococcum* and/or *Alcaligenes faecalis* inoculation on canola (*Brassica napus* L.) plants grown in saline soil. *Plants, 10*, 110.
75. Shaik, M. R., Albalawi, G. H., Khan, S. T., Khan, M., Adil, S. F., Kuniyil, M., & Khan, M. (2016). "Miswak" based green synthesis of silver nanoparticles: Evaluation and comparison of their microbicidal activities with the chemical synthesis. *Molecules, 21*(11), 1478.
76. Lade, B. D., & Gogle, D. P. (2019). Nano-biopesticides: Synthesis and applications in plant safety. In K. A. Abd-Elsalam & R. Prasad (Eds.), *Nanobiotechnology Applications in Plant Protection* (Vol. 2, pp. 169–189). Springer.

77. Manivannan, N., Aswathy, S., Malaikozhundan, B., & Boopathi, T. (2021). Nano-zinc oxide synthesized using diazotrophic Azospirillum improves the growth of mung bean, *Vigna radiata*. *International Nano Letters, 11*, 405–415.
78. Haghighi, M., & Pessarakli, M. (2013). Influence of silicon and nano-silicon on salinity tolerance of cherry tomatoes (*Solanum lycopersicum* L.) at early growth stage. *Scientia Horticulturae, 161*, 111–117.
79. Haghighi, M., Abolghasemi, R., & Silva, J. A. T. (2014). Low and high temperature stress affect the growth characteristics of tomato in hydroponic culture with se and nano-se amendment. *Scientia Horticulturae, 178*, 231–240.
80. Qados, A. M. A., & Moftah, A. E. (2015). Influence of silicon and nano-silicon on germination, growth, and yield of faba bean (*Vicia faba* L.) under salt stress conditions. *Journal of Experimental Agriculture International, 5*, 509–524.
81. Arafat, A. H. A., Ashish, K. S., Mahmmoud, S. A., Mojtaba, K., & Lam-Son, P. T. (2017). Titanium dioxide nanoparticles improve growth and enhance tolerance of broad bean plants under saline soil conditions. *Land Degradation and Development, 29*, 1065–1073.
82. Ejaz, M., Raja, N. I., Mashwani, Z. U. R., Sheeraz, A. M., Mubashir, H., & Iqbal, M. (2018). Effect of silver nanoparticles and silver nitrate on growth of rice under biotic stress. *IET Nanobiotechnology, 12*(7), 927–932.
83. Pérez-Labrada, F., López-Vargas, E. R., Ortega-Ortiz, H., Cadenas-Pliego, G., Benavides-Mendoza, A., & Juárez-Maldonado, A. (2019). Responses of tomato plants under saline stress to foliar application of copper nanoparticles. *Plants, 8*(6), 151.
84. Behboudi, F., Tahmasebi-Sarvestani, Z., Kassaee, M. Z., Modarres-Sanavy, S. A. M., Sorooshzadeh, A., & Mokhtassi-Bidgoli, A. (2019). Evaluation of chitosan nanoparticles effects with two application methods on wheat under drought stress. *Journal of Plant Nutrition, 42*(13), 1439–1451.
85. Ogunkunle, C. O., Bornmann, B., Wagner, R., Fatoba, P. O., Frahm, R., & Lützenkirchen-Hecht, D. (2019). Copper uptake, tissue partitioning and biotransformation evidence by XANES in cowpea (*Vigna unguiculata* L) grown in soil amended with nanosized copper particles. *Environmental Nanotechnology, Monitoring and Management, 12*, 100231.
86. Salajegheh, M., Yavarzadeh, M., Payandeh, A., & Akbarian, M. M. (2020). Effects of titanium and silicon nanoparticles on antioxidant enzymes activity and some biochemical properties of *Cuminum cyminum* L. under drought stress. Banats. *Journal of Biotechnology, 11*(21), 19–25.
87. Siddiqui, M. H., Al-Whaibi, M. H., Faisal, M., & Al Sahli, A. A. (2014). Nano-silicon dioxide mitigates the adverse effects of salt stress on *Cucurbita pepo* L. *Environmental Toxicology and Chemistry, 33*(11), 2429–2437.
88. Youssef, M. S., & Elamawi, R. M. (2020). Evaluation of phytotoxicity, cytotoxicity, and genotoxicity of ZnO nanoparticles in *Vicia faba*. *Environmental Science and Pollution Research, 27*(16), 18972–18984.
89. Noori, A. T., Donnelly, J., Colbert, W., Cai, L. A., & Newman, J. C. (2020). White exposure of tomato (Lycopersicon esculentum) to silver nanoparticles and silver nitrate: Physiological and molecular response int. *Journal of Phytoremediation, 22*, 40–51.
90. Prabha, A. S., Dorothy, R., Jancirani, S., Rajendran, S., Singh, G., & Kumaran, S. S. (2020). Recent advances in the study of toxicity of polymer-based nanomaterials. *Nanotoxicity*, 143–165.
91. Shukla, S. K., Mishra, A. K., Arotiba, O. A., & Mamba, B. B. (2013). Chitosan-based nanomaterials: A state-of-the-art 7 review. *International Journal of Biological Macromolecules, 59*, 46–58.
92. Solima, W. S., & El-Shaieny, A. H. A. (2014). Effect of saline water on germination and early growth stage of five Apiaceae species. *African Journal of Agricultural Research, 9*(7), 713–719.
93. Singh, A. (2015). Soil salinization and waterlogging: A threat to environment and agricultural sustainability. *Ecological Indicators, 57*, 128–130.

94. Oliveira, H. C., Gomes, B. C., Pelegrino, M. T., & Seabra, A. B. (2016). Nitric oxide-releasing chitosan nanoparticles alleviate the effects of salt stress in maize plants. *Nitric Oxide, 61*, 10–19.
95. Sheikhalipour, M., Esmaielpour, B., Gohari, G., Haghighi, M., Jafari, H., Farhadi, H., & Kalisz, A. (2021). Salt stress mitigation via the foliar application of chitosan-functionalized selenium and anatase titanium dioxide nanoparticles in stevia (*Stevia rebaudiana* Bertoni). *Molecules, 26*(13), 4090.
96. Sen, S. K., Chouhan, D., Das, D., Ghosh, R., & Mandal, P. (2020). Improvisation of salinity stress response in mung bean through solid matrix priming with normal and nano-sized chitosan. *International Journal of Biological Macromolecules, 145*, 108–123.
97. Zayed, M. M., Elkafafi, S. H., Zedan, A. M., & Dawoud, S. F. (2017). Effect of nano chitosan on growth, physiological and biochemical parameters of Phaseolus vulgaris under salt stress. *Journal of Plant Production, 8*(5), 577–585.
98. Hassan, F. A. S., Ali, E., Gaber, A., Fetouh, M. I., & Mazrou, R. (2021). Chitosan nanoparticles effectively combat salinity stress by enhancing antioxidant activity and alkaloid biosynthesis in *Catharanthus roseus* (L.) G. Don. *Plant Physiology and Biochemistry, 162*, 291–300.
99. Ali, E. F., El-Shehawi, A. M., Ibrahim, O. H. M., Abdul-Hafeez, E. Y., Moussa, M. M., & Hassan, F. A. S. (2021). A vital role of chitosan nanoparticles in improvisation the drought stress tolerance in *Catharanthus roseus* (L.) through biochemical and gene expression modulation. *Plant Physiology and Biochemistry, 161*, 166–175.
100. Silveira, N. M., Seabra, A. B., Marcos, F. C., Pelegrino, M. T., Machado, E. C., & Ribeiro, R. V. (2019). Encapsulation of S-nitrosoglutathione into chitosan nanoparticles improves drought tolerance of sugarcane plants. *Nitric Oxide, 84*, 38–44.
101. Sarhan, M. G., Bashanday, S., & El-Gayed, A. (2021). Water stress on cotton plants as affected by potassium and nano chitosan-NPK fertilizer. *Scientific Journal of Agricultural Sciences, 3*, 236–250.
102. Faizan, M., Rajput, V. D., Al-Khuraif, A. A., Arshad, M., Minkina, T., Sushkova, S., & Yu, F. (2021). Effect of foliar fertigation of chitosan nanoparticles on cadmium accumulation and toxicity in *Solanum lycopersicum*. *Biology, 10*(7), 666.
103. Azimi, F., Oraei, M., Gohari, G., Panahirad, S., & Farmarzi, A. (2021). Chitosan-selenium nanoparticles (cs–se NPs) modulate the photosynthesis parameters, antioxidant enzymes activities and essential oils in *Dracocephalum moldavica* L. under cadmium toxicity stress. *Plant Physiology and Biochemistry, 167*, 257–268.
104. Shirkhani, Z., Rad, A. C., & Mohsenzadeh, F. (2021). Improving Cd-phytoremediation ability of *Datura stramonium* L. by Chitosan and Chitosan nanoparticles. *Biologia, 76*, 2161–2171.
105. Saharan, V., Sharma, G., Yadav, M., Choudhary, M. K., Sharma, S. S., Pal, A., & Biswas, P. (2015). Synthesis and in vitro antifungal efficacy of cu–chitosan nanoparticles against pathogenic fungi of tomato. *International Journal of Biological Macromolecules, 75*, 346–353.
106. Chun, S. C., & Chandrasekaran, M. (2019). Chitosan and chitosan nanoparticles induced expression of pathogenesis-related proteins genes enhances biotic stress tolerance in tomato. *International Journal of Biological Macromolecules, 125*, 948–954.
107. Sathiyabama, M., & Parthasarathy, R. (2016). Biological preparation of chitosan nanoparticles and its in vitro antifungal efficacy against some phytopathogenic fungi. *Carbohydrate Polymers, 151*, 321–325.
108. Choudhary, R. C., Kumaraswamy, R. V., Kumari, S., Sharma, S. S., Pal, A., Raliya, R., Biswas, P., & Saharan, V. (2017). Cu-chitosan nanoparticle boost defense responses and plant growth in maize (*Zea mays* L.). *Scientific Reports, 7*(1), 1–11.
109. Kheiri, A., Jorf, S. M., Malihipour, A., Saremi, H., & Nikkhah, M. (2016). Application of chitosan and chitosan nanoparticles for the control of fusarium head blight of wheat (*fusarium graminearum*) in vitro and greenhouse. *International Journal of Biological Macromolecules, 93*, 1261–1272.

110. Safikhan, S., Khoshbakht, K., Chaichi, M. R., Amini, A., & Motesharezadeh, B. (2018). Role of chitosan on the growth, physiological parameters and enzymatic activity of milk thistle (*Silybum marianum* (L.) Gaertn.) in a pot experiment. *Journal of Applied Research on Medicinal and Aromatic Plants, 10*, 49–58.
111. Hernández-Hernández, H., Juárez-Maldonado, A., Benavides-Mendoza, A., Ortega-Ortiz, H., Cadenas-Pliego, G., Sánchez-Aspeytia, D., & González-Morales, S. (2018). Chitosan-PVA and copper nanoparticles improve growth and overexpress the SOD and JA genes in tomato plants under salt stress. *Agronomy, 8*(9), 175.
112. Abdullah, D. Z., Khan, S. A., Jebran, K., & Ali, A. (2015). Agricultural credit in Pakistan: Past trends and future prospects. *Journal of Applied Environmental and Biological Sciences, 5*, 178–188.
113. Razak, A. A., Ismail, M. R., Karim, M. F., Wahab, P. E. M., Abdullah, S. N. K., & H. (2013). Changes in leaf gas exchange, biochemical properties, growth and yield of chilli grown under soilless culture subjected to deficit fertigation. *Australian Journal of Crop Science, 7*, 1582–1589.
114. Leung, J., & Giraudat, J. (1998). Abscisic acid signal transduction. *Annual Review of Plant Biology, 49*(1), 199–222.
115. Priyaadharshini, M., Sritharan, N., Senthil, A., & Marimuthu, S. (2019). Physiological studies on effect of chitosan nanoemulsion in pearl millet under drought condition. *Journal of Pharmacognosy and Phytochemistry, 8*, 3304–3307.
116. Yadav, S. K. (2010). Heavy metal toxicity in plants: An overview on the role of glutathione and phytochelatins in heavy metal stress tolerance of plants. *South African Journal of Botany, 76*, 167–179.
117. Rizwan, M., Ali, S., Adrees, M., Rizvi, H., Zia-ur-Rehman, M., Hannan, F., Qayyum, M. F., Hafeez, F., & Ok, Y. S. (2016). Cadmium stress in rice: Toxic effects, tolerance mechanisms, and management: A critical review. *Environmental Science and Pollution Research, 23*(18), 17859–17879.
118. Irfan, M., Ahmad, A., & Hayat, S. (2014). Effect of cadmium on the growth and antioxidant enzymes in two varieties of *Brassica juncea*. *Saudi Journal of Biological Sciences, 21*(2), 125–131.
119. Mostofa, M. G., Rahman, M., Ansary, M., Uddin, M., Fujita, M., & Tran, L.-S. P. (2019). Interactive effects of salicylic acid and nitric oxide in enhancing rice tolerance to cadmium stress. *International Journal of Molecular Sciences, 20*, 5798.
120. Qu, D. Y., Gu, W. R., Zhang, L. G., Li, C. F., Chen, X. C., Li, J., Li, L. J., Xie, T. L., & Wei, S. (2019). Role of chitosan in the regulation of the growth, antioxidant system and photosynthetic characteristics of maize seedlings under cadmium stress. *Russian Journal of Plant Physiology, 66*(1), 140–151.
121. Ferreira, R. B., Monteiro, S., Freitas, R., Santos, C. N., Chen, Z., Batista, L. M., Duarte, J., Borges, A., & Teixeira, A. R. (2007). The role of plant defence proteins in fungal pathogenesis. *Molecular Plant Pathology, 8*, 677–700.
122. Moura, J. C. M. S., Bonine, C. A. V., Oliveira, F. V. J., Dornelas, M. C., & Mazzafera, P. (2010). Abiotic and biotic stresses and changes in the lignin content and composition in plants. *Journal of Integrative Plant Biology, 52*(4), 360–376.
123. Sasidharan, R., Voesenek, L. A., & Pierik, R. (2011). Cell wall modifying proteins mediate plant acclimatization to biotic and abiotic stresses. *Critical Reviews in Plant Sciences, 30*(6), 548–562.
124. Chen, L., Meng, J., He, X. L., Zhang, M., & Luan, Y. S. (2019). *Solanum lycopersicum* micro RNA1916 targets multiple target genes and negatively regulates the immune response in tomato. *Plant, Cell & Environment, 42*(4), 1393–1407.
125. Mitra, D., Djebaili, R., Pellegrini, M., Mahakur, B., Sarker, A., Chaudhary, P., & Mohapatra, P. K. D. (2021). Arbuscular mycorrhizal symbiosis: Plant growth improvement and induction of resistance under stressful conditions. *Journal of Plant Nutrition, 44*, 1–37.

126. Li, J., & Zhuang, S. (2020). Antibacterial activity of chitosan and its derivatives and their interaction mechanism with bacteria: Current state and perspectives. *European Polymer Journal, 138*, 109984.
127. Rouphael, Y., & Colla, G. (2020). Biostimulants in Agriculture. *Frontiers in Plant Science, 11*.
128. Colla, G., & Rouphael, Y. (2015). Biostimulants in horticulture. *Scientia Horticulturae, 196*, 1–2.
129. ParaĎiković, N., Teklić, T., Zeljković, S., Lisjak, M., & Špoljarević, M. (2019). Biostimulants research in some horticultural plant species—A review. *Food and Energy Security, 8*(2), e00162.
130. Yakhin, O. I., Lubyanov, A. A., Yakhin, I. A., & Brown, P. H. (2017). Biostimulants in plant science: A global perspective. *Frontiers in Plant Science, 7*, 2049.
131. Gopalakrishnan, S., Sathya, A., Vijayabharathi, R., Varshney, R. K., Gowda, C. L., & Krishnamurthy, L. (2015). Plant growth promoting rhizobia: Challenges and opportunities. *3Biotech, 5*(4), 355–377.
132. Luginbuehl, L. H., & Oldroyd, G. E. (2017). Understanding the arbuscule at the heart of endomycorrhizal symbioses in plants. *Current Biology, 27*(17), R952–R963.
133. Ramírez-Valdespino, C. A., Casas-Flores, S., & Olmedo-Monfil, V. (2019). Trichoderma as a model to study effector-like molecules. *Frontiers in Microbiology, 10*, 1030.
134. Ameen, F., Alsamhary, K., Alabdullatif, J. A., & ALNadhari, S. (2021). A.: Review on metal-based nanoparticles and their toxicity to beneficial soil bacteria and fungi. *Ecotoxicology and Environmental Safety, 213*, 112027.
135. Guilger, M., Pasquoto-Stigliani, T., Bilesky-Jose, N., Grillo, R., Abhilash, P. C., Fraceto, L. F., & De Lima, R. (2017). Biogenic silver nanoparticles based on Trichoderma harzianum: Synthesis, characterization, toxicity evaluation and biological activity. *Scientific Reports, 7*(1), 1–13.
136. Boruah, S., & Dutta, P. (2021). Fungus mediated biogenic synthesis and characterization of chitosan nanoparticles and its combine effect with Trichoderma asperellum against *fusarium oxysporum, sclerotium rolfsii* and *rhizoctonia solani. Indian Phytopathology, 74*(1), 81–93.
137. Guilger-Casagrande, M., Germano-Costa, T., Pasquoto-Stigliani, T., Fraceto, L. F., & de Lima, R. (2019). Biosynthesis of silver nanoparticles employing *Trichoderma harzianum* with enzymatic stimulation for the control of *sclerotinia sclerotiorum. Scientific Reports, 9*(1), 1–9.
138. Saravanakumar, K., Chelliah, R., Mubarak Ali, D., Jeevithan, E., Oh, D. H., Kathiresan, K., & Wang, M. H. (2018). Fungal enzyme-mediated synthesis of chitosan nanoparticles and its biocompatibility, antioxidant and bactericidal properties. *International Journal of Biological Macromolecules, 118*, 1542–1549.
139. El-Gazzar, N., El-Bakery, A. M., & Ata, A. A. (2018). Influence of some bioagents and chitosan nanoparticles on controlling maize late wilt and improving plants characteristics. *Egyptian Journal of Phytopathology, 46*(2), 243–264.
140. Zavala-González, E. A., Lopez-Moya, F., Aranda-Martinez, A., Cruz-Valerio, M., Lopez-Llorca, L. V., & Ramírez-Lepe, M. (2016). Tolerance to chitosan by Trichoderma species is associated with low membrane fluidity. *Journal of Basic Microbiology, 56*(7), 792–800.
141. Palma-Guerrero, J., Jansson, H. B., Salinas, J., & Lopez-Llorca, L. V. (2008). Effect of chitosan on hyphal growth and spore germination of plant pathogenic and biocontrol fungi. *Journal of Applied Microbiology, 104*(2), 541–553.
142. Lopez-Moya, F., Suarez-Fernandez, M., & Lopez-Llorca, L. V. (2019). Molecular mechanisms of chitosan interactions with fungi and plants. *International Journal of Molecular Sciences, 20*(2), 332.
143. Tian, H., Kah, M., & Kariman, K. (2019). Are nanoparticles a threat to mycorrhizal and rhizobial symbioses? A critical review. *Frontiers in Microbiology, 10*, 1660.
144. El Amerany, F., Meddich, A., Wahbi, S., Porzel, A., Taourirte, M., Rhazi, M., & Hause, B. (2020). Foliar application of chitosan increases tomato growth and influences

mycorrhization and expression of endochitinase-encoding genes. *International Journal of Molecular Sciences, 21*(2), 535.
145. Khati, P., Chaudhary, P., Gangola, S., Bhatt, P., & Sharma, A. (2017). Nanochitosan supports growth of Zea mays and also maintains soil health following growth. *3Biotech, 7*(1), 81.
146. Maruyama, C. R., Guilger, M., Pascoli, M., Bileshy-José, N., Abhilash, P. C., Fraceto, L. F., & De Lima, R. (2016). Nanoparticles based on chitosan as carriers for the combined herbicides imazapic and imazapyr. *Scientific Reports, 6*(1), 1–15.
147. Atalla, S. M., Abdel-Kader, M. M., El-Gamal, N. G., & El-Mougy, N. S. (2020). Using maize wastes, fermented by co-cultures of *Trichoderma harzianum* and *Pseudomonas fluorescens*, as grain dressing against m maize diseases under field conditions. *Egyptian Journal of Biological Pest control, 30*(1), 1–8.
148. Young, C. C., Rekha, P. D., Lai, W. A., & Arun, A. B. (2006). Encapsulation of plant growth-promoting bacteria in alginate beads enriched with humic acid. *Biotechnology and Bioengineering, 95*(1), 76–83.
149. Moradi Pour, M., Saberi-Riseh, R., Mohammadinejad, R., & Hosseini, A. (2019). Nano-encapsulation of plant growth-promoting rhizobacteria and their metabolites using alginate-silica nanoparticles and carbon nanotube improves UCB1 pistachio micropropagation. *Journal of Microbiology and Biotechnology, 29*(7), 1096–1103.
150. Del Val, C., Barea, J. M., & Azcon-Aguilar, C. (1999). Diversity of arbuscular mycorrhizal fungus populations in heavy-metal-contaminated soils. *Applied and Environmental Microbiology, 65*(2), 718–723.
151. El Amerany, F., Rhazi, M., Wahbi, S., Taourirte, M., & Meddich, A. (2020). The effect of chitosan, arbuscular mycorrhizal fungi, and compost applied individually or in combination on growth, nutrient uptake, and stem anatomy of tomato. *Scientia Horticulturae, 261*, 109015.
152. Stewart-Wade, S. M. (2020). Efficacy of organic amendments used in containerized plant production: Part 1–compost-based amendments. *Scientia Horticulturae, 266*, 108856.
153. Amir, S., Jouraiphy, A., Meddich, A., El Gharous, M., Winterton, P., & Hafidi, M. (2010). Structural study of humic acids during composting of activated sludge-green waste: Elemental analysis, FTIR and ^{13}C NMR. *Journal of Hazardous Materials, 177*(1–3), 524–529.
154. Wong, W. S., Zhong, H. T., Cross, A. T., & Yong, J. W. H. (2020). Plant biostimulants in vermicomposts: Characteristics and plausible mechanisms. In *The Chemical Biology of Plant Biostimulants* (pp. 155–180). John Wiley & Sons.
155. Guo, X. X., Liu, H. T., & Wu, S. B. (2019). Humic substances developed during organic waste composting: Formation mechanisms, structural properties, and agronomic functions. *Science of the Total Environment, 662*, 501–510.
156. Chen, K. L., & Elimelech, M. (2008). Interaction of fullerene (C60) nanoparticles with humic acid and alginate coated silica surfaces: Measurements, mechanisms, and environmental implications. *Environmental Science & Technology, 42*(20), 7607–7614.
157. Ibrahim, M. F. (2020). The role of vermicompost and chitosan nanoparticles as foliar application to enhancing growth, yield and oil of black cumin (*Nigella sativa* L.) plants. *Archives of Agriculture and Science Journal, 3*(2), 205–223.
158. Yang, F. M., Li, H. M., Li, F., Xin, Z. H., Zhao, L. Y., Zheng, Y. H., & Hu, Q. H. (2010). Effect of nano-packing on preservation quality of fresh strawberry (*Fragaria ananassa Duch.* cv Fengxiang) during storage at 4 °C. *Journal of Food Science, 75*(3), C236–C240.
159. Van Hung, D., Tong, S., Tanaka, F., Yasunaga, E., Hamanaka, D., Hiruma, N., & Uchino, T. (2011). Controlling the weight loss of fresh produce during postharvest storage under a nano-size mist environment. *Journal of Food Engineering, 106*(4), 325–330.
160. Song, H., Yuan, W., Jin, P., Wang, W., Wang, X., Yang, L., & Zhang, Y. (2016). Effects of chitosan/nano-silica on postharvest quality and antioxidant capacity of loquat fruit during cold storage. *Postharvest Biology and Technology, 119*, 41–48.
161. Chandra, S., Chakraborty, N., Dasgupta, A., Sarkar, J., Panda, K., & Acharya, K. (2015). Chitosan nanoparticles: A positive modulator of innate immune responses in plants. *Scientific Reports, 5*(1), 1–14.

162. Kumaraswamy, R. V., Kumari, S., Choudhary, R. C., Sharma, S. S., Pal, A., Raliya, R., Biswas, P., & Saharan, V. (2019). Salicylic acid functionalized chitosan nanoparticle: A sustainable biostimulant for plant. *International Journal of Biological Macromolecules, 123*, 59–69.
163. Choudhary, R. C., Kumaraswamy, R. V., Kumari, S., Sharma, S. S., Pal, A., Raliya, R., Biswas, P., & Saharan, V. (2019). Zinc encapsulated chitosan nanoparticle to promote maize crop yield. *International Journal of Biological Macromolecules, 127*, 126–135.
164. Kumaraswamy, R. V., Saharan, V., Kumari, S., Choudhary, R. C., Pal, A., Sharma, S. S., Rakshit, S., Raliya, R., & Biswas, P. (2021). Chitosan-silicon nanofertilizer to enhance plant growth and yield in maize (Zea mays L.). *Plant Physiology and Biochemistry, 159*, 53–66.
165. Esyanti, R. R., Zaskia, H., & Amalia, A. (2019). Chitosan nanoparticle-based coating as post-harvest technology in banana. *Journal of Physics: Conference Series, 1204*, 012109.
166. Lustriane, C., Dwivany, F. M., Suendo, V., & Reza, M. (2018). Effect of chitosan and chitosan-nanoparticles on post harvest quality of banana fruits. *Journal of Plant Biotechnology, 45*(1), 36–44.
167. Melo, N. F. C. B., de MendonçaSoares, B. L., Diniz, K. M., Leal, C. F., Canto, D., Flores, M. A., da Costa Tavares-Filho, J. H., Galembeck, A., Stamford, T. L. M., Stamford-Arnaud, T. M., et al. (2018). Effects of fungal chitosan nanoparticles as eco-friendly edible coatings on the quality of postharvest table grapes. *Postharvest Biology and Technology, 139*, 56–66.
168. Ishkeh, S. R., Shirzad, H., Asghari, M. R., Alirezalu, A., Pateiro, M., & Lorenzo, J. M. (2021). Effect of chitosan nanoemulsion on enhancing the phytochemical contents, health-promoting components, and shelf life of raspberry (*Rubus sanctus Schreber*). *Applied Sciences, 11*(5), 2224.
169. Abdel-Aziz, H. M. A., Hasaneen, M. N., & Omer, A. M. (2016). Nano chitosan-NPK fertilizer enhances the growth and productivity of wheat plants grown in sandy soil. *Spanish Journal of Agricultural Research, 14*(1), 17.
170. Abdel-Aziz, H. M. M., Hasaneen, M. N. A. G., & Omer, A. M. (2018). Foliar application of nano chitosan NPK fertilizer improves the yield of wheat plants grown on two different soils. *Egyptian Journal of Experimental Biology, 14*(1), 63–72.
171. Santo Pereira, A. D. E., Oliveira, H. C., & Fraceto, L. F. (2019). Polymeric nanoparticles as an alternative for application of gibberellic acid in sustainable agriculture: A field study. *Scientific Reports, 9*(1), 1–10.
172. Picchi, V., Gobbi, S., Fattizzo, M., Zefelippo, M., & Faoro, F. (2021). Chitosan nanoparticles loaded with N-acetyl cysteine to mitigate ozone and other possible oxidative stresses in durum wheat. *Plants, 10*(4), 691.
173. Nasr, F., Pateiro, M., Rabiei, V., Razavi, F., Formaneck, S., Gohari, G., & Lorenzo, J. M. (2021). Chitosan-phenylalanine nanoparticles (cs-Phe Nps) extend the postharvest life of persimmon (*Diospyros kaki*) fruits under chilling stress. *Coatings, 11*(7), 819.
174. Hmmam, I., Zaid, N. M., Mamdouh, B., Abdallatif, A., Abd-Elfattah, M., & Ali, M. (2021). Storage behavior of "Seddik" mango fruit coated with CMC and guar gum-based silver nanoparticles. *Horticulturae, 7*(3), 44.
175. Gutiérrez-Jara, C., Bilbao-Sainz, C., McHugh, T., Chiou, B. S., Williams, T., & Villalobos-Carvajal, R. (2021). Effect of Cross-linked alginate/oil Nanoemulsion coating on cracking and quality parameters of sweet cherries. *Food, 10*(2), 449.
176. Sasson, E., Pinhasi, R. V. O., Margel, S., & Klipcan, L. (2020). Engineering and use of proteinoid polymers and nanocapsules containing agrochemicals. *Scientific Reports, 10*(1), 1–13.
177. Zambrano-Zaragoza, M. L., Quintanar-Guerrero, D., Del Real, A., González-Reza, R. M., Cornejo-Villegas, M. A., & Gutiérrez-Cortez, E. (2020). Effect of nano-edible coating based on beeswax solid lipid nanoparticles on strawberry's preservation. *Coatings, 10*(3), 253.
178. Zambrano-Zaragoza, M. L., Mercado-Silva, E., Ramirez-Zamorano, P., Cornejo-Villegas, M. A., Gutiérrez-Cortez, E., & Quintanar-Guerrero, D. (2013). Use of solid lipid nanoparticles

(SLNs) in edible coatings to increase guava (*Psidium guajava* L.) shelf-life. *Food Research International, 51*(2), 946–953.
179. García-Betanzos, C. I., Hernández-Sánchez, H., Bernal-Couoh, T. F., Quintanar-Guerrero, D., & de la Luz Zambrano-Zaragoza, M. (2017). Physicochemical, total phenols and pectin methylesterase changes on quality maintenance on guava fruit (*Psidium guajava* L.) coated with candeuba wax solid lipid nanoparticles-xanthan gum. *Food Research International, 101*, 218–227.
180. Hurd, R. G., & Purvis, O. N. (1964). The effect of gibberellic acid on the flowering of spring and winter rye. *Annals of Botany, 28*(1), 137–151.
181. Guan, Y. R., Xue, J. Q., Xue, Y. Q., Yang, R. W., Wang, S. L., & Zhang, X. X. (2019). Effect of exogenous GA3 on flowering quality, endogenous hormones, and hormone-and flowering-associated gene expression in forcing-cultured tree peony (*Paeonia suffruticosa*). *Journal of Integrative Agriculture, 18*(6), 1295–1311.
182. Moneruzzaman, K. M., Hossain, A. B. M. S., Normaniza, O., & Boyce, A. N. (2011). Growth, yield and quality responses to gibberellic acid (GA3) of wax apple *Syzygium samarangense* var. Jambu air madu fruits grown under field conditions. *African Journal of Biotechnology, 10*(56), 11911–11918.
183. Mahajan, P. V., Caleb, O. J., Singh, Z., Watkins, C. B., & Geyer, M. (2014). Postharvest treatments of fresh produce. *Philosophical Transactions of the Royal Society A, 372*, 20130309.
184. Park, H. J. (1999). Development of advanced edible coatings for fruits. *Trends in Food Science and Technology, 10*(8), 254–260.
185. De Moura, M. R., Avena-Bustillos, R. J., McHugh, T. H., Krochta, J. M., & Mattoso, L. H. C. (2008). Properties of novel hydroxypropyl methylcellulose films containing chitosan nanoparticles. *Journal of Food Science, 73*(7), N31–N37.
186. Khan, I., Saeed, K., & Khan, I. (2019). Nanoparticles: Properties, applications and toxicities. *Arabian Journal of Chemistry, 12*(7), 908–931.
187. Xu, L., Liang, H. W., Yang, Y., & Yu, S. H. (2018). Stability and reactivity: Positive and negative aspects for nanoparticle processing. *Chemical Reviews, 118*(7), 3209–3250.
188. Avellan, A., Yun, J., Zhang, Y., Spielman-Sun, E., Unrine, J. M., Thieme, J., Li, J., Lombi, E., Bland, G., & Lowry, G. V. (2019). Nanoparticle size and coating chemistry control foliar uptake pathways, translocation, and leaf-to-rhizosphere transport in wheat. *ACS Nano, 13*(5), 5291–5305.
189. Aziz, H. M. A., Hasaneen, M. N., & Omer, A. M. (2016). Nano chitosan-NPK fertilizer enhances the growth and productivity of wheat plants grown in sandy soil. *Spanish Journal of Agricultural Research, 14*(1), 17.

Plant-Mediated Eco-Friendly Synthesis of Platinum Nanoparticles and Their Applications

Canh Minh Vu, Suresh Ghotekar, Thanh-Dong Pham, Nguyen Minh Viet, Rajeshwari Oza, Muhammad Bilal, and Arpita Roy

Abstract A critical milestone in the discipline of nanotechnology is the progress of trustworthy and eco-benevolent pathways for the production of diverse metallic nanoparticles (NPs). Among all noble metal NPs, platinum nanoparticles (PtNPs) are gaining popularity due to their biocompatibility and catalytic characteristics. Numerous natural active biomolecules detected in plant broths, such as enzymes, coumarins, steroids, alkaloids, polyphenols, terpenoids, flavonoids, proteins, and vitamins, may have a role in PtNPs bioreduction, production, and stability. In addition, several contributions have been made in the recent decade to produce eco-accommodating synthesis processes that prevent toxic intermediates. This chapter presented an insight into recent advancements in biogenically assisted PtNPs production, characterization tools, and implications in medicine and the pharmaceutical industry.

C. M. Vu
Advanced Institue of Science and Technology, The University of Da Nang, Da Nang, Vietnam

S. Ghotekar (✉)
Department of Chemistry, Smt. Devkiba Mohansinhji Chauhan College of Commerce and Science, Silvassa 396 230, University of Mumbai, Dadra and Nagar Haveli (UT), India

T.-D. Pham · N. M. Viet
VNU Key Laboratory of Advanced Materials for Green Growth, Faculty of Chemistry, University of Science, Vietnam National University, Hanoi, Hanoi, Vietnam

R. Oza
Department of Chemistry, S.N. Arts, D.J.M. Commerce and B.N.S. Science College, Sangamner 422 605, Savitribai Phule Pune University, Pune, Maharashtra, India

M. Bilal
School of Life Science and Food Engineering, Huaiyin Institute of Technology, Huaian, China

A. Roy
Department of Biotechnology, School of Engineering & Technology, Sharda University, Greater Noida, India

J.-T. Chen (ed.), *Plant and Nanoparticles*,
https://doi.org/10.1007/978-981-19-2503-0_6

1 Introduction

The enormous proliferation of advanced nanotechnology in daily life, such as agriculture, energy, catalysis, food, optics, medicine, sensors, defense, cosmetics, medicines, and textile, has been experienced worldwide, and nanotechnology is now considered an essential need of advanced technology [1–5]. Nevertheless, synthesizing materials with precise properties through synthetic strategies remains an arduous issue. The usage of living organisms for NP synthesis has been revealed to be a cost-effective and eco-accommodating approach. The elemental composition, precise size, and morphologies of the NPs are altered as required by controlling the metrics. In this modern era, eco-benevolent fabrication of NPs is emphasizing viable chemical strategies to accomplish reliable progressive aims, and interest in eco-benign production of NPs is rapidly expanding globally [6–11]. Considering the state of environmental contamination [12–14], eco-benign strategies for preparing NPs must be developed to prevent the use of chemicals and minimize waste formation [15–18]. Renewable resources and environmentally friendly solvents are used in biogenic synthesis [19–25].

Moreover, nanobiotechnology is a concept used to describe the effective connection between modern nanotechnology and biology [1–11]. Modern nanobiotechnology is a versatile and intriguing discipline of modern nanotechnology that involves a divergent study segment, including biology, material chemistry, advanced engineering, medicine, and physics [21, 22]. Because biological entities may access more components for the production of NPs, advanced nanotechnology has much more effective merits than the other traditional strategy. Advanced bionanomaterials should be produced employing the rich biodiversity of related biological entities [18].

Due to their peculiar crystalline, optical, and catalytic characteristics, platinum nanoparticles (PtNPs) are especially useful for effective biomedical and catalytic uses [26, 27]. During the preliminary stages of PtNPs synthesis, chemical and physical approaches were extensively applied [28, 29]. The physicochemical approaches ensure that the crystallographic structure of the NPs can be modified to produce the desired topology and size [30, 31]. Despite this, nasty chemicals and rigorous synthesis conditions eventually result in health and environmental problems. As a result, "green chemistry" is required to produce environmentally beneficial materials [1, 8, 32]. As a result, plant-mediated fabrication has been included in the manufacturing of PtNPs. For the effective extracellular production of NPs, microbial biomass and plant broths are versatile but efficient sources [33]. Furthermore, bioactive compounds stimulate metal NP bioreduction, reducing or stabilizing the manufacturing process [1, 22]. However, research into the biosynthesis of PtNPs is restricted. No study has looked at the bioreductive mechanism included in the plant-assisted fabrication of PtNPs, as well as the factors that influence it.

This chapter covers a comprehensive description of the methodologies for synthesizing PtNPs and the parameters that affect the biogenic production of PtNPs.

2 Synthesis of PtNPs

The topology, composition, size, and crystal structure of PtNPs and the existence of a capping agent all affect their biomedical and industrial implications [26, 31, 34, 35]. The optimization of these intrinsic characteristics has demanded the development of advanced fabrication techniques. Previously obtained data has guided the alteration and functionalization of PtNPs for biomedical applications, illustrating that the physicochemical characteristics and stability of the multifunctional NPs in a biological culture perform critical involvements in deciding their harmfulness or safety levels. Biocompatibility, carefully defined characteristics, and pollutant processing are major obstacles to the prospective use of PtNPs as antioxidant materials and drug carriers [35, 36]. The diverse fabrication approaches are discussed and highlighted in the underlying sections.

2.1 Chemical Approaches

Among chemical approaches, electrochemical reduction (ECR) [37, 38], wet chemical reduction (WCR) [39], chemical vapor deposition (CVD) [40], and galvanic displacement [41] have recently sparked interest in controlling the physicochemical features of PtNPs in chemical strategies. WCR is frequently employed because it helps control NP characteristics. To manufacture PtNPs in solution from Pt salts, the WCR process primarily requires reducing agents [39], and the size and topology of PtNPs could be controlled by altering the composition of the Pt metal salt and the experimental temperature [42]. A solid study was also given to the development of WCR techniques to create PtNPs with higher catalytic performance. To promote the growth of PtNPs, various shape-directing agents were utilized. Many capping agents, surfactants, and polymers have also been utilized to improve the function of PtNPs [43, 44].

To obtain greater control of the reaction parameters, multiphase fabrication methodologies have also been devised, involving the employment of reducing reagents in the gaseous state [45]. Furthermore, because of its large-scale development, it harms the environment. Glycerol and microwave heating have been advised to prevent adverse environmental impacts and enhance industrial scale-up [44–46].

Surface characteristics and functionalization serve a big role in the creation of PtNPs for bioengineering implications. Therefore, synthetic strategies based on "natural reagents," such as sodium citrate and ascorbic acid, appear particularly appealing because they allow for precise control of some key features for pharmaceutical uses, such as size, topology, catalytic properties, durability, and development efficacy, as well as the possibility of surface functionalization after synthesis.

2.2 Physical Approaches

Physical approaches such as aerosol-induced deposition [47], electron beam-assisted reduction [48], flame fabrication [49], and laser ablation [50] are currently receiving a lot of interest as a way to get around some of the shortcomings of chemical approaches. A considerable intensity laser beam is applied as a prime source in the laser ablation process. A laser beam could be employed in both constant and pulsed modes. To acquire specific PtNPs qualities, this flexible technique relies on vibrations, temperature modulation, and ambient pressure fluctuations [51]. This approach of PtNPs fabrication is not well known, and its application is restricted due to significant dilution and complexities that could cause issues with tuning the scale, yield, topology, and size of the PtNPs [52, 53].

Another physical approach for producing PtNPs is called cathodic corrosion, and it includes transferring a material electrode into a suspension of NPs [54]. Nevertheless, the latter protocol has several limitations, like insufficient profit fabrication and shape, size, and scale tenability.

2.3 Biological Approaches

The biological approaches were suggested as alternatives to physical and chemical protocols because they avoid using harmful toxic solvents in the reaction. In the literature, only a few papers mention the synthesis of PtNPs. Bacteria [55], cyanobacteria [56], fungi [57], plants [58], seaweeds [59], and biomaterials such as aqueous honey solutions and egg yolk [60] have all been shown to have efficient PtNPs production procedures. Several investigations have proposed the conversion of Pt(IV) into PtNPs in sulfate-reducing bacteria [61]. Pt metal salts and protein levels, like WCR, serve an essential involvement in regulating the size and topology of PtNPs in the green fabrication approach. Fungi like *Fusarium oxysporum* and *Neurospora crassa* have also been recognized as a useful "scale-up" protocol for PtNPs fabrication [62, 63]. Moreover, metal NPs were eco-benevolently synthesized using broths from medicinal plants with biomolecule constituents serving as capping agents [58].

The number of research demonstrating various microorganism-mediated PtNPs synthesis is steadily increasing. All of these techniques revealed a plethora of low-toxic, natural, and cost-effective ways to process NPs, with the majority of them avoiding expensive laboratory sets. However, the presence of undesired chemicals such as microbial remnants with undesirable biological efficacy, which need sophisticated and time-consuming refining techniques, may limit their massive-scale use for NPs implications [64, 65]. Furthermore, while bio-assisted techniques are intriguing, they have yet to achieve good control over the characteristics of NPs fully. To summarize, the range of current technologies for PtNPs fabrication makes it challenging to adopt a typical protocol for optimizing stability, biocompatibility, and productivity [64]. As a result, these characteristics must be addressed to obtain biogenically produced PtNPs with the required stability and size for biomedical purposes using green synthesis approaches.

3 Green Synthesis of PtNPs from Plant Extracts

Plant-mediated NPs synthesis has recently garnered considerable attention, and diverse medicinal plant species have been investigated to produce NPs of various sizes and morphologies for diverse purposes [66]. Researchers have emphasized plant-mediated NPs synthesis above other biological approaches since the microorganisms employed for NPs synthesis took a lot of effort to develop microbe's cultures, and maintaining these cultures is still a challenge. On the other hand, plants are readily available and do not require any of the growing material required for microbial growth cultures. As a result, several medicinal plant species have been prominently exploited for NPs production, notably for biomedical uses, due to their eco-benign nature, easy availability, cost-effectiveness, and great biodiversity [1–5].

For a particular plant-assisted fabrication of PtNPs, dried material of diverse parts of plants is ground, filtered to discard unwanted debris, and aqua is mixed to form an aqueous extract. At ambient temperature, a mixture of aqueous metal precursor and plant broth is allowed to react. To synthesize PtNPs of various forms, different concentrations and amounts of metal salt solution and plant extract can be used. The produced PtNPs are separated from the active bioconstituents in the plant broth by sonicating the reduced solution for some time. The solution is centrifuged and rinsed multiple times with distilled water to eliminate any remaining contaminants following sonication. The as-prepared PtNPs are dried, collected, and stored for subsequent examination [72]. A protocol for plant-mediated PtNPs production is shown in Fig. 1.

Bio-inspired fabrication of PtNPs using medicinal plant extracts with their structure and estimated size are presented in Table 1.

Diospyros kaki leaves extract was employed for the facile production of PtNPs. UV-visible spectroscopy analyzed color changes at 95 °C [78]. The fabrication of hexagonal and pentagonal form of PtNPs was seen at 50 °C using *Fumariae herba*

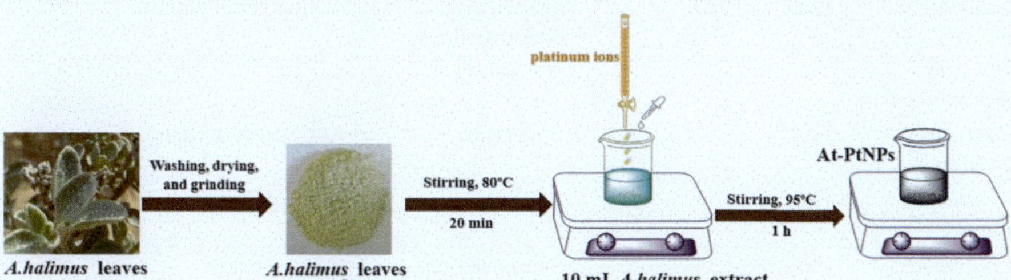

Fig. 1 Protocol of green fabrication of PtNPs. (Reproduced from Ref. 72)

Table 1 Green fabrication of PtNPs utilizing diverse therapeutic plant broths with their estimated size and topology

Name of the plants	Part of plant	Size (nm)	Shape	Applications	Ref.
Ajwa	–	1.1–2.5	Spherical	Anticancer and antibacterial study	67
AjwaBarni	–	1.1–2.5	Spherical	Anticancer and antibacterial study	68
Alchornea laxiflora	Bark	3.68–8.77	Spherical	Catalytic activity	69
Anacardium occidentale	Leaf	–	Irregular rod	Catalytic reduction of 4-nitrophenol	70
Antigonon leptopus	–	5	Spherical	–	71
Atriplex halimus	Leaves	1–3	Spherical	Photocatalytic and antibacterial activity	72
Azadirachta indica	Leaves	5–50	Spheres	–	73
Bacopa monnieri	Leaf	5–20	Spherical	Neuroprotective study	58
Bidens tripartita	–	10	Irregular rod	–	74
Nigella sativa	Seeds	1–6	Spherical	Antimicrobial and anti-cancer study	75
Cacumen platycladi	–	2.4 ± 0.8	Spherical	–	76
Combretum erythrophyllum	Leaf	1.04 ± 0.26	Spherical	Antibacterial activity	77
Diospyros kaki	Leaf	2–20	Spheres and plates	–	78
Fumariae herba	–	30	Hexagonal and pentagonal	Photocatalytic activity	79
Garcinia mangostana	Fruit	20–25	Spherical	Antibacterial activity	80
Gloriosa superba	Tuber	10	Spherical	Anticancer activity	81
Heterotheca inuloides	–	7.1	–	Hydrogenation	82
Jatropha gossypifolia *Jatropha glandulifera*	Leaf	20 100	Spherical, dodecahedron, and cubic	Antibacterial activity	83
Maytenus royleanus	Leaf	5	Spherical	Anticancer activity	84
Mentha piperita	Leaf	54	Spherical	Anticancer activity	85
Nymphaea alba	Flower	35	Semispherical	H_2O_2 sensing	86

(continued)

Table 1 (continued)

Name of the plants	Part of plant	Size (nm)	Shape	Applications	Ref.
Ocimum sanctum	Leaf	23	Irregular	Water electrolysis study	87
Ononidis radix	–	20	Spherical and hexagonal	Anticancer activity	88
Orange	Peel	23	–	Antimicrobial study and p-nitrophenol reduction	89
Orange	Peel	1.6–4.0	Spherical	–	90
Ocimum sanctum	Leaf	20–60	Irregular	–	91
Peganum harmala	Seed	20.3 ± 1.9	Spherical	Antioxidant and anticancer activity	92
Prosopis farcta	Fruit	3.8	–	Cytotoxicity study	93
Prosopis farcta	Fruit	1.6–5	Spherical	–	94
Psidium guajava	Leaf	113	Spherical	Anticancer and antibacterial activity	95
Punica granatum	Peel	20.12	Spherical	Cytotoxicity study	96
Quercus glauca	Leaves	5–15	Spherical	Electrochemical study	97
Salix tetrasperma	Leaf	12	Spherical	Anticancer and photocatalytic activity	98
Sapindus mukorossi	Fruit pericarp	2–19	–	–	99
Taraxacum laevigatum	–	2–7	Spherical	Antibacterial activity	100
Terminalia chebula	Fruit pericarp	4	Cubic and spherical	–	101
Tragia involucrata	Leaf	10	Spherical	Anticancer and antibacterial activity	102
Water hyacinth	Leaves	3.74	Spherical	–	60

extract, and the fabrication took 4 h. The change in color from yellow to brown displayed that platinum ions had been reduced to PtNPs with a median size of 10–30 nm, and the consequent peaks were seen using a UV-visible spectrometer [79].

PtNPs were produced from *Anacardium occidentale* leaf extract at various pH levels. The optimal pH range for NPs synthesis is 6 to 9. TEM images demonstrated the production of irregular rod-shaped PtNPs. Moreover, secondary metabolites in the leaf [70] showed that produced PtNPs had a good capability to reduce aromatic nitro compounds. Another experiment indicated the production of PtNPs from neem leaf, with color shifts (yellow to brown) confirming PtNPs synthesis and further verified by UV-visible spectrum study. TEM examination reveals the fabrication of polydisperse NPs in the 5–50 nm size range [73].

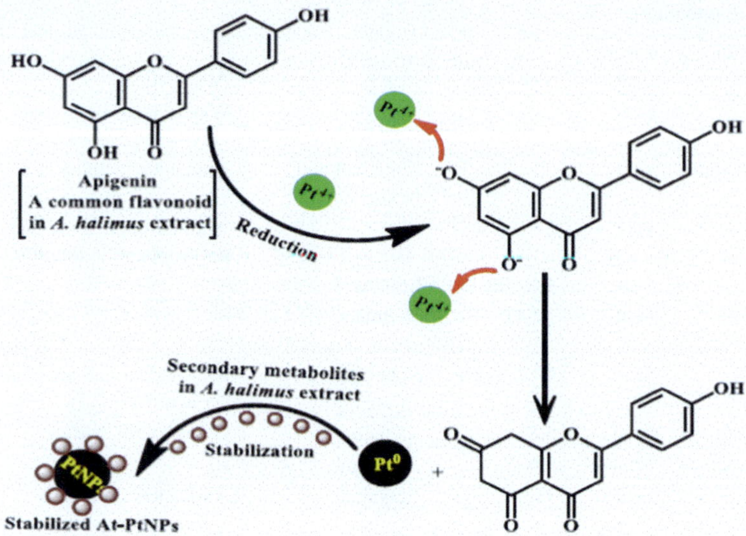

Fig. 2 A plausible mechanism for the green fabrication of PtNPs utilizing plant extracts. (Reproduced from ref. 72)

The utilization of *Ocimum sanctum* leaves for the facile production of PtNPs at higher temperatures was reported in a study, indicating that the higher temperature is optimal for the swift creation of multifunctional PtNPs. As-synthesized PtNPs was 23 nm in diameter and had an uneven shape. Several phenolic chemicals, antioxidants, terpenoids, amino acids, some proteins, flavonoids, and ascorbic acid are found in *Ocimum sanctum* leaf extract [87]. Debility, worm plague, arthritis, dyspepsia, intermittent fever, scrofula, inflammation, hemorrhoids, flatulence, piles, ulcers, leprosy, and snake poison are all treated using *Gloriosa superba* roots extract [81]. Spherical NPs were produced at high temperatures. At 90 °C, *Cacumen platycladi* produced PtNPs of similar size (0.8–2 nm) [76].

Another study used date extracts to synthesize homogeneous spherical-shaped PtNPs with diameters 1.3 to 6 nm. The fabricated PtNPs were employed as an antibacterial agent for *B. subtilis* and *E. coli* [68]. Furthermore, the leaf extract of *Atriplex halimus* was used to produce PtNPs with diameters 1–3 nm in recent work. The eco-benevolent fabrication of PtNPs was affirmed at a high temperature of 95 °C with a rapid color shift. *Atriplex halimus* extract was utilized as a bioreductants and bio-stabilizer in the effective preparation of PtNPs [72]. Furthermore, the plausible mechanism for the formation of PtNPs using *Atriplex halimus* extract is depicted in Fig. 2.

In another study, PtNPs were prepared using *Ajwa, Alchornea laxiflora, Antigonon leptopus, Bidens tripartita, Nigella sativa, Combretum erythrophyllum, Garcinia mangostana, Heterotheca inuloides, Jatropha gossypifolia, Maytenus royleanus, Mentha piperita, Nymphaea alba, Ononidis radix, Orange, Peganum harmala, Prosopis farcta, Psidium guajava, Punica granatum, Quercus glauca,*

Salix tetrasperma, Sapindus mukorossi, Taraxacum laevigatum, Terminalia chebula, Tragia involucrata, and *Water hyacinth* extract [60, 67–102].

4 Applications

4.1 Antibacterial Efficacy of PtNPs

The antibacterial performance of bio-inspired PtNPs was also thoroughly examined, yielding a promising result. PtNPs made from *J. gossypifolia* and *J. glandulifera*, as well as PtNPs, demonstrated tremendous antibacterial efficacy against pathogenic bacteria [83]. Date-derived PtNPs were also discovered to be highly effective bactericidal agents against a panel of harmful microorganisms. At ambient circumstances, a date (*Ajwa* and *Barni*) was employed to synthesize and stabilize PtNPs, and performance was evaluated against *E. coli* and *B. subtilis*, with PtNPs antibacterial response being highly promising [68]. The PtNPs made from *G. mangostana* fruit rind extract also exhibited potential antibacterial activity against *Bacillus sp*. [80]. PtNPs were recently synthesized employing black cumin seed extract, and their bactericidal efficacy was evaluated against a few selected pathogens [75].

4.2 Anticancer Efficacy of PtNPs

PtNPs made by green techniques were used as an anticancer drug, and the results were promising. At ambient temperatures, *Ajwa* and *Barni* dates were employed to make PtNPs. Their anticancer potential was assessed in MCF-7, HCT-116, and HePG-2 cancer cells, and their efficacy was compared to that of a regularly used effective anticancer treatment (Doxorubicin HCl). The effectiveness of PtNPs prepared by the green approach was encouraging [68]. Similarly, the toxicity of PtNPs made from *M. piperita* leaves extract was tested against the HCT-116 cell line, and it was discovered that the PtNPs inhibited the growth of cancer cells at minimum concentrations, with an IC_{50} value of 20 g/mL [85]. In addition, the anticancer potential of *G. superba* tuber extract-mediated PtNPs was assessed. MTT experiments revealed that monodispersed spherical NPs of 10 nm had potent anticancer efficacy for MCF-7 cell lines. PtNPs had antitumor activity of 49.65 ± 1.99 percent [81].

In addition, PtNPs were synthesized using *B. monnieri* leaf extract, and their toxicity was also assessed [58]. PtNPs were recently synthesized utilizing black cumin extract, and their anticancer efficacy was tested against HeLa and MDA-MB-231 breast cell lines. PtNPs were found to be toxic to cancer cell lines in a dose-dependent manner, with IC_{50} values of 19.83 g/mL and 36.86 g/mL, respectively [75].

4.3 Catalytic and Photocatalytic Performance of PtNPs

The bio-inspired PtNPs catalytic performance was also investigated and potent capacity was noticed, i.e., the leaf extracts of *F. herba* were employed for PtNPs production, and their photocatalytic performance was studied in the source of light. In the degradation of MB and CV dyes, 30 nm PtNPs with hexagonal and pentagonal shapes showed outstanding catalytic capabilities [79]. In another study, *O. sanctum* leaf extract was employed to reduce chloroplatinic acid to create PtNPs and then examined for hydrogen-generating capabilities. The hydrogen evolution potential of biosynthesized PtNPs was comparable to that of pure Pt, indicating that they might be used in the water electrolysis process [87]. The catalytic activity of PtNPs made from dried leaves of *A. occidentale* was also investigated. In the reduction of aromatic nitro compounds, PtNPs show good catalytic activity [70]. Furthermore, in another investigation, PtNPs were made employing peel extract as a function of starting pH, and the NPs evinced outstanding catalytic performance [89].

5 Concluding Remarks and Future Direction

PtNPs are well known as important essential players in a variety of scientific and technological fields. They have sparked much attention in the biomedical area due to their strange features, which suggest that they could be used as medications, nano-vehicles, and nano-diagnostic tools for targeted drug administration. The current literature on diverse PtNPs synthesis procedures, including physical and chemical approaches, is discussed in this review and a detailed interpretation of biological protocol. When weighing the benefits and drawbacks of various fabrication methods, physical and chemical routes as traditional PtNPs are deemed hazardous and less suited for biomedical uses. However, because of their eco-friendly and harmless nature, the biological fabrication approach of PtNPs is now regarded as an alternative to chemical and physical protocols, thereby promoting their effective use for biomedical uses.

In addition, we discussed the prospective medicinal properties of PtNPs as a significant anticancer and antibacterial agent and their toxicological effects. Furthermore, looking into the possibilities of manufacturing metallic nanomaterials (such as Pt-Cu, Pt-Au, Pt-Au-ZnO, Pt-Ag, and so on) with multifunctional modalities may aid in the development of new ways to use PtNPs as diagnostic agents.

References

1. Cuong, H. N., Pansambal, S., Ghotekar, S., Oza, R., Hai, N. T. T., Viet, N. M., & Nguyen, V. H. (2022). New frontiers in the plant extract mediated biosynthesis of copper oxide (CuO)

nanoparticles and their potential applications: A review. *Environmental Research, 203*, 111858.

2. Dabhane, H., Ghotekar, S., Tambade, P., Pansambal, S., Murthy, H. A., Oza, R., & Medhane, V. (2021). A review on environmentally benevolent synthesis of CdS nanoparticle and their applications. *Environmental Chemistry and Ecotoxicology, 3*, 209–219.
3. Dabhane, H., Ghotekar, S. K., Tambade, P. J., Pansambal, S., Ananda Murthy, H. C., Oza, R., & Medhane, V. (2021). Cow urine mediated green synthesis of nanomaterial and their applications: A state-of-the-art review. *Journal of Water and Environmental Nanotechnology, 6*(1), 81–91.
4. Ghotekar, S. (2019). A review on plant extract mediated biogenic synthesis of CdO nanoparticles and their recent applications. *Asian Journal of Green Chemistry, 3*(2), 187–200.
5. Nasrollahzadeh, M., Sajjadi, M., Dadashi, J., & Ghafuri, H. (2020). Pd-based nanoparticles: Plant-assisted biosynthesis, characterization, mechanism, stability, catalytic and antimicrobial activities. *Advances in Colloid and Interface Science, 276*, 102103.
6. Ghotekar, S., Dabhane, H., Pansambal, S., Oza, R., Tambade, P., & Medhane, V. (2020). A review on biomimetic synthesis of Ag2O nanoparticles using plant extract, characterization and its recent applications. *Advanced Journal of Chemistry-Section B, 2*(3), 102–111.
7. Ghotekar, S., Pagar, K., Pansambal, S., Murthy, H. A., & Oza, R. (2021). Biosynthesis of silver sulfide nanoparticle and its applications. In *Handbook of Greener Synthesis of Nanomaterials and Compounds* (pp. 191–200). Elsevier.
8. Ghotekar, S., Pansambal, S., Bilal, M., Pingale, S. S., & Oza, R. (2021). Environmentally friendly synthesis of Cr2O3 nanoparticles: Characterization, applications and future perspective— a review. *Case Studies in Chemical and Environmental Engineering, 3*, 100089.
9. Pagar, T., Ghotekar, S., Pagar, K., Pansambal, S., & Oza, R. (2019). A review on bio-synthesized Co3O4 nanoparticles using plant extracts and their diverse applications. *Journal of Chemical Reviews, 1*(4), 260–270.
10. Pagar, T., Ghotekar, S., Pansambal, S., Oza, R., & Marasini, B. P. (2020). Facile plant extract mediated eco-benevolent synthesis and recent applications of CaO-NPs: A state-of-the-art review. *Journal of Chemical Reviews, 2*(3), 201–210.
11. Soni, V., Raizada, P., Singh, P., Cuong, H. N., Rangabhashiyam, S., Saini, A., ... Nguyen, V. H. (2021). Sustainable and green trends in using plant extracts for the synthesis of biogenic metal nanoparticles toward environmental and pharmaceutical advances: A review. *Environmental Research, 202*, 111622.
12. Nikam, A., Pagar, T., Ghotekar, S., Pagar, K., & Pansambal, S. (2019). A review on plant extract mediated green synthesis of zirconia nanoparticles and their miscellaneous applications. *Journal of chemical reviews, 1*(3), 154–163.
13. Khan, I., Saeed, K., & Khan, I. (2019). Nanoparticles: Properties, applications and toxicities. *Arabian Journal of Chemistry, 12*(7), 908–931.
14. Matussin, S., Harunsani, M. H., Tan, A. L., & Khan, M. M. (2020). Plant-extract-mediated SnO_2 nanoparticles: Synthesis and applications. *ACS Sustainable Chemistry & Engineering, 8*(8), 3040–3054.
15. Korde, P., Ghotekar, S., Pagar, T., Pansambal, S., Oza, R., & Mane, D. (2020). Plant extract assisted eco-benevolent synthesis of selenium nanoparticles-a review on plant parts involved, characterization and their recent applications. *Journal of Chemical Reviews, 2*(3), 157–168.
16. Ghotekar, S., Pagar, K., Pansambal, S., Murthy, H. A., & Oza, R. (2020). A review on eco-friendly synthesis of BiVO4 nanoparticle and its eclectic applications. *Advanced Journal of Science and Engineering, 1*, 106–112.
17. Ghotekar, S. (2019). Plant extract mediated biosynthesis of Al2O3 nanoparticles-a review on plant parts involved, characterization and applications. *Nanochemistry Research, 4*(2), 163–169.
18. Gawande, M. B., Goswami, A., Felpin, F. X., Asefa, T., Huang, X., Silva, R., ... Varma, R. S. (2016). Cu and cu-based nanoparticles: Synthesis and applications in catalysis. *Chemical Reviews, 116*(6), 3722–3811.

19. Dabhane, H., Ghotekar, S., Tambade, P., & Medhane, V. (2020). Plant mediated green synthesis of lanthanum oxide (La_2O_3) nanoparticles: A review. *Asian Journal of Nanosciences and Materials, 3*(4), 291–299.
20. Ghotekar, S., Pagar, T., Pansambal, S., & Oza, R. (2020). A review on green synthesis of sulfur nanoparticles via plant extract, characterization and its applications. *Advanced Journal of Chemistry-Section B*, 128–143.
21. Bandeira, M., Giovanela, M., Roesch-Ely, M., Devine, D. M., & da Silva Crespo, J. (2020). Green synthesis of zinc oxide nanoparticles: A review of the synthesis methodology and mechanism of formation. *Sustainable Chemistry and Pharmacy, 15*, 100223.
22. Basnet, P., Chanu, T. I., Samanta, D., & Chatterjee, S. (2018). A review on bio-synthesized zinc oxide nanoparticles using plant extracts as reductants and stabilizing agents. *Journal of Photochemistry and Photobiology B: Biology, 183*, 201–221.
23. Naikoo, G. A., Mustaqeem, M., Hassan, I. U., Awan, T., Arshad, F., Salim, H., & Qurashi, A. (2021). Bioinspired and green synthesis of nanoparticles from plant extracts with antiviral and antimicrobial properties: A critical review. *Journal of Saudi Chemical Society, 25*(9), 101304.
24. Shreyash, N., Bajpai, S., Khan, M. A., Vijay, Y., Tiwary, S. K., & Sonker, M. (2021). Green synthesis of nanoparticles and their biomedical applications: A review. *ACS Applied Nano Materials, 4*(11), 11428–11457.
25. Abinaya, S., Kavitha, H. P., Prakash, M., & Muthukrishnaraj, A. (2021). Green synthesis of magnesium oxide nanoparticles and its applications: A review. *Sustainable Chemistry and Pharmacy, 19*, 100368.
26. Chen, A., & Holt-Hindle, P. (2010). Platinum-based nanostructured materials: Synthesis, properties, and applications. *Chemical Reviews, 110*(6), 3767–3804.
27. Pedone, D., Moglianetti, M., De Luca, E., Bardi, G., & Pompa, P. P. (2017). Platinum nanoparticles in nanobiomedicine. *Chemical Society Reviews, 46*(16), 4951–4975.
28. Murphin Kumar, P. S., Thiripuranthagan, S., Imai, T., Kumar, G., Pugazhendhi, A., Vijayan, S. R., Esparza, R., Abi, H., & Krishnan, S. K. (2017). Pt nanoparticles supported on mesoporous CeO_2 nanostructures obtained through green approach for efficient catalytic performance toward ethanol electro-oxidation. *ACS Sustainable Chemistry & Engineering, 5*(12), 11290–11299.
29. Nakano, S., Akedo, J., & Ogiso, H. (2007). Platinum nanoparticle catalyst support technique by ion implantation and surface etching (IISE) method. *Surface and Coatings Technology, 201*(19–20), 8539–8541.
30. Zhang, X., Xia, Z., Huang, Y., Jia, Y., Sun, X., Li, Y., . . . Wen, W. (2016). Shape-controlled synthesis of Pt nanopeanuts. *Scientific Reports, 6*(1), 1–7.
31. Leong, G. J., Schulze, M. C., Strand, M. B., Maloney, D., Frisco, S. L., Dinh, H. N., Pivovar, B., & Richards, R. M. (2014). Shape-directed platinum nanoparticle synthesis: Nanoscale design of novel catalysts. *Applied Organometallic Chemistry, 28*(1), 1–17.
32. Duan, H., Wang, D., & Li, Y. (2015). Green chemistry for nanoparticle synthesis. *Chemical Society Reviews, 44*(16), 5778–5792.
33. Zheng, B., Kong, T., Jing, X., Odoom-Wubah, T., Li, X., Sun, D., . . . Li, Q. (2013). Plant-mediated synthesis of platinum nanoparticles and its bioreductive mechanism. *Journal of Colloid and Interface Science, 396*, 138–145.
34. Papst, S., Brimble, M. A., Evans, C. W., Verdon, D. J., Feisst, V., Dunbar, P. R., Tilley, R. D., & Williams, D. E. (2015). Cell-targeted platinum nanoparticles and nanoparticle clusters. *Organic & Biomolecular Chemistry, 13*(23), 6567–6572.
35. Jameel, M. S., Aziz, A. A., & Dheyab, M. A. (2020). Green synthesis: Proposed mechanism and factors influencing the synthesis of platinum nanoparticles. *Green Processing and Synthesis, 9*(1), 386–398.
36. Crist, R. M., Grossman, J. H., Patri, A. K., Stern, S. T., Dobrovolskaia, M. A., Adiseshaiah, P. P., Clogston, J. D., & McNeil, S. E. (2013). Common pitfalls in nanotechnology: Lessons

learned from NCI's nanotechnology characterization laboratory. *Integrative Biology, 5*(1), 66–73.
37. Mahima, S., Kannan, R., Komath, I., Aslam, M., & Pillai, V. K. (2008). Synthesis of platinum Y-junction nanostructures using hierarchically designed alumina templates and their enhanced electrocatalytic activity for fuel-cell applications. *Chemistry of Materials, 20*(3), 601–603.
38. Raoof, J. B., Ojani, R., & Hosseini, S. R. (2012). Electrocatalytic oxidation of methanol onto platinum particles decorated nanostructured poly (1, 5-diaminonaphthalene) film. *Journal of Solid State Electrochemistry, 16*(8), 2699–2708.
39. Bönnemann, H., & Richards, R. M. (2001). Nanoscopic metal particles– synthetic methods and potential applications. *European Journal of Inorganic Chemistry, 2001*(10), 2455–2480.
40. Shafiei, M., Riahi, A. R., Sen, F. G., & Alpas, A. T. (2010). Improvement of platinum adhesion to carbon surfaces using PVD coatings. *Surface and Coatings Technology, 205*(2), 306–311.
41. Mahmoud, M. A., & El-Sayed, M. A. (2012). Metallic double shell hollow nanocages: The challenges of their synthetic techniques. *Langmuir, 28*(9), 4051–4059.
42. Miyabayashi, K., Nakamura, S., & Miyake, M. (2011). Synthesis of small platinum cube with less than 3 nm by the control of growth kinetics. *Crystal Growth & Design, 11*(10), 4292–4295.
43. Kang, Y., Ye, X., & Murray, C. B. (2010). Size-and shape-selective synthesis of metal nanocrystals and nanowires using CO as a reducing agent. *Angewandte Chemie, 122*(35), 6292–6295.
44. Jan, H., Gul, R., Andleeb, A., Ullah, S., Shah, M., Khanum, M., ... Abbasi, B. H. (2021). A detailed review on biosynthesis of platinum nanoparticles (PtNPs), their potential antimicrobial and biomedical applications. *Journal of Saudi Chemical Society, 25*(8), 101297.
45. Zhou, W., Wu, J., & Yang, H. (2013). Highly uniform platinum icosahedra made by hot injection-assisted GRAILS method. *Nano Letters, 13*(6), 2870–2874.
46. Kou, J., Bennett-Stamper, C., & Varma, R. S. (2013). Green synthesis of noble nanometals (au, Pt, pd) using glycerol under microwave irradiation conditions. *ACS Sustainable Chemistry & Engineering, 1*(7), 810–816.
47. Paschos, O., Choi, P., Efstathiadis, H., & Haldar, P. (2008). Synthesis of platinum nanoparticles by aerosol assisted deposition method. *Thin Solid Films, 516*(12), 3796–3801.
48. Ke, X., Bittencourt, C., Bals, S., & Van Tendeloo, G. (2013). Low-dose patterning of platinum nanoclusters on carbon nanotubes by focused-electron-beam-induced deposition as studied by TEM. *Beilstein Journal of Nanotechnology, 4*(1), 77–86.
49. Choi, I. D., Lee, H., Shim, Y. B., & Lee, D. (2010). A one-step continuous synthesis of carbon-supported Pt catalysts using a flame for the preparation of the fuel electrode. *Langmuir, 26*(13), 11212–11216.
50. Rakshit, R. K., Bose, S. K., Sharma, R., Budhani, R. C., Vijaykumar, T., Neena, S. J., & Kulkarni, G. U. (2008). Correlations between morphology, crystal structure, and magnetization of epitaxial cobalt-platinum films grown with pulsed laser ablation. *Journal of Applied Physics, 103*(2), 023915.
51. Dhand, C., Dwivedi, N., Loh, X. J., Ying, A. N. J., Verma, N. K., Beuerman, R. W., Lakshminarayanan, R., & Ramakrishna, S. (2015). Methods and strategies for the synthesis of diverse nanoparticles and their applications: A comprehensive overview. *RSC Advances, 5*(127), 105003–105037.
52. Scaramuzza, S., Zerbetto, M., & Amendola, V. (2016). Synthesis of gold nanoparticles in liquid environment by laser ablation with geometrically confined configurations: Insights to improve size control and productivity. *The Journal of Physical Chemistry C, 120*(17), 9453–9463.
53. Correard, F., Maximova, K., Estève, M. A., Villard, C., Roy, M., Al-Kattan, A., Gingras, M., Kabashin, A. V., & Braguer, D. (2014). Gold nanoparticles prepared by laser ablation in aqueous biocompatible solutions: Assessment of safety and biological identity for nanomedicine applications. *International Journal of Nanomedicine, 9*, 5415.

54. Yanson, A. I., Rodriguez, P., Garcia-Araez, N., Mom, R. V., Tichelaar, F. D., & Koper, M. T. (2011). Cathodic corrosion: A quick, clean, and versatile method for the synthesis of metallic nanoparticles. *Angewandte Chemie International Edition, 50*(28), 6346–6350.
55. Rashamuse, K. J., Mutambanengwe, C. C. Z., & Whiteley, C. G. (2008). Enzymatic recovery of platinum (IV) from industrial wastewater using a biosulphidogenic hydrogenase. *African Journal of Biotechnology, 7*(8).
56. Brayner, R., Barberousse, H., Hemadi, M., Djedjat, C., Yéprémian, C., Coradin, T., Livage, J., Fiévet, F., & Couté, A. (2007). Cyanobacteria as bioreactors for the synthesis of au, ag, pd, and Pt nanoparticles via an enzyme-mediated route. *Journal of Nanoscience and Nanotechnology, 7*(8), 2696–2708.
57. Govender, Y., Riddin, T., Gericke, M., & Whiteley, C. G. (2009). Bioreduction of platinum salts into nanoparticles: A mechanistic perspective. *Biotechnology Letters, 31*(1), 95–100.
58. Nellore, J., Pauline, C., & Amarnath, K. (2013). Bacopa monnieri phytochemicals mediated synthesis of platinum nanoparticles and its neurorescue effect on 1-methyl 4-phenyl 1, 2, 3, 6 tetrahydropyridine-induced experimental parkinsonism in zebrafish. *Journal of Neurodegenerative Diseases, 2013*, 972391.
59. Shiny, P. J., Mukherjee, A., & Chandrasekaran, N. (2016). DNA damage and mitochondria-mediated apoptosis of A549 lung carcinoma cells induced by biosynthesised silver and platinum nanoparticles. *RSC Advances, 6*(33), 27775–27787.
60. Leo, A. J., & Oluwafemi, O. S. (2017). Plant-mediated synthesis of platinum nanoparticles using water hyacinth as an efficient biomatrix source–an eco-friendly development. *Materials Letters, 196*, 141–144.
61. Riddin, T. L., Govender, Y., Gericke, M., & Whiteley, C. G. (2009). Two different hydrogenase enzymes from sulphate-reducing bacteria are responsible for the bioreductive mechanism of platinum into nanoparticles. *Enzyme and Microbial Technology, 45*(4), 267–273.
62. Castro-Longoria, E., Moreno-Velasquez, S. D., Vilchis-Nestor, A. R., Arenas-Berumen, E., & Avalos-Borja, M. (2012). Production of platinum nanoparticles and nanoaggregates using neurospora crassa. *Journal of Microbiology and Biotechnology, 22*(7), 1000–1004.
63. Syed, A., & Ahmad, A. (2012). Extracellular biosynthesis of platinum nanoparticles using the fungus fusarium oxysporum. *Colloids and Surfaces B: Biointerfaces, 97*, 27–31.
64. Stephen, A. J., Rees, N. V., Mikheenko, I., & Macaskie, L. E. (2019). Platinum and palladium bio-synthesized nanoparticles as sustainable fuel cell catalysts. *Frontiers in Energy Research, 7*, 66.
65. Kavitha, K. S., Baker, S., Rakshith, D., Kavitha, H. U., Yashwantha Rao, H. C., Harini, B. P., & Satish, S. (2013). Plants as green source towards synthesis of nanoparticles. *International Research Journal of Biological Sciences, 2*(6), 66–76.
66. Kuppusamy, P., Yusoff, M. M., Maniam, G. P., & Govindan, N. (2016). Biosynthesis of metallic nanoparticles using plant derivatives and their new avenues in pharmacological applications–an updated report. *Saudi Pharmaceutical Journal, 24*(4), 473–484.
67. Ismail, E. H., & Al-Radadi, N. S. (2017). An eco-friendly synthesis of platinum nanoparticles and their applications on the cancer cell treatments. *Journal of Computational and Theoretical Nanoscience, 14*(12), 6044–6052.
68. Al-Radadi, N. S. (2019). Green synthesis of platinum nanoparticles using Saudi's dates extract and their usage on the cancer cell treatment. *Arabian Journal of Chemistry, 12*(3), 330–349.
69. Olajire, A. A., Adeyeye, G. O., & Yusuf, R. A. (2017). Alchornea laxiflora bark extract assisted green synthesis of platinum nanoparticles for oxidative desulphurization of model oil. *Journal of Cluster Science, 28*(3), 1565–1578.
70. Sheny, D. S., Philip, D., & Mathew, J. (2013). Synthesis of platinum nanoparticles using dried Anacardium occidentale leaf and its catalytic and thermal applications. *Spectrochimica Acta Part A: Molecular and Biomolecular Spectroscopy, 114*, 267–271.
71. Ganaie, S. U., Abbasi, T., & Abbasi, S. A. (2018). Biomimetic synthesis of platinum nanoparticles utilizing a terrestrial weed Antigonon leptopus. *Particulate Science and Technology, 36*(6), 681–688.

72. Eltaweil, A. S., Fawzy, M., Hosny, M., Abd El-Monaem, E. M., Tamer, T. M., & Omer, A. M. (2022). Green synthesis of platinum nanoparticles using Atriplex halimus leaves for potential antimicrobial, antioxidant, and catalytic applications. *Arabian Journal of Chemistry, 15*(1), 103517.
73. Thirumurugan, A., Aswitha, P., Kiruthika, C., Nagarajan, S., & Christy, A. N. (2016). Green synthesis of platinum nanoparticles using Azadirachta indica–an eco-friendly approach. *Materials Letters, 170*, 175–178.
74. Dobrucka, R. (2016). Synthesis and structural characteristic of platinum nanoparticles using herbal bidens tripartitus extract. *Journal of Inorganic and Organometallic Polymers and Materials, 26*(1), 219–225.
75. Aygun, A., Gülbagca, F., Ozer, L. Y., Ustaoglu, B., Altunoglu, Y. C., Baloglu, M. C., Atalar, M. N., Alma, M. H., & Sen, F. (2020). Biogenic platinum nanoparticles using black cumin seed and their potential usage as antimicrobial and anticancer agent. *Journal of Pharmaceutical and Biomedical Analysis, 179*, 112961.
76. Zheng, B., Kong, T., Jing, X., Odoom-Wubah, T., Li, X., Sun, D., Lu, F., Zheng, Y., Huang, J., & Li, Q. (2013). Plant-mediated synthesis of platinum nanoparticles and its bioreductive mechanism. *Journal of Colloid and Interface Science, 396*, 138–145.
77. Fanoro, O. T., Parani, S., Maluleke, R., Lebepe, T. C., Varghese, R. J., Mgedle, N., Mavumengwana, V., & Oluwafemi, O. S. (2021). Biosynthesis of smaller-sized platinum nanoparticles using the leaf extract of Combretum erythrophyllum and its antibacterial activities. *Antibiotics, 10*(11), 1275.
78. Song, J. Y., Kwon, E. Y., & Kim, B. S. (2010). Biological synthesis of platinum nanoparticles using Diopyros kaki leaf extract. *Bioprocess and Biosystems Engineering, 33*(1), 159–164.
79. Dobrucka, R. (2019). Biofabrication of platinum nanoparticles using Fumariae herba extract and their catalytic properties. *Saudi Journal of Biological Sciences, 26*(1), 31–37.
80. Nishanthi, R., Malathi, S., & Palani, P. (2019). Green synthesis and characterization of bioinspired silver, gold and platinum nanoparticles and evaluation of their synergistic antibacterial activity after combining with different classes of antibiotics. *Materials Science and Engineering: C, 96*, 693–707.
81. Rokade, S. S., Joshi, K. A., Mahajan, K., Patil, S., Tomar, G., Dubal, D. S., Parihar, V. S., Kitture, R., Bellare, J. R., & Ghosh, S. (2018). Gloriosa superba mediated synthesis of platinum and palladium nanoparticles for induction of apoptosis in breast cancer. *Bioinorganic Chemistry and Applications, 2018*, 4924186.
82. Gama-Lara, S. A., Natividad, R., Vilchis-Nestor, A. R., López-Castañares, R., García-Orozco, I., Gonzalez-Pedroza, M. G., & Morales-Luckie, R. A. (2019). Ultra-small platinum nanoparticles with high catalytic selectivity synthesized by an eco-friendly method supported on natural hydroxyapatite. *Catalysis Letters, 149*(12), 3447–3453.
83. Jeyapaul, U., Kala, M. J., Bosco, A. J., Piruthiviraj, P., & Easuraja, M. (2018). An eco-friendly approach for synthesis of platinum nanoparticles using leaf extracts of Jatropa gossypifolia and Jatropa glandulifera and its antibacterial activity. *Oriental Journal of Chemistry, 34*(2), 783.
84. Ullah, S., Ahmad, A., Wang, A., Raza, M., Jan, A. U., Tahir, K., Rahman, A. U., & Qipeng, Y. (2017). Bio-fabrication of catalytic platinum nanoparticles and their in vitro efficacy against lungs cancer cells line (A549). *Journal of Photochemistry and Photobiology B: Biology, 173*, 368–375.
85. Yang, C., Wang, M., Zhou, J., & Chi, Q. (2017). Bio-synthesis of peppermint leaf extract polyphenols capped nano-platinum and their in-vitro cytotoxicity towards colon cancer cell lines (HCT 116). *Materials Science and Engineering: C, 77*, 1012–1016.
86. Mohamadi, A. R., Nami, N., & Norouzi, B. (2020). Bio-directed synthesis of platinum nanoparticles by Nymphaea alba extract: Fabrication of a novel non-enzymatic hydrogen peroxide sensor. *Journal of Materials Science: Materials in Electronics, 31*(21), 18721–18731.
87. Soundarrajan, C., Sankari, A., Dhandapani, P., Maruthamuthu, S., Ravichandran, S., Sozhan, G., & Palaniswamy, N. (2012). Rapid biological synthesis of platinum nanoparticles using

Ocimum sanctum for water electrolysis applications. *Bioprocess and Biosystems Engineering, 35*(5), 827–833.
88. Dobrucka, R., Romaniuk-Drapała, A., & Kaczmarek, M. (2019). Evaluation of biological synthesized platinum nanoparticles using Ononidis radix extract on the cell lung carcinoma A549. *Biomedical Microdevices, 21*(3), 1–10.
89. Castro, L., Blázquez, M. L., González, F., Muñoz, J. Á., & Ballester, A. (2015). Biosynthesis of silver and platinum nanoparticles using orange peel extract: Characterisation and applications. *IET Nanobiotechnology, 9*(5), 252–258.
90. Karim, N. A., Rubinsin, N. J., Burukan, M. A. A., & Kamarudin, S. K. (2019). Sustainable route of synthesis platinum nanoparticles using orange peel extract. *International Journal of Green Energy, 16*(15), 1518–1526.
91. Prabhu, N., & Gajendran, T. (2017). Green synthesis of noble metal of platinum nanoparticles from Ocimum sanctum (Tulsi) plant-extracts. *IOSR Journal of Biotechnology and Biochemistry, 3*, 107–112.
92. Fahmy, S. A., Fawzy, I. M., Saleh, B. M., Issa, M. Y., Bakowsky, U., & Azzazy, H. M. E. S. (2021). Green synthesis of platinum and palladium nanoparticles using Peganum harmala L. Seed Alkaloids: Biological and Computational Studies. *Nanomaterials, 11*(4), 965.
93. Jameel, M. S., Aziz, A. A., Dheyab, M. A., Mehrdel, B., Khaniabadi, P. M., & Khaniabadi, B. M. (2021). Green sonochemical synthesis platinum nanoparticles as a novel contrast agent for computed tomography. *Materials Today Communications, 27*, 102480.
94. Jameel, M. S., Aziz, A. A., & Dheyab, M. A. (2021). Impacts of various solvents in ultrasonic irradiation and green synthesis of platinum nanoparticle. *Inorganic Chemistry Communications, 128*, 108565.
95. Manzoor, S., Bashir, D. J., Imtiyaz, K., Rizvi, M. M. A., Ahamad, I., Fatma, T., Agarwal, N. B., Arora, I., & Samim, M. (2021). Biofabricated platinum nanoparticles: Therapeutic evaluation as a potential nanodrug against breast cancer cells and drug-resistant bacteria. *RSC Advances, 11*(40), 24900–24916.
96. Şahin, B., Aygün, A., Gündüz, H., Şahin, K., Demir, E., Akocak, S., & Şen, F. (2018). Cytotoxic effects of platinum nanoparticles obtained from pomegranate extract by the green synthesis method on the MCF-7 cell line. *Colloids and Surfaces B: Biointerfaces, 163*, 119–124.
97. Karthik, R., Sasikumar, R., Chen, S. M., Govindasamy, M., Kumar, J. V., & Muthuraj, V. (2016). Green synthesis of platinum nanoparticles using quercus glauca extract and its electrochemical oxidation of hydrazine in water samples. *International Journal of Electrochemical Science, 11*, 8245–8255.
98. Ramachandiran, D., Elangovan, M., & Rajesh, K. (2021). Structural, optical, biological and photocatalytic activities of platinum nanoparticles using Salix tetrasperma leaf extract via hydrothermal and ultrasonic methods. *Optik, 244*, 167494.
99. Kumar, M. N., Govindh, B. O. D. D. E. T. I., & Annapurna, N. O. W. D. U. R. I. (2017). Green synthesis and characterization of platinum nanoparticles using sapindus mukorossi Gaertn. Fruit Pericarp. *Asian Journal of Chemistry, 29*, 2541–2544.
100. Tahir, K., Nazir, S., Ahmad, A., Li, B., Khan, A. U., Khan, Z. U. H., . . . Rahman, A. U. (2017). Facile and green synthesis of phytochemicals capped platinum nanoparticles and in vitro their superior antibacterial activity. *Journal of Photochemistry and Photobiology B: Biology, 166*, 246–251.
101. Kumar, K. M., Mandal, B. K., & Tammina, S. K. (2013). Green synthesis of nano platinum using naturally occurring polyphenols. *RSC Advances, 3*(12), 4033–4039.
102. Selvi, A. M., Palanisamy, S., Jeyanthi, S., Vinosha, M., Mohandoss, S., Tabarsa, M., .Sang Guan You, Ethiraj Kannapiran Prabhu, N. M. (2020). Synthesis of Tragia involucrata mediated platinum nanoparticles for comprehensive therapeutic applications: Antioxidant, antibacterial and mitochondria-associated apoptosis in HeLa cells. Process Biochemistry, 98, 21–33.

Foliar Application of Metallic Nanoparticles on Crops Under Field Conditions

Martin Šebesta, Ľuba Ďurišová, Dávid Ernst, Samuel Kšiňan, Ramakanth Illa, B. Ratna Sunil, Avinash P. Ingle, Yu Qian, Martin Urík, and Marek Kolenčík

M. Šebesta (✉) · M. Urík
Institute of Laboratory Research on Geomaterials, Faculty of Natural Sciences, Comenius University in Bratislava, Bratislava, Slovakia
e-mail: martin.sebesta@uniba.sk; martin.urik@uniba.sk

Ľ. Ďurišová · S. Kšiňan
Institute of Plant and Environmental Sciences, Faculty of Agrobiology and Food Resources, Slovak University of Agriculture in Nitra, Nitra, Slovakia
e-mail: luba.durisova@uniag.sk; samuel.ksinan@uniag.sk

D. Ernst
Institute of Agronomic Sciences, Faculty of Agrobiology and Food Resources, Slovak University of Agriculture in Nitra, Nitra, Slovakia
e-mail: david.ernst@uniag.sk

R. Illa
Department of Chemistry, Rajiv Gandhi University of Knowledge Technologies, AP IIIT, Nuzvid, India
e-mail: iramakanth@rguktn.ac.in

B. R. Sunil
Department of Mechanical Engineering, Bapatla Engineering College, Bapatla, Andhra Pradesh, India
e-mail: bratnasunil@becbapatla.ac.in

A. P. Ingle
Biotechnology Centre, Department of Agricultural Botany, Dr. Panjabrao Deshmukh Krishi Vidyapeeth, Akola, Maharashtra, India

Y. Qian
School of Ecology and Environmental Science, Yunnan University, Kunming, China
e-mail: qianyu@ynu.edu.cn

M. Kolenčík
Institute of Agronomic Sciences, Faculty of Agrobiology and Food Resources, Slovak University of Agriculture in Nitra, Nitra, Slovakia

Nanotechnology Centre, VŠB Technical University of Ostrava, Ostrava, Poruba, Czech Republic
e-mail: marek.kolencik@uniag.sk

J.-T. Chen (ed.), *Plant and Nanoparticles*,
https://doi.org/10.1007/978-981-19-2503-0_7

Abstract Various industries, including agriculture, are increasingly interested in inorganic nanomaterials. Their smaller size (<100 nm) and tunability give rise to different chemical, physical, and biological properties compared with conventionally applied products. The study of the nanoworld with its unexplored processes and scientific challenges led to a new generation of more beneficial and efficient tools such as nanofertilizer and growth enhancers with either direct or gradual and well-regulated effects. The novelty of nanoparticles consists mainly in the reduction of inputs with the same function, for example, fertilizers in crop production, while also improving food quality, safety within the food chain, and the overall environmental impact. The aim of this chapter includes (1) effects of foliar application of engineered nanoparticles (mainly metallic) as new generation agrochemicals on plant production with (2) description of interactions in nanoparticle-plant systems, specifically their quantitative and qualitative nutritional parameters, (3) analysis of transfer and interaction of nanoparticles in leaves, and (4) palynological analysis in the context of impact on pollen quality to quantify distribution, transport, bioavailability, and potential toxicity of nanoparticles or their residues in field experimental setups.

Keywords Foliar application · Metals and metal oxides nanoparticles · Field studies · Plant production · Quantitative and nutritional parameters · Physiological response · Palynological analysis · Environmental safety

1 Introduction

Engineered nanomaterials have been extensively studied for the last three decades. Nanomaterials are materials that have tunable and size-dependent properties and are defined as materials with one, two, or three dimensions below 100 nm. Based on their shape, nanomaterials are divided into materials with one dimension in nanometer-sized scale, for example, nanosheets; two dimensions in nanoscale range, for example, nanorods or nanotubes; and nanoparticles (NPs) with all three dimensions below 100 nm [1]. Recently, engineered nanomaterials have been used in many areas associated with human life, such as electronics [2]; energy solutions [3]; additives in metals [4]; concrete [5], ceramic, and plastic composites [6–9]; inks [10]; and paints and coatings [11, 12]. Their catalytic and photocatalytic properties [13–15] are utilized in various chemical processes. In medicine also, delivery systems and biosensors have been developed using different types of engineered nanomaterials [16]. Overall, these nanomaterials can be effectively used in medical, cosmetic, and food products [16–18]. For example, in the food industry, they are used in food packaging [19]. Cleaning the tap water or environmental waters, remediation of contaminated sites, and environmental sensing are another avenues where nanoparticles are used, or their use is heavily researched [20, 21].

The use of engineered nanomaterials, especially engineered nanoparticles (ENPs), also captured the interest of many scientists in agricultural fields [22–25]. The research captured interest because the ENPs showed both effects found in their ionic counterparts and nano-specific effects in some cases. Appropriate application of ENPs also led to increases in the efficiency of crops to absorb fertilizer micronutrients [22, 24]. At the same time, several different ENPs have shown antibacterial and antifungal properties that were also utilized in other studies. In addition, ENPs served as catalyzers that break up pesticides after their initial use is not needed or as biosensors providing easily accessed information in situ [26]. Yet the buildup of ENPs in soils or their other effects are not fully investigated, and hence, we need to proceed cautiously with their application [27, 28]. Foliar application is often used under the field conditions to supplement crops with the least amount of ENPs with the most significant effect [29–33] and was also observed to be effective to a greater extent than the application to the soil in certain situations [34].

Agriculture faces many challenges, such as climate change, including more frequent extreme weather, urbanization, and declining oil and gas reserves. Problems also lie in the high environmental burden of agriculture, especially in the ever-increasing application of industrial fertilizers and pesticides, the reduction of biodiversity, and the eutrophication of ground and surface. These problems are also compounded by an exponentially growing population, with 9.6 billion people expected to live on Earth by 2050 [22]. Great emphasis is now placed on sustainable agriculture and its practices to eliminate these problems. Sustainable agriculture is essential for protecting and enhancing natural resources such as soil, water, biodiversity, vegetation, renewable energy, and other ecosystem services, guaranteeing a growing population's security. In the interests of sustainable agriculture, nanotechnologies are considered an innovative and promising technology that will provide sufficient food resources for an increasing population and reduce the burden on the environment [35, 36]. In addition, nanotechnology can improve our understanding of plant functioning, leading to higher crop yields of the desired quality, better nutrient utilization, and overall less use of pesticides [37, 38]. The significant imbalance between the number of nutrients applied and plants absorbed in crop production is reflected both in the economy of cultivation and in the adverse effects on the environment. It turns out that with standard elements, such as N, P, and K, for nanofertilizers, 40% of the delivered amount was sufficient in the case of nanofertilizers when compared to the commonly used fertilizers [39]. However, extensive studies are required on the evaluation of the effects of nanofertilizers on plants especially due to their small size [40]. The main goal in the development of these technologies must be the protection of human health and the environment [35, 41]. Therefore, the use of nanotechnology in agriculture is closely tied to the principles of precision agriculture, where the concept of crop health is essential, and nanotechnologies enable more effective measurement of temporal, spatial, and individual data.

Effects of ENPs on crop health and production range from improved physiological growth parameters such as seed germination, fresh and dry weight, shoot and root length of plants, improved photosynthetic parameters, the activity of antioxidants, and improved uptake of micro- and macronutrients [42–46]. In crop production improvements, parameters such as the production of oils, seeds, and protein, as well as other nutritional parameters and agriculturally important phytochemicals such as flavonols were observed [31, 32, 44, 47]. Most of these effects were observed in both field and laboratory studies. Also, the concentration-dependent effects were comparable, and higher concentrations negatively affected plants [32, 44, 48]. The improvement of the grain yields and nutritional parameters after the foliar application were observed even for low concentrations of ENPs, for example, 2.6 mg/L of ZnO or TiO_2 NPs in field experiments. These results show that we can reduce the amount of already applied micronutrients by using nano-enhanced formulae or using low concentrations of growth-enhancing ENPs that do not have commonly used ionic counterparts [31, 32]. Moreover, the foliar application of ENPs was shown to decrease the content of potentially toxic heavy metals, such as Cd, in grains and edible parts of crops grown on contaminated sites, and ENPs can help to alleviate problems connected to the agricultural management of these sites [29, 49, 50].

This chapter aims to provide up-to-date information on the application of metallic nanoparticles on agricultural plants under field conditions and how the properties of ENPs or the concentrations and methods of application may affect the influence they have on the plants under these conditions. The experiments with ENPs under the field conditions are relatively sparse. Therefore, many unknowns are connected to potential best practices that can be utilized when ENPs are used as either nanofertilizer, growth promoters, or protection against biotic and abiotic stresses. Yet a lot can be surmised from more prevalent studies of either short-term or life cycle pot experiments in laboratory or greenhouse conditions and from the best practices when the ionic counterparts of ENPs were used similarly to the studied ENPs.

2 Different Roles of Engineered Nanomaterials in Agricultural Fields

The development of effective nanoscale tools in agricultural applications is based on logical premises often solved in medical areas such as sensors and drug delivery. The premise is that smart technology delivers an active substance to a targeted site, and then the substance is gradually released to provide a promising effect. In the context of agriculture and agroecology, this will ensure a better ecological balance in the long run and offer more sustainable solutions despite gradual climate change and environmental pollution [51]. Other agronomic benefits include reducing inputs and thus lowering financial costs, improving soil health, and maintaining a more natural biogeochemical cycle, which is then reflected in higher food quality and plant production. In the broader context of conventional fertilization, engineered nanomaterials are designed to increase the efficiency of uptake and loss to the environment since up to 50–70% of substances of conventional fertilizers are leached, mineralized, or transformed by microorganisms or other organisms present in the agricultural fields [52]. This inefficiency has a wide area of environmental impacts, including human health, local ecosystem damage to soil microbial flora, a

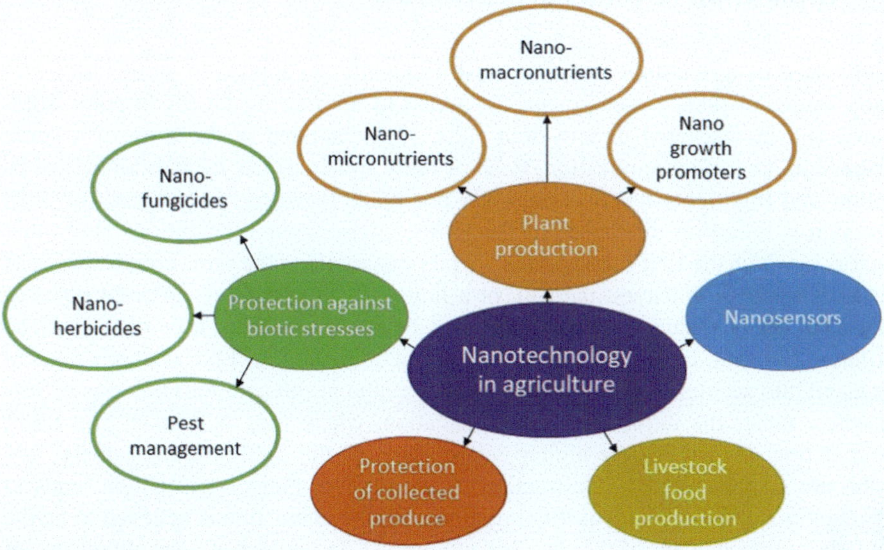

Fig. 1 Application of nanomaterials in different fields of agriculture

proliferation of parasites, and leaching that results in eutrophication processes affecting ecosystems downstream, i.e., freshwater and marine environments [53].

Based on these aspects, new types of nanomaterials have been developed for agricultural purposes. Singh et al. [54] defined three broad areas of application of nanotechnologies in agriculture: (1) application of nanomaterials for more effective quality management of pre- and postharvest products, for example, in the protection, improvement of their taste, maintenance of nutritional values, and possible identification of pathogens or toxins using nano-biosensors; (2) research and development of nanomaterials effects in laboratory and pilot plant conditions for photocatalysis, interactions, and behavior; the bioremediation of persistent pesticides, disinfectants, and targeted types of nano-macronutrients (such as nanocarriers and nano-cellular materials); or visualization of pathogens and nano-biosensors; and (3) direct application of nanomaterials in real agricultural conditions ensuring desalinization, photocatalysis of pollutants, removal of potentially toxic elements, and sensing of plant health, pathogens, toxicants, and levels of nutrients. Nanomaterials are also divided based on the type of application to several categories: (1) nanomaterials for protection against biotic stresses such as insects and pest repellants and biocides, nanoherbicides, and nanofungicides, where nanomaterials play roles of active substances or supporting chemicals that better their attachment and efficacy or catalyze the active substances after their intended application length; (2) plant production with nanomicronutrient, nanomacronutrients, and nano growth promoters; (3) nanosensors; (4) livestock food production with nutrient supplementation and enhanced drug delivery; and (5) protection and evaluation of the collected produce (Fig. 1).

3 Application of Nanoagrochemicals

ENPs used in agricultural areas are most frequently (1) applied in germination and early stages of plant development, (2) added to nutrient media (hydroponic solutions) and incorporated directly into soils, (3) or applied in the form of a spray dispersion, i.e., foliar application. Each of these strategies has its agronomic justification and highlights nanotechnology's potential. However, in general, the final decision is based on the nature of the ENPs, environmental conditions, and other factors to which the ENPs are applied [55]. For example, if the same concentration of ENPs is applied on leaves, the soil, or a liquid growth medium, different types of reactions occur within these systems, and hence, effects that take place will be different [34, 56]. In this context, the presence and nature of ENP coatings need to be taken into account. These coatings may be stable in soil solutions but may be less effective under the influence of solar radiation, where they will undergo a faster degradation process, where higher energy parts of the light spectra, namely, UV light, are responsible. The destruction of the ENP coatings by UV light leads to aggregation, and the strong oscillating dipole-dipole interaction is believed to be the underlying mechanism of the destabilization [57]. In context of agricultural use of ENPs, this aggregation may lead to slower release of the ions from the ENPs that are supposed to supplement the plant with micronutrients or to lower absorption of ENPs to subcutaneous tissues of plants, and thus, this process may limit their effectiveness.

4 Role of Properties of Engineered Nanoparticles in Crop-Nanoparticle Interaction

The interaction of ENPs with crops and their plant surfaces is theoretically linked to the ENPs' (1) surface properties, such as moieties attached on the surface that have hydrophobic or hydrophilic properties, shape, charge, and energy of the surface; (2) the chemistry and crystallinity of ENPs; and (3) the physicochemical properties of the environment the ENPs interact with, including chemistry, input energy, such as sunlight or other types of radiation, and temperature changes [58–61].

4.1 Surface Modification

The interaction of ENPs with crops was found to be largely dependent on the multiple properties of ENPs. The most influential properties are the chemical composition, size and shape, and surface properties of the ENPs. Inorganic nanoforms are often coated with functional groups by the process of encapsulation to alter adhesion, absorbability, transport, and targeted effect on vegetative organs, for example, entrance to the leaves, stems, and roots. Often, these nanoforms are then called core-shell nanostructures or nanoparticles with surface coatings. Core-structure or surface coating can be split into several groups based on the composition (1) organogenic, containing organic molecules of active, industrially prepared substances, for example, pesticides and stabilizers [62]; (2) biogenic, for example, plant extract coatings (residues after nanomaterial biosynthesis) [46, 54]; or (3) metal or inorganic coating [63]. Industrially prepared organogenic and biogenic ENP coatings often take the form of polymers, emulsions, foams, or other surfactant forms from colloidal dispersions [46, 54]. To date, no long-term study has been directly focused on the role of adjuvants and additives in nano-enabled pesticides or fertilizers.

Molecules that affect surface charge are often used to adjust the ENP properties like aggregation and attachment to solid particles in the system. ENPs with negative charge were much more mobile in porous media than ENPs with the same composition but with surface modification with molecules with a positive charge, which was retained to a much higher degree [64]. Surface modification often changes the toxicity toward organisms and can lead to both lower and higher toxicity [65, 66]. For example, when both bare and Z-COTE HP1®-coated ZnO NPs were applied on beans through soil application, coated ZnO NPs stimulated more root growth and increased concentration of nutrient compared to bare ZnO NPs [67, 68]. It was also found that surface defects may improve reactive oxygen species generation and thus have higher toxicity [69].

Biosynthesis is often used to create more benign ENPs that may have additional effects based on the active substances present, for example, in the leaves of a plant [70, 71]. Different types of biosynthesis were used, and biologically induced synthesis with plant, fungi, or microbial extracts was used in studies [70, 71]. There is some evidence that ENPs may benefit from synergistic effects of ENPs and their surface-bound molecules of organic extracts [70]. In a study by Irshad et al. [29], both solgel-synthesized and biosynthesized TiO_2 NPs were applied on wheat, and biosynthesized TiO_2 NPs had a better effect on plant height, straw and grain yield, and reduced Cd toxicity. The application of biosynthesized ENPs often leads to better crop production with lower side effects, and thus, their qualities may be important for the future use of ENPs in agriculture.

4.2 Size

The size of ENPs is one of the most important properties that play a key role during crop-nanoparticle interaction. Properties such as their dissolution rate, absorption, and translocation in plants are directly affected by the size of ENPs [72–77]. Smaller ENPs dissolve more readily and can pass the pores of plant cell walls. The size limit for the passage of ENPs to apoplast was reported to be 5 nm [78] or 20 nm [25]. Size is, therefore, also crucial in toxicity toward plants since more significant dissolution and translocation of small ENPs into plant tissues renders them potentially more toxic [66, 73, 79]. In ZnO NPs, the higher toxicity of smaller nanoparticles toward fava beans was linked to their 30% greater dissolution, and reactive oxygen species generation was similar in 25 nm- and 70 nm-sized ZnO NPs [80]. Thwala et al. [81] also arrived at a similar conclusion with Ag NPs sized 10 and 40 nm, where the smaller Ag NPs dissolved more readily, and their dissolution was linked to their higher toxicity to *Salvinia minima*. Also, Cu NPs toxicity was found to be size dependent when applied on *Glycine max* cv. Kowsar via soil application [82]. However, the smallest 25 nm Cu NPs were even more toxic than Cu^{2+}, indicating that the toxicity of these ENPs is related to some other mechanisms than just dissolution and may be tied to their potential to generate reactive oxygen species, which is a size-dependent process.

4.3 Shape

Some of the ENPs have highly tunable shape and form structures with one, two, or three dimensions in the nanoscale (1–100 nm) [83]. For example, ZnO nanomaterials can be synthesized in many different shapes such as (1) structures with one dimension in nanoscale, including nanoplates, nanosheets, and nanopellets [84, 85]; (2) structures that have two dimensions within 100 nm, such as belts, combs, helixes, nanorods, needles, ribbons, rings, springs, tubes, and wires [86–94]; (3) nanoparticles with hexagonal, oval, pyramidal, or spherical shape [9]; and (4) more complicated shapes with their parts having a nanoscale size, such as urchin-like forms, dandelions, flower, and snowflakes [95–97]. Spherical nanoparticles may have more compensated surface energies and, thus, have been found to coagulate/aggregate at higher concentrations compared to rods and platelets [98]. Also, nanorods were observed to dissolve faster compared to the spherical ENPs of similar volume since they have a higher active surface that reacts with the water solvent [99]. This may affect the interaction of ENPs with plants; for example, hexagonal ZnO NPs were shown to have a slightly higher positive impact on the growth of tomato plants compared with spherical ZnO NPs [100].

4.4 Chemical Composition and Crystal Structure

The chemical composition of ENPs is essential as it changes important properties, such as their dissolution, surface reactivity, and effect on crops, since they may affect their metabolism directly or indirectly by releasing the ions [76, 101]. Doping of ENPs may result in a heightened ability to create reactive oxygen species. For example, ZnO NPs doped with Mn and Co showed higher toxicity compared to undoped ZnO NPs, and Fe-doped ZnO NPs had a similar effect [65]. However, Se-doped ZnO NPs exhibited decreased toxicity toward *Escherichia coli* in a study by Dutta et al. [102] even though they were shown to produce a higher number of reactive oxygen species due to Se leaching from ZnO NPs.

The crystal structure is important for ENPs that exist in more than one form of crystal structure, and it may have a direct effect on the surface properties of ENPs and, thus, on the interaction with plants. Also, additionally, to creating polymorphs, some ENPs may lack a long-range order and be amorphous. Good examples of polymorphous nanomaterials are TiO_2 NPs that are synthesized as rutile, anatase, or, less commonly, brookite [103], and they can also be synthesized in amorphous form [104]. The highest photocatalytic activity, production of reactive oxygen species, and, thus, toxicological effects were observed for amorphous TiO_2 NPs, followed by anatase and rutile $\approx$ brookite [104, 105]. In this context, the modification of the crystal structure of ENPs by doping with other elements is a strategy that creates agrochemicals and nanofertilizers that can directly affect the efficiency of photosynthesis [106]. For example, the modification of TiO_2 NPs by doping their structure with silver and nitrogen can increase their antimicrobial effect [107].

5 Interaction of Metallic Nanoparticles with Plants After Foliar Application

During foliar application, several types of effects are expected depending on environmental factors and conditions and the species, health condition, and developmental stage of a plant. After application of ENPs, plants were observed to be more resistant to pathogens [108] and contain more nutrients capable of redistribution within the plant [31, 60] and higher content of chlorophyll [109–111]. Their overall vitality was higher, and there was also a positive effect on crop production [31, 32]. The mechanisms that affect the foliar application of ENPs on plant species are shown in Fig. 2.

The properties and composition of colloidal dispersions in which ENPs are suspended affect the adhesion to the leaf surface, absorption, translocation, transport of ENPs, accumulation, and behavior at a specific site within the plant [60]. The adhesion of the applied foliar colloidal solution itself depends on the proportion of the individual components, their overall charge, hydrophobicity, or hydrophily [112, 113]. This depends not only on ENPs' ability to be absorbed but also on the properties of additives in the colloidal solution and the health of the leaves.

Also, the surface of the leaf is important with its many structures such as trichomes, vents, the content and composition of the waxes, and the shape of the epidermal cells [114–116]. Because of that, faster absorption kinetics and a more targeted effect were observed for hydrophilic ENPs against specific leaf organelles [117]. ENPs in pure aqueous solution affect the redistribution of trichomes, or upon absorption, they can form localized pools in the lower parts of trichomes [32]. Moreover, a synergistic or antagonistic effect can be expected depending on the compounds added to colloidal dispersion. If the solution contains other readily

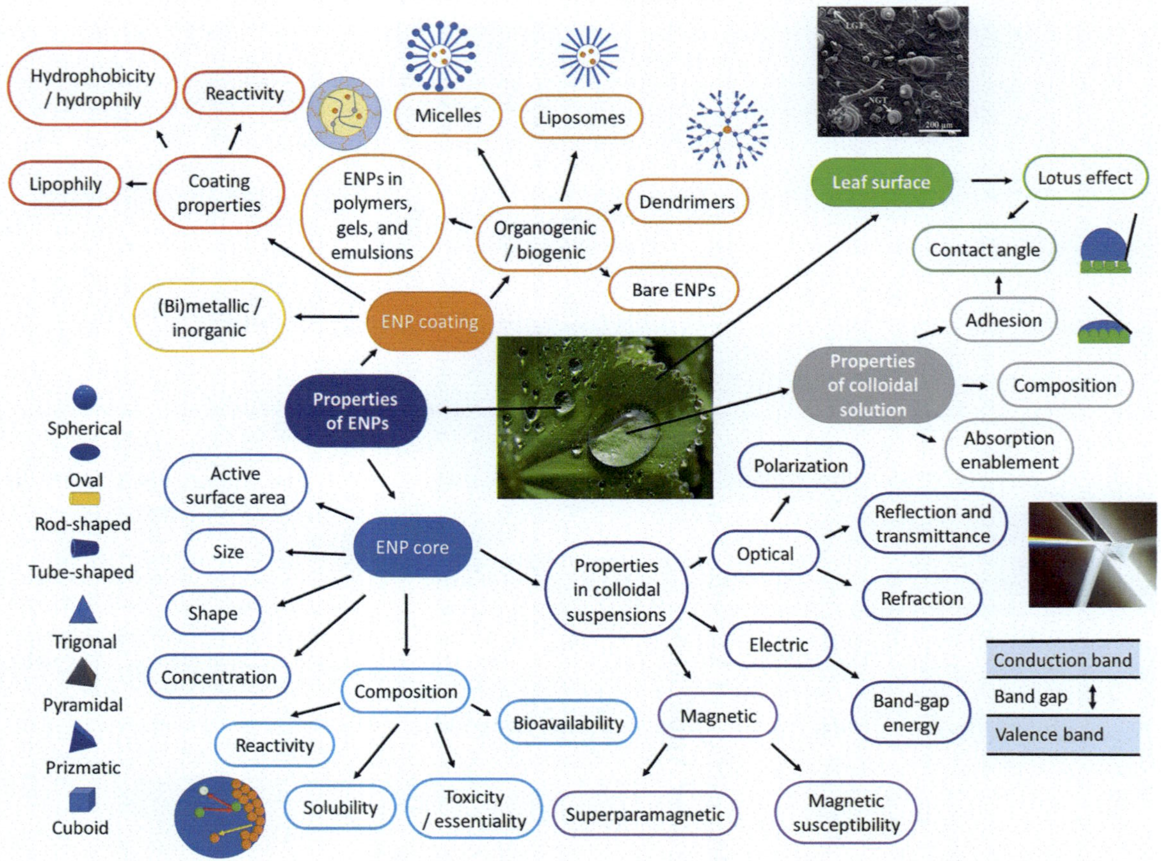

Fig. 2 Selected types of interactions of ENPs with the surface of a leaf upon foliar application

bioavailable forms of macro- and micronutrients in a suitable ratio, they may have synergistic positive effects. Unfortunately, there are not many studies on the application of the combined effect of ionic and ENP forms to date.

Surface coating of the ENPs, or in the case of core-shell ENPs, shell, is largely responsible for the fate and behavior of the ENPs in the environment since it is a surface responsible for the interaction with the outside environment. When the colloidal solutions with ENPs are applied on leaves, sunlight, mainly its UV component of the spectrum, can damage the coating, and the interaction with plant surface can change. Also, it should be taken into account that in the case of biosynthetically created ENPs, the coating itself may have antimicrobial effects if the extracts used in the synthesis are from medicinal plants containing molecules with such properties [118]. This combination of antimicrobial effects of the ENP coatings and ENPs themselves can prevent the emergence or manifestation of microbial pathogens [108]. Similarly, in the case of inorganic coatings, we expect different antimicrobial behavior in comparison with the individual effectiveness of bare ENP alone [119]. From this point of view, the physicochemical properties including photostability, electro-optical properties, their exposure and transition within the ENPs, such as the production of surface plasmons in the case of Au ENPs [120], or the formation of reactive oxygen species and simply overall effect on photochemical centers of plant leaves are of interest when ENPs are surface modified [109–111, 121].

One of the decisive criteria of using the ENPs is that they can be applied at much lower concentrations when compared to their bulk or conventional counterparts to have similar production and physiological effects [31, 32]. At higher doses, ENPs often have antimicrobial effects. They can be applied alone or in combination with other active substances, and they may be a more environment-friendly way of management of pathogen weeds and pests [122]. Another influencing factor is the chemical and physical stability of the ENPs. Stable, insoluble ENPs are often less bioavailable to plants. For more soluble ENPs, their chemical composition plays an important role. Currently, the trend is to apply soluble organic and inorganic ENPs that are based on organic carbon or contain macronutrients (nitrogen, phosphorus, or potassium), for example, modifications of apatite (containing Ca, P, also inorganic C) [123]. Research also focuses on ZnO NPs that are dissolving on the leaf surface, and their dissolution is enhanced by sunlight, where, in addition to photocatalytic properties of ENPs, it is expected that the effects also come from released ionic Zn^{2+} and also from Zn bound to phytic acid or tricarboxylic acids [124]. When entering a plant, Zn supports the enzymatic, metabolic, and physiological functions of the plant [125]. On the other side of the ENP spectrum are nanoparticles such as TiO_2 NPs, where Ti is not conventionally classified as an essential element, but both Ti and TiO_2 NPs still have a positive influence on the development and health of plants [126]. Ti in plants stimulates the activity of selected enzymes and nutrients and partly intensifies photosynthesis that includes increasing chlorophyll content and similar effects [127]. The proposed mechanisms by which TiO_2 NPs affect the health and growth of plants lie in their ability to protect the chlorophyll against degradation, provide surface sites that decrease the local energy of electron transfers in photosynthesis, and help activate specific stress response mechanisms in plants, and both TiO_2 NPs and Ti may specifically or nonspecifically bind with certain proteins in plant bodies [60, 128]. It was proposed that TiO_2 NPs may specifically interact with Fe and the proteins and metabolic systems the Fe is part of [127].

As mentioned briefly in the last paragraph, some research suggests that the conversion of light quanta by ENPs, such as TiO_2 NPs, to electron energy, is activating or lowering the energy requirements of chemical processes in plants, for example, electron transport, oxygen evolution, and chloroplast photophosphorylation [109, 110, 129]. TiO_2 NPs and similar metal oxide ENPs are semiconductors, and their dielectric properties arise from their semiconductive properties, i.e., their bandgap energy. On the nanoscale, the bandgap energy can be manipulated by reducing the size of ENPs, doping other elements into the crystal lattice of the reactive surface and introducing crystal lattice deformations on the nanoparticle surfaces [130, 131]. Moreover, other ENPs, such as electron-conducting Au NPs, possess surface plasmons that may also enhance photosynthetic processes in plants [132]. When adjusting the electrical properties of ENPs, we must also consider the negligible but existing effect of the magnetic fields. For example, an effective improvement in the quality and yield of maize (*Zea mays* L.) was observed when the magnetic field was combined with Ag NPs [133]. Moreover, a positive effect on the yield and nutrient intake of ajwain (*Carum copticum* L.) has also been demonstrated [134]. Otherwise, magnetic ENPs or static magnetic fields are used in plant sciences for microscopic visualization of plant cell tissues [135] or gene transformation strategies [136].

Current research in agriculture is concerned chiefly with the size and shape control of ENPs and adjustment of their surface properties [101]. Modification of these physicochemical properties follows a logic that the point of entry and immediate effect on the plant are bottlenecked by the size, shape, and properties of cuticle or the size of leaf stomata, where the upper limit of ENP entry is about 5–20 nm [25, 78]. In this context, the hydrodynamic size is the most critical property of ENPs, but surface characteristics such as hydrophilicity or hydrophobicity and charge of the nanoparticle are also important. For example, the study by Hu et al. [117] suggests that ENPs with a positive or negative zeta potential higher than 30 mV have improved transport to lower parts of plant epidermis compared with more charge-neutral ENPs, which were less able to penetrate the lipid bilayer.

6 Impact of Engineered Nanoparticles on the Leaf Traits

The foliar application exposes leaves directly to sprayed ENPs, and the leaf-ENP interaction is important for the distribution of ENPs and ions released from them throughout the plants. Spraying of the plant's leaves by ENPs may lead to a rise in the number of leaves per plant and overall leaves' area [137]. In addition, depending on the composition of the ENPs, they may have various antibacterial, antifungal, or UV protective properties that enhance the leaf health and growth, and they can also

Table 1 Impact of various applications of engineered nanoparticles on the leaf traits

Type of NPs or conventional spray	Concentration and type of application	Species	Effect	References
CuO NPs	2.5, 10, 50, 100, and 1000 mg/L	*Oryza sativa* L.	Decreased size and lower number of stomata, increased size, and higher number of trichomes	Costa and Sharma [138]
Fe_2O_3 NPs	10, 20, and 30 μm, root application	*Mentha piperita* L.	Decrease in density and number of trichomes	Askary et al. [139]
GA-Ag NPs (gum arabic-coated silver NPs)	5, 10, 20, and 60 mg/L, foliar application	*Phaseolus vulgaris* L. (Bronco and Nebraska varieties)	Significant rise in the number of leaves per plant and leaves' area	El-Batal et al. [137]
TiO_2 NPs	50, 100, 150, 200, and 250 mg/L	*Mentha piperita* L.	Significant improvement in the diameter and density of peltate and glandular trichomes	Ahmad et al. [140]
TiO_2 NPs	2.6 mg/L	*Helianthus annuus* L.	Different ratio of nonglandular trichomes: linear glandular trichomes	Kolenčík et al. [32]
ZnO NPs	300 and 2000 mg/L	*Hordeum sativum* distichum cv. Travnik	Deformation of stomatal and trichome morphology (increased area of stomata and trichomes treated with 300 mg/L ZnO NP, size reduction in leaves treated at higher concentrations)	Rajput et al. [141]
ZnO NPs	2.6 mg/L	*Helianthus annuus* L.	The appearance of capitate glandular trichomes, different ratios of nonglandular trichomes: linear glandular trichomes	Kolenčík et al. [32]

supplement needed nutrients to the plant. These properties are derived either from their chemical composition [137] or based on their photocatalytic properties [47] or other properties related to their reactive surface. And similar to conventional sprays or conventional soil-applied fertilizers that mostly contain ionic forms of chemicals, ENPs have various effects on the leaf traits whether they are applied directly on the leaf or to the root area (Table 1).

Applications of ionic forms of fertilizers or growth promoters were found to have various effects on leaves, for example, a significant rise in the number of leaves per plant, leaves' area for ionic Ag applied to leaves of *Phaseolus vulgaris* [137], decrease in density and number of trichomes by chelated Fe applied to roots of *Mentha piperita* [139], or small, dark-brown/black spots, which were smaller than 0.5 mm in width, on the petioles and blades of the lower leaves along with dark-brown lesions smaller than 2 mm on the lower leaves spotted in *Helianthus annuus* L. after the application of Mn [142].

When both conventional forms and nanoparticulate forms were compared, their effects were similar, but the effects' strength was different for conventional forms and ENPs. In a study by Askary et al. [139], where Fe_2O_3 ENPs were applied to the root area of *Mentha piperita* via hydroponics, they had a more pronounced effect than the iron chelate at the same concentration. The application of Fe_2O_3 ENPs resulted in a decrease in density and number of trichomes of different types, including peltate, glandular, and nonglandular trichomes. Two mechanisms were assumed by Askary et al. [139] for reducing trichomes density: Either (1) the Fe_2O_3 NPs had just adverse effects on the formation of trichomes or (2) they can reduce the effect of salt stress and, thus, the formation of certain types of trichomes that form as a response to that stress. In another study [137], Ag NPs surface modified with gum arabic were compared with $AgNO_3$. The ENPs have shown a better ability to increase the number of leaves and leaves' area than $AgNO_3$ in *Phaseolus vulgaris* L. (Bronco and Nebraska varieties).

ENPs usually have positive effects on leaves at lower concentrations and negative effects at high concentrations. For example, 2000 mg L of ZnO NPs caused size reduction of leaves [141]. Similarly, CuO NPs at concentrations above 50 mg/L reduced the number of thylakoids per granum and distorted the thylakoid membranes in *Oryza sativa*, and swelling of the intrathylakoidal space was noted at 1000 mg/L. Also, above 10 mg/L, a decline in size and a lower number of stomata were found along with a bigger size and higher number of trichomes after CuO NPs treatment [138]. However, they can increase the number of leaves per plant and overall leaves' area at lower concentrations [137]. Reduction of certain structures of leaves was also observed, and in many cases, it is not yet fully understood if the effects are benign toward the plants or not [139]. The application of TiO_2 NPs to leaves at a concentration of 100–150 mg/L was found to be optimal for *Mentha piperita* [140]. At 100 mg/L, increase of 62.5% and 77.8% was recorded for the peltate and glandular trichome density and diameter, respectively. An increase in peltate and glandular trichomes is a way of increasing the production of essential oils in *Mentha piperita* [140]. A deformation of trichome and stomatal morphology was found in *Hordeum sativum* distichum cv. Travnik after ZnO hydroponic root application [141]. Also, a high concentration of ZnO NPs disorder thylakoids, expand interthylakoidal gaps, heightened the size of starch granules, and injure the outer and inner membranes of chloroplasts in leaves. The analysis by cytomorphometry identified a decrease in the number and size of plastoglobules and chloroplasts per cell of the leaf.

7 Metal and Metal Oxide Nanoparticles Affect Plant Yields and Nutritional Parameters

From the point of view of plant production and agricultural yields, we can consider ENPs made of several metals, such as ZnO, TiO_2, Ag, Fe_2O_3, CeO_2, and MoO_3, effective [25, 143]. Most of the ENPs affect the yields and health of the grown crops positively at suitable concentrations. The effect of ENPs on crops is both in the improvement of quantity, i.e., higher yields, and the quality of the product with increased content of essential nutrients and elements (Fig. 3). A summary of some of the effects of different inorganic nanoparticles on plants is given in Table 2.

7.1 *Quantitative Improvements in Plant Yields*

The advantage of ZnO NP fertilizers is their ability to apply them up to 10 times lower concentrations when compared to traditional $ZnSO_4$ fertilizer while producing the same or similar effects [154, 155]. The influence of ZnO NPs on crop production parameters in field conditions is relatively well developed for crops such as wheat, corn, or rice. Rizwan et al. [147] noted that in field conditions, ZnO NPs had affected positively the quantitative parameters of winter wheat, and Rizwan et al. [153] found similar positive effects in maize. The impact of zinc-based ENPs on millet, most often in the form of ZnO, was primarily observed in laboratory studies. Nandhini et al. [156] confirmed the positive influence of ZnO NPs, especially on the germination and emergence of millet. Only a few studies on the effect of ZnO NPs on

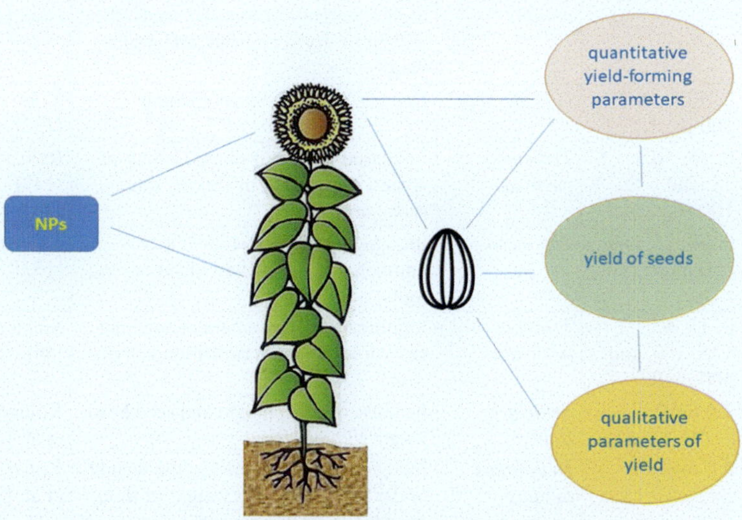

Fig. 3 The application of ENPs has effects on qualitative and quantitative parameters of yield

Table 2 Effects of different ENPs on yield and growth parameters of crops

Type of NPs	Concentration	Species	Effect	References
Ag	0 and 44 mg/L	*Solanum lycopersicum*	Increased height, shoot length, and plant biomass	Noshad et al. [144]
Ag	0 and 2000 mg/L	*Triticum aestivum*	Decreased height, grain weight, and plant biomass	Yang et al. [145]
CeO_2	0, 100, 200, 400, and 800 mg/L	*Helianthus annuus*	Accumulation of Ce in sunflower roots, very low translocation to the aboveground parts of the plant, and negligible effects on production parameters	Tassi et al. [146]
Fe_3O_4	0, 5, 10, 15, and 20 mg/L	*Triticum aestivum*	Increased height of plants, length of spikes, the weight of spikes, shoots, roots, and grains	Rizwan et al. [147]
Silica NPs	0, 250, and 1000 mg/L	*Arabidopsis thaliana*	Size-dependent uptake by roots; no toxicity even at higher doses	Slomberg and Schoenfisch [148]
TiO_2	0 and 750 mg/L	*Oryza sativa*	Increased length of shoot and roots and higher yield of grains	Zahra et al. [149]
TiO_2	0 and 400 mg/L	*Raphanus sativus*	Increased germination parameters	Haghighi and Teixeira da Silva [150]
TiO_2	0, 40, 80, 120, 160, 200, and 240 mg/L	*Brassica oleracea*	Higher concentrations impacted negatively the shoot length and positively the root length.	Singh et al. [151]
TiO_2	0 and 2.6 mg/L	*Helianthus annuus*	Head diameter increased and the higher weight of thousand seeds, the weight of dry seed head, and grain yield	Kolenčík et al. [32]
ZnO	0, 50, and 500 mg/L	*Glycine max*	Decreased stem and root length	Yoon et al. [152]
ZnO	0, 25, 50, 75, and 100 mg/L	*Triticum aestivum*	Increased height of plants, length of spikes, the weight of spikes, shoots, roots, and grains	Rizwan et al. [147]
ZnO	0, 50, 75, and 100 mg/L	*Zea mays*	Increased length of shoots, leaves' number, and dry weight of shoots and roots	Rizwan et al. [153]
ZnO	0, 15, 62, 125, 250, and 500 mg/L	*Triticum aestivum*	Significant enhancement of seedling growth and seed germination activity	Singh et al. [54]
ZnO	2.6 mg/L	*Setaria italica*	Increase of plant height and seed head length	Kolenčík et al. [31]
ZnO	2.6 mg/L	*Helianthus annuus*	Increase of head diameter, the weight of dry seed head, the weight of thousand seeds, and grain yield	Kolenčík et al. [32]

millet in field conditions have been published to this date. Kolenčík et al. [31] described the positive influence of the foliar application of ZnO NPs on the foxtail millet plant height and length at a concentration of 2.6 mg/L.

Tarafdar et al. [157], in turn, confirmed the positive effects of foliar application of ZnO NPs on millet yields. Sunflower has been shown to respond very well to lower concentrations, 2.6 mg/L, of ZnO NPs [32]. The foliar application was observed to significantly increase the values of production parameters, namely, the head diameter, the weight of the head, the weight of a thousand seeds, and the seed yield. Due to the better absorption of ZnO NPs and nano-derived effects, there was also an increase in the yield parameters of peanuts compared to both control and the ionic form of $ZnSO_4$, for example, 34% improvement in pod yield per plant reported in contrast to ionic form [48]. In another study, Mahajan et al. [158] showed that the ZnO NP at an optimal concentration at the level of 20 mg/L significantly improved beans' growth parameters and yield. They recorded a 42% increase in root length and stem length up to 98% at said dose. Rezaei and Abbasi [159] found that usage of Zn-based NPs in chelated form stimulated the production of cotton by increasing the number and weight of capsules per plant. The addition of ZnO NPs at a concentration of 40 mg/L significantly promoted parameters affecting the yield of rice and yield itself [160]. It is also interesting to note that the combination of Zn- and B-based NPs increased via the foliar application the yields of pomegranate fruit [161]. The ZnO nanoparticles can act as plant growth stimulators and increase their yields. However, it must be said that nanoparticles, like any fertilizer, can be toxic to the plant when incorrectly dosed. It has been found that the concentration at which ZnO NPs are phytotoxic depends primarily on the specific plant species. Most often, adverse effects of foliar application of ZnO NPs are occurring at concentrations exceeding 1500–2000 mg/L [22].

TiO_2 NPs can have substantial positive as well as negative effects on crops. However, it all depends on the used concentration and growth phase when these ENPs are applied. At suitable concentrations, TiO_2 NPs are regarded to be plant growth promoters. Zahra et al. [162, 163] confirmed the vital role of TiO_2 NPs in increasing plant phosphorus uptake and promoting growth in soil moisture deficiency conditions. TiO_2 NPs also play an important role in nitrogen metabolism, which has a significant effect on the final crop yield, and it was confirmed in a study by Kolenčík et al. [32]. The positive impact of foliar application of TiO_2 NPs on the diameter of the head, the weight of the head, the weight of a thousand seeds, and the yield of seeds of annual sunflower was observed at a concentration of 2.6 mg/L. Similar effects have been reported by several studies in other plants, such as barley [164], wheat [165], and cowpea [166].

Similar to Ti-based ENPs, such as TiO_2 NPs, Ag NPs can have a different nature of actions in their nanoparticulate form compared to free ions. Ag NPs are attractive for agronomical use primarily for their antibacterial and antifungal effects. Therefore, it is suitable to use these ENPs in crops susceptible to bacterial or fungal diseases, such as potatoes. Tahmasbi et al. [167] observed that foliar application of Ag NPs at 50 mg/L increased the yield of potato tubers per plant by 86.9% compared to untreated control, mainly due to their improved health.

The role of Fe in plant photosynthesis is well known. Its application in the form of ENPs significantly supports physiological development, which is also reflected in the favorable values of yield-improving parameters and increased yields of crops. Positive effects in soybean [168] or peas [169] have been observed after foliar application. In both cases, there were observed increases in pods per plant and the weight of a thousand seeds. Soybean had 48% higher yields than control when 500 mg/L of Fe-based NPs were applied on its leaves at three plant growth stages [168]. Peas were also positively affected by the same concentration of Fe-based NPs, where 1000 seed weight and number of seeds per pods were 196.3% and 21.5% higher than control, respectively [169]. Similar effects of Fe-based NPs have been observed in maize [170]. The application of magnetite NPs supported the yield and growth parameters of mung beans.

Strategies of applying combinations of metal nanoparticles on crops are also known: Fe and Zn [170], Fe and Mg [169], or ZnO and $MnFe_2O_4$ [171]. For example, the combination of ZnO and $MnFe_2O_4$ significantly increased the yield of peanuts due to better nutrient utilization and more intense photosynthesis [171]. Similarly, Drostkar et al. [172] reported that foliar application of Zn and Fe ENPs had a positive effect on chickpea production, which was reflected in the positive impact on parameters connected to higher yields and higher yield.

ENPs created with other elements have also been used in crop production. The positive effects of Mo-based NPs are also known. Manjili et al. [173] reported their positive influence on peanuts via foliar application. They improved plant height and branching, number of pods per plant, the weight of thousands of seeds, number of seeds per plant, seed size, and seed yield. The chelated form of Mo nanoparticles was used in this study.

7.2 *Qualitative Improvements in Plant Yields*

Inorganic ENPs such as Zn- and Fe-based NPs affect the quality of field crop production most significantly. In plants, Zn has an important function in root growth and absorption of essential nutrients, notably N, required for the synthesis of proteins [174]. Several studies [54, 147, 153] in field conditions confirmed that ZnO nanoparticles positively affected the quality parameters of winter wheat production and winter corn. Studies also confirmed similar positive effects by Kolenčík et al. [31] on foxtail millet and Kolenčík et al. [32] on sunflower. The millet had an increased content of total nitrogen, oil, starch, and grain dry matter. However, the crude protein content decreased [31]. In sunflower, a significant increase in oil content was observed after the application of ZnO nanoparticles [32]. Both crops had a ZnO nanoparticle concentration of 2.6 mg/L. Singh [175] and Patel [176] also described the beneficial effect of low concentrations of Zn-based ENPs on sunflower oil contents. Nandhini et al. [156] found that ZnO nanoparticles promoted the activity of important millet enzymes under laboratory conditions, leading to higher plant resistance to fungal diseases. It has been proven that plants without fungal infestation have a higher quality of final products [177]. In the studies by Kolenčík et al. [31, 32], it was found that the plants treated with ZnO NPs did not suffer from any fungal diseases and achieved a higher quality than the control variant. However, this was not always statistically significant. Rizwan et al. [147, 153] found that ZnO NPs positively affected the concentration of chlorophylls and, thus, the intensity of photosynthesis of wheat and maize. Schmidt et al. [177], Bellesi et al. [178], and Matzen et al. [179] state that a higher intensity of photosynthesis is a prerequisite for increasing the qualitative parameters of production, such as the content of starch, oil, dry matter, or crude protein. Higher intensity of photosynthesis is directly connected to lower water stress [180, 181]. This is consistent with the results of the study by Kolenčík et al. [31] as it was noted that crops treated with ZnO NPs showed lower water stress and had a higher content of nitrogenous substances, oil, starch, and dry matter. Similar results were reported by Kirnak et al. [182], who found that the content of oil in pumpkin seeds increased when water stress values decreased. The same trend was observed by Candogan et al. [183], where lower values of water stress in soybean crops resulted in a higher content of protein and oil in its seeds. Higher contents of proteins, starch, sugars, and oil [171] were also reported in peanuts after the application of zinc nanofertilizers. Raigond et al. [184] reported increased content of anthocyanin, starch, and soluble sugars in potatoes. Leaf application of nano-chelated Zn and Fe also improved the concentration of crude protein and carbohydrates in maize grains [170]. Furthermore, higher protein content was also recorded in pea seeds after the application of Fe NPs compared to the classical macroforms [169].

Growth promoters based on TiO_2 NPs can also contribute to increasing the quality parameters of field crops. There was a significant increase in carotenoids in crops such as rapeseed, mung beans, and tomatoes [185]. It can be stated that concentrations of TiO_2 NPs (2.6 mg/L) also have the potential to increase the oil content in oilseeds, for example, in sunflowers [32]. Other ENPs, such as Ag NPs in fenugreek, increase the content of proteins, phenolic, and flavonoid compounds [186]. The influence of different metal nanoparticles on the qualitative production parameters of plants is given in Table 3.

In general, the long-term application of engineered nanoparticles in agriculture comes with several advantages and disadvantages. Therefore, their ecological and environmental impact is studied in several ways, including the toxicity to soil organisms other than the plants they are applied to and their impact on the soil [187]. However, effects to plant reproductive organs are relatively less studied, and this field of study also needs our attention.

Table 3 The influence of different metal nanoparticles on the qualitative production parameters of plants

Type of NPs	Concentration	Species	Effect	References
Ag	0 and 60 mg/L	*Trigonella foenum-graecum*	Increase of the content of carbon, protein, phenolic, and flavonoid acid	Sadak [186]
TiO_2	0 and 2.6 mg/L	*Helianthus annuus*	Increase of oil content	Kolenčík et al. [32]
ZnO	400, 1000, and 2000 mg/L	*Arachis hypogaea*	Higher N, P, Fe, Mn, and Zn contents in both seeds and straw; higher chlorophyll, carotenoids, total carbohydrate, total soluble sugars, total proteins, and oil percentages content in seeds	El-Metwally et al. [171]
ZnO	0 and 500 mg/L	*Solanum tuberosum*	Increase of the content of anthocyanin, starch, and soluble sugars	Raigond et al. [184]
ZnO	0 and 250 mg/L	*Helianthus annuus*	Increase of the content of sugars, proteins, and oil	Rajiv et al. [143]
ZnO	0 and 2.6 mg/L	*Setaria italica*	Increase of the content of total nitrogen, oil, starch, and dry mass	Kolenčík et al. [31]
			Decrease of total proteins	
ZnO	0 and 2.6 mg/L	*Helianthus annuus*	Increase of oil content	Kolenčík et al. [32]

8 Evaluation of Nanoparticles in the Context of Reproductive and Environmental Safety Through the Palynological Analysis

The reproductive phase is the most crucial period in the life cycle of the plants. It includes processes such as the development of the flowering parts, differentiation of the gametophytes, as well as pollination, fertilization, and embryogenesis (Fig. 4). Some of the elements used in nanoparticles (for example, B, Fe, Zn) represent microelements necessary for the normal development of reproductive parts [188–191]. In the case of the supply of these elements in the form of nanoparticles, the stimulatory or inhibitory effect on the reproductive organs and generative cells could be achieved by a well-balanced concentration of these elements and suitable form in the applied ENPs [192]. The surface charge of the particles and their treatment is also important. It was proved that negatively charged Fe_2O_3 NPs cause a decrease in pollen germination [193].

Particle size seems to be an essential parameter for the transport of nanoparticles into cells. In addition to the entrance pathways through the cell wall of pollen grains, smaller ENPs can enter the large pollen grains through the pores. The developmental phase of the plant in which the application of ENPs takes place is also important. It is necessary to demonstrate whether the application of ENPs on vegetative parts of crops, especially in field conditions, sufficiently supports the morphogenesis of

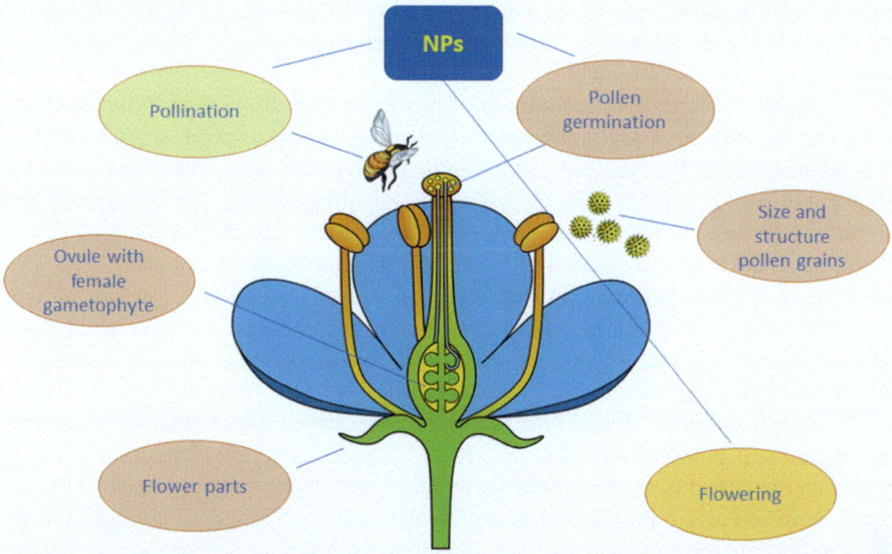

Fig. 4 Impact of the nanoparticles on the reproductive parts of plants

generative organs and the successful development of gametes. A positive correlation was found between better vegetative growth and increased flower production after ZnO application [194].

It is known that the distribution of elements in plant organs and tissues is not as efficient, and the accumulation of substances in different parts of the plant varies; therefore, the method of application is essential to achieve the expected effect [34]. However, there is a lack of knowledge on in situ experiments since most of the data are obtained from in vitro experiments [192]. Furthermore, there is a lack of comprehensive knowledge on the impact of ENPs on the whole reproductive cycle of the plants, from the establishment of flowers through the differentiation of flower structures to the final development of seeds and fruit.

The application of ENPs affects the onset and time of the flowering process. In some cases, the application of ENPs accelerates the flowering process. An earlier flowering process was noted after applying Ag NPs in tulip *Tulipa gesneriana* [195] and lily *Lilium* cv. Mona Lisa [196]. The same effect was observed when ZnO NPs were applied on onion, which reduced the number of days required for the flowering process [194]. However, in the other species, ENPs may have the opposite effect. Flowering time was significantly delayed after exposure of plants to carbon nanotubes in tomatoes [197]. The delayed flowering process of *Arabidopsis thaliana* after the application of Ag NPs was caused by a changed photoperiod, autonomous, and vernalization pathways [198]. However, the application of ENPs may not affect flowering time. CeO_2 NP did not cause any abnormalities in flowering time in the bean [199]. The prolongation of flowering time of ornamental plants is a desirable aim. A longer flowering process was observed after the application of Ag NPs in two cultivars of lily, even if the bulbs were soaked in ENP solution [196]. The timing of the flowering process is also important from the perspective of the pollination conditions.

The impact of ENPs on the number of flowers is directly related to the fruit set. Spraying Ag NPs at 3 mg/L increased flower percentage as well as fruit yield in *Prunus persica* [200]. Similarly, the treatment of onion with ZnO NPs caused a higher number of flowers and seeded fruits per umbel [194]. An increase in the number of flowers in lily *Lilium* cv Mona Lisa was achieved after applying Ag NPs [196], and ZnO NPs increased the number of flowers in tomatoes [201]. The application of ENPs has a significant effect on flower formation. It turned out that TiO_2 NP foliar application helped to form flowers in tomatoes at considerably low concentrations, but plants growing in treated soil produced a similar number of flowers at higher concentrations.

The improvement of flower characteristics is economically important, especially for ornamental plants. TiO_2 NPs positively affected morphological traits such as flower diameter and fresh flower weight in widely cultivated species *Petunia hybrida* [202]. Similarly, Ag NPs affected positively the width and length of petals in *Lilium* [196] and improved the quality of cut flowers in tulip *Tulipa gesneriana* [195]. The opposite effect was observed in *A. thaliana* when the application of Ag NPs caused a decrease in flower and calyx size as well as petal viability. Both the pistil and stamens were also adversely affected [198]. The response to the concentration of applied ENPs may be variety specific. A positive effect on the flower characteristics of the *Lilium* cv Mona Lisa was achieved at 50 mg/L Ag NPs while in cv Little John at 100 mg/L [196].

Although there are no apparent phenotypic changes in flower structures, the disturbance in gamete development within the reproductive organs can occur [193]. The normal development of male and female gametophytes is a crucial factor in successful generative reproduction. The haploid phase is a particularly sensitive indicator of various environmental changes [203, 204]. Disturbance during sporogenesis and gametogenesis due to adverse conditions results in undeveloped or degenerated gametophytes unable to participate in the formation of generative diaspores. From this point of view, the male gametophyte is more sensitive to various negative effects. Therefore, pollen is used as a bioindicator of environmental pollution [205]. The increasing use of ENPs in various areas of industry and agriculture requires increased attention to the impact of these particles on plants, especially on the reproductive organs of the crops grown for seeds or fruits. Therefore, it would be appropriate to test the impact of nanoparticles on the model and sensitive species and subsequently verify it on the crops. The species *Arabidopsis thaliana* is the most used object in plant biology. It was observed that ENPs inhibited pollen development [198], and Fe_2O_3 NPs increased the pollen abortion of this model species [193]. It seems that these findings could be applied to the other group of plants, including crops. It was found that disturbance during microsporogenesis after the application of CeO_2 NPs in beans led to irregular chromosome pairing, formation of laggard chromosomes, development of abnormal or degenerated tetrads, and collapsed or degenerated microspores. Pollen grains of the beans were degenerated or undeveloped, and the shape of the pollen grain was variable [199]. Likewise, different types of pollen aberrations were detected in peach after application of Zn-based NPs [200], as well as kiwifruit *Actinidia deliciosa* var. *deliciosa* pollen affected by Pd NPs [206].

An important indicator of pollen quality is the size and structure of pollen grains. The Ag NPs increased substantially the size of pollen in peach [200], and selenium NPs with supplementation at 3 mg/L increased the pollen size in tomatoes [197]. However, higher doses of Se NPs reduced not only the size of the pollen but also the density of the pollen grains in microsporangium, drawing attention to their potential toxicity [197].

The effect of ENPs on the male reproduction phase can also be manifested by changes in the microscopic and submicroscopic structure of pollen grains. Although pollen grains are coated with a very resistant lipid wall of sporopollenin, after contact of the ENPs with the surface of the pollen, the exine of the pollen could be damaged. It was found that Ag NPs caused extensive alterations in the muri and lumen of the exine pollen grain in *Peltophorum pterocarpum* [207]. Fullerenes adhered to pollen grains and damaged the plasma membrane of *Cryptomeria japonica*, *Chamaecyparis obtusa*, and *Camellia japonica* [208]. Changes in the ultrastructural level of the kiwifruit *Actinidia deliciosa* pollen grains such as plasmalemma invagination, vacuolization, and degenerative changes were observed after the application of Ag NPs [206]. For the same species, Pd NPs showed deeper cell penetration into cytoplasm and vacuoles [206]. In addition to the uptake of ENPs through cell walls, there is the possibility of small NPs penetration (5–10 nm) via germinating pores, especially with large pollen grains.

The most important indicator of the pollen grains quality is the viability of pollen. Current findings suggest that the positive or negative influence of ENPs on pollen viability depends on the type of nanoparticle, the concentration or size of the particles, the method of preparation, and the method of application. In addition, ENPs affect various stages during pollen germination, such as the emergence of the pollen tube, germination rate, and length of the pollen tube.

Application of Fe_2O_3 and Ag NPs decreased the pollen viability in model species *A. thaliana* [193, 199]. In addition, the accumulation of Ag NPs in the proximity of the germination apertures in kiwifruit [206] and blockage of germ pore of the pollen in *Peltophorum pterocarpum* [207] disrupted the emergence of the pollen tube and caused a substantial decrease in pollen viability.

Adhesion of ENPs to the surface of pollen grains hurts pollen germination. Fine ZnO NPs adhered to lily pollen grains and decreased germination rate. The continuous absorption of zinc from ZnO NPs by pollen grains and pollen tubes was observed; thereby, it was increasing its intracellular concentration and inhibiting pollen germination [209]. Fullerenes captured on the surface of the exine had the same effect on the pollen germination of *Camellia japonica* [208]. Graphene oxide decreased pollen germination and tube elongation at concentrations over 50 mg/L in tobacco *Nicotiana tabacum* and *Corylus avellana* [210]. It appears that the effect of ENPs on pollen may be specific to different species. For example, graphene oxide ENPs induced acidification of the media, which is the primary toxicity mechanism causing the impairment of pollen performance in *C. avellana* but not in *N. tabacum* [210]. The acidic properties of graphene oxide ENPs were manifested at 100 mg/L. Low doses did have any effect on the viability of tobacco pollen, yet they doubled the frequency of bent tubes after disruption of the intracellular pH homeostasis.

The decrease in pollen viability is usually due to the increasing concentration of ENPs. It was observed that increasing concentrations of Ag NPs reduced pollen germination and tube growth in *Peltophorum pterocarpum* and impaired pollen performance more severely at the stage of the emergence of the pollen tube. The negative effect also showed a decreased pollen tube growth rate and shorter length of the pollen tubes. The cause of the pollen tube growth disturbance was a lower amount of Ca^{2+} in the apex pollen tube [207]. A similar effect on pollen viability and performance in *Actinidia deliciosa* pollen had an application of the same type of nanoparticles. Ag NPs decreased pollen viability, and tube elongation was much more sensitive to Ag NPs than just the tube emergence [211]. Also, Pd NPs decreased significantly the pollen viability in kiwifruit. In this case, inhibition of pollen germination, cessation of pollen tube emergence, and elongation were due to the perturbation in redox balance and loss of endogenous calcium [206].

In some cases, positive effects of nanoparticles on pollen germination were reported. Commercial Ag NPs showed a significant increase in pollen viability without any aberrations in peach [200], and a positive effect of organic nanofertilizers Lithovit ($CaCO_3$, $MgCO_3$, Fe) and Nargo (N, P, K, Mg, Zn, Fe, Cu, Mo, B, Ca, Se) on in vitro pollen germination and pollen tube elongation manifested in higher pollen viability in peas [212].

A lot of ENPs obtained by organic synthesis appears to have a positive effect on pollen germination. For example, Au NPs green synthesized with *Terminalia arjuna* extract increased pollen germination in onion *Allium cepa* and *Gloriosa superba* [213, 214], and nettle-derived carbon nanosheets had no inhibitory effects on tobacco pollen germination [215].

Knowledge of the impact of ENPs on pollen viability in field crops is very rare. In this context, it was found that CeO_2 NPs decreased germination in bean plants [199]. In addition, most of the data come from controlled greenhouse or laboratory conditions. Also, most palynological experiments were performed by pollen germination on artificial media with added ENPs. Therefore, it will be necessary to focus on field experiments to verify the mechanism of impact of the nanoparticles in complex conditions. The first mention is that CeO_2 NPs increased pollen germination, and their foliar spray protected pollen germination under drought conditions in *Sorghum bicolor*. The mechanism responsible for this positive effect is the antioxidant properties of NPs, which reduce oxidative stress from drought [216].

Effective pollination is crucial to achieving a high-quality yield of most crops. ENPs released into the environment can be a potential risk to pollinators [217, 218]. In particular, ENPs that adhere to the pollen surface of entomophilic species affect pollen performance and can be transported on the bodies and in the digestive tracts of pollinators and cause contamination of bee products. This pathway poses a risk of transmission to the food chain. It was found that Fe released from

ENPs was transported on the body surface of worker bees, and Pb was found in their digestive tract [219].

In addition to the normal development of the male gametophyte, the second necessary condition for reproductive success is the normal development of the female gametophyte. Disturbance during megagametogenesis resulted in the formation of the incomplete or degenerate female gametophyte. Knowledge of the impact of ENPs on this process is absent. There is a mention of disturbance of this process in bean plants. CeO_2 NPs caused megagametogenesis disturbance in the stage of two-celled female gametophyte due to apoptosis nuclei [199]. However, further research on other crops is necessary. The impact of ENPs on the reproduction of plants can be seen in Table 4.

Interaction between pollen and stigma after the impact of ENPs on these structures is insufficiently examined, and the influence on fertilization and the process of embryo development. Some research showed that although ENPs did not affect crop yields, the nutrient content of the grains was affected [220]. Further research on the transmission of ENPs from seeds to offspring in crops is needed. The assessment of food safety risks is also necessary. So far, experimental data for the model species showed that this transport is possible [198]. Experiments with magnetofection of pollen gave inconsistent results. Zhao et al. [221] described a new method that used magnetic nanoparticles to deliver DNA into pollen grains of some dicots, which was not confirmed in monocots [222].

9 Conclusion and Future Perspectives and Development

Plant growth and health are of utmost importance as far as agriculture is concerned. Inexpensive alternatives to conventional fertilizers that can reduce the number of applied materials are needed. Nanofertilizer and growth promoters have a great potential for success in this area. These ENPs offer an effective strategy to prevent malnutrition via increased uptake of micronutrients to edible plant parts and their better yields. However, further research is needed to fully utilize their potential in large-scale agricultural applications that will be in line with the environmental goals of soil protection and climate crisis mitigation. Further extensive studies are urgently required to fully comprehend the scale and various aspects that are related to their potential to improve agriculture since field experiments with ENPs are sparse. Many ENPs are photoreactive when applied to leaves. Dosage of ENPs and high solar exposure may boost antioxidant systems in leaves at appropriate concentrations. Still, they may also overwhelm the defense systems and result in leaf damage at higher doses. ENPs exhibit the potential to significantly improve agricultural yields even at relatively low concentrations without significant risk to the consumers or the sprayed plants.

Future experiments with ENPs will need to address their specific influence by comparing the applications of ENPs with their ionic counterparts and the specificities of ENP application to find the best practices that provide the highest yield at the

Table 4 Impact of NPs on reproduction of the plants

Type of NPs	Size of NPs/concentration	Plant species	Flower parts/effects	References
Ag	10 nm 2 and 20 pollen 0, 5, 10, 15, and 20 mg/L	*Actinidia deliciosa*	Decreased pollen viability and performance, early changes in calcium content, specific ultrastructural alterations, imbalance of redox status, damaged pollen membranes, inhibited germination disruption of both the tube elongation process and pollen tube emergence, pollen grains dark and shrunken (at 20 mg/L), anomalous ultrastructures accumulated beneath the inner cell wall of the membranous structures at 20 mg/L, plasmalemma invagination, vacuolization, and large empty cavities in pollen grains	Speranza et al. [211]
Ag	12.5 mg/kg	*Arabidopsis thaliana*	Decreased expression of floral integrators, delayed flowering time, inhibition of pollen development, flower and calyx size decrease, decreased petal viability, stamen and pistil adversely affected, decreased pollen viability, thinner and shorter pods, decreased weight and length of pods, reduced pod production, small and unfilled seeds, inhibition of pod growth, and reduced seed number in pods	Ke et al. [198]

(continued)

Table 4 (continued)

Type of NPs	Size of NPs/concentration	Plant species	Flower parts/effects	References
Ag	0, 25, 50, 100, and 150 mg/L	*Lilium* "Mona Lisa"	"Mona Lisa"	Salachna et al. [196]
		Lilium "Little John"	Positive effect on morphological parameters, enhanced fresh weight of bulb (23.6–50.5%), the earlier flowering process by 2 or 3 days, and longer flowering time.	
			Soaking bulbs: Increased flower production, accelerated flowering, and increased number of flowers without affecting the flower longevity	
			"Little John"	
			Soaking bulbs: Stimulatory effect on flowering, greater fresh weight of bulbs (40.6–56.5%), increased flower production with longer tepals, and accelerated flowering by 2–4 days	
Ag	5, 10, 20, and 25 mg/L	*Peltophorum pterocarpum*	Increasing concentrations reduced pollen germination, tube growth, impaired pollen performance at the stage of the emergence of the pollen tube, changes in overall morphology of the pollen grains, extensive changes in the muri and lumen of exine and blockage of germ pore, decreased rate of pollen tube growth, decreased length of pollen tubes, and a lower amount of Ca^{2+} in apex pollen tube	Dutta Gupta et al. [207]

(continued)

Table 4 (continued)

Type of NPs	Size of NPs/concentration	Plant species	Flower parts/effects	References
Ag	2, 2.5, and 3 mg/L	*Prunus persica* "Florida Prince"	Increased flower percentage, fruit physical and chemical characteristics, and fruit yield.	Mosa et al. [200]
Zn	0.5, 1, and 1.5 mg/L		Different types of pollen aberrations during Zn NPs treatment (stickiness in content, ultrastructural changes in the exine and interior walls of pollen grains, increase in ultrastructural changes, partially or fully degenerated content, and shrunken pollen content with a big gap in capacity).	
			Ag NPs increased pollen viability without any aberrations and increased the pollen size and pores.	
Ag	25, 50, 100, and 150 mg/L	*Tulipa gesneriana* "Pink Impression"	Earlier flowering, increased length and fresh weight of cut flowers, no effect on the postharvest longevity of cut flowers, a higher number of daughter bulbs, enhanced fresh weight of daughter bulbs, shortened production cycle, improved quality of cut flowers, and stimulation effect at 50 and 100 mg/L	Byczyńska et al. [195]
Au	Size 10–20 nm	*Allium cepa*	Increments in the frequency of mitotic index (most significant at 1000 μM), increased mitotic index of *A. cepa* root tip cells, no cytotoxic effect in cell cycle, no endocytosis, and increased pollen germination in *G. superba*	Gopinath et al. [214], Alharbi et al. [213]
	0, 100, 500, and 1000 μM	*Gloriosa superba*		

(continued)

Table 4 (continued)

Type of NPs	Size of NPs/concentration	Plant species	Flower parts/effects	References
CeO_2	250–2000 mg/L	*Phaseolus vulgaris*	Disturbance during microsporogenesis at the chromosome pairing stage, formation of laggard chromosomes, collapsed and degenerated microspores, abnormal and degenerated tetrads, degenerated and misshapen pollen, and decreased germination.	Salehi et al. [199]
			Megagametogenesis disturbance in the stage of two-celled female gametophyte	
CeO_2	10 mg/L	*Sorghum bicolor*	Increased pollen germination. Foliar spray improved pollen germination (10%) under drought conditions.	Djanaguiraman et al. [216]
Fe_2O_3	3 and 25 mg/L	*Arabidopsis thaliana*	3 mg/L increased in the abortion of pollen and reduction in pollen viability	Bombin et al. [193]
Pd	5–10 nm 0.1–0.4 mg/L	*Actinidia deliciosa* "Tomuri" male genotype	Pd NPs altered kiwifruit pollen morphology (shape, wrinkled appearance, largely undulated exine, anomalous changes in pollen ultrastructure, damage of plasma membrane), immediate accumulation of NPs behind the wall, primarily in the proximity of the germination apertures, deeper cell penetration of NPs (in cytoplasm and vacuoles), decrease in pollen viability, inhibition of pollen germination and cessation of pollen tube emergence, perturbation in redox balance, and decrease and loss of endogenous calcium	Speranza et al. [211]

(continued)

Table 4 (continued)

Type of NPs	Size of NPs/concentration	Plant species	Flower parts/effects	References
Se and NaSe	0, 3, and 10 mg/L Foliar application	*Solanum lycopersicum*	The size and density of pollen grains affected in a dose- and material type-dependent manners, decreased density of pollen grains in pollen sacs, the size of pollen grains increased, and reduction in both the size and the density of pollen grains	Neysanian et al. [197]
TiO_2	0, 5, 10, 15, 20, and 40 mg/L	*Petunia hybrida*	Best flower diameter in 5 ppm and positive effect on morphological traits (flower diameter, fresh flower weight)	Kamali et al. [202]
TiO_2	0–1000 mg/L (kg)	*Solanum lycopersicum*	TiO_2 foliar application improved flowering at low concentrations. Soil-treated plants—a similar amount of flowers at the higher concentrations	
ZnO			ZnO increased the number of flowers, and the foliar application was more effective.	
			Higher accumulation of NPs—retarded plant growth and development	
ZnO	10, 20, 30, and 40 mg/L	*Allium cepa*	Reduced number of days required for flowering, a higher number of seeded fruits per umbel, higher number of flowers and seeded fruits per umbel, higher seed weight per umbel, and 1000 seed weight	Laware and Raskar [194]
ZnO	100 mg/L	*Lilium longiflorum*	Fine ZnO adhered to pollen grains. A decrease of the germination rate of pollen	Yoshihara et al. [209]

(continued)

Table 4 (continued)

Type of NPs	Size of NPs/concentration	Plant species	Flower parts/effects	References
Fullerenes	F1 C_{60}/C_{70} (20% C_{70} + 1% higher fullerenes)	*Cryptomeria japonica*	Fullerenes adhered to pollen grains (F1 more than F2), autofluorescence of all species decreased, germination ratio of *Camelina japonica* pollen grains with F1 adhesion decreased by one-third, viability reached 90%, and adhesion of F1 to pollen grains caused damage to the plasma membrane.	Aoyagi and Ugwu [208]
	F2 (C_{60} 99%)	*Chamaecyparis obtusa*		
		Camellia japonica		
GO	25, 50, and 100 mg/L	*Nicotiana tabacum*	Decreased pollen germination and tube elongation at GO concentrations higher than 50 mg/L by 20% and 19% in *N. tabacum* and 68% and 58% in *C. avellana*.	Candotto Carniel et al. [210]
GBMs	3 and 9 nm	*Corylus avellana*	Adhesion to the pollen grains and the pollen tube surfaces in both species.	
			GO did not affect the viability of *N. tabacum* pollen yet doubled the frequency of bent tubes.	
			At 100 mg/L, GO-induced acidification of the media is the main toxicity mechanism causing the impairment of pollen performance in *C. avellana*.	
	0.2% Lithovit	*Pisum sativum* "Pleven 4"	Positive effect on in vitro pollen germination and pollen tube elongation, higher pollen viability, and significantly highest values had Nargo.	Georgieva et al. [212]
	0.05% Nargo			

(continued)

Table 4 (continued)

Type of NPs	Size of NPs/concentration	Plant species	Flower parts/effects	References
CNS	10, 50, and 100 mg/L	*Nicotiana tabacum*	No inhibitory effect on pollen germination	Shah et al. [215]
CNTs	1 and 10 mg/kg	*Solanum lycopersicum* L. "Ailsa Craig"	Delayed flowering time. Plants exposed to CNTs treatments flowered 2–3 days later.	

lowest used concentrations. Furthermore, more studies into their use when applied at low concentrations where their plant growth and health improvements are combined with the economic viability of the application are also needed. One of the avenues to decrease the risk of ENPs is their green synthesis with chemicals produced by bacteria, algae, fungi, or plants. However, only a few studies compared such effects with inorganically synthesized ENPs. Studies on multiple generations of plants should be undertaken to assess the potential safety risks in the long term or in the future. One of the crucial strategies to evaluate the impact on the environment after the agricultural application of ENPs in the study of reproductive and environmental safety through the palynological analysis and more emphasis should be put on this research in the future.

Acknowledgment This research was supported by the Science Grant Agency of the Ministry of Education, Science, Research, and Sports of the Slovak Republic and the Slovak Academy of Sciences via grants VEGA No. VEGA 2 1/0175/22, by the project from the Grant Agency of the Slovak University of Agriculture in Nitra No. 04-GASPU-2021, and by the project supported by MHRD, New Delhi, titled "Design of piezoresistive micro cantilever beam for Bio/Agro applications," by the Design Innovation Center (DIC), Rajiv Gandhi University of Knowledge Technologies (RGUKT), Nuzvid, Andhra Pradesh, India.

References

1. Rumble, J. (2016). *Uniform description system for materials on the nanoscale. Version 2.0.* CODATA-VAMAS Working Group on the Description of Nanomaterials, Zenodo.
2. Tan, H. W., An, J., Chua, C. K., & Tran, T. (2019). Metallic nanoparticle inks for 3D printing of electronics. *Advanced Electronic Materials, 5*, 1800831. https://doi.org/10.1002/aelm.201800831
3. Selvakumar, D., Nagaraju, P., Arivanandhan, M., & Jayavel, R. (2021). Metal oxide–grafted graphene nanocomposites for energy storage applications. *Emergent Materials, 4*, 1143. https://doi.org/10.1007/s42247-021-00215-4
4. Zhou, D., Qiu, F., Wang, H., & Jiang, Q. (2014). Manufacture of nano-sized particle-reinforced metal matrix composites: A review. *Acta Metallurgica Sinica (English Letters), 27*, 798–805. https://doi.org/10.1007/s40195-014-0154-z

5. Li, Z., Ding, S., Yu, X., et al. (2018). Multifunctional cementitious composites modified with nano titanium dioxide: A review. *Composites. Part A, Applied Science and Manufacturing, 111*, 115–137. https://doi.org/10.1016/j.compositesa.2018.05.019
6. Antunes, A., Popelka, A., Aljarod, O., et al. (2020). Accelerated weathering effects on poly (3-hydroxybutyrate-co-3-hydroxyvalerate) (PHBV) and PHBV/TiO2 nanocomposites. *Polymers (Basel), 12*, 1743. https://doi.org/10.3390/polym12081743
7. Hirota, K., Sugimoto, M., Kato, M., et al. (2010). Preparation of zinc oxide ceramics with a sustainable antibacterial activity under dark conditions. *Ceramics International, 36*, 497–506. https://doi.org/10.1016/j.ceramint.2009.09.026
8. Luyt, A. S., Antunes, A., Popelka, A., et al. (2021). Effect of poly(ε-caprolactone) and titanium (IV) dioxide content on the UV and hydrolytic degradation of poly(lactic acid)/poly (ε-caprolactone) blends. *Journal of Applied Polymer Science, 138*, 51266. https://doi.org/10.1002/app.51266
9. Moezzi, A., McDonagh, A. M., & Cortie, M. B. (2012). Zinc oxide particles: Synthesis, properties and applications. *Chemical Engineering Journal, 185–186*, 1–22. https://doi.org/10.1016/j.cej.2012.01.076
10. Grant, C. A., Twigg, P. C., Baker, R., & Tobin, D. J. (2015). Tattoo ink nanoparticles in skin tissue and fibroblasts. *Beilstein Journal of Nanotechnology, 6*, 1183–1191. https://doi.org/10.3762/bjnano.6.120
11. Jašková, V., Hochmannová, L., & Vytřasová, J. (2013). TiO 2 and ZnO nanoparticles in photocatalytic and hygienic coatings. *International Journal of Photoenergy, 2013*, 1–6. https://doi.org/10.1155/2013/795060
12. Larue, C., Castillo-Michel, H., Sobanska, S., et al. (2014). Fate of pristine TiO2 nanoparticles and aged paint-containing TiO2 nanoparticles in lettuce crop after foliar exposure. *Journal of Hazardous Materials, 273*, 17–26. https://doi.org/10.1016/j.jhazmat.2014.03.014
13. Sarina, S., Waclawik, E. R., & Zhu, H. (2013). Photocatalysis on supported gold and silver nanoparticles under ultraviolet and visible light irradiation. *Green Chemistry, 15*, 1814–1833. https://doi.org/10.1039/C3GC40450A
14. Vinitha, V., Preeyanghaa, M., Vinesh, V., et al. (2021). Two is better than one: Catalytic, sensing and optical applications of doped zinc oxide nanostructures. *Emergent Materials, 4*, 1093. https://doi.org/10.1007/s42247-021-00262-x
15. Yan, N., Yuan, Y., & Dyson, P. J. (2013). Nanometallic chemistry: Deciphering nanoparticle catalysis from the perspective of organometallic chemistry and homogeneous catalysis. *Dalton Transactions, 42*, 13294–13304. https://doi.org/10.1039/C3DT51180D
16. Klębowski, B., Depciuch, J., Parlińska-Wojtan, M., & Baran, J. (2018). Applications of noble metal-based nanoparticles in medicine. *International Journal of Molecular Sciences, 19*, 4031. https://doi.org/10.3390/ijms19124031
17. De la Calle, I., Menta, M., & Séby, F. (2016). Current trends and challenges in sample preparation for metallic nanoparticles analysis in daily products and environmental samples: A review. *Spectrochimica Acta Part B: Atomic Spectroscopy, 125*, 66–96. https://doi.org/10.1016/j.sab.2016.09.007
18. Lu, P.-J., Huang, S.-C., Chen, Y.-P., et al. (2015). Analysis of titanium dioxide and zinc oxide nanoparticles in cosmetics. *Journal of Food and Drug Analysis, 23*, 587–594. https://doi.org/10.1016/j.jfda.2015.02.009
19. Souza, V. G. L., & Fernando, A. L. (2016). Nanoparticles in food packaging: Biodegradability and potential migration to food—A review. *Food Packaging and Shelf Life, 8*, 63–70. https://doi.org/10.1016/J.FPSL.2016.04.001
20. Saravanan, A., Kumar, P. S., Karishma, S., et al. (2021). A review on biosynthesis of metal nanoparticles and its environmental applications. *Chemosphere, 264*, 128580. https://doi.org/10.1016/J.CHEMOSPHERE.2020.128580
21. Xie, J., Zhang, X., Wang, H., et al. (2012). Analytical and environmental applications of nanoparticles as enzyme mimetics. *TrAC Trends in Analytical Chemistry, 39*, 114–129. https://doi.org/10.1016/J.TRAC.2012.03.021

22. Ditta, A., & Arshad, M. (2016). Applications and perspectives of using nanomaterials for sustainable plant nutrition. *Nanotechnology Reviews, 5*, 209–229. https://doi.org/10.1515/ntrev-2015-0060
23. Lombi, E., Donner, E., Dusinska, M., & Wickson, F. (2019). A one health approach to managing the applications and implications of nanotechnologies in agriculture. *Nature Nanotechnology, 14*, 523–531. https://doi.org/10.1038/s41565-019-0460-8
24. Monreal, C. M., DeRosa, M., Mallubhotla, S. C., et al. (2016). Nanotechnologies for increasing the crop use efficiency of fertilizer-micronutrients. *Biology and Fertility of Soils, 52*, 423–437. https://doi.org/10.1007/s00374-015-1073-5
25. Wang, P., Lombi, E., Zhao, F.-J., & Kopittke, P. M. (2016). Nanotechnology: A new opportunity in plant sciences. *Trends in Plant Science, 21*, 699–712. https://doi.org/10.1016/j.tplants.2016.04.005
26. Servin, A., Elmer, W., Mukherjee, A., et al. (2015). A review of the use of engineered nanomaterials to suppress plant disease and enhance crop yield. *Journal of Nanoparticle Research, 17*, 92. https://doi.org/10.1007/s11051-015-2907-7
27. Navarro, E., Baun, A., Behra, R., et al. (2008). Environmental behavior and ecotoxicity of engineered nanoparticles to algae, plants, and fungi. *Ecotoxicology, 17*, 372–386. https://doi.org/10.1007/s10646-008-0214-0
28. Ranjan, A., Rajput, V. D., Minkina, T., et al. (2021). Nanoparticles induced stress and toxicity in plants. *Environmental Nanotechnology, Monitoring and Management, 15*, 100457. https://doi.org/10.1016/J.ENMM.2021.100457
29. Irshad, M. A., ur Rehman, M. Z., Anwar-ul-Haq, M., et al. (2021). Effect of green and chemically synthesized titanium dioxide nanoparticles on cadmium accumulation in wheat grains and potential dietary health risk: A field investigation. *Journal of Hazardous Materials, 415*, 125585. https://doi.org/10.1016/J.JHAZMAT.2021.125585
30. Khattak, A., Ullah, F., Shinwari, Z. K., & Mehmood, S. (2021). The effect of titanium dioxide nanoparticles and salicylic acid on growth and biodiesel production potential of sunflower (helianthus annuus l.) Under water stress. *Pakistan Journal of Botany, 53*, 1987–1995.
31. Kolenčík, M., Ernst, D., Komár, M., et al. (2019). Effect of foliar spray application of zinc oxide nanoparticles on quantitative, nutritional, and physiological parameters of foxtail millet (Setaria italica l.) under field conditions. *Nanomaterials, 9*, 1559. https://doi.org/10.3390/nano9111559
32. Kolenčík, M., Ernst, D., Urík, M., et al. (2020). Foliar application of low concentrations of titanium dioxide and zinc oxide nanoparticles to the common sunflower under field conditions. *Nanomaterials, 10*, 1619. https://doi.org/10.3390/nano10081619
33. Umar, W., Hameed, M. K., Aziz, T., et al. (2021). Synthesis, characterization and application of ZnO nanoparticles for improved growth and Zn biofortification in maize. *Archives of Agronomy and Soil Science, 67*, 1164–1176. https://doi.org/10.1080/03650340.2020.1782893
34. Lian, J., Zhao, L., Wu, J., et al. (2020). Foliar spray of TiO2 nanoparticles prevails over root application in reducing Cd accumulation and mitigating Cd-induced phytotoxicity in maize (Zea mays L.). *Chemosphere, 239*, 124794. https://doi.org/10.1016/j.chemosphere.2019.124794
35. Karunanayaka, M. P. (2021). Nanofertilizers use for modern agriculture. *Journal of Research Technology and Engineering, 2*, 86–91.
36. Shaban, E. E., Elbakry, H. F. H., Ibrahim, K. S., et al. (2019). The effect of white kidney bean fertilized with nano-zinc on nutritional and biochemical aspects in rats. *Biotechnology Reports, 23*, e00357. https://doi.org/10.1016/J.BTRE.2019.E00357
37. Ndlovu, N., Tatenda, M., Clemence, M., & Munyengwa, N. (2020). Nanotechnology applications in crop production and food systems. *International Journal of Plant Breeding and Crop Science, 7*, 603–613.
38. Veronica, N., Tulasi, G., Ramesh, T., & Narender Reddy, S. (2015). Role of nano fertilizers in agricultural farming. *International journal of Environmental Science and Technology, 1*, 1–3.

39. Zulfiqar, F., Navarro, M., Ashraf, M., et al. (2019). Nanofertilizer use for sustainable agriculture: Advantages and limitations. *Plant Science, 289*, 110270. https://doi.org/10.1016/J.PLANTSCI.2019.110270
40. Javed, Z., Dashora, K., Mishra, M., et al. (2019). Effect of accumulation of nanoparticles in soil health - A concern on future. *Frontiers in Nanoscience and Nanotechnology, 5*, 1–9.
41. Saini, S., Kumar, P., Sharma, N. C., et al. (2021). Nano-enabled Zn fertilization against conventional Zn analogues in strawberry (Fragaria × ananassa Duch.). *Scientia Horticulturae, 282*, 110016. https://doi.org/10.1016/J.SCIENTA.2021.110016
42. Faizan, M., Faraz, A., Yusuf, M., et al. (2018). Zinc oxide nanoparticle-mediated changes in photosynthetic efficiency and antioxidant system of tomato plants. *Photosynthetica, 56*, 678–686. https://doi.org/10.1007/s11099-017-0717-0
43. García-López, J. I., Zavala-García, F., Olivares-Sáenz, E., et al. (2018). Zinc oxide nanoparticles boosts phenolic compounds and antioxidant activity of Capsicum annuum L. during germination. *Agronomy, 8*, 215. https://doi.org/10.3390/agronomy8100215
44. Peralta-Videa, J. R., Hernandez-Viezcas, J. A., Zhao, L., et al. (2014). Cerium dioxide and zinc oxide nanoparticles alter the nutritional value of soil cultivated soybean plants. *Plant Physiology and Biochemistry, 80*, 128–135. https://doi.org/10.1016/j.plaphy.2014.03.028
45. Salama, D. M., Osman, S. A., Abd El-Aziz, M. E., et al. (2019). Effect of zinc oxide nanoparticles on the growth, genomic DNA, production and the quality of common dry bean (Phaseolus vulgaris). *Biocatalysis and Agricultural Biotechnology, 18*, 101083. https://doi.org/10.1016/j.bcab.2019.101083
46. Singh, J., Kumar, S., Alok, A., et al. (2019a). The potential of green synthesized zinc oxide nanoparticles as nutrient source for plant growth. *Journal of Cleaner Production, 214*, 1061–1070. https://doi.org/10.1016/j.jclepro.2019.01.018
47. Kőrösi, L., Bouderias, S., Csepregi, K., et al. (2019). Nanostructured TiO2-induced photocatalytic stress enhances the antioxidant capacity and phenolic content in the leaves of Vitis vinifera on a genotype-dependent manner. *Journal of Photochemistry and Photobiology B: Biology, 190*, 137–145. https://doi.org/10.1016/J.JPHOTOBIOL.2018.11.010
48. Prasad, T. N. V. K. V., Sudhakar, P., Sreenivasulu, Y., et al. (2012). Effect of nanoscale zinc oxide particles on the germination, growth and yield of peanut. *Journal of Plant Nutrition, 35*, 905–927. https://doi.org/10.1080/01904167.2012.663443
49. Kurtinová, S., & Šebesta, M. (2022). Heavy metal stress alleviation in plants by ZnO and TiO2 nanoparticles. In A. P. Ingle (Ed.), *Nanotechnology in agro-ecosystem*. Elsevier.
50. Šebesta, M., Kurtinová, S., Kolenčík, M., & Illa, R. (2021b). Enhancement of stress tolerance of crop plants by ZnO nanoparticles. In M. Faizan, S. Hayat, & F. Yu (Eds.), *Sustainable agriculture reviews. Nanoparticles: A new tool to enhance stress tolerance* (Vol. 53). Springer International Publishing.
51. Solanki, P., Bhargava, A., Chhipa, H., et al. (2015). Nano-fertilizers and their smart delivery system. In M. Rai, C. Ribeiro, L. Mattoso, & N. Duran (Eds.), *Nanotechnologies in food and agriculture* (pp. 81–101). Springer International Publishing.
52. Bollag, J.-M., Myers, C. J., & Minard, R. D. (1992). Biological and chemical interactions of pesticides with soil organic matter. *Science of the Total Environment, 123–124*, 205–217. https://doi.org/10.1016/0048-9697(92)90146-J
53. Chhipa, H. (2017). Nanofertilizers and nanopesticides for agriculture. *Environmental Chemistry Letters, 15*, 15–22. https://doi.org/10.1007/s10311-016-0600-4
54. Singh, M., Srivastava, M., Kumar, A., & Pandey, K. D. (2019b). 12 - Biosynthesis of nanoparticles and applications in agriculture. In A. Kumar, A. K. Singh, & K. K. Choudhary (Eds.), *Role of plant growth promoting microorganisms in sustainable agriculture and nanotechnology* (pp. 199–217). Woodhead Publishing.
55. Chugh, G., Siddique, K. H. M., & Solaiman, Z. M. (2021). Nanobiotechnology for agriculture: smart technology for combating nutrient deficiencies with nanotoxicity challenges. *Sustainability, 13*, 1781. https://doi.org/10.3390/su13041781

56. Nath, J., Dror, I., Landa, P., et al. (2019). Isotopic labelling for sensitive detection of nanoparticle uptake and translocation in plants from hydroponic medium and soil. *Environment and Chemistry, 16*, 391–400.
57. Cheng, Y., Yin, L., Lin, S., et al. (2011). Toxicity reduction of polymer-stabilized silver nanoparticles by sunlight. *Journal of Physical Chemistry C, 115*, 4425–4432. https://doi.org/10.1021/jp109789j
58. Holišová, V., Konvičková, Z., Kratošová, G., et al. (2019). Phytosynthesis of Au and Au/ZrO 2 bi-phasic system nanoparticles with evaluation of their colloidal stability. *Journal of Nanoscience and Nanotechnology, 19*, 2807–2813. https://doi.org/10.1166/jnn.2019.15851
59. Holišová, V., Urban, M., Konvičková, Z., et al. (2021). Colloidal stability of phytosynthesised gold nanoparticles and their catalytic effects for nerve agent degradation. *Scientific Reports, 11*, 4071. https://doi.org/10.1038/s41598-021-83460-1
60. Kolenčík, M., Nemček, L., Šebesta, M., et al. (2021). Effect of TiO2 as plant growth-stimulating nanomaterial on crop production. In V. P. Singh, S. Singh, S. M. Prasad, et al. (Eds.), *Plant responses to nanomaterials* (pp. 129–144). Springer Nature Switzerland AG.
61. Konvičková, Z., Holišová, V., Kolenčík, M., et al. (2018). Phytosynthesis of colloidal Ag-AgCl nanoparticles mediated by Tilia sp. leachate, evaluation of their behaviour in liquid phase and catalytic properties. *Colloid & Polymer Science, 296*, 677–687. https://doi.org/10.1007/s00396-018-4290-2
62. Gahukar, R. T., & Das, R. K. (2020). Plant-derived nanopesticides for agricultural pest control: Challenges and prospects. *Nanotechnology for Environmental Engineering, 5*, 3. https://doi.org/10.1007/s41204-020-0066-2
63. Ghosh Chaudhuri, R., & Paria, S. (2012). Core/shell nanoparticles: Classes, properties, synthesis mechanisms, characterization, and applications. *Chemical Reviews, 112*, 2373–2433. https://doi.org/10.1021/cr100449n
64. Yecheskel, Y., Dror, I., & Berkowitz, B. (2016). Transport of engineered nanoparticles in partially saturated sand columns. *Journal of Hazardous Materials, 311*, 254–262. https://doi.org/10.1016/j.jhazmat.2016.03.027
65. Le, T. C., Yin, H., Chen, R., et al. (2016). An experimental and computational approach to the development of ZnO nanoparticles that are safe by design. *Small, 12*, 3568–3577. https://doi.org/10.1002/smll.201600597
66. Nair, S., Sasidharan, A., Divya Rani, V. V., et al. (2009). Role of size scale of ZnO nanoparticles and microparticles on toxicity toward bacteria and osteoblast cancer cells. *Journal of Materials Science. Materials in Medicine, 20*, 235–241. https://doi.org/10.1007/s10856-008-3548-5
67. López-Moreno, M. L., Cedeño-Mattei, Y., Bailón-Ruiz, S. J., et al. (2018). Environmental behavior of coated NMs: Physicochemical aspects and plant interactions. *Journal of Hazardous Materials, 347*, 196–217. https://doi.org/10.1016/j.jhazmat.2017.12.058
68. Medina-Velo, I. A., Barrios, A. C., Zuverza-Mena, N., et al. (2017). Comparison of the effects of commercial coated and uncoated ZnO nanomaterials and Zn compounds in kidney bean (Phaseolus vulgaris) plants. *Journal of Hazardous Materials, 332*, 214–222. https://doi.org/10.1016/j.jhazmat.2017.03.008
69. Persaud, I., Raghavendra, A. J., Paruthi, A., et al. (2020). Defect-induced electronic states amplify the cellular toxicity of ZnO nanoparticles. *Nanotoxicology, 14*, 145–161. https://doi.org/10.1080/17435390.2019.1668067
70. Gebre, S. H., & Sendeku, M. G. (2019). New frontiers in the biosynthesis of metal oxide nanoparticles and their environmental applications: An overview. *SN Applied Sciences, 1*, 928. https://doi.org/10.1007/s42452-019-0931-4
71. Huang, Y., Haw, C. Y., Zheng, Z., et al. (2021). Biosynthesis of zinc oxide nanomaterials from plant extracts and future green prospects: A topical review. *Advanced Sustainable Systems, 5*, 2000266. https://doi.org/10.1002/adsu.202000266
72. Bian, S.-W. W., Mudunkotuwa, I. A., Rupasinghe, T., & Grassian, V. H. (2011). Aggregation and dissolution of 4 nm ZnO nanoparticles in aqueous environments: Influence of pH, ionic

strength, size, and adsorption of humic acid. *Langmuir, 27*, 6059–6068. https://doi.org/10.1021/la200570n
73. Chang, Y.-N., Zhang, M., Xia, L., et al. (2012). The toxic effects and mechanisms of CuO and ZnO nanoparticles. *Materials (Basel), 5*, 2850–2871. https://doi.org/10.3390/ma5122850
74. Dietz, K.-J., & Herth, S. (2011). Plant nanotoxicology. *Trends in Plant Science, 16*, 582–589. https://doi.org/10.1016/j.tplants.2011.08.003
75. Meulenkamp, E. A. (1998). Size dependence of the dissolution of ZnO nanoparticles. *The Journal of Physical Chemistry. B, 102*, 7764–7769. https://doi.org/10.1021/jp982305u
76. Molnárová, M., Filová, A., & Peško, M. (2015). *Iónové a nanočasticové formy ťažkých kovov v prostredí a ich interakcia s fotosyntetizujúcimi organizmami*. Vysoká škola báňská - Technická univerzita Ostrava, Hornicko-geologická fakulta, Institut environmentálního inženýrství.
77. Mudunkotuwa, I. A., Rupasinghe, T., Wu, C.-M., & Grassian, V. H. (2012). Dissolution of ZnO nanoparticles at circumneutral pH: A study of size effects in the presence and absence of citric acid. *Langmuir, 28*, 396–403. https://doi.org/10.1021/la203542x
78. Gogarten, J. P. (1988). Physical properties of the cell wall of photoautotrophic suspension cells from Chenopodium rubrum L. *Planta, 174*, 333–339. https://doi.org/10.1007/bf00959518
79. Nemček, L., Šebesta, M., Urík, M., et al. (2020). Impact of bulk ZnO, ZnO nanoparticles and dissolved Zn on early growth stages of barley—A pot experiment. *Plants, 9*, 1365. https://doi.org/10.3390/plants9101365
80. Pedruzzi, D. P., Araujo, L. O., Falco, W. F., et al. (2020). ZnO nanoparticles impact on the photosynthetic activity of Vicia faba: Effect of particle size and concentration. *NanoImpact, 19*, 100246. https://doi.org/10.1016/j.impact.2020.100246
81. Thwala, M., Klaine, S., & Musee, N. (2021). Exposure media and nanoparticle size influence on the fate, bioaccumulation, and toxicity of silver nanoparticles to higher plant Salvinia minima. *Molecules, 26*, 2305. https://doi.org/10.3390/molecules26082305
82. Yusefi-Tanha, E., Fallah, S., Rostamnejadi, A., & Pokhrel, L. R. (2020). Particle size and concentration dependent toxicity of copper oxide nanoparticles (CuONPs) on seed yield and antioxidant defense system in soil grown soybean (Glycine max cv. Kowsar). *Science of the Total Environment, 715*, 136994. https://doi.org/10.1016/j.scitotenv.2020.136994
83. Thorny Chanu, T., & Upadhyaya, H. (2019). Chapter 3 - Zinc oxide nanoparticle-induced responses on plants: a physiological perspective. In D. K. Tripathi, P. Ahmad, S. Sharma, et al. (Eds.), *Nanomaterials in plants, algae and microorganisms* (pp. 43–64). Academic Press.
84. Chiu, W. S., Khiew, P. S., Cloke, M., et al. (2010). Photocatalytic study of two-dimensional ZnO nanopellets in the decomposition of methylene blue. *Chemical Engineering Journal, 158*, 345–352. https://doi.org/10.1016/j.cej.2010.01.052
85. José-Yacamán, M., Gutierrez-Wing, C., Miki, M., et al. (2005). Surface diffusion and coalescence of mobile metal nanoparticles. *The Journal of Physical Chemistry. B, 109*, 9703–9711. https://doi.org/10.1021/jp0509459
86. Chen, W. J., Liu, W. L., Hsieh, S. H., & Tsai, T. K. (2007). Preparation of nanosized ZnO using α brass. *Applied Surface Science, 253*, 6749–6753. https://doi.org/10.1016/j.apsusc.2007.01.091
87. Frade, T., Melo Jorge, M. E., & Gomes, A. (2012). One-dimensional ZnO nanostructured films: Effect of oxide nanoparticles. *Materials Letters, 82*, 13–15. https://doi.org/10.1016/j.matlet.2012.05.028
88. Huang, Y., He, J., Zhang, Y., et al. (2006). Morphology, structures and properties of ZnO nanobelts fabricated by Zn-powder evaporation without catalyst at lower temperature. *Journal of Materials Science, 41*, 3057–3062. https://doi.org/10.1007/s10853-006-6978-9
89. Kong, X. Y., Ding, Y., Yang, R., & Wang, Z. L. (2004). Single-crystal nanorings formed by epitaxial self-coiling of polar nanobelts. *Science, 303*, 1348–1351. https://doi.org/10.1126/science.1092356

90. Liu, J., Huang, X., Duan, J., et al. (2005). A low-temperature synthesis of multiwhisker-based zinc oxide micron crystals. *Materials Letters, 59*, 3710–3714. https://doi.org/10.1016/j.matlet.2005.06.043
91. Nikoobakht, B., Wang, X., Herzing, A., & Shi, J. (2013). Scalable synthesis and device integration of self-registered one-dimensional zinc oxide nanostructures and related materials. *Chemical Society Reviews, 42*, 342–365. https://doi.org/10.1039/C2CS35164A
92. Pan, Z. W., Dai, Z. R., & Wang, Z. L. (2001). Nanobelts of semiconducting oxides. *Science, 291*, 1947–1949. https://doi.org/10.1126/science.1058120
93. Wahab, R., Ansari, S. G., Kim, Y.-S., et al. (2007). Room temperature synthesis of needle-shaped ZnO nanorods via sonochemical method. *Applied Surface Science, 253*, 7622–7626. https://doi.org/10.1016/j.apsusc.2007.03.060
94. Xu, T., Ji, P., He, M., & Li, J. (2012). Growth and structure of pure ZnO micro/nanocombs. *Journal of Nanomaterials, 2012*, 797935. https://doi.org/10.1155/2012/797935
95. Bitenc, M., & Crnjak Orel, Z. (2009). Synthesis and characterization of crystalline hexagonal bipods of zinc oxide. *Materials Research Bulletin, 44*, 381–387. https://doi.org/10.1016/j.materresbull.2008.05.005
96. Liu, J., Huang, X., Li, Y., et al. (2006). Selective growth and properties of zinc oxide nanostructures. *Scripta Materialia, 55*, 795–798. https://doi.org/10.1016/j.scriptamat.2006.07.010
97. Polshettiwar, V., Baruwati, B., & Varma, R. S. (2009). Self-assembly of metal oxides into three-dimensional nanostructures: Synthesis and application in catalysis. *ACS Nano, 3*, 728–736. https://doi.org/10.1021/nn800903p
98. Zhou, D., & Keller, A. A. (2010). Role of morphology in the aggregation kinetics of ZnO nanoparticles. *Water Research, 44*, 2948–2956. https://doi.org/10.1016/j.watres.2010.02.025
99. Joo, S. H., & Zhao, D. (2017). Environmental dynamics of metal oxide nanoparticles in heterogeneous systems: A review. *Journal of Hazardous Materials, 322*, 29–47. https://doi.org/10.1016/j.jhazmat.2016.02.068
100. Pérez Velasco, E. A., Betancourt Galindo, R., Valdez Aguilar, L. A., et al. (2020). Effects of the morphology, surface modification and application methods of ZnO-NPs on the growth and biomass of tomato plants. *Molecules, 25*, 1282. https://doi.org/10.3390/molecules25061282
101. Mahil, E. I. T., & Kumar, B. N. A. (2019). Foliar application of nanofertilizers in agricultural crops–A review. *Journal of Farm Science, 32*, 239–249.
102. Dutta, R. K., Nenavathu, B. P., & Talukdar, S. (2014). Anomalous antibacterial activity and dye degradation by selenium doped ZnO nanoparticles. *Colloids and Surfaces. B, Biointerfaces, 114*, 218–224. https://doi.org/10.1016/j.colsurfb.2013.10.007
103. Baranowska-Wójcik, E., Szwajgier, D., Oleszczuk, P., & Winiarska-Mieczan, A. (2020). Effects of titanium dioxide nanoparticles exposure on human health—A review. *Biological Trace Element Research, 193*, 118–129. https://doi.org/10.1007/s12011-019-01706-6
104. Jiang, J., Oberdörster, G., Elder, A., et al. (2008). Does nanoparticle activity depend upon size and crystal phase? *Nanotoxicology, 2*, 33–42. https://doi.org/10.1080/17435390701882478
105. Zhang, J., Zhou, P., Liu, J., & Yu, J. (2014). New understanding of the difference of photocatalytic activity among anatase, rutile and brookite TiO2. *Physical Chemistry Chemical Physics, 16*, 20382–20386. https://doi.org/10.1039/C4CP02201G
106. Kamal, R., & Mogazy, A. M. (2021). Effect of doping on TiO2 nanoparticles characteristics: Studying of fertilizing effect on cowpea plant growth and yield. *Journal of Soil Science and Plant Nutrition*. https://doi.org/10.1007/s42729-021-00648-0
107. Yuan, Y., Ding, J., Xu, J., et al. (2010). TiO2 nanoparticles co-doped with silver and nitrogen for antibacterial application. *Journal of Nanoscience and Nanotechnology, 10*, 4868–4874. https://doi.org/10.1166/jnn.2010.2225
108. El-Gazzar, N., & Ismail, A. M. (2020). The potential use of Titanium, Silver and Selenium nanoparticles in controlling leaf blight of tomato caused by Alternaria alternata. *Biocatalysis and Agricultural Biotechnology, 27*, 101708. https://doi.org/10.1016/J.BCAB.2020.101708

109. Hong, F., Yang, F., Liu, C., et al. (2005a). Influences of nano-TiO2 on the chloroplast aging of spinach under light. *Biological Trace Element Research, 104*, 249–260. https://doi.org/10.1385/BTER:104:3:249
110. Hong, F., Zhou, J., Liu, C., et al. (2005b). Effect of nano-TiO2 on photochemical reaction of chloroplasts of spinach. *Biological Trace Element Research, 105*, 269–279. https://doi.org/10.1385/BTER:105:1-3:269
111. Zheng, L., Hong, F., Lu, S., & Liu, C. (2005). Effect of nano-TiO2 on strength of naturally aged seeds and growth of spinach. *Biological Trace Element Research, 104*, 83–91.
112. Nairn, J. J., Forster, W. A., & van Leeuwen, R. M. (2016). Effect of solution and leaf surface polarity on droplet spread area and contact angle. *Pest Management Science, 72*, 551–557. https://doi.org/10.1002/ps.4022
113. Papierowska, E., Szporak-Wasilewska, S., Szewińska, J., et al. (2018). Contact angle measurements and water drop behavior on leaf surface for several deciduous shrub and tree species from a temperate zone. *Trees, 32*, 1253–1266. https://doi.org/10.1007/s00468-018-1707-y
114. He, Y., Xiao, S., Wu, J., & Fang, H. (2019). Influence of multiple factors on the wettability and surface free energy of leaf surface. *Applied Sciences, 9*, 593. https://doi.org/10.3390/app9030593
115. Schönherr, J., & Bukovac, M. J. (1972). Penetration of stomata by liquids: Dependence on surface tension, wettability, and stomatal morphology 1. *Plant Physiology, 49*, 813–819. https://doi.org/10.1104/pp.49.5.813
116. Weiss, A. (1988). Contact angle of water droplets in relation to leaf water potential. *Agricultural and Forest Meteorology, 43*, 251–259. https://doi.org/10.1016/0168-1923(88)90053-6
117. Hu, P., An, J., Faulkner, M. M., et al. (2020). Nanoparticle charge and size control foliar delivery efficiency to plant cells and organelles. *ACS Nano, 14*, 7970–7986. https://doi.org/10.1021/acsnano.9b09178
118. Mirza, A. U., Kareem, A., Nami, S. A. A., et al. (2019). Malus pumila and Juglen regia plant species mediated zinc oxide nanoparticles: Synthesis, spectral characterization, antioxidant and antibacterial studies. *Microbial Pathogenesis, 129*, 233–241. https://doi.org/10.1016/J.MICPATH.2019.02.020
119. Chen, Y., Gao, N., & Jiang, J. (2013). Surface matters: Enhanced bactericidal property of core-shell Ag-Fe2O3 nanostructures to their heteromer counterparts from one-pot synthesis. *Small, 9*, 3242. https://doi.org/10.1002/smll.201300543
120. Shah, M., Badwaik, V., Kherde, Y., et al. (2014). Gold nanoparticles: Various methods of synthesis and antibacterial applications. *Frontiers in Bioscience, 19*, 1320–1344.
121. Ehsan, M., Raja, N. I., Mashwani, Z.-R., et al. (2021). Responses of bimetallic Ag/ZnO alloy nanoparticles and urea on morphological and physiological attributes of wheat. *IET Nanobiotechnology, 15*, 602–610. https://doi.org/10.1049/nbt2.12048
122. Pérez-de-Luque, A., & Rubiales, D. (2009). Nanotechnology for parasitic plant control. *Pest Management Science, 65*, 540–545. https://doi.org/10.1002/ps.1732
123. Liu, R., & Lal, R. (2015). Potentials of engineered nanoparticles as fertilizers for increasing agronomic productions. *Science of the Total Environment, 514*, 131–139. https://doi.org/10.1016/j.scitotenv.2015.01.104
124. Li, C., Wang, P., van der Ent, A., et al. (2019). Absorption of foliar-applied Zn in sunflower (Helianthus annuus): Importance of the cuticle, stomata and trichomes. *Annals of Botany, 123*, 57–68. https://doi.org/10.1093/aob/mcy135
125. Sturikova, H., Krystofova, O., Huska, D., & Adam, V. (2018). Zinc, zinc nanoparticles and plants. *Journal of Hazardous Materials, 349*, 101–110. https://doi.org/10.1016/j.jhazmat.2018.01.040
126. Šebesta, M., Illa, R., Zvěřina, O., et al. (2021a). Positive effects of TiO2 nanomaterials on crop growth. In A. P. Ingle (Ed.), *Nanotechnology for plant diseases* (pp. 17–44). Wiley.
127. Lyu, S., Wei, X., Chen, J., et al. (2017). Titanium as a beneficial element for crop production. *Frontiers in Plant Science, 8*, 597.

128. Zierden, M. R., & Valentine, A. M. (2016). Contemplating a role for titanium in organisms. *Metallomics, 8*, 9–16. https://doi.org/10.1039/C5MT00231A
129. Lei, Z., Mingyu, S., Chao, L., et al. (2007). Effects of nanoanatase TiO2 on photosynthesis of spinach chloroplasts under different light illumination. *Biological Trace Element Research, 119*, 68–76. https://doi.org/10.1007/s12011-007-0047-3
130. Chen, X., & Mao, S. S. (2007). Titanium dioxide nanomaterials: Synthesis, properties, modifications, and applications. *Chemical Reviews, 107*, 2891–2959. https://doi.org/10.1021/cr0500535
131. Weldegebrieal, G. K. (2020). Synthesis method, antibacterial and photocatalytic activity of ZnO nanoparticles for azo dyes in wastewater treatment: A review. *Inorganic Chemistry Communications, 120*, 108140. https://doi.org/10.1016/j.inoche.2020.108140
132. Govorov, A. O., & Carmeli, I. (2007). Hybrid structures composed of photosynthetic system and metal nanoparticles: Plasmon enhancement effect. *Nano Letters, 7*, 620–625. https://doi.org/10.1021/nl062528t
133. Berahmand, A. A., Ghafariyan Panahi, A., Sahabi, H., et al. (2012). Effects silver nanoparticles and magnetic field on growth of fodder maize (Zea mays L.). *Biological Trace Element Research, 149*, 419–424. https://doi.org/10.1007/s12011-012-9434-5
134. Seghatoleslami, M., Feizi, H., Mousavi, G., & Berahmand, A. (2015). Effect of magnetic field and silver nanoparticles on yield and water use efficiency of Carum copticum under water stress conditions. *Polish Journal of Chemical Technology, 17*, 110–114. https://doi.org/10.1515/pjct-2015-0016
135. González-Melendi, P., Fernández-Pacheco, R., Coronado, M. J. E., et al. (2008). Nanoparticles as smart treatment-delivery systems in plants: assessment of different techniques of microscopy for their visualization in plant tissues. *Annals of Botany, 101*, 187–195. https://doi.org/10.1093/aob/mcm283
136. Lv, Z., Jiang, R., Chen, J., & Chen, W. (2020). Nanoparticle-mediated gene transformation strategies for plant genetic engineering. *The Plant Journal, 104*, 880–891. https://doi.org/10.1111/tpj.14973
137. El-Batal, A. I., Gharib, F. A. E.-L., Ghazi, S. M., et al. (2016). Physiological responses of two varieties of common bean (Phaseolus Vulgaris L.) to foliar application of silver nanoparticles. *Nanomaterials and Nanotechnology, 6*, 13. https://doi.org/10.5772/62202
138. Costa, M. V. J., & Sharma, P. K. (2016). Effect of copper oxide nanoparticles on growth, morphology, photosynthesis, and antioxidant response in Oryza sativa. *Photosynthetica, 54*, 110–119. https://doi.org/10.1007/s11099-015-0167-5
139. Askary, M., Talebi, S. M., Amini, F., & Bangan, A. D. B. (2016). Effects of stress on foliar trichomes plasticity in Mentha piperita. *Nusantara Bioscience, 8*, 32–38. https://doi.org/10.13057/nusbiosci/n080107
140. Ahmad, B., Shabbir, A., Jaleel, H., et al. (2018). Efficacy of titanium dioxide nanoparticles in modulating photosynthesis, peltate glandular trichomes and essential oil production and quality in Mentha piperita L. *Current Plant Biology, 13*, 6–15. https://doi.org/10.1016/j.cpb.2018.04.002
141. Rajput, V. D., Minkina, T., Fedorenko, A., et al. (2021). Effects of zinc oxide nanoparticles on physiological and anatomical indices in spring barley tissues. *Nanomaterials, 11*, 1722. https://doi.org/10.3390/nano11071722
142. Blamey, F. P. C., Joyce, D. C., Edwards, D. G., & Asher, C. J. (1986). Role of trichomes in sunflower tolerance to manganese toxicity. *Plant and Soil, 91*, 171–180. https://doi.org/10.1007/BF02181785
143. Rajiv, P., Vanathi, P., & Thangamani, A. (2018). An investigation of phytotoxicity using Eichhornia mediated zinc oxide nanoparticles on Helianthus annuus. *Biocatalysis and Agricultural Biotechnology, 16*, 419–424. https://doi.org/10.1016/J.BCAB.2018.09.017
144. Noshad, A., Hetherington, C., & Iqbal, M. (2019). Impact of AgNPs on seed germination and seedling growth: A focus study on its antibacterial potential against *Clavibacter michiganensis*

subsp. *michiganensis* infection in *Solanum lycopersicum*. *Journal of Nanomaterials, 2019*, 6316094. https://doi.org/10.1155/2019/6316094
145. Yang, J., Jiang, F., Ma, C., et al. (2018). Alteration of crop yield and quality of wheat upon exposure to silver nanoparticles in a life cycle study. *Journal of Agricultural and Food Chemistry, 66*, 2589–2597. https://doi.org/10.1021/acs.jafc.7b04904
146. Tassi, E., Giorgetti, L., Morelli, E., et al. (2017). Physiological and biochemical responses of sunflower (Helianthus annuus L.) exposed to nano-CeO2 and excess boron: Modulation of boron phytotoxicity. *Plant Physiology and Biochemistry, 110*, 50–58. https://doi.org/10.1016/J.PLAPHY.2016.09.013
147. Rizwan, M., Ali, S., Ali, B., et al. (2019a). Zinc and iron oxide nanoparticles improved the plant growth and reduced the oxidative stress and cadmium concentration in wheat. *Chemosphere, 214*, 269–277. https://doi.org/10.1016/J.CHEMOSPHERE.2018.09.120
148. Slomberg, D. L., & Schoenfisch, M. H. (2012). Silica nanoparticle phytotoxicity to Arabidopsis thaliana. *Environmental Science & Technology, 46*, 10247–10254. https://doi.org/10.1021/es300949f
149. Zahra, Z., Waseem, N., Zahra, R., et al. (2017). Growth and metabolic responses of rice (Oryza sativa L.) cultivated in phosphorus-deficient soil amended with TiO2 nanoparticles. *Journal of Agricultural and Food Chemistry, 65*, 5598–5606. https://doi.org/10.1021/acs.jafc.7b01843
150. Haghighi, M., & Teixeira da Silva, J. A. (2014). The effect of N-TiO2 on tomato, onion, and radish seed germination. *Journal of Crop Science and Biotechnology, 17*, 221–227. https://doi.org/10.1007/s12892-014-0056-7
151. Singh, D., Kumar, S., Singh, S. C., et al. (2012). Applications of liquid assisted pulsed laser ablation synthesized TiO2 nanoparticles on germination, growth and biochemical parameters of brassica oleracea var. Capitata. *Science of Advanced Materials, 4*, 522–531. https://doi.org/10.1166/sam.2012.1313
152. Yoon, S. J., Kwak, J. I., Lee, W. M., et al. (2014). Zinc oxide nanoparticles delay soybean development: A standard soil microcosm study. *Ecotoxicology and Environmental Safety, 100*, 131–137. https://doi.org/10.1016/J.ECOENV.2013.10.014
153. Rizwan, M., Ali, S., ur Rehman, M. Z., et al. (2019b). Alleviation of cadmium accumulation in maize (Zea mays L.) by foliar spray of zinc oxide nanoparticles and biochar to contaminated soil. *Environmental Pollution, 248*, 358–367. https://doi.org/10.1016/j.envpol.2019.02.031
154. Dapkekar, A., Deshpande, P., Oak, M. D., et al. (2018). Zinc use efficiency is enhanced in wheat through nanofertilization. *Scientific Reports, 8*, 6832. https://doi.org/10.1038/s41598-018-25247-5
155. Khanm, H., Vaishnavi, B. A., & Shankar, A. G. (2018). Raise of nano-fertilizer era: Effect of nano scale zinc oxide particles on the germination, growth and yield of tomato (Solanum lycopersicum). *International Journal of Current Microbiology and Applied Sciences, 7*, 1861–1871. https://doi.org/10.20546/ijcmas.2018.705.219
156. Nandhini, M., Rajini, S. B., Udayashankar, A. C., et al. (2019). Biofabricated zinc oxide nanoparticles as an eco-friendly alternative for growth promotion and management of downy mildew of pearl millet. *Crop Protection, 121*, 103–112. https://doi.org/10.1016/J.CROPRO.2019.03.015
157. Tarafdar, J. C., Raliya, R., Mahawar, H., & Rathore, I. (2014). Development of zinc nanofertilizer to enhance crop production in pearl millet (Pennisetum americanum). *Agricultural Research, 3*, 257–262. https://doi.org/10.1007/s40003-014-0113-y
158. Mahajan, P., Dhoke, S. K., & Khanna, A. S. (2011). Effect of nano-ZnO particle suspension on growth of mung (*Vigna radiata*) and gram (*Cicer arietinum*) seedlings using plant agar method. *Journal of Nanotechnology, 2011*, 696535. https://doi.org/10.1155/2011/696535
159. Rezaei, M., & Abbasi, H. (2014). Foliar application of nanochelate and non-nanochelate of zinc on plant resistance physiological processes in cotton (Gossipium hirsutum L.). *Iranian Journal of Plant Physiology, 4*, 1137–1144.

160. Ghasemi, M., Ghorban, N., Madani, H., et al. (2017). Effect of foliar application of zinc nano oxide on agronomic traits of two varieties of rice (Oryza sativa L.). *Crop Research, 52*, 195–2001. https://doi.org/10.5958/2454-1761.2017.00017.1
161. Davarpanah, S., Tehranifar, A., Davarynejad, G., et al. (2016). Effects of foliar applications of zinc and boron nano-fertilizers on pomegranate (Punica granatum cv. Ardestani) fruit yield and quality. *Scientia Horticulturae, 210*, 57–64. https://doi.org/10.1016/j.scienta.2016.07.003
162. Zahra, Z., Ali, M. A., Parveen, A., et al. (2019). Exposure–response of wheat cultivars to TiO2 nanoparticles in contrasted soils. *Soil and Sediment Contamination: An International Journal, 28*, 184–199. https://doi.org/10.1080/15320383.2018.1561650
163. Zahra, Z., Arshad, M., Ali, M. A., et al. (2020). Phosphorus phytoavailability upon nanoparticle application. In S. Hayat, J. Pichtel, M. Faizan, & Q. Fariduddin (Eds.), *Sustainable agriculture reviews: Nanotechnology for plant growth and development* (Vol. 41, pp. 41–61). Springer International Publishing.
164. Janmohammadi, M., Amanzadeh, T., Sabaghnia, N., & Dashti, S. (2016). Impact of foliar application of nano micronutrient fertilizers and titanium dioxide nanoparticles on the growth and yield components of barley under supplemental irrigation. *Acta Agriculturae Slovenica, 107*, 265–276.
165. Mustafa, H., Ilyas, N., Akhtar, N., et al. (2021). Biosynthesis and characterization of titanium dioxide nanoparticles and its effects along with calcium phosphate on physicochemical attributes of wheat under drought stress. *Ecotoxicology and Environmental Safety, 223*, 112519. https://doi.org/10.1016/J.ECOENV.2021.112519
166. Owolade, O., & Ogunleti, D. (2008). Effects of titanium dioxide on the diseases, development and yield of edible cowpea. *Journal of Plant Protection Research, 48*, 329–336.
167. Tahmasbi, D., Zarghami, R., Azghandi, A. V., et al. (2011). Effects of nanosilver and nitroxin biofertilizer on yield and yield components of potato minitubers. *International Journal of Agriculture and Biology, 13*, 986–990.
168. Sheykhbaglou, R., Sedghi, M., Tajbakhsh Shishevan, M., & Seyed Sharifi, R. (2010). Effects of nano-iron oxide particles on agronomic traits of soybean. *Notulae Scientia Biologicae, 2*, 112–113.
169. Delfani, M., Firouzabadi, M. B., Farrokhi, N., & Makarian, H. (2014). Some physiological responses of black-eyed pea to iron and magnesium nanofertilizers. *Communications in Soil Science and Plant Analysis, 45*, 530–540. https://doi.org/10.1080/00103624.2013.863911
170. Sharifi, R., Mohammadi, K., & Rokhzadi, A. (2016). Effect of seed priming and foliar application with micronutrients on quality of forage corn (Zea mays). *Environmental and Experimental Biology, 14*, 151–156. https://doi.org/10.22364/eeb.14.21
171. El-Metwally, I. M., Abo-Basha, D. M. R., & Abd El-Aziz, M. E. (2018). Response of peanut plants to different foliar applications of nano- iron, manganese and zinc under sandy soil conditions. *Middle East Journal of Applied Sciences, 8*, 474–482.
172. Drostkar, E., Talebi, R., & Kanounim, H. (2016). Foliar application of Fe, Zn and NPK nano-fertilizers on seed yield and morphological traits in chickpea under rainfed condition. *Research Letters in Ecology, 4*, 221–228.
173. Manjili, M., Bidarigh, S., & Amiri, E. (2014). Study the effect of foliar application of nano chelate molybdenum fertilizer on the yield and yield components of peanut. *Egyptian Academic Journal of Biological Sciences, H. Botany, 5*, 67–71. https://doi.org/10.21608/eajbsh.2014.16829
174. Safyan, N., Naderidarbaghshahi, M. R., & Bahari, B. (2012). The effect of microelements spraying on growth, qualitative and quantitative grain corn in Iran. *International Research Journal of Applied and Basic Sciences, 3*, 2780–2784.
175. Singh, M. D. (2015). *Studies on the effect of time of application and concentration of nano zinc sulphide (nZS) on the growth and yield of sunflower (Helianthus annuus L.)*. University of Agricultural Sciences, Dharwad, Karnataka, India.
176. Patel, S. S., Kumar, B. N. A., Singh, M. D., et al. (2019). Foliar application of green synthesized zinc sulphide and zinc oxide nano particles enhances growth, root attributes,

yield and oil quality of sunflower (Helianthus annuus L.). *Global Journal of Science Frontier Research, 19*, 11–19. https://doi.org/10.34257/GJSFRDVOL19IS4PG11
177. Schmidt, M., Horstmann, S., De Colli, L., et al. (2016). Impact of fungal contamination of wheat on grain quality criteria. *Journal of Cereal Science, 69*, 95–103. https://doi.org/10.1016/j.jcs.2016.02.010
178. Bellesi, F. J., Arata, A. F., Martínez, M., et al. (2019). Degradation of gluten proteins by Fusarium species and their impact on the grain quality of bread wheat. *Journal of Stored Products Research, 83*, 1–8. https://doi.org/10.1016/j.jspr.2019.05.007
179. Matzen, N., Ravn Jørgensen, J., Holst, N., & Nistrup Jørgensen, L. (2019). Grain quality in wheat—Impact of disease management. *European Journal of Agronomy, 103*, 152–164. https://doi.org/10.1016/j.eja.2018.12.007
180. Taghvaeian, S., Comas, L., DeJonge, K. C., & Trout, T. J. (2014). Conventional and simplified canopy temperature indices predict water stress in sunflower. *Agricultural Water Management, 144*, 69–80. https://doi.org/10.1016/j.agwat.2014.06.003
181. Zhang, Q., Chen, M. J., Ju, W., et al. (2017). Improving the ability of the photochemical reflectance index to track canopy light use efficiency through differentiating sunlit and shaded leaves. *Remote Sensing of Environment, 194*, 1–15. https://doi.org/10.1016/j.rse.2017.03.012
182. Kirnak, H., Irik, H. A., & Unlukara, A. (2019). Potential use of crop water stress index (CWSI) in irrigation scheduling of drip-irrigated seed pumpkin plants with different irrigation levels. *Scientia Horticulturae, 256*, 108608. https://doi.org/10.1016/j.scienta.2019.108608
183. Candogan, B. N., Sincik, M., Buyukcangaz, H., et al. (2013). Yield, quality and crop water stress index relationships for deficit-irrigated soybean [Glycine max (L.) Merr.] in sub-humid climatic conditions. *Agricultural Water Management, 118*, 113–121. https://doi.org/10.1016/j.agwat.2012.11.021
184. Raigond, P., Raigond, B., Kaundal, B., et al. (2017). Effect of zinc nanoparticles on antioxidative system of potato plants. *Journal of Environmental Biology, 38*, 435–439. https://doi.org/10.22438/jeb/38/3/MS-209
185. Morteza, E., Moaveni, P., Farahani, H. A., & Kiyani, M. (2013). Study of photosynthetic pigments changes of maize (Zea mays L.) under nano Tio2 spraying at various growth stages. *Springerplus, 2*, 247. https://doi.org/10.1186/2193-1801-2-247
186. Sadak, M. S. (2019). Impact of silver nanoparticles on plant growth, some biochemical aspects, and yield of fenugreek plant (Trigonella foenum-graecum). *Bulletin of the National Research Centre, 43*, 38. https://doi.org/10.1186/s42269-019-0077-y
187. Rajput, V. D., Minkina, T. M., Behal, A., et al. (2018). Effects of zinc-oxide nanoparticles on soil, plants, animals and soil organisms: A review. *Environmental Nanotechnology, Monitoring and Management, 9*, 76–84.
188. Hafeez, B., Khanif, Y. M., & Saleem, M. (2013). Role of zinc in plant nutrition - A review. *American Journal of Experimental Agriculture, 3*, 374–391. https://doi.org/10.9734/AJEA/2013/2746
189. Pandey, N., Pathak, G. C., & Sharma, C. P. (2006). Zinc is critically required for pollen function and fertilisation in lentil. *Journal of Trace Elements in Medicine and Biology, 20*, 89–96. https://doi.org/10.1016/j.jtemb.2005.09.006
190. Roschzttardtz, H., Conéjéro, G., Divol, F., et al. (2013). New insights into Fe localization in plant tissues. *Frontiers in Plant Science, 4*, 350. https://doi.org/10.3389/fpls.2013.00350
191. Wang, Q., Lu, L., Wu, X., et al. (2003). Boron influences pollen germination and pollen tube growth in Picea meyeri. *Tree Physiology, 23*, 345–351. https://doi.org/10.1093/treephys/23.5.345
192. Zhao, L., Lu, L., Wang, A., et al. (2020). Nano-biotechnology in agriculture: Use of nanomaterials to promote plant growth and stress tolerance. *Journal of Agricultural and Food Chemistry, 68*, 1935–1947. https://doi.org/10.1021/acs.jafc.9b06615
193. Bombin, S., LeFebvre, M., Sherwood, J., et al. (2015). Developmental and reproductive effects of iron oxide nanoparticles in Arabidopsis thaliana. *International Journal of Molecular Sciences, 16*, 24174–24193. https://doi.org/10.3390/ijms161024174

194. Laware, S. L., & Raskar, S. (2014). Influence of zinc oxide nanoparticles on growth, flowering and seed productivity in onion. *International Journal of Current Microbiology and Applied Sciences, 3*, 874–881.
195. Byczyńska, A., Zawadzińska, A., & Salachna, P. (2019). Silver nanoparticles preplant bulb soaking affects tulip production. *Acta Agriculturae Scandinavica Section B Soil and Plant Science, 69*, 250–256. https://doi.org/10.1080/09064710.2018.1545863
196. Salachna, P., Byczyńska, A., Zawadzińska, A., et al. (2019). Stimulatory effect of silver nanoparticles on the growth and flowering of potted oriental lilies. *Agronomy, 9*, 610. https://doi.org/10.3390/agronomy9100610
197. Neysanian, M., Iranbakhsh, A., Ahmadvand, R., et al. (2020). Comparative efficacy of selenate and selenium nanoparticles for improving growth, productivity, fruit quality, and postharvest longevity through modifying nutrition, metabolism, and gene expression in tomato; Potential benefits and risk assessment. *PLoS One, 15*, 1–17. https://doi.org/10.1371/journal.pone.0244207
198. Ke, M., Li, Y., Qu, Q., et al. (2020). Offspring toxicity of silver nanoparticles to Arabidopsis thaliana flowering and floral development. *Journal of Hazardous Materials, 386*, 121975. https://doi.org/10.1016/j.jhazmat.2019.121975
199. Salehi, H., Chehregani Rad, A., Raza, A., & Chen, J.-T. (2021). Foliar application of CeO2 nanoparticles alters generative components fitness and seed productivity in bean crop (Phaseolus vulgaris L.). *Nanomaterials, 11*, 862. https://doi.org/10.3390/nano11040862
200. Mosa, W. F. A., El-Shehawi, A. M., Mackled, M. I., et al. (2021). Productivity performance of peach trees, insecticidal and antibacterial bioactivities of leaf extracts as affected by nanofertilizers foliar application. *Scientific Reports, 11*, 10205. https://doi.org/10.1038/s41598-021-89885-y
201. Raliya, R., Nair, R., Chavalmane, S., et al. (2015). Mechanistic evaluation of translocation and physiological impact of titanium dioxide and zinc oxide nanoparticles on the tomato (Solanum lycopersicum L.) plant. *Metallomics, 7*, 1584–1594. https://doi.org/10.1039/C5MT00168D
202. Kamali, M., Shoor, M., & Feizi, H. (2018). Impacts of nanosized and bulk titanium dioxide on flowering and morphophysiological traits of petunia (Petunia hybrida) under salinity stress. *Journal of Horticultural Sciences, 32*, 199–212.
203. Gray, S. B., & Brady, S. M. (2016). Plant developmental responses to climate change. *Developmental Biology, 419*, 64–77. https://doi.org/10.1016/j.ydbio.2016.07.023
204. Hedhly, A. (2011). Sensitivity of flowering plant gametophytes to temperature fluctuations. *Environmental and Experimental Botany, 74*, 9–16. https://doi.org/10.1016/j.envexpbot.2011.03.016
205. Pogányova, A., Mičieta, K., & Dušička, J. (2019). Genotoxic assessment of selected native plants to deferentially exposed urban ecosystems. *Environmental Science and Pollution Research, 26*, 9055–9064. https://doi.org/10.1007/s11356-019-04230-1
206. Speranza, A., Leopold, K., Maier, M., et al. (2010). Pd-nanoparticles cause increased toxicity to kiwifruit pollen compared to soluble Pd(II). *Environmental Pollution, 158*, 873–882. https://doi.org/10.1016/j.envpol.2009.09.022
207. Dutta Gupta, S., Saha, N., Agarwal, A., & Venkatesh, V. (2020). Silver nanoparticles (AgNPs) induced impairment of in vitro pollen performance of Peltophorum pterocarpum (DC.) K. Heyne. *Ecotoxicology, 29*, 75–85. https://doi.org/10.1007/s10646-019-02140-z
208. Aoyagi, H., & Ugwu, C. U. (2011). Fullerene fine particles adhere to pollen grains and affect their autofluorescence and germination. *Nanotechnology, Science and Applications, 4*, 67–71. https://doi.org/10.2147/NSA.S14263
209. Yoshihara, S., Hirata, S., Yamamoto, K., et al. (2021). ZnO nanoparticles effect on pollen grain germination and pollen tube elongation. *Plant Cell, Tissue and Organ Culture, 145*, 405–415. https://doi.org/10.1007/s11240-021-02017-2
210. Candotto Carniel, F., Gorelli, D., Flahaut, E., et al. (2018). Graphene oxide impairs the pollen performance of Nicotiana tabacum and Corylus avellana suggesting potential negative effects

on the sexual reproduction of seed plants. *Environmental Science. Nano, 5*, 1608–1617. https://doi.org/10.1039/C8EN00052B
211. Speranza, A., Crinelli, R., Scoccianti, V., et al. (2013). In vitro toxicity of silver nanoparticles to kiwifruit pollen exhibits peculiar traits beyond the cause of silver ion release. *Environmental Pollution, 179*, 258–267. https://doi.org/10.1016/j.envpol.2013.04.021
212. Georgieva, N., Nikolova, I., Kosev, V., & Naydenova, Y. (2017). In vitro germination and viability of pea pollen grains after application of organic nano-fertilizers. *Pestic i fitomedicina, 32*, 61–65. https://doi.org/10.2298/PIF1701061G
213. Alharbi, N. S., Bhakyaraj, K., Gopinath, K., et al. (2017). Gum-mediated fabrication of eco-friendly gold nanoparticles promoting cell division and pollen germination in plant cells. *Journal of Cluster Science, 28*, 507–517. https://doi.org/10.1007/s10876-016-1130-8
214. Gopinath, K., Venkatesh, K. S., Ilangovan, R., et al. (2013). Green synthesis of gold nanoparticles from leaf extract of Terminalia arjuna, for the enhanced mitotic cell division and pollen germination activity. *Industrial Crops and Products, 50*, 737–742. https://doi.org/10.1016/j.indcrop.2013.08.060
215. Shah, S. S., Qasem, M. A. A., Berni, R., et al. (2021). Physico-chemical properties and toxicological effects on plant and algal models of carbon nanosheets from a nettle fibre clone. *Scientific Reports, 11*, 6945. https://doi.org/10.1038/s41598-021-86426-5
216. Djanaguiraman, M., Nair, R., Giraldo, J. P., & Prasad, P. V. V. (2018). Cerium oxide nanoparticles decrease drought-induced oxidative damage in sorghum leading to higher photosynthesis and grain yield. *ACS Omega, 3*, 14406–14416. https://doi.org/10.1021/acsomega.8b01894
217. Kos, M., Jemec Kokalj, A., Glavan, G., et al. (2017). Cerium(iv) oxide nanoparticles induce sublethal changes in honeybees after chronic exposure. *Environmental Science. Nano, 4*, 2297–2310. https://doi.org/10.1039/C7EN00596B
218. Milivojević, T., Glavan, G., Božič, J., et al. (2015). Neurotoxic potential of ingested ZnO nanomaterials on bees. *Chemosphere, 120*, 547–554. https://doi.org/10.1016/j.chemosphere.2014.07.054
219. Wang, T.-H., Jian, C.-H., Hsieh, Y.-K., et al. (2013). Spatial distributions of inorganic elements in honeybees (Apis mellifera L.) and possible relationships to dietary habits and surrounding environmental pollutants. *Journal of Agricultural and Food Chemistry, 61*, 5009–5015. https://doi.org/10.1021/jf400695w
220. Du, W., Gardea-Torresdey, J. L., Ji, R., et al. (2015). Physiological and biochemical changes imposed by CeO2 nanoparticles on wheat: A life cycle field study. *Environmental Science & Technology, 49*, 11884–11893. https://doi.org/10.1021/acs.est.5b03055
221. Zhao, X., Meng, Z., Wang, Y., et al. (2017). Pollen magnetofection for genetic modification with magnetic nanoparticles as gene carriers. *Nature Plants, 3*, 956–964. https://doi.org/10.1038/s41477-017-0063-z
222. Vejlupkova, Z., Warman, C., Sharma, R., et al. (2020). No evidence for transient transformation via pollen magnetofection in several monocot species. *Nature Plants, 6*, 1323–1324. https://doi.org/10.1038/s41477-020-00798-6

Phytotoxic Effects of Nanoparticles and Defense Mechanisms in Plants

Muhammad Adil, Amar Nasir, Noor Muhammad Khan, and Arbab Sikandar

Abstract Nanoparticles have drawn considerable attention on account of their unique physicochemical characteristics and valuable applications in various sectors. Nevertheless, the rapid development and widespread utilization of nanotechnology may cause the inevitable dispersion of nanoparticles into the environment and potential ecotoxicological consequences. Contrary to relatively more common nanotoxicological investigations on humans and animals, the phytotoxicity of nanoparticles still requires substantial investigation. Being the end recipients of environmental contaminants, plants are not only affected directly but also determine the ecological fate of nanoparticles and the nature of exposure for higher species. Therefore, elucidating the interaction between nanoparticles and plants is critical for highlighting the environmental impact of nanotechnology. This chapter describes the phytotoxic potential of nanoparticles emphasizing the uptake, translocation, accumulation, toxicological and detoxification mechanisms, phytotoxicity assessment assays, toxicity-affecting factors, and spectrum of detrimental effects in plants.

Keywords Phytotoxic effects · Nanoparticles · Defense mechanisms · Plants

M. Adil (✉)
Pharmacology and Toxicology Section, University of Veterinary & Animal Sciences, Lahore, Jhang Campus, Pakistan
e-mail: muhammad.adil@uvas.edu.pk

A. Nasir
Department of Clinical Sciences, University of Veterinary and Animal Sciences, Lahore, Jhang Campus, Pakistan

N. M. Khan
Physiology and Biochemistry Section, University of Veterinary and Animal Sciences, Lahore, Jhang Campus, Pakistan

A. Sikandar
Anatomy and Histology Section, University of Veterinary and Animal Sciences, Lahore, Jhang Campus, Pakistan

J.-T. Chen (ed.), *Plant and Nanoparticles*,
https://doi.org/10.1007/978-981-19-2503-0_8

1 Introduction

Nanotechnology constitutes an emerging discipline of science and nanoparticles have a variety of applications, particularly in the agricultural, biomedical, engineering, and industrial sectors [1, 2]. However, nanoparticles can enter the soil, air and water through wastewater, sewage sludge and landfills during the processes of synthesis, handling and application [3–6]. Agricultural application, surface runoff, rain erosion and atmospheric deposition are the major pathways for the entry of nanoparticles into the soil. The high reactivity and unique physicochemical profile of nanoparticles are attributed to considerable differences from their bulk counterparts in terms of size, structure, surface characteristics and shape. The constant interaction of plants with water, air, and water facilitates the uptake, storage and transfer of nanoparticles.

Although most of the nanoparticles are known to induce phytotoxic effects in the form of morphological, biochemical, physiological, or genetic perturbations, plants also possess certain innate defense mechanisms to tackle these stressful conditions [7, 8]. However, plant cells are subjected to apoptosis when the extent of phytotoxic damage exceeds the capacity of inherent detoxification potential [9]. Besides deleteriously affecting the soil organisms and plants, the deposited nanoparticles can be further transported by the plants to animals and humans, representing the high trophic levels of the ecosystem [10–12]. Consequently, the toxicological interaction between nanoparticles and plants requires thorough investigation for the assessment of risk to environmental well-being, food safety, biodiversity conservation and public health.

2 Methods for the Assessment of Nanoparticle-Induced Phytotoxicity

Generally, phytotoxicity tests are conducted using certain crops including both the Monocotyledoneae and Dicotyledoneae species [13]. Commonly recommended species for this purpose *are* maize (*Zea mays* subsp. mays), wheat (*Triticum aestivum*), cabbage (*Brassica oleracea*), bean (*Phaseolus vulgaris*), rice (*Oryza sativa*), radish (*Raphanus sativus*), carrot (*Daucus carota* subsp. sativus), cucumber (*Cucumis sativus*), onion (*Allium cepa*), tomato (*Solanum lycopersicum*), oat (*Avena sativa*), soybean (*Glycine max*), lettuce (*Lactuca sativa*) and ryegrass (*Lolium perenne L.*). Recently, characteristic model species like *Arabidopsis thaliana* have also been successfully used for phytotoxic examination [14].

Phytotoxicity assays are carried out during the germination of plants for measuring the germination percentage that requires the targeting of seeds by the test solution for about 4 days [13] and at the time of growth of the seedlings, using shoot/root elongation and dry weight as variables in order to gauge the impact of plant exposure to detrimental substances [15]. Different media are used for the growth of plants during phytotoxicity testing and water is the simplest and cheapest medium for this purpose. Besides, soft gels or agars, as well as soil are also commonly employed [14]. Frequently evaluated parameters in this context are germination, root/stem growth rates, leaf count and chlorophyll content [16, 17].

Microscopic examination has been described in several studies wherein nanoparticles were mapped inside the plant cells. For instance, the presence of selenium nanoparticles in root tissues of exposed plants has been described as bright green spots in fluorescent microscopy [18]. Comparable findings were also recorded in soybean plants exposed to magnetite nanoparticles [19]. Likewise, the penetration of super magnetic oxide nanoparticles inside the cells of soybean was also visualized under the fluorescent microscope [19]. Cell viability studies constitute an integral approach of nanotoxicity assessment in plants by determining their viability using vital stains like Evan's blue, in addition to measuring cellular metabolic activity markers including triphenyltetrazolium chloride. The dye is expelled out by viable cell membranes but retained by the dead cells because of the existing membrane permeability [20]. On the other hand, the performance of metabolic activity markers depends upon the activity of mitochondrial enzymes wherein it is transformed to peculiar red formazan products under the action of mitochondrial oxidoreductase enzymes [21]. Likewise, fluorescein diacetate/propidium iodide dual staining also provides a reliable outcome ensuring cell viability. Living cells metabolize fluorescein diacetate resulting in the emission of green fluorescence, while in nonliving or dying cells, propidium iodide is taken up via ruptured cell membranes causing intercalation within the DNA, emitting a red fluorescence [22]. Comet assay, a microgel electrophoresis technique, is another approach for the interpretation of tissue damage at the single-cell level [23].

In plants, generally, the levels of flavonoids and phenols undergo variation, in response to stresses such as exposure to ultraviolet B due to drought and heavy metals culminating in the removal of reactive oxygen species [24]. It occurs due to the peroxidation of unsaturated fatty acids in phospholipids [25]. The reactive oxygen species are produced by nanoparticles targeting the cell membranes resulting in lipid peroxidation. Consequently, the DNA and proteins are denatured by lipid-derived free radicals [26]. Malondialdehyde content is used to measure the integrity of the membranes in plants as it has a direct correlation with the accumulation of reactive oxygen species [11, 27].

3 Factors Influencing the Phytotoxicity of Nanoparticles

The phytotoxicity of nanoparticles depends upon their physicochemical features (including size, concentration, surface area and stability), the type and growth stage of target plants, culture medium and environmental factors (Fig. 1).

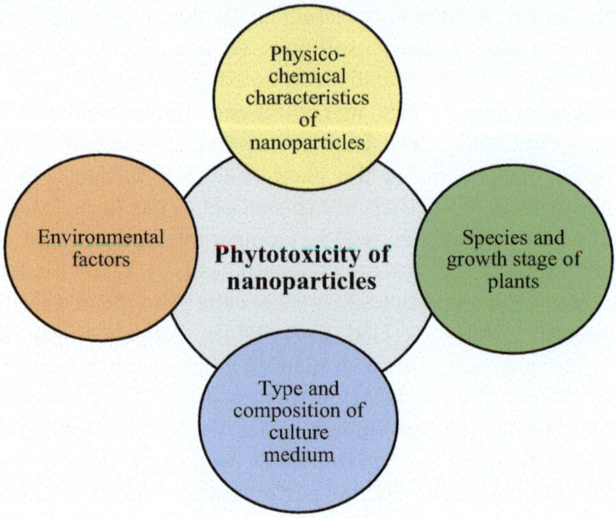

Fig. 1 Factors affecting the phytotoxicity of nanoparticles

3.1 Physicochemical Characteristics of Nanoparticles

Nanoparticles are relatively more harmful than their bulk counterparts on account of high reactivity and ease of intracellular diffusion [28]. Contrary to the smaller gold nanoparticles (3.5 nm), the larger ones (18 nm) were unable to induce leaf necrosis [29]. However, the conversion of certain bulk substances into nanoparticles results in the loss of phytotoxic potential. Unlike their bulk counterparts, the silver nanoparticles were devoid of inhibitory effect on castor beans' growth [30]. Moreover, the spectrum of phytotoxic effects also differed with the concentrations of cerium oxide nanoparticles, nano zerovalent iron and titanium dioxide nanoparticles [31–33]. Titanium dioxide nanoparticles exerted dose-dependent inhibitory effects on the germination, growth, root length and biomass of exposed plants [34]. Besides, surface modification also influences the detrimental impact of nanoparticles on plants and polymeric coating of copper oxide nanoparticles reduced the growth of *Lemna gibba,* even at smaller concentrations [35]. Furthermore, gibberellic acid-coated silver nanoparticles exhibited prolonged phytotoxicity in comparison with un-coated nanoparticles of silver nitrate [36]. Conversely, zinc sulfide coating enhanced biocompatibility and reduced the toxicity of quantum dots [37].

3.2 Species and Growth Stage of Target Plants

The nature and intensity of phytotoxic insult differ among plants on account of species-specific variations in the antioxidant defense mechanism. Besides, xylem structure and seed size have also been implicated in governing the susceptibility of bifoliate plants to nanoparticle-mediated toxicity [38]. In contrast to alfalfa, tomato, cucumber and corn exhibited a reduction in growth rate on exposure to cerium oxide nanoparticles [39]. Likewise, the phytotoxic impact of nanoparticles was significantly different on conventional and transgenic seeds [40]. Immature plants are relatively more vulnerable to the detrimental effects of nanoparticles, owing to lack of properly functional detoxification system.

3.3 Type and Composition of Culture Medium

The toxicity profile of nanoparticles also depends on the type and composition of the growth medium. Accordingly, several ingredients of soil including organic matter, nutrients and colloids are known to influence the impact of the nanoparticles on plants. For instance, the exclusion of organic matter from soil enhanced the accumulation of cerium in bean plants treated with cerium oxide nanoparticles [41]. Contrary to organic carbon, the pH and clay content of soil inversely affected the phytotoxic capacity of silver nanoparticles [42]. In addition to organic carbon and pH value, ionic strength and salt type also determine the phytotoxic effectiveness of nanoparticles in an aqueous medium.

3.4 Environmental Factors

Changes in environmental temperature and light conditions may interfere with the phytotoxicity of nanoparticles by altering the biochemical and physiological attributes of plants. Illumination changed the phytotoxic effect of silver nanoparticles on *Wolffia globosa* through the modification of protein content and photosynthetic pigments [43]. In contrast to light conditions, increased temperature and aging reduced the phytotoxic impact of nanoparticles [44].

4 Uptake and Translocation of Nanoparticles in Plants

4.1 Nanoparticles Uptake

Plants take up the nanoparticles primarily through two portals. One way is the foliar entry where nanoparticles make their way through cuticles, stomata, and hydathodes and the other is soil entry via irrigation where nanoparticles enter the plants through wounds, roots, root tips, or root hairs [45]. Nanoparticles undergo penetration across the cell wall and cell membrane and thereby interact with the different processes in plants. In the first instance, nanoparticles get penetrated through the roots or parts above the ground such as root junctions. During the process of uptake and translocation, nanoparticles encounter several physiological barriers of the plants, for example, the cell wall. In plants the cell wall consists of cellulose, which allows the passage of small nanoparticles and opposes the entry of larger particles. The plant cell wall has its exclusion size limit that ranges from 5 to 20 nm, so it is presumed that particles falling within this range can easily get across the cell wall and penetrate the plant cell [46]. Certain nanoparticles possess the potential to generate larger pores in the cell wall and therefore provide an enhanced passage to the larger particles [47]. Having high reactivity and increased surface area, the nanoparticles contained in the soil are subsequently transported into the plants by interacting with root exudates and membrane transporters [48]. Soon after penetration across the cell wall and subsequent uptake, the nanoparticles are transported into the cell by means of endocytosis and ultimately distributed among the plant tissues with the help of plasmodesmata through a symplastic pathway [49]. Nanoparticles are translocated to the aerial parts on sneaking in the roots and ultimately accumulated in cellular or subcellular organelles. Soil to roots adsorption of nanoparticles in plants is the first step that occurs in bioaccumulation [50].

4.2 Translocation of Nanoparticles in Plants

The smaller nanoparticles (3–5 nm in diameter) pass through the pores easily as compared to larger ones. After being penetrated through the cell wall, nanoparticles are apoplastically driven via extracellular spaces to the central vascular cylinder, permitting their unidirectional ascending movement by means of xylem. Nanoparticles symplastically pass through the Casparian strip barrier to reach the central vascular cylinder. It occurs through the binding of nanoparticles to certain carrier proteins of the endodermal cell membrane using endocytosis, forming of pores, and transport [51, 52]. The nanoparticles remain internalized in the cytoplasm for cell-to-cell transport process. Nanoparticles not getting internalized become accumulated on the Casparian strip, and only those reaching xylem are transported to the shoots and redistributed to the roots via the phloem. The taken-up nanoparticles in plants exist in the epidermal cell wall, cortical cell cytoplasm and nuclei. Nanoparticles not getting transported in plants through the soil, accumulate on the surface of roots and interfere with the absorption of nutrients from the soil [53]. Direct absorption of nanoparticles inside seeds can happen after their entry into the seed coat through parenchymatic intercellular spaces coupled with diffusion in the cotyledon. Moreover, nanoparticles can enter the leaves via the portals of cuticles or stomata. Plant cuticle serves as the prime fence in limiting the stomatal transport of nanoparticles measuring 10 nm or above, and thus, their cellular transport occurs using symplastic and apoplastic portals into the vascular system of the plant [54]. Passage of nanoparticles ranging from 10 to 50 nm is facilitated by the cytoplasm of adjacent cells through symplastic route. Nanoparticles ranging from 50 to 200 nm are transferred between the cells by means of apoplastic route. The transit of internalized nanoparticles occurs alongside the sugar flow via phloem sieve tubes [55]. Consequent to phloem vascular transport, nanoparticles move in two directions and gather variably in roots, stems, young leaves, grains and fruits, owing to the role of described organs as effective sinks for the sap [56]. The apoplastic transport is termed as a nonselective process of minimum resistance. Majority of water-based nutrients and nonessential metals interact with nanoparticles and facilitate the apoplastic portals for their translocation. The transfer and buildup of nanoparticles in plants rely upon the physiology and structure of their cells, interface of nanomaterials with soil, and type as well as the stability of nanoparticles.

5 Intrinsic Detoxification and Defense Mechanisms in Plants

Plants being frequently exposed to environmental toxicants are vulnerable to phytotoxicity and they have evolved specialized defense mechanisms to neutralize the injurious effects of those toxins. Detoxifications constitutes on of the most important and common defense mechanisms present in plants. Plants are fortunate in possessing the detoxification pathways through which the deleterious effects of toxic substances are minimized. Scavenging of reactive oxygen species is an example of such pathways for combating the oxidative stress induced by environmental toxins, for example, nanoparticles. Reactive oxygen species have been implicated to enact in the intercellular signaling pathways as second messengers, the closing of stomata [57], apoptosis [58] and gravitropism [59]. Under equilibrium, the marked damaging effects are not much pronounced; however, excess generation of reactive oxygen species associated with environmental chemicals such as nanoparticles can lead to oxidative stress. Plant cells have developed antioxidant defense mechanisms in their cell organelles including peroxisomes, chloroplast and mitochondria, which shields them against the toxic hazards of chemical exposure [60]. Plant antioxidant defense systems consist of both enzymatic (such as superoxide dismutase, catalase dehydroascorbate reductase, monodehydroascorbate reductase, glutathione reductase, glutathione S-transferases and glutathione peroxidase) and nonenzymatic antioxidants including thiols, glutathione, ascorbic acid and phenolics [61].

5.1 Enzymatic Antioxidant Defense System

The enzymatic defense system is comprised of antioxidant enzymes having a vital role against the stressors inducing oxidative stress. Superoxide dismutase is one of the antioxidant enzymes that catalyze the conversion of highly toxic reactive oxygen species; O_2^- into less toxic O_2 and H_2O_2. It has three different isozymes, i.e., superoxide dismutase-iron, superoxide dismutase-copper and superoxide dismutase-Zinc [62]. These isozymes are effective against reactive oxygen species, in aerobic organisms and at subcellular levels [63]. Superoxide dismutase exerts its potent antioxidant effect in various plant species. It is now clear that increased levels of superoxide dismutase have been associated with a more pronounced antioxidant response against the oxidative stress caused by reactive oxygen species and nanoparticles-mediated toxicity [64]. The antioxidant activity under nanoparticles' exposure was enhanced by superoxide dismutase in tomatoes, and comparable effects have also been demonstrated in onion and rice [65].

Another antioxidant enzyme is catalase which plays an indispensable role in the detoxification of reactive oxygen species during stress conditions [66]. It combats reactive oxygen species by converting H_2O_2 into water and oxygen. Ascorbate peroxidase is another well-known and highly potent scavenger of reactive oxygen species, that converts H_2O_2 into H_2O and O_2. Ascorbate peroxidase has a high affinity for H_2O_2 as compared to catalase. Dose-dependent catalase activity was observed in cucumber leaves post-exposure to titanium dioxide nanoparticles while no quantifiable variation in ascorbate peroxidase activity was observed [67]. Glutathione reductase acts as an antioxidant in both enzymatic and nonenzymatic defense systems. It is a reduced nicotinamide adenine dinucleotide phosphate-dependent enzyme that catalyzes the oxidation of glutathione to glutathione disulfide [68]. Altogether, these enzymes provide an efficient antioxidant defense against the oxidative stress induced by nanoparticles.

5.2 *Nonenzymatic Antioxidant Defense System*

The nonenzymatic antioxidant system is composed of chemical substances like ascorbic acid, thiols and glutathione, which exert their antioxidant effects through neutralizing the reactive oxygen species. Ascorbic acid protects the cell membrane against oxidative damage by scavenging reactive oxygen species as it possesses the ability to transfer its free electrons in nonenzymatic and enzymatic reactions [69]. Both thiols and ascorbic acid exert a high antioxidant activity in nanoparticles-exposed plants. However, exposure at greater levels of nanoparticles demonstrated a significant decline in ascorbic acid and thiols activity leading to enhanced oxidative damage and reduced photosynthetic activity [70]. Glutathione is an important thiol of nonprotein nature and plays its role in antioxidant defense mechanism against the reactive oxygen species-induced oxidative stress. It is essentially found in different compartments of the cell such as mitochondria, chloroplast, peroxisomes, cytosol, endoplasmic reticulum and vacuoles [71]. The antioxidant potential of glutathione is attributed to its ability to donate protons in the organic free radicals and scavenging reactive oxygen species with ultimately being reduced to glutathione disulfide. Recently, it has been reported that nanoparticles of indium oxide and cerium oxide caused glutathione biosynthesis and sulfur assimilation gene regulation in Arabidopsis [72]. Phenolic compounds due to their antioxidant potential can form chelates, scavenge reactive oxygen species, and minimize lipid peroxidation through the entrapment of lipid alkoxyl polyphenols. Carotenoids can detoxify the reactive oxygen species and are generally classified as lipophilic antioxidants [73]. Tocopherols (α, β, γ, and δ) are classified as lipophilic phenols, involved in the scavenging of oxygen free radicals, lipid peroxidation radicals, and $^1O_2$ [74]. However, the potential of phenolic compounds including tocopherols and carotenoids against exposure to nanoparticles has been very rarely investigated, and their role in nanoparticles-induced oxidative stress remains elusive.

6 Mechanisms of Nanoparticle-Induced Phytotoxicity

After the nanoparticles have been absorbed and taken up by the plant cells, they may interact with the various physiological activities and disrupt the growth and repair processess. It is assumed that nanoparticles may interfere with the regulation of genes and oxidative mechanisms leading to oxidative burst [75]. Nanoparticles induce oxidative damage through the generation of reactive oxygen species and thereby lead to cell death.

It has also been established that nanoparticles further interfere with the mitochondria and chloroplast, culminating in the oxidative burst of the respective cell [65]. Damage to the cellular proteins, DNA and lipid peroxidation ascribed to nanoparticles has also been investigated [76]. Exposure to nanoparticles increases the levels of reactive oxygen species that leads to oxidative burst and the death of plant cells that can be either programmed (apoptosis) or necrosis. Again, the damaging role of reactive oxygen species is associated with the equilibrium between the production of reactive oxygen species and their scavenging by the antioxidant enzymes. During nanoparticle-induced oxidative stress, plants tend to increase their production of antioxidant enzymes for the trapping and scavenging of reactive oxygen species [77]. It is also interesting to know that exposure to nanoparticles also gives rise to up- and downregulation of different plant hormones (phytohormones). An increased cytokine level following the exposure of cotton plants to silver nanoparticles reveals that nanoparticles interfere with the hormonal balance through reactive oxygen species and therefore affect the growth and development of plants.

As we are aware that plants prepare their food by the process of photosynthesis, it is imperative for their growth and repair that the photosynthetic activity operates without any interruption. Nanoparticles cause the synthesis of reactive oxygen species and influence the activity of photosynthetic pigments and their concentration [78]. This provides an understanding regarding the mechanism of plant's death due to altered metabolism and growth pathways following exposure to the nanoparticle. In plants, nanoparticles affect root elongation, seed germination and biomass. Nanoparticles of iron oxide, carbon and titanium oxide adversely affected the seed root elongation, germination rate and germination index in cucumber plants [79]. Although nanoparticles toxicity has been extensively evaluated at different angles, its mechanisms and effects at the proteomic levels are still not clear and offer a gap that requires to be filled through future research.

7 Phytotoxic Effects of Nanoparticles

Nanoparticles are capable to exert a variety of phytotoxic effects ranging from morphological, physiological, genetic and transgenerational perturbations to productivity deficits (Fig. 2).

7.1 *Effects of Nanoparticles on Seed Germination*

The germination of a seed is the initiation of the physiological process for a new life. The outer seed coat serves to safeguard the developing embryo [80]. After cracking of seed coat, the emerging radicle is the foremost tissue of the plant coming in direct contact with metal particles. The seed germination and growth of many edible crops have been reported to be badly affected by exposure to nanoparticles [81]. It has already been explored that copper oxide nanoparticles affect the seed germination and growth of the seedling [82]. In addition, these can severely compromise the plant elongation and even may lead to the death of the plant.

Many of the earlier studies demonstrated the retardation of certain species of plants like soybean (*Glycine max*), maize (*Zea mays* subsp. Mays), ryegrass, wheat (*Triticum aestivum*) and barley when exposed to silver, iron, and zinc oxide nanoparticles. Many facets of the growth of a plant have been reported to be affected such as seed germination, biomass, shoot length, gene expression and biomass [83, 84]. Growth suppression was noticed in *Bacillus thuringiensis* (Bt)-transgenic cotton in response to silicon dioxide nanoparticle exposure [85]. Likewise, the growth of wheat plants, grown on sand matrix, was suppressed by copper oxide nanoparticles along with a change in the structure of roots [86]. Moreover, reduction in the root lengths and fresh weights of Arabidopsis seedlings, as well as the biomass and germination rate of rice seeds have also been described [87].

7.2 *Influence of Nanoparticles on Plant Hormones and Growth*

Plant hormones are the products of plant metabolism and active organic materials involved in regulating various physiological activities throughout plant growth and coordinating reactions to encounters [88]. The composition and activity profile of the hormones are the key determinants of phytotoxicity index. Nanoparticles considerably impact the synthesis of plant hormones. Even small levels of nanoparticles can perform significant activity as copper ions in low concentrations enhance the growth of plants acting as microelements [89]. Besides, exposure to carbon nanotubes reduced the concentrations of phytohormones in rice seedlings [90]. Another

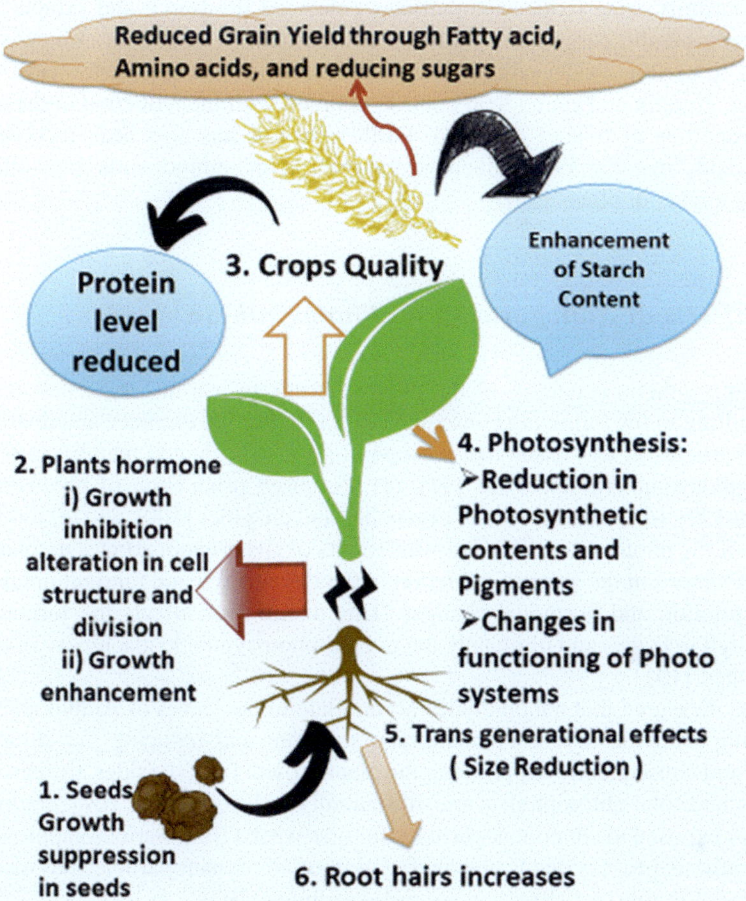

Fig. 2 Spectrum of phytotoxic effects associated with nanoparticles

investigation [91] performed on irrigation water having copper nanoparticles described a decrease in the lengths of shoots and root of Spinach plant (*Spinacia oleracea*).

7.3 Impact of Nanoparticles on Grain Quality and Yield

Earlier studies carried out on hydroponic plants reported that environmental aggregation of nanoparticles may drastically alter the quality and yield of food crops grown on soil [92]. The effect on the quality of food crops was reported in one such study where the protein was not influenced by silver nanoparticles in comparison to carbohydrates except only at a much higher level (100 mg/L) of silver nanoparticles

[93]. Contrarily, zinc oxide nanoparticles enhanced the starch and protein levels, while, reduced the copper and molybdenum content in cucumber. Likewise, rice exposed to cerium oxide nanoparticles had less iron, sulfur, valeric acids, lauric, prolamin, glutelin and starch than control, coupled with the reduction in the antioxidant properties of treated rice [94]. Furthermore, it has also been recorded that cerium oxide nanoparticles influenced the fatty acids, amino acids, phenolics and reducing sugars in plants [95].

7.4 *Effects of Nanoparticles on Photosynthesis*

The photosynthetic apparatus of the plant is struck by various toxic nanoparticles [96] resulting in the following consequences: (a) undesired entrance and distribution of nanoparticles in leaf tissue like mesophyll [97], (b) changed membrane physiology of photosynthetic apparatus [98], (c) decreased production of photosynthetic pigments [99], (d) altered cytosolic organelles and enzymes [100], and (e) modulated activity of the photosystem. The harmful effects of silver nanoparticles on plants can also be foreseen in the form of reduction in the chlorophyll and nutrient uptake, rate of transpiration and hormonal changes. This disturbance affects the formation of chlorophyll in leaves and ultimately the overall photosynthetic machinery of plant is deteriorated [101].

It was measured that nanoparticles accumulate in the leaves of Arabidopsis plant interfering with the thylakoid membrane structure and reducing the chlorophyll content and consequently suppressing the plant growth [78]. Besides, the carotenoid contents and total chlorophyll were substantially reduced in rice (*Oryza sativa* L.) seedlings exposed to silver nanoparticles for one-week [102]. Serious suppression of photosynthesis process due to the aggregation of silver nanoparticles has also been documented in mustard (Brassica species) seedlings [103].

Furthermore, the exposure of silver nanoparticles to the leafy gametophytes altered the thylakoid, reduced the extent of chlorophyll b, and thereby influenced the balance of certain essential elements in *Physcomitrella patens* [104]. The nanoparticles also led to a drastic decline in the transpiration rate of *Cucurbita pepo* [105]. Likewise, a 10-day exposure of *Lupinus termis* L. seedlings to nanoparticles caused the elongation of shoot and root, while, the fresh weights, total chlorophyll and total protein contents got substantially lowered [106].

7.5 *Cytotoxic and Genotoxic Effects of Nanoparticles in Plants*

Nanoparticles enter into the plants through a smart delivery system mediated by attacking the genes or using DNA of specific plant organelles [107]. Their entry involves different pathways, one of which is via transporters [108]. For example, the genes acting as transporters for the nanoparticles of silicon; Lsi1, Lsi2, and Lsi6 exist in *Oryza sativa* (rice) roots [109]. Upon entry of the particles into the cell, a cascade of macromolecular interactions sets in, having varied manifestations. For instance, the upregulation of many genes like those associated with water channels and stress has been reported. Contrarily, titanium dioxide nanoparticles damaged the genomic DNA in *Cucurbita pepo* [110].

The deleterious effects of titanium dioxide nanoparticles were also observed in a range of plants like *Nicotiana tabacum, Allium cepa* and *Zea mays.* The DNA damage occurred at 4 mM, 2 mM, and 10 mM levels, respectively [111, 112]. Moreover, negative impacts pertinent to chromosome structures of *Zea mays* were also evident [111]. Similarly, these particles have led to an increase in the number of tubulin monomers, ultimately affecting the proteosome system in *Arabidopsis thaliana*. Likewise, the exposure of rice seedlings to silver nanoparticles resulted in the differential expressions of genes associated with the tolerance of oxidative stress [102].

In *Lolium multiflorum* (Italian ryegrass), the seedlings were largely vacuolated coupled with the damaged epidermis, and could not grow root hair with their cells on exposure to 40 mg/L silver nanoparticles [36]. However, in a similar experiment these particles reduced the vacuole size and led to decreased cell turgidity in cabbage (*Brassica oleracea*) and maize (*Zea mays*) plants [101].

Besides instigating the morphological and physiological changes, nanoparticles also exert their effects at cellular and molecular levels, and may influence the cell structure and cell division. Likewise, the seedlings were unable to grow root hair, cortical cells were significantly vacuolated and distorted, whereas, the root cap and epidermis were also affected [36]. A decrease in cell size and turgidity in maize plants exposed to nanoparticles was also recorded [113]. Similarly, the cell wall integrity and vacuoles were damaged following the penetration of silver nanoparticles into the plant cells [114]. Additionally, the exposure of *Allium cepa* to silver nanoparticles caused a drastic decline in mitotic index and impaired cell division, ensuing chromatin bridge, cell disintegration, disturbed metaphase, stickiness, and multiple chromosomal breaks [115].

7.6 Transgenerational Effects of Nanoparticles in Plants

The aggregation of nanoparticles in plants can occur within various tissues including fruits, seeds, leaves and roots. The entry and buildup of nanoparticles in seeds lead to transgenerational effects in plants [116]. The absorption of nanoparticles by plants not only carries the jeopardy of harming the plants but also the congregation of pollutants and producing polluted progeny acting as a medium to transmit toxic substances in the naïve and clear environment [11]. In the absence of extraneous sources of exposure, nanoparticles can be transferred to plant generations via seeds [117].

Few beneficial effects of cerium oxide nanoparticles have also been reported by authors on the first-generation seedlings, nevertheless, harmful effects were noticed on the growth of second progeny. It was evident that the second progeny of tomato plants grown from seeds obtained from parent plants with exposure to cerium oxide nanoparticles were characterized by decreased biomass, lesser potential of water transpiration, and a higher proportion of reactive O_2 [116]. Further investigations on the impact of cerium oxide nanoparticles using decreased concentrations (10 mg L^{-1} nanoparticles) of these particles during the initial life stages of the tomato plants still showed untoward effects. The seedlings of the second progeny of the parent plants were relatively smaller, contained lesser biomass and higher reactive oxygen species, as recorded earlier. Albeit, the growth of root hairs and gathering of nanoparticles were more pronounced than controls and parent generation. These findings may help in devising future studies on the transgenerational impact of nanoparticles in plants [116].

8 Conclusions and Future Perspectives

The interaction between nanoparticles and plants is a multifaceted phenomenon that depends upon the physicochemical attributes of nanoparticles, the inherent susceptibility of plants, route of exposure and the nature of soil or growth medium [101]. Although, nanoparticles consisting of essential heavy metals and metal oxides (such as manganese oxides and iron oxides) are relatively harmless for agricultural applications [118], most of the typical nanoparticles have been linked with a variety of deleterious effects on several species of plants. Accordingly, the underlying mechanisms of nanoparticle-induced phytotoxicity require inclusive comprehension before their application in the field. Nevertheless, the existing data on phytotoxic effects of nanoparticles has been primarily obtained through controlled greenhouse studies or laboratory experiments, and therefore, the predicted responses may differ in field conditions [119]. Likewise, the potential role of soil characteristics and microorganisms in nanoparticle-induced phytotoxicity cannot be determined using hydroponic systems, synthetic soil and potting soil [118]. Besides investigating the detrimental impact of nanoparticles on plants in soil and quartz sand, the phytotoxic screening of nanoparticles should be extended to adverse conditions such as salinity, drought and floods [60].

In contrast to conventional nanoparticles, the phytotoxic impact of newly engineered coated nanoparticles has not been completely investigated so far. Even though, the transgenerational and genotoxic effects of coated nanoparticles have been evaluated, the estimation of threshold concentrations is still to be determined [119]. Likewise, the interaction of weathered nanoparticles with plants and the signaling mechanisms underlying the association of nanoparticles with reactive oxygen species have not been fully elucidated [101, 119]. Moreover, the chronic exposure-response relationship and resultant phytotoxic effects of nanoparticles require consideration to facilitate the subsequent long-term risk assessment [60].

Being smaller in size and lighter in weight, the nanoparticles are projected to become airborne and thereby disseminate to other plants resulting in probable ill effects [120]. The movement of nanoparticles across the different portions of plants indicates their likely transfer to various ecological trophic levels, thereby influencing the soil microbiota, animals and humans [101]. Accordingly, the uptake and accumulation of nanoparticles by edible plants reflect serious concerns about food safety and public health [119, 121]. Prolonged exposure of freshwater habitats to nanoparticles may result in increased pH and salt concentration, thus affecting the inhabiting aquatic organisms. Therefore, properly designed, life cycle-based investigations are critically needed for explicating the transfer of nanoparticles across the food chain and evaluating their cumulative impact on the ecosystem [101].

More effective and convenient techniques should be devised for in situ analysis of the detrimental effects caused by nanoparticles in plants. Detailed mechanistic investigations are critically required for targeting DNA damage repair and cell death by exposing edible and model plants to environmentally feasible concentrations of nanoparticles. Plant cell suspensions can be effectively used for in vitro nanotoxicity analyses due to high susceptibility and larger surface area for exposure. Novel tools of microscopy and next-generation sequencing can be employed to localize the nanoparticles in plants and explore their phytotoxic effects at genetic and proteomic levels [46]. Besides, the species-specific and genetic-based variations in the susceptibility of plants to nanoparticle-induced toxicity as well as the potential role of nanoparticles in translocating pollutants also need further investigation. Finally, effective measures are also requisite for precluding or diminishing the phytotoxic effects attributed to nanoparticles and regulating their agro-industrial applications [60].

References

1. Nel, A., Xia, T., Madler, L., & Li, N. (2006). Toxic potential of materials at the nanolevel. *Science, 311*(5761), 622–627.
2. Peralta-Videa, J. R., Zhao, L., Lopez-Moreno, M. L., de la Rosa, G., Hong, J., & Gardea-Torresdey, J. L. (2011). Nanomaterials and the environment: A review for the biennium 2008–2010. *Journal of Hazardous Materials, 186*(1), 1–15.
3. Aziz, N., Faraz, M., Pandey, R., Shakir, M., Fatma, T., Varma, A., Barman, I., & Prasad, R. (2015). Facile algae-derived route to biogenic silver nanoparticles: Synthesis, antibacterial, and photocatalytic properties. *Langmuir, 31*(42), 11605–11612.
4. Cornelis, G., Hund-Rinke, K., Kuhlbusch, T., Van den Brink, N., & Nickel, C. (2014). Fate and bioavailability of engineered nanoparticles in soils: A review. *Critical Reviews in Environmental Science and Technology, 44*(24), 2720–2764.
5. Gottschalk, F., Sonderer, T., Scholz, R. W., & Nowack, B. (2009). Modeled environmental concentrations of engineered nanomaterials (TiO2, ZnO, Ag, CNT, fullerenes) for different regions. *Environmental Science & Technology, 43*(24), 9216–9222.
6. Prasad, R., Pandey, R., & Barman, I. (2016). Engineering tailored nanoparticles with microbes: Quo vadis? *Wiley Interdisciplinary Reviews: Nanomedicine and Nanobiotechnology, 8*(2), 316–330.

7. Rico, C. M., Peralta-Videa, J., & Gardea-Torresdey, J. (2015). Chemistry, biochemistry of nanoparticles, and their role in antioxidant defense system in plants. In *Nanotechnology and plant sciences* (pp. 1–17). Springer.
8. Tripathi, D., Singh, V., Swati, S., Prasad, S., Chauhan, D., & Dubey, N. (2016). Effect of silicon and silicon nanoparticle (SiNp) on seedlings of maize cultivar and hybrid differing in arsenate tolerance. *Frontiers in Environmental Science, 4*, 46.
9. Hossain, Z., Mustafa, G., Sakata, K., & Komatsu, S. (2016). Insights into the proteomic response of soybean towards Al2O3, ZnO, and Ag nanoparticles stress. *Journal of Hazardous Materials, 304*, 291–305.
10. Hong, J., Peralta-Videa, J. R., Rico, C., Sahi, S., Viveros, M. N., Bartonjo, J., Zhao, L., & Gardea-Torresdey, J. L. (2014). Evidence of translocation and physiological impacts of foliar applied CeO2 nanoparticles on cucumber (Cucumis sativus) plants. *Environmental Science & Technology, 48*(8), 4376–4385.
11. Rico, C. M., Majumdar, S., Duarte-Gardea, M., Peralta-Videa, J. R., & Gardea-Torresdey, J. L. (2011). Interaction of nanoparticles with edible plants and their possible implications in the food chain. *Journal of Agricultural and Food Chemistry, 59*(8), 3485–3498.
12. Zhu, H., Han, J., Xiao, J. Q., & Jin, Y. (2008). Uptake, translocation, and accumulation of manufactured iron oxide nanoparticles by pumpkin plants. *Journal of Environmental Monitoring, 10*(6), 713–717.
13. Wang, W., & Freemark, K. (1995). The use of plants for environmental monitoring and assessment. *Ecotoxicology and Environmental Safety, 30*(3), 289–301.
14. Miralles, P., Church, T. L., & Harris, A. T. (2012). Toxicity, uptake, and translocation of engineered nanomaterials in vascular plants. *Environmental Science & Technology, 46*(17), 9224–9239.
15. Wang, W. (1991). Literature review on higher plants for toxicity testing. *Water, Air, and Soil Pollution, 59*(3), 381–400.
16. Lee, C. W., Mahendra, S., Zodrow, K., Li, D., Tsai, Y. C., Braam, J., & Alvarez, P. J. (2010). Developmental phytotoxicity of metal oxide nanoparticles to *Arabidopsis thaliana*. *Environmental Toxicology and Chemistry: An International Journal, 29*(3), 669–675.
17. Parsons, J. G., Lopez, M. L., Gonzalez, C. M., Peralta-Videa, J. R., & Gardea-Torresdey, J. L. (2010). Toxicity and biotransformation of uncoated and coated nickel hydroxide nanoparticles on mesquite plants. *Environmental Toxicology and Chemistry, 29*(5), 1146–1154.
18. Tarrahi, R., Khataee, A., Movafeghi, A., Rezanejad, F., & Gohari, G. (2017). Toxicological implications of selenium nanoparticles with different coatings along with Se4+ on Lemna minor. *Chemosphere, 181*, 655–665.
19. Ghafariyan, M. H., Malakouti, M. J., Dadpour, M. R., Stroeve, P., & Mahmoudi, M. (2013). Effects of magnetite nanoparticles on soybean chlorophyll. *Environmental Science & Technology, 47*(18), 10645–10652.
20. Baker, C. J., & Mock, N. M. (1994). An improved method for monitoring cell death in cell suspension and leaf disc assays using Evans blue. *Plant Cell Tissue and Organ Culture, 39*(1), 7–12.
21. Chang, C., Larsen, P., Clark, K., Wen, C.-K., Ding, W., Shockey, J., & Pan, Z. (1999). Protein-protein interactions in ethylene signal transduction in Arabidopsis. In *Biology and biotechnology of the plant hormone ethylene II* (pp. 65–70). Springer.
22. Babula, P., Masarik, M., Adam, V., Eckschlager, T., Stiborova, M., Trnkova, L., Skutkova, H., Provaznik, I., Hubalek, J., & Kizek, R. (2012). Mammalian metallothioneins: properties and functions. *Metallomics, 4*(8), 739–750.
23. Santos, C. L., Pourrut, B., & Oliveira, J. M. P. (2015). The use of comet assay in plant toxicology: Recent advances. *Frontiers in Genetics, 6*, 216.
24. Winkel-Shirley, B. (2002). Biosynthesis of flavonoids and effects of stress. *Current Opinion in Plant Biology, 5*(3), 218–223.

25. Tanou, G., Molassiotis, A., & Diamantidis, G. (2009). Induction of reactive oxygen species and necrotic death-like destruction in strawberry leaves by salinity. *Environmental and Experimental Botany, 65*(2–3), 270–281.
26. Sharma, D., Kanchi, S., & Bisetty, K. (2019). Biogenic synthesis of nanoparticles: A review. *Arabian Journal of Chemistry, 12*(8), 3576–3600.
27. Montillet, J.-L., Chamnongpol, S., Rustérucci, C., Dat, J., Van De Cotte, B., Agnel, J.-P., Battesti, C., Inzé, D., Van Breusegem, F., & Triantaphylidès, C. (2005). Fatty acid hydroperoxides and H_2O_2 in the execution of hypersensitive cell death in tobacco leaves. *Plant Physiology, 138*(3), 1516–1526.
28. Oberdürster, G. (2000). Toxicology of ultrafine particles: *In vivo* studies. *Philosophical Transactions of the Royal Society of London Series A: Mathematical, Physical and Engineering Sciences, 358*(1775), 2719–2740.
29. Sabo-Attwood, T., Unrine, J. M., Stone, J. W., Murphy, C. J., Ghoshroy, S., Blom, D., Bertsch, P. M., & Newman, L. A. (2012). Uptake, distribution and toxicity of gold nanoparticles in tobacco (*Nicotiana xanthi*) seedlings. *Nanotoxicology, 6*(4), 353–360.
30. Yasur, J., & Rani, P. U. (2013). Environmental effects of nanosilver: Impact on castor seed germination, seedling growth, and plant physiology. *Environmental Science and Pollution Research, 20*(12), 8636–8648.
31. El-Temsah, Y. S., & Joner, E. J. (2012). Impact of Fe and Ag nanoparticles on seed germination and differences in bioavailability during exposure in aqueous suspension and soil. *Environmental Toxicology, 27*(1), 42–49.
32. Morales, M. I., Rico, C. M., Hernandez-Viezcas, J. A., Nunez, J. E., Barrios, A. C., Tafoya, A., Flores-Marges, J. P., Peralta-Videa, J. R., & Gardea-Torresdey, J. L. (2013). Toxicity assessment of cerium oxide nanoparticles in cilantro (*Coriandrum sativum* L.) plants grown in organic soil. *Journal of Agricultural and Food Chemistry, 61*(26), 6224–6230.
33. Rafique, R., Arshad, M., Khokhar, M., Qazi, I., Hamza, A., & Virk, N. (2014). Growth response of wheat to titania nanoparticles application. *NUST Journal of Engineering Sciences, 7*(1), 42–46.
34. Frazier, T. P., Burklew, C. E., & Zhang, B. (2014). Titanium dioxide nanoparticles affect the growth and microRNA expression of tobacco (*Nicotiana tabacum*). *Functional & Integrative Genomics, 14*(1), 75–83.
35. Perreault, F., Popovic, R., & Dewez, D. (2014). Different toxicity mechanisms between bare and polymer-coated copper oxide nanoparticles in *Lemna gibba*. *Environmental Pollution, 185*, 219–227.
36. Yin, L., Cheng, Y., Espinasse, B., Colman, B. P., Auffan, M., Wiesner, M., Rose, J., Liu, J., & Bernhardt, E. S. (2011). More than the ions: The effects of silver nanoparticles on *Lolium multiflorum*. *Environmental Science & Technology, 45*(6), 2360–2367.
37. Su, Y., He, Y., Lu, H., Sai, L., Li, Q., Li, W., Wang, L., Shen, P., Huang, Q., & Fan, C. (2009). The cytotoxicity of cadmium based, aqueous phase–synthesized, quantum dots and its modulation by surface coating. *Biomaterials, 30*(1), 19–25.
38. Lee, S., Cha, E. J., Park, K., Lee, S. Y., Hong, J. K., Sun, I. C., Kim, S. Y., Choi, K., Kwon, I. C., & Kim, K. (2008). A nearinfrared-fluorescence-quenched gold-nanoparticle imaging probe for *in vivo* drug screening and protease activity determination. *Angewandte Chemie, 120*(15), 2846–2849.
39. López-Moreno, M. L., de la Rosa, G., Hernández-Viezcas, J. A., Peralta-Videa, J. R., & Gardea-Torresdey, J. L. (2010). X-ray absorption spectroscopy (XAS) corroboration of the uptake and storage of CeO_2 nanoparticles and assessment of their differential toxicity in four edible plant species. *Journal of Agricultural and Food Chemistry, 58*(6), 3689–3693.
40. Li, X., Gui, X., Rui, Y., Ji, W., Yu, Z., & Peng, S. (2014). Bt-transgenic cotton is more sensitive to CeO_2 nanoparticles than its parental non-transgenic cotton. *Journal of Hazardous Materials, 274*, 173–180.
41. Majumdar, S., Peralta-Videa, J. R., Trujillo-Reyes, J., Sun, Y., Barrios, A. C., Niu, G., Flores-Margez, J. P., & Gardea-Torresdey, J. L. (2016). Soil organic matter influences cerium

translocation and physiological processes in kidney bean plants exposed to cerium oxide nanoparticles. *Science of the Total Environment, 569*, 201–211.

42. Schlich, K., & Hund-Rinke, K. (2015). Influence of soil properties on the effect of silver nanomaterials on microbial activity in five soils. *Environmental Pollution, 196*, 321–330.
43. Zou, X., Li, P., Huang, Q., & Zhang, H. (2016). The different response mechanisms of *Wolffia globosa*: Light-induced silver nanoparticle toxicity. *Aquatic Toxicology, 176*, 97–105.
44. Jośko, I., & Oleszczuk, P. (2013). Manufactured nanomaterials: The connection between environmental fate and toxicity. *Critical Reviews in Environmental Science and Technology, 43*(23), 2581–2616.
45. Ali, S., Mehmood, A., & Khan, N. (2021). Uptake, translocation, and consequences of nanomaterials on plant growth and stress adaptation. *Journal of Nanomaterials*, 1–17.
46. Dietz, K.-J., & Herth, S. (2011). Plant nanotoxicology. *Trends in Plant Science, 16*(11), 582–589.
47. Navarro, E., Piccapietra, F., Wagner, B., Marconi, F., Kaegi, R., Odzak, N., Sigg, L., & Behra, R. (2008). Toxicity of silver nanoparticles to *Chlamydomonas reinhardtii. Environmental Science & Technology, 42*(23), 8959–8964.
48. Kurepa, J., Paunesku, T., Vogt, S., Arora, H., Rabatic, B. M., Lu, J., Wanzer, M. B., Woloschak, G. E., & Smalle, J. A. (2010). Uptake and distribution of ultrasmall anatase TiO_2 alizarin red S nanoconjugates in *Arabidopsis thaliana. Nano Letters, 10*(7), 2296–2302.
49. Ma, X., Geiser-Lee, J., Deng, Y., & Kolmakov, A. (2010). Interactions between engineered nanoparticles (ENPs) and plants: Phytotoxicity, uptake and accumulation. *Science of the Total Environment, 408*(16), 3053–3061.
50. Zhang, W., Musante, C., White, J. C., Schwab, P., Wang, Q., Ebbs, S. D., & Ma, X. (2017b). Bioavailability of cerium oxide nanoparticles to *Raphanus sativus* L. in two soils. *Plant Physiology and Biochemistry, 110*, 185–193.
51. Khan, I., Saeed, K., & Khan, I. (2019). Nanoparticles: Properties, applications and toxicities. *Arabian Journal of Chemistry, 12*(7), 908–931.
52. Syu, Y.-y., Hung, J.-H., Chen, J.-C., & Chuang, H.-w. (2014). Impacts of size and shape of silver nanoparticles on Arabidopsis plant growth and gene expression. *Plant Physiology and Biochemistry, 83*, 57–64.
53. Chen, X., Zhou, Y., Han, H., Wang, X., Zhou, L., Yi, Z., Fu, Z., Wu, X., Li, G., & Zeng, L. (2021). Optical and magnetic properties of small-size core–shell Fe_3O_4@C nanoparticles. *Materials Today Chemistry, 22*, 100556.
54. Avellan, A., Yun, J., Morais, B. P., Clement, E. T., Rodrigues, S. M., & Lowry, G. V. (2021). Critical review: Role of inorganic nanoparticle properties on their foliar uptake and in planta translocation. *Environmental Science & Technology, 55(20)*, 13417–13431.
55. Schwab, F., Zhai, G., Kern, M., Turner, A., Schnoor, J. L., & Wiesner, M. R. (2016). Barriers, pathways and processes for uptake, translocation and accumulation of nanomaterials in plants–critical review. *Nanotoxicology, 10*(3), 257–278.
56. Achari, G. A., & Kowshik, M. (2018). Recent developments on nanotechnology in agriculture: Plant mineral nutrition, health, and interactions with soil microflora. *Journal of Agricultural and Food Chemistry, 66*(33), 8647–8661.
57. Kwak, J. M., Mori, I. C., Pei, Z. M., Leonhardt, N., Torres, M. A., Dangl, J. L., Bloom, R. E., Bodde, S., Jones, J. D., & Schroeder, J. I. (2003). NADPH oxidase AtrbohD and AtrbohF genes function in ROS-dependent ABA signaling in Arabidopsis. *The EMBO Journal, 22*(11), 2623–2633.
58. Mittler, R. (2002). Oxidative stress, antioxidants and stress tolerance. *Trends in Plant Science, 7*(9), 405–410.
59. Joo, J. H., Bae, Y. S., & Lee, J. S. (2001). Role of auxin-induced reactive oxygen species in root gravitropism. *Plant Physiology, 126*(3), 1055–1060.
60. Yang, J., Cao, W., & Rui, Y. (2017). Interactions between nanoparticles and plants: Phytotoxicity and defense mechanisms. *Journal of Plant Interactions, 12*(1), 158–169.

61. Singh, V. P., Singh, S., Kumar, J., & Prasad, S. M. (2015). Investigating the roles of ascorbate-glutathione cycle and thiol metabolism in arsenate tolerance in ridged luffa seedlings. *Protoplasma, 252*(5), 1217–1229.
62. Fridovich, I. (1989). Superoxide dismutases: an adaptation to a paramagnetic gas. *Journal of Biological Chemistry, 264*(14), 7761–7764.
63. Scandalios, J. G. (1993). Oxygen stress and superoxide dismutases. *Plant Physiology, 101*(1), 7.
64. Zhang, P., Ma, Y., Liu, S., Wang, G., Zhang, J., He, X., Zhang, J., Rui, Y., & Zhang, Z. (2017a). Phytotoxicity, uptake and transformation of nano-CeO_2 in sand cultured romaine lettuce. *Environmental Pollution, 220*, 1400–1408.
65. Faisal, M., Saquib, Q., Alatar, A. A., Al-Khedhairy, A. A., Hegazy, A. K., & Musarrat, J. (2013). Phytotoxic hazards of NiO-nanoparticles in tomato: A study on mechanism of cell death. *Journal of Hazardous Materials, 250*, 318–332.
66. Garg, N., & Manchanda, G. (2009). ROS generation in plants: Boon or bane? *Plant Biosystems, 143*(1), 81–96.
67. Servin, A. D., Morales, M. I., Castillo-Michel, H., Hernandez-Viezcas, J. A., Munoz, B., Zhao, L., Nunez, J. E., Peralta-Videa, J. R., & Gardea-Torresdey, J. L. (2013). Synchrotron verification of TiO_2 accumulation in cucumber fruit: A possible pathway of TiO2 nanoparticle transfer from soil into the food chain. *Environmental Science & Technology, 47*(20), 11592–11598.
68. Rao, K. M., Raghavendra, A., & Reddy, K. J. (2006). *Physiology and molecular biology of stress tolerance in plants*. Springer Science & Business Media.
69. Noctor, G., & Foyer, C. H. (1998). Ascorbate and glutathione: Keeping active oxygen under control. *Annual Review of Plant Biology, 49*(1), 249–279.
70. Rico, C. M., Morales, M. I., Barrios, A. C., McCreary, R., Hong, J., Lee, W.-Y., Nunez, J., Peralta-Videa, J. R., & Gardea-Torresdey, J. L. (2013b). Effect of cerium oxide nanoparticles on the quality of rice (*Oryza sativa* L.) grains. *Journal of Agricultural and Food Chemistry, 61* (47), 11278–11285.
71. Foyer, C. H., & Noctor, G. (2003). Redox sensing and signalling associated with reactive oxygen in chloroplasts, peroxisomes and mitochondria. *Physiologia Plantarum, 119*(3), 355–364.
72. Ma, C., Chhikara, S., Xing, B., Musante, C., White, J. C., & Dhankher, O. P. (2013). Physiological and molecular response of *Arabidopsis thaliana* (L.) to nanoparticle cerium and indium oxide exposure. *ACS Sustainable Chemistry & Engineering, 1*(7), 768–778.
73. Young, A. J. (1991). The photoprotective role of carotenoids in higher plants. *Physiologia Plantarum, 83*(4), 702–708.
74. Diplock, A., Machlin, L., Packer, L., & Pryor, W. (1989). Vitamin E: Biochemistry and health aspects. *Annals of the New York Academy of Sciences, 570*, 1–555.
75. Hossain, Z., Mustafa, G., & Komatsu, S. (2015). Plant responses to nanoparticle stress. *International Journal of Molecular Sciences, 16*(11), 26644–26653.
76. Van Breusegem, F., & Dat, J. F. (2006). Reactive oxygen species in plant cell death. *Plant Physiology, 141*(2), 384–390.
77. Mustafa, G., & Komatsu, S. (2016). Toxicity of heavy metals and metal-containing nanoparticles on plants. *Biochimica et Biophysica Acta (BBA)-Proteins and Proteomics, 1864*(8), 932–944.
78. Qian, H., Peng, X., Han, X., Ren, J., Sun, L., & Fu, Z. (2013). Comparison of the toxicity of silver nanoparticles and silver ions on the growth of terrestrial plant model *Arabidopsis thaliana*. *Journal of Environmental Sciences, 25*(9), 1947–1956.
79. Mushtaq, Y. K. (2011). Effect of nanoscale Fe_3O_4, TiO_2 and carbon particles on cucumber seed germination. *Journal of Environmental Science and Health, Part A, 46*(14), 1732–1735.
80. Wierzbicka, M., & Obidzińska, J. (1998). The effect of lead on seed imbibition and germination in different plant species. *Plant Science, 137*(2), 155–171.

81. Yan, A., & Chen, Z. (2019). Impacts of silver nanoparticles on plants: A focus on the phytotoxicity and underlying mechanism. *International Journal of Molecular Sciences, 20*(5), 1003.
82. Zafar, H., Ali, A., & Zia, M. (2017). CuO nanoparticles inhibited root growth from *Brassica nigra* seedlings but induced root from stem and leaf explants. *Applied Biochemistry and Biotechnology, 181*(1), 365–378.
83. Ghosh, M., Bhadra, S., Adegoke, A., Bandyopadhyay, M., & Mukherjee, A. (2015). MWCNT uptake in *Allium cepa* root cells induces cytotoxic and genotoxic responses and results in DNA hyper-methylation. *Mutation Research/Fundamental and Molecular Mechanisms of Mutagenesis, 774*, 49–58.
84. Yan, S., Zhao, L., Li, H., Zhang, Q., Tan, J., Huang, M., He, S., & Li, L. (2013). Single-walled carbon nanotubes selectively influence maize root tissue development accompanied by the change in the related gene expression. *Journal of Hazardous Materials, 246*, 110–118.
85. Le, V. N., Rui, Y., Gui, X., Li, X., Liu, S., & Han, Y. (2014). Uptake, transport, distribution and bio-effects of SiO_2 nanoparticles in Bt-transgenic cotton. *Journal of Nanobiotechnology, 12*(1), 1–15.
86. Tang, Y., He, R., Zhao, J., Nie, G., Xu, L., & Xing, B. (2016). Oxidative stress-induced toxicity of CuO nanoparticles and related toxicogenomic responses in *Arabidopsis thaliana*. *Environmental Pollution, 212*, 605–614.
87. Shaw, A. K., & Hossain, Z. (2013). Impact of nano-CuO stress on rice (*Oryza sativa* L.) seedlings. *Chemosphere, 93*(6), 906–915.
88. Santner, A., Calderon-Villalobos, L. I. A., & Estelle, M. (2009). Plant hormones are versatile chemical regulators of plant growth. *Nature Chemical Biology, 5*(5), 301–307.
89. Karlsson, H. L., Gustafsson, J., Cronholm, P., & Möller, L. (2009). Size-dependent toxicity of metal oxide particles—A comparison between nano-and micrometer size. *Toxicology Letters, 188*(2), 112–118.
90. Hao, Y., Yu, F., Lv, R., Ma, C., Zhang, Z., Rui, Y., Liu, L., Cao, W., & Xing, B. (2016). Carbon nanotubes filled with different ferromagnetic alloys affect the growth and development of rice seedlings by changing the C: N ratio and plant hormones concentrations. *PLoS One, 11*(6), e0157264.
91. Singh, D., & Kumar, A. (2016). Impact of irrigation using water containing CuO and ZnO nanoparticles on *spinach oleracea* grown in soil media. *Bulletin of Environmental Contamination and Toxicology, 97*(4), 548–553.
92. Priester, J. H., Ge, Y., Mielke, R. E., Horst, A. M., Moritz, S. C., Espinosa, K., Gelb, J., Walker, S. L., Nisbet, R. M., & An, Y.-J. (2012). Soybean susceptibility to manufactured nanomaterials with evidence for food quality and soil fertility interruption. *Proceedings of the National Academy of Sciences, 109*(37), E2451–E2456.
93. Rani, P. U., Yasur, J., Loke, K. S., & Dutta, D. (2016). Effect of synthetic and biosynthesized silver nanoparticles on growth, physiology and oxidative stress of water hyacinth: *Eichhornia crassipes* (Mart) Solms. *Acta Physiologiae Plantarum, 38*(2), 58.
94. Rico, C. M., Hong, J., Morales, M. I., Zhao, L., Barrios, A. C., Zhang, J.-Y., Peralta-Videa, J. R., & Gardea-Torresdey, J. L. (2013a). Effect of cerium oxide nanoparticles on rice: A study involving the antioxidant defense system and *in vivo* fluorescence imaging. *Environmental Science & Technology, 47*(11), 5635–5642.
95. Rico, C. M., Lee, S. C., Rubenecia, R., Mukherjee, A., Hong, J., Peralta-Videa, J. R., & Gardea-Torresdey, J. L. (2014). Cerium oxide nanoparticles impact yield and modify nutritional parameters in wheat (*Triticum aestivum* L.). *Journal of Agricultural and Food Chemistry, 62*(40), 9669–9675.
96. Sardoiwala, M. N., Kaundal, B., & Choudhury, S. R. (2018). Toxic impact of nanomaterials on microbes, plants and animals. *Environmental Chemistry Letters, 16*(1), 147–160.
97. Xiong, T., Dumat, C., Dappe, V., Vezin, H., Schreck, E., Shahid, M., Pierart, A., & Sobanska, S. (2017). Copper oxide nanoparticle foliar uptake, phytotoxicity, and consequences for sustainable urban agriculture. *Environmental Science & Technology, 51*(9), 5242–5251.

98. Rajput, V. D., Minkina, T., Sushkova, S., Tsitsuashvili, V., Mandzhieva, S., Gorovtsov, A., Nevidomskyaya, D., & Gromakova, N. (2018b). Effect of nanoparticles on crops and soil microbial communities. *Journal of Soils and Sediments, 18*(6), 2179–2187.
99. Rajput, V. D., Minkina, T., Fedorenko, A., Mandzhieva, S., Sushkova, S., Lysenko, V., Duplii, N., Azarov, A., & Chokheli, V. (2018a). Destructive effect of copper oxide nanoparticles on ultrastructure of chloroplast, plastoglobules and starch grains in spring barley (*Hordeum sativum distichum*). *International Journal of Agriculture and Biology, 21*, 171–174.
100. Tighe-Neira, R., Carmora, E., Recio, G., Nunes-Nesi, A., Reyes-Diaz, M., Alberdi, M., Rengel, Z., & Inostroza-Blancheteau, C. (2018). Metallic nanoparticles influence the structure and function of the photosynthetic apparatus in plants. *Plant Physiology and Biochemistry, 130*, 408–417.
101. Tripathi, D. K., Singh, S., Singh, S., Pandey, R., Singh, V. P., Sharma, N. C., Prasad, S. M., Dubey, N. K., & Chauhan, D. K. (2017). An overview on manufactured nanoparticles in plants: Uptake, translocation, accumulation and phytotoxicity. *Plant Physiology and Biochemistry, 110*, 2–12.
102. Nair, P. M. G., & Chung, I. M. (2014). Physiological and molecular level effects of silver nanoparticles exposure in rice (*Oryza sativa* L.) seedlings. *Chemosphere, 112*, 105–113.
103. Vishwakarma, K., Upadhyay, N., Singh, J., Liu, S., Singh, V. P., Prasad, S. M., Chauhan, D. K., Tripathi, D. K., & Sharma, S. (2017). Differential phytotoxic impact of plant mediated silver nanoparticles (AgNPs) and silver nitrate ($AgNO_3$) on Brassica sp. *Frontiers in Plant Science, 8*, 1501.
104. Liang, L., Tang, H., Deng, Z., Liu, Y., Chen, X., & Wang, H. (2018). Ag nanoparticles inhibit the growth of the bryophyte, *Physcomitrella patens*. *Ecotoxicology and Environmental Safety, 164*, 739–748.
105. Musante, C., & White, J. C. (2012). Toxicity of silver and copper to Cucurbita pepo: Differential effects of nano and bulk-size particles. *Environmental Toxicology, 27*(9), 510–517.
106. Al-Huqail, A. A., Hatata, M. M., Al-Huqail, A. A., & Ibrahim, M. M. (2018). Preparation, characterization of silver phyto nanoparticles and their impact on growth potential of *Lupinus termis* L. seedlings. *Saudi Journal of Biological Sciences, 25*(2), 313–319.
107. Siddiqui, M. H., Al-Whaibi, M. H., Firoz, M., & Al-Khaishany, M. Y. (2015). Role of nanoparticles in plants. In *Nanotechnology and plant sciences* (pp. 19–35). Springer.
108. Sahebi, M., Hanafi, M. M., Siti Nor Akmar, A., Rafii, M. Y., Azizi, P., Tengoua, F., Nurul Mayzaitul Azwa, J., & Shabanimofrad, M. (2015). Importance of silicon and mechanisms of biosilica formation in plants. *BioMed Research International, 2015*, 396010.
109. Ma, J. F., & Yamaji, N. (2015). A cooperative system of silicon transport in plants. *Trends in Plant Science, 20*(7), 435–442.
110. Moreno-Olivas, F., Gant, V. U., Johnson, K. L., Peralta-Videa, J. R., & Gardea-Torresdey, J. L. (2014). Random amplified polymorphic DNA reveals that TiO_2 nanoparticles are genotoxic to *Cucurbita pepo*. *Journal of Zhejiang University Science A, 15*(8), 618–623.
111. Castiglione, M. R., Giorgetti, L., Geri, C., & Cremonini, R. (2011). The effects of nano-TiO_2 on seed germination, development and mitosis of root tip cells of *Vicia narbonensis* L. and *Zea mays* L. *Journal of Nanoparticle Research, 13*(6), 2443–2449.
112. Ghosh, M., Bandyopadhyay, M., & Mukherjee, A. (2010). Genotoxicity of titanium dioxide (TiO_2) nanoparticles at two trophic levels: Plant and human lymphocytes. *Chemosphere, 81* (10), 1253–1262.
113. Pokhrel, L. R., & Dubey, B. (2013). Evaluation of developmental responses of two crop plants exposed to silver and zinc oxide nanoparticles. *Science of the Total Environment, 452*, 321–332.
114. Abdelsalam, N. R., Abdel-Megeed, A., Ali, H. M., Salem, M. Z., Al-Hayali, M. F., & Elshikh, M. S. (2018). Genotoxicity effects of silver nanoparticles on wheat (*Triticum aestivum* L.) root tip cells. *Ecotoxicology and Environmental Safety, 155*, 76–85.

115. Kumari, M., Mukherjee, A., & Chandrasekaran, N. (2009). Genotoxicity of silver nanoparticles in *Allium cepa*. *Science of the Total Environment, 407*(19), 5243–5246.
116. Wang, Q., Ebbs, S. D., Chen, Y., & Ma, X. (2013). Trans-generational impact of cerium oxide nanoparticles on tomato plants. *Metallomics, 5*(6), 753–759.
117. Lin, S., Reppert, J., Hu, Q., Hudson, J. S., Reid, M. L., Ratnikova, T. A., Rao, A. M., Luo, H., & Ke, P. C. (2009). Uptake, translocation, and transmission of carbon nanomaterials in rice plants. *Small, 5*(10), 1128–1132.
118. Ruttkay-Nedecky, B., Krystofova, O., Nejdl, L., & Adam, V. (2017). Nanoparticles based on essential metals and their phytotoxicity. *Journal of Nanobiotechnology, 15*(1), 1–19.
119. Zuverza-Mena, N., Martínez-Fernández, D., Du, W., Hernandez-Viezcas, J. A., Bonilla-Bird, N., López-Moreno, M. L., Komárek, M., Peralta-Videa, J. R., & Gardea-Torresdey, J. L. (2017). Exposure of engineered nanomaterials to plants: Insights into the physiological and biochemical responses-a review. *Plant Physiology and Biochemistry, 110*, 236–264.
120. Pacheco, I., & Buzea, C. (2018). Nanoparticle uptake by plants: Beneficial or detrimental? In *Phytotoxicity of nanoparticles* (pp. 1–61). Springer.
121. Millán-Chiu, B. E., Md, P. R.-T., & Loske, A. M. (2020). Nanotoxicology in plants. In *Green Nanoparticles* (pp. 43–76). Springer.

Plant Molecular Responses to Nanoparticle Stress

Ilham Khan, Murtaza Hasan, Rehana Kausar, Junaid Shehzad, and Ghazala Mustafa

Abstract Adverse environmental conditions such as global warming, water shortage, flooding, and salinity are the most challenging tasks for agriculture and threaten global food security. To address these food security issues, it is imperative to develop better agricultural technologies as well as stress-tolerant cultivars. Nanomaterials are now being used as a vital tool for improving the growth and productivity of crops under abiotic stresses. Recently, managing crop stress using nanomaterials has been utilized to lessen the negative impacts caused by salinity, drought, flooding, and temperature. Current findings have suggested that nanoparticles might help plants to overcome abiotic stresses at lower toxicity and higher effectiveness as compared to their bulk or ionic counterparts. These nanomaterials can easily penetrate plant cells and are readily taken up, subsequently influencing various biological functions. Even though studies have shown that nanoparticles boost the growth of crops and alleviate abiotic stress, many questions regarding the nanoparticle-dependent regulation processes still need to be answered. This chapter attempts to summarize the plant responses to several metal oxide nanoparticles and their impact on gene expression and metabolite regulation. A general mechanism through which the metal oxide nanoparticles may cause an impact on the plant and organelle-specific response will be discussed.

Keywords Nanoparticles · Plants · Genomics · Metabolomics

I. Khan · J. Shehzad · G. Mustafa (✉)
Department of Plant Sciences, Quaid-i-Azam University, Islamabad, Pakistan
e-mail: mghazala@qau.edu.pk

M. Hasan
College of Chemistry and Chemical Engineering, Zhongkai University of Agriculture and Engineering, Guangzhou, China

Department of Biotechnology, The Institute of Biochemistry, Biotechnology and Bioinformatics, The Islamia University of Bahawalpur, Bahawalpur, Pakistan

R. Kausar
Department of Botany, The University of Azad Jammu and Kashmir, Muzaffarabad, Pakistan

J.-T. Chen (ed.), *Plant and Nanoparticles*,
https://doi.org/10.1007/978-981-19-2503-0_9

1 Introduction

Production of world food should rise by 70% to satisfy the growing demand by the year 2050, as per FAO [1]. In the world, agronomy is believed as the primary mainstay to produce food. Because of increasing pressure on land, for use of nano agriculture, escalating land yield in food production is a test as the cultivating area is going to remain consistent or even reduce [2]. With the environmental change, abiotic stresses are the chief limitation for supporting yield production [3]. According to one of the assessments, roughly 70% of yield decrease is directly or indirectly affected by abiotic stresses [4]. Abiotic stress prompts a sequence of morphological, physiological, chemical, and molecular changes that negatively affect the productivity and growth of plants. Salinity, drought, and temperature are the most predominant abiotic stresses, intimidating worldwide food security.

Common breeding techniques have met with restricted achievement in further developing the stress tolerance of crop plants including intergeneric hybridization. It is significant to search for other strategies to foster stress-tolerant crop plants [5]. These days, the worldwide interest is to enhance the production of food with contained accessible resources and least yet effective utilization of compost and pesticides to reshape current agriculture. Among the most recent innovation, nanotechnology is the most encouraging one in plant biotechnology and agriculture [6]. Plant growth at various stages is affected both positively and negatively using nanodevice or nanoparticles (NPs). Nanotechnology contains unique properties of nanomaterial that make it affluent for agricultural research in crop enhancement programs as well as mitigation to stresses [7]. NPs prompted different significant impacts on the physiological, morphological, and chemical properties of plants.

NPs are made of components of extremely small size (up to 100 nm), and these factors affect the properties at the cellular level. Nanomaterials (NMs) have a moderately bigger surface area when contrasted with a similar mass of material in larger form [8]. NPs can modify the chemical properties and make materials more chemically reactive. NMs have a high surface-to-volume ratio that boosts their reactivity and achievable biochemical activity [9]. In many sectors, the use of nanotechnology has been followed such as energy production, water treatment, health and medicine, agriculture, and food production [10]. The present status of information regarding NMs and plants interaction at both molecular and physiological levels, including toxicity, absorption, and cell compartmentalization, has been checked [11, 12]. In the last 10 years, the use of NPs rises exponentially, with even more use is expected in the future as it improves the crop yield more effectively.

There are currently some conceptions that which organs of the plant, cells, tissues, and organelles are related to response, as well as the molecular pathways related to toxicity (e.g., ROS production, DNA damage, protein misfolding, and so on) [13]. In any case, given the wide scope of NMs utilized (size, shape, composition, coating) and their effects, it remains hard to focus on constant endpoints common in response to various classes of these materials. Estimation of metabolites or other physiological parameters indicates the biological dose and impact on exposure to NPs, showing indirectly or directly the effects on physiology [14]. Biomarkers of impacts show changes at the molecular and cellular levels and indicate the protein's abundance or genes expression under controlled conditions. Because of this potential, biomarkers of impacts can be conveniently applied as a tool for the evaluation of toxicity [15]. Therefore, the present chapter attempts to summarize the plant responses to several metal oxide nanoparticles and their impact on gene expression and metabolite regulation.

2 Mechanism of Plant-Nanoparticle Interaction

Identifying the nature of interactions among plants and NPs is vital for evaluating their uptake and distribution. However, not much advancement has been done toward understanding the effect of NMs at the molecular level [16]. To understand the molecular response and phytotoxicity of plants under NP stress, evaluation of gene expression through transcriptomics is a potent method. NPs enter largely through leaves and roots to plant systems [17]. After the entrance, NPs interface with plants at the cell and subcellular levels, expediating changes to morphological and physiological levels, which might be stimulatory or oppressive relying upon the properties of NPs.

Chemical nature, reactivity, size, and, particularly, the NP amount present in or on the plant define their impact on plant systems [18]. To comprehend the interaction between plant and NP and to analyze the threats or advantages of agriculture, we consider the positive as well as negative effects of NPs on the development and growth of plants [19]. Specialists have utilized various strategies for NP application, viz., seed treatment, soil application, or foliar spray, while assessing the impact of nanoparticles on seed germination or plant development. Nonetheless, seed treatment has been utilized in the larger part of the assessments, definitely because of its practicality.

2.1 *Metal Containing Nanoparticles*

Interaction of NPs-crop plants and agriculture systems can have a huge impact on human well-being [20]. Therefore, NPs in crops should be entirely researched considering crop nutrients, molecular biology, biochemistry, and human health. Thus, inside crop visualization of NP's behavior is made possible through progressed instruments. Utilizing the macerozyme (R-10 enzyme) SP–ICP–MS gave direct data on Au NP phytoaccumulation in *Solanum lycopersicum* by explaining the concentration and size of the particle [21]. Yet there is a deficiency of detailed study on absorption, accumulation, transformation, and threats of NPs for crops. Moreover, besides influencing the nutritional values in crops/plants, NPs are broadly shown to have different biochemical and morphological effects [22]. These impacts are related to both transgenic and natural crops. Through decreased concentration of phytohormones, the conferred nanotoxicity to plants might be visualized after interaction with NPs, such as cerium and titanium NPs (Ce NPs and Ti NPs). Further, the nutrients of the crop (fats, proteins, and sugars) are also influenced by different NPs [23, 24]. After phytoaccumulation, the effects of NPs might be negative, positive, or impartial on crops.

Due to the human health associations, as well as the impacts on the productivity of plants, the uptake of NPs in consumable plants is a significant subject [25]. In agriculture, crops uptake of several NPs has been studied. Surprisingly, a few NPs had positive results; for instance, Ti and Cs NPs phytoaccumulated in tomato, rice, cucumber, and soybean [26, 27]. Under iron (Fe NPs) stress, the yield of soybean and nut seedlings and the dry weight of its pod and leaf have been improved [28, 29]. Similarly, Prasad et al. [30] reported that Cu and Cs NPs improved the growth of shoot and root in mung bean and improved the pod dry weight in peanuts, respectively. Moreover, in wheat crop, Cu NP increases tolerance against stress. Copper and iron NPs also increased the yield of wheat through physiological effects, which were shown through enhanced content of sugar and SOD activity on proteomics study [31]. In plants, because of aerobic metabolism, reactive oxygen species (ROS) are produced which acts as signaling molecules [32]. ROS have both free radicals (OH, O_2^-) and non-radical molecules ($^1O_2$, H_2O_2) [33]. When the level of ROS exceeds the defense mechanism, it causes adverse impacts such as oxidative stress in plants. The excess level of ROS can cause a threat to cells through membrane and DNA damage and electrolyte leakage, and at last, cell death occurs [34]. Previously reported research proposed that the phytotoxicity of metal and metal-based NPs can cause oxidative stress in plants [35–37]. Lin and Xing [38] reported that Zn NP phytotoxicity in ryegrass is due to the lipid peroxidation and particle-dependent ROS formation on cellular membrane surface.

Moreover, in *Lemna gibba,* Ag NP phytotoxicity was caused due to a high level of ROS production after being exposed to Ag NP stress for 7 days [39]. In general, H_2O_2 (ROS) might be transformed into more lethal •OH, where no significant details exist on the determination of •OH in plants under NP stress, and cannot be detoxified through an enzymatic system (cellular damage unavoidable) [13]. Among all ROS, •OH is highly reactive; hence, it can cause damage to proteins and membrane, lipid peroxidation, and other cellular injuries by reaction with all biological molecules [40]. Oxidative stress is intensified by the generation of lipid-derived radicals, when the level of ROS increases, which not only disturb the normal cellular functions but also damage DNA and proteins [41, 42]. In all living organisms, maximum damaging processes are due to lipid peroxidation, and under stress conditions, membrane injury typically acts as a signal of lipid peroxidation. Cell membrane damage is due to MDA (malondialdehyde), which is one of the end products of unsaturated fatty acids peroxidation [43, 44]. It has been proved that due to ROS production, lipid peroxidation increases in plants when exposed to various stresses such as metal-containing NPs.

2.2 Metal Oxide Nanoparticles

In plant cells, the metal oxide nanoparticle-induced toxicity is mostly mediated by the production of ROS [45]. In chloroplast, ROS is produced as a by-product of metabolic pathways, responsible for the degradation of chlorophyll [46]. So under MO-NPs, plants produce ROS due to disturbance in photosynthetic activity, which stimulates the defense mechanisms to prevent the plants from oxidative stress damage. Enzymes (SOD (superoxide dismutase), CAT (catalase), GPX (guaiacol peroxidase), POD (peroxidase), and ascorbate peroxidase (APX)), MDA, and thiols contents are usually changed in response to variation in ROS concentration [47]. 3,3-′-diaminobenzidine (DAB) (ROS-sensitive dye) can be used to observe H_2O_2 accumulation in *Arabidopsis* roots exposed to La_2O_3 and CeO_2 NPs.

Results revealed that under NP stress, deep brown color was produced, which can be easily seen with naked eyes [48]. Also, *Zea mays* exposed to CeO_2 NPs (800 mg/kg) result in the accumulation of H_2O_2, which is 10 times greater than the control plants. However, the results showed that no ion leakage and lipid oxidation occurred due to oxidative stress [49]. Variation in antioxidant and ROS responses depends on plant species, type of MO-NPs, and exposure conditions. When cucumber plants were treated with La_2O_3-NPs, at 2000 mg/L, the ROS production was higher in stressed plants [50]. Similarly, the POD, CAT, and SOD activities were also enhanced in *Cucumis sativus* under ZnO and CuO NP stress [51]. TiO_2 NPs improve the CAT activity at a concentration of 250–750 mg/kg while reducing the APX activity at 500 mg/kg concentration in *C. sativus* [52]. Moreover, CeO_2 enhanced the production of H_2O_2 in *Brassica rapa* and corn but reduces it in rice plants [53–55]. CuO NPs induced oxidative stress in rice, wheat, and soybean plants (elevated MDA content and ROS) and improved the antioxidant enzymes activities [36, 56, 57], and reduced activity of APX in *B. juncea* [58]. As previously mentioned, on plants, NPs have both positive and adverse effects. Zheng et al. [59] reported the positive effect of TiO_2 NPs on cucumber and spinach seed. The positive impact of TiO_2 depends on its different phases such as in spinach plants, anatase and rutile TiO_2 increase the formation of chlorophyll by 19% and 28%, correspondingly [60]. TiO_2 enhanced the photosynthetic activity by 30% and improved the germination of spinach seed.

Furthermore, anatase TiO_2 increases the fresh and dry weight of the spinach plant by 58.2 and 69.8%, though rutile TiO_2 increased the growth of *Spinacia oleracea* by 63–76%. This improvement in spinach growth is due to the capability of TiO_2 NPs to convert the N (nitrogen) to NH_3 [61]. In soybean, the antioxidant system and nitrate reductase activity are enhanced by applying a mixture of SiO_2 and TiO_2 NPs [62]. TiO_2 NPs boost up the formation of chlorophyll and Rubisco activities to improve plant growth [63]. ZnO NPs increase the root length of soybean while iron oxide NPs enhanced the dry weight of its leaf and pod [28, 64]. Maize seed when exposed to SiO_2 improves its germination rate by providing better nutrients to seeds [65]. SiO_2 improves plant growth by enhancing electron transport rate, transpiration rate, photosynthesis, and other physiological factors [66, 67]. Wheat germination is enhanced when under ZnO NPs [68]. NPs stimulate the growth of plants under certain concentrations, which seems very important. Additional studies are necessary to investigate the enhancing effects of these NPs on important crop plants. Moreover, it is important to discover the mode of action of these NPs.

3 Nanoparticle's Role in Stress Mitigation

Nanotechnology can reduce environmental contamination and assist plants to tolerate biotic and abiotic stress conditions. The green-synthesized NPs from plants are environment-friendly and economical [69]. Nanoparticle applications can assist with minimizing the use of harsh, harmful, and expensive chemicals, normally utilized in plant production [70]. Salinity has appeared as a worldwide interest due to constant rises inland impacted with salt. Salinity stress involves several harmful effects on a plant's biochemical, physiological, and molecular features and decreases crop productivity [71]. When compared with other metal oxide nanoparticles, ZnO is reliable and safe. Nanoparticles regulate the ion balance to reduce the effects of salt stress.

Zinc oxide nanoparticles (ZnO NPs) have many advantages on plant production, soil fertility, and zinc source, which is a major microelement for improving plant protection and development [72, 73]. ZnO nanoparticles can mitigate the salinity stress in *Abelmoschus esculentus* by enhancing the activity of CAT and SOD, increasing photosynthetic contents, and decreasing total soluble sugar and proline accumulation [74]. Similarly, in salinity stress, foliar spray of ZnO NPs improves chlorophyll content along with the performance of E, gs, ci, and PN [75]. Rubisco, which is directly linked with photosynthetic activity, is enhanced through NPs [76]. Latef et al. [77] described that in lupine (*L. termis*), the photosynthetic pigments, growth parameters, Zn content, and CAT activity decreased during salinity stress as compared to control plants. Moreover, in salinized plants, the contents of MDA, total phenols, organic solutes, sodium (Na), and ascorbic acids increased [78], though priming seeds with Zn NPs stimulated the growth of salinized plants by fortification in the level of photosynthetic pigments, growth parameters, Zn content, as well as CAT, SOD, POD, and APX enzyme activities. Also, in stressed plants, Zn NPs decrease the sodium and MDA contents [79, 80], thus increasing the salt tolerance of lupin plants. In comparison with NaCl stress, the plants treated with Zn NPs have high nonenzymatic activity and antioxidant enzymes. Seed priming with Zn NPs increases Zn content and decreases the accumulation of Na, which is a sign of salt resistance. Foliar application of cerium dioxide (CeO_2) NPs on Moldavian balm under NaCl stress significantly enhanced the agronomic traits, antioxidant enzymes, SPAD, and chlorophyll pigments. Also, CeO_2 NPs decrease the H_2O_2, electrolyte leakage, proline content, and MDA content.

While there was a significant reduction in agronomic traits and chlorophyll contents in salinized soil, CeO_2 NPs improve the defense mechanism against NaCl stress [81]. According to Rossi et al. [82] in *Brassica,* Ce NPs shortened the apoplastic barrier in roots, which enhanced the movement of Na^+ to shoots whereas less accumulation in roots of Na+ ions. Ce also exhibits antioxidant abilities due to the occurrence of both Ce_3^+ and Ce_4^+ on the surface of NPs [83]. The distinct redox chemistry on the CeO_2 NP surface provided NPs for superoxide dismutase (SOD) or catalase (CAT) mimetic activities. Foliar spray of iron oxide (IO-NPs) considerably enhanced the leaf area and enzymes activity in *Dracocephalum moldavica* under salt stress as compared to control ones [84]. Silicon is considered as one of the highly important elements for the plant, which occurs in three forms such as solid fractions, adsorbed, and liquid [85, 86]. Adsorbed and liquid forms of silicon contain mono silicic acid, due to which it is available to plants and an important form of Si [87]. Silicon NPs activate the defense and protection systems of plants, enhance their growth and yield, and act as a specific ion scavenger [88]. Also, silicon dioxide NPs increase the activity of antioxidant enzymes under salinity stress, act as growth inducers in plants, and enhance the tolerance to combat abiotic stresses [89, 90]. Silicon NPs have been suggested to enhance the chlorophyll pigments, reduce cell wall damage, regulate Na^+ levels, and stimulate K^+ uptake in wheat (*Triticum aestivum* L.) under salt stress. Si NPs might have the ability to enhance detoxification of Na^+ in plants by boosting Na^+ binding to the cell wall [91]. In sorghum, drought stress reduced the yield of grains (76%) and delayed the leaf and grain head emergence. ZnO NPs successfully alleviate the drought stress by increasing the enzyme's activity and can improve the grain yield (22–183%) [92]. Drought stress causes accretion of osmolytes and malondialdehyde and changes in the structure of subcellular organelles.

Thus, drought induces damage to the chloroplast, and mitochondria are lessened through the application of Zn NPs (activate antioxidant enzymes and promote the synthesis of melatonin) [93]. Iron is the fundamental component for the development and growth of living creatures (plants, human beings, and animals). Iron performs a significant role in different physiological and biochemical processes. For several enzymes, iron acts as a co-factor that acts as catalyst for several biochemical reactions [94]. As compared to iron-based fertilizers, iron nanoparticles (Fe NPs) improve the production and growth of plants [95]. For mitigation of adverse effects of HMs stress, iron oxide NPs are more desirable due to their large surface area to weight ratio [96]. Adrees et al. [97] reported that Fe NPs enhanced the process of photosynthesis, crop yield, and iron concentration in wheat while decreasing the oxidative stress because of drought and cadmium (Cd) stress. The efficiency of iron nanoparticles depends on their concentration in soil. The mitigation effects of IO-NPs were also investigated in rice under drought stress [98]. The results showed that iron oxide NPs elevate antioxidant enzyme activity, biomass, and photosynthesis and a decrease in ROS. Increased Zn content using ZnO NPs considerably enhanced the expression of SOD in plant leaves and is more efficient to reduce oxidative stress [99]. Foliar spray of ZnO NPs mitigates the chilling stress in rice seedlings, enhancing plant growth parameters and reducing the level of proline, H_2O_2, and MDA [100]. Venkatachalam et al. [101] confirmed the enhancing effects of ZnO NPs by elevating the level of antioxidant enzymes and reducing MDA content in leaves of *L. leucocephala* during oxidative stress. Other studies also revealed a potential increase in soluble protein contents and photosynthetic pigments by ZnO NPs.

Cerium oxide (CeO_2) NPs, because of their distinct optical, thermal, and electrical properties, have been widely utilized in various attributes of agriculture and plant science [102]. For example, CeO_2 NPs enhanced the antioxidant enzymes and plant growth and protect the membrane from peroxidation leakage [103]. Liable on the size, pH, surface charge, concentration, and subcellular localization, CeO_2 NPs may act as an antioxidant or induce oxidative stress [104–106]. Ce NPs can protect the chloroplast structure and cell wall. In low concentration, Ce acts as a catalyst in the synthesis of chlorophyll and in ROS scavenging and helps to maintain the cell structure [107, 108]. Thus, the application of metal NPs is a safe method to mitigate the negative effects of abiotic stresses such as salinity and drought in different crops.

4 Subcellular Transport and Mobilization of Nanoparticles

Data from experiments confirm that the symplast pathway is the predominant and efficient pathway for the transport of NPs inside plants. Symplast pathway is assisted by a vast range of transporting protein, endocytosed pathway, interconnected particle channels, aquaporins, or novel pores for the passage of NPs [109]. The most favored plant tissue is the xylem for NP assimilation and translocation alongside the stomata and phloem. Carbon-based and designed NPs exhibited various responses in plants for their application and passage [110]. Designed NMs are translocated and aggregated contrarily inside the leaves, trichomes, petioles, fruits, and stems of various plants. In different cellular organelles, designed NMs are aggregated into the cell wall, plastids, cytoplasm, small vesicles, and nuclei. Certainly, there is not one transportation mechanism, but a different range of mechanisms at physiological, biochemical, and molecular levels is included for perforation and accretion of NPs [111]. Compared to animal cells, the cell wall of the plant is the foremost barrier for nanotechnology.

The size of NPs and water content are the most basic parameters for infiltration and movement of NPs in the plant cell [112]. Size is very important not only for diffusion through cuticle and stomata of leaf but also for the movement of NPs in cell wall matrix [113]. There are a significant number of reports that have documented the infiltration of NPs into plants generally through young roots, tissue culture, parenchyma cells of stem, and protoplasts [114, 115]. The translocation of NPs from leaf to root reveals that NPs travel by the phloem transport mechanism. The shape, application technique, and nature of plant tissues are the main factors on which the transportation and accumulation of NPs depend. NPs in the soil go through a series of bioconversion, which defines the toxicity and bioavailability of NPs. The NPs interact with the roots of plants before translocation to the aerial part and then accumulate in cellular organelles. In bioaccumulation, adsorption of NPs through roots of the plant is the first step [116]. Moreover, the transport of NPs into plasmodesmata of cell or other cellular organelles is also defined by the size of NPs [117]. The accumulation and reactivity of NPs in plant structures or on the surface of the cell are associated with its shape [118]. Additionally, the charge on NPs is imported for attachment to the cell wall, which is negatively charged, after this hydrophobicity on the surface of plants performs a vital role in the ingestion and translocation of NPs [119]. Du et al. [120] reported that small size NPs penetrate the root with capillary forces or osmotic pressure or through epidermal cells of root directly. Epidermal cells restrict the NPs with large size due to their semipermeable nature, but some nanoparticles generated new pores in the epidermal cell wall to enable its entrance [120]. NPs that once cross the cell wall are carried apoplastically by extracellular spaces and move to the central vascular cylinder, facilitating the xylem to move upward in one direction. Nanoparticles bind to transporter proteins (in the endodermal membrane of a cell) due to pore formation, endocytosis, and transport to cross the Casparian strip barrier through the symplast pathway.

Nanoparticles are internalized in the cytoplasm and move from cell to cell through plasmodesmata [117]. Those NPs that are incapable of internalization form aggregation on the Casparian strip [121]. NPs carried up by plants might be found inside the cortical cell cytoplasm, epidermal cell wall, and nucleus. In seeds, direct assimilation of NPs can happen by passing into the coat utilizing intercellular spaces of parenchyma, supplemented by dispersion in the cotyledon [117]. The NPs applied through a foliar spray can enter the leaves through cuticles of stomatal pores [122]. The cuticular layer limits the entry of NPs to a size of 10 nm, acts as an essential leaf barrier, enters through stomatal pores, and transfers to the vascular bundle of plants through apoplastic and symplastic pathways.

The exchange of NPs (10–50 nm) is supported through the nearby cell-to-cell cytoplasm (symplastic pathway). In this manner, NPs of large sizes (50–200 nm) are moved between the cells (apoplastic pathway). NPs that are internalized are carried through the sieve tubes in phloem along with the sugar movement. Because of vascular transport through the phloem, NPs accumulate in fruits, root, leaves, stems, and grains since these organs serve as effective sinks for the sap and can move bidirectionally [117, 118]. Morphology of the leaf and its composition are fundamental factors that influence the catching of NPs on the outer layer of the leaf. Cell wall pore size is the main factor for NP's entrance into the plant cell. NPs that can penetrate the cell wall from the surface of the plant could be 40 nm to 50 nm [123]. Zhu et al. [124] revealed that gold NPs cannot accumulate in the shoot of *Raphanus raphanistrum* and *Cucurbita pepo* but accumulate in *Oryza sativa* shoots. Besides, positively charged gold (Au NPs) are taken more rapidly.

Interestingly, negatively charged Au NPs are moved from the roots more efficiently into plant shoots. According to literature, silicon dioxide (SiO_2) and titanium dioxide (TiO_2) are the most stable NPs and can be found in their impeccable speciation in plant tissues [125]. Maximum Zn accumulates in shoots and roots of *Zea mays* in various forms like Zn phosphate. This might be due to absorption of plant, Zn translocation in ionic form, and enhanced dissolution in the rhizosphere [126]. It is also reported previously that CuO NPs are transported from roots to shoots through the xylem and back from shoots to roots by phloem in *Z. mays*. Hernandez-Viezcas et al. [25] observed the translocation of CeO_2 and ZnO NPs in *Glycine max*. Zn biotransformed into Zn citrus inside plant tissue, while CeO2 NPs translocated in the form of nanoparticles.

5 Gene Expression Analyses in Response to Nanoparticle Stress

Determination of genes expression is an effective approach to find out how plants respond to specific environmental stress. In various photosynthetic species like green algae (unicellular) to vascular plants (rice, *A. thaliana*, or tomato), transcriptomic studies have been performed to investigate the nano-impact [127]. Therefore, to evaluate the nano-impact on plants, transcriptional data including a variety of species are accessible with completely developed genomics tools. Omics advances can shift the research on plant-NP associations from low-throughput to high-throughput revelation [128]. Whereas data on the transcriptional impacts of NP exposure is accessible, the outcomes are to some extent conflicting. Few investigations show a solid impact on the transcript of genes related to stress and propose high-level toxicity for the plant (Table 1), whereas other researchers did not discover major changes in transcription and have reasoned that NPs are probably not going to deliver any adverse impacts for the plant [143]. Also, a small part of plant genes is utilized as markers for toxicological examinations, and they are generally involved in response to oxidative stress, which is not very useful at the morphological level.

Kaveh et al. [146] investigated the response of *A. thaliana* when exposed to silver nanoparticles (Ag NPs). The results show that 446 genes revealed steady expression levels out of which 375 considerably expressed at different levels. Various genes differentially expressed in response to Ag NPs were observed to be involved in many processes such as cellular and metabolic processes, biological regulation, and in response to the stimulus. Various genes that are downregulated are involved in the cellular response to hormone stimuli and hormone signaling pathways [129]. Subsequently, the upregulated genes are mostly involved in response to abiotic stresses (salts, metal ions, radiation, starvation, light, oxidative and osmotic stress). Genes that are significantly upregulated by Ag NPs are involved in salt stress: one gene involved in defense mechanism against biotic stress and encoding myrosinase binding protein, three genes involved in the biosynthesis pathway of thalianol, and one gene involved in response to wounding that encodes MLP. Other studies show that Ag NPs reduce the expression of ACO2 and ACS7, signifying that it acts as an ethylene inhibitor [127]. According to Landa et al. [130], ZnO and Ag NPs upregulated the genes AT1G08830 (SOD), two peroxidases (AT3G21770 and AT2G18150) that are produced in response to oxidative stress, and one gene (AT3G49780) phytosulfokine-β growth factor that is involved in wounding stress. Plants need to survive in certain conditions (e.g., new environment exploitation) for which plants produced stress-stimulated secondary metabolites in which these clusters are involved. Cu-Zn SOD expression levels were stimulated by treating *C. sativus* plants with CuNPs [143]. The result shows that Cu-Zn SOD expression level was enhanced sixfold in treated plants by reference to untreated plants, which indicates the increase in ROS production and conversion of superoxide to H_2O_2. Similarly, Nair and Chung [147] also reported the increase in gene expression of SOD in *P. sativum* on exposure to CuO NPs. The genes upregulated and downregulated in plants due to TiO_2 (80 up- and 74 downregulated) and ZnO

Table 1 Expression of different genes in response to NP stress and their functions

Nanoparticles	Plant species	Gene expression ↑↓	Function	Reference
Ag NPs	*A. thaliana*	AT3G28220 ↑ AT1G52000 ↑ AT1G52040 ↑ AT5G48010 ↑ AT5G48000 ↑ AT5G47990 ↑ AT2G01520 ↑	Defense mechanism Biosynthesis pathway	Bari and Jones [129]
		ACC oxidase 2 ↓ ACC synthase 7 ↓	Ethylene inhibitor	Syu et al. [127]
ZnO and Ag NPs	*A. thaliana*	AT1G08830 ↑ AT3G21770 ↑ AT2G18150 ↑ AT3G49780 ↑	Combat oxidative and wounding stress	Landa et al. [130]
ZnO NPs	*O. sativa*	OsWRKY76↓ OsbZIP52 ↓ OsMYB30 ↓ OsMYB4 ↓ OsNAC5 ↓ OsWRKY94 ↓	Protect cells from injuries due to chilling stress	Song et al. [100]
	Hordeum vulgare	miR156a ↑ miR159a ↑	Produce in response to abiotic stress	Plaksenkova et al. [131]
	Brassica napus	ARP↑ MYC↑ SnRK2D ↓ MAPK3↑ MAPK4 ↑	Enhance plant growth and stress tolerance	Hezaveh et al. [132]
	Hordeum vulgare	*HvNM1* ↓ *HvNM2* ↓ *HvNM3* ↓ *HvNM4* ↓ *HvNM5* ↓ *HvNM6* ↓	Regulate stress response and zinc transporter	Dong et al. [128]
	Solanum lycopersicum	Cu-ZnSOD ↑ GSH1 ↑ GR1↑ CAT1 ↑	Genes related to antioxidant activity	Li et al. [133]
	O. sativa	APX ↑ CAT ↑ SOD ↑	Defense mechanism	Salah et al. [134]
	O. sativa	N-acetylserotonin methyltransferase↓ Peroxidase ↓ Caffeic acid O-methyltransferase ↑ Tryptophan decarboxylase ↑	Alleviate phytotoxicity	Huang et al. [135]
Ag NPs	*A. thaliana*	IAA8 ↑ NCED3 ↑ RD22 ↑ ACS7 ↓ ACO2 ↓	Involve in metabolism and signaling pathways	Syu et al. [127]
	O. sativa	CAT gene↑ APX gene↑ CuZnSOD gene↓	Protection of cell from damage	Gupta et al. [136]
	Chironomus riparius	CuZnSOD↑ Catalase↑ PHGPx1 ↑	Produce in response to oxidative stress and detoxification of metals	Nair et al. [137]

(continued)

Table 1 (continued)

Nanoparticles	Plant species	Gene expression ↑↓	Function	Reference
		Glutathione S-transferases ↑ Thioredoxin reductase 1 ↑		
	Cajanus cajan	Lipoxygenase ↓ NADPH oxidase↓ P5CS (stress-responsive gene) ↑	Reduce the level of ROS	Yadu et al. [138]
	T. aestivum	Metallothionein genes ↑	Protect cells from toxic effects of metals	Dimkpa et al. [139]
CuO NPs	*A. thaliana*	RHL41 ↑ MSRB7 ↑ BCB ↑ PRXCA ↑ MC8 ↑	Oxidative stress-related genes	Tang et al. [140]
		SLR1/IAA14 ↑ AXR3/IAA17 ↑ AXR2/AA7 ↑	Lateral root development	Wang et al. [141]
SiO_2 NPs	*Hyoscyamus*	h6h gene ↑	Alkaloid production	Hedayati et al. [142]
Cu NPs	*C. sativus*	Cu-Zn SOD ↑	Reduce oxidative stress	Mosa et al. [143]
Al_2O_3 NPs	*N. tabacum*	miR397, miR399, miR395, miR398 ↑	Mitigate NP stress	Burklew et al. [144]
		miRNA ↑	Defense mechanism	Burklew et al. [144]
TiO_2 NPs	*A. thaliana*	LHCII *b* ↑	Enhance absorption of light	Ze et al. [145]

(660 up- and 826 downregulated) NPs are mainly stress-responsive genes, both biotic (defense to pathogen and wounding) and abiotic (salinity, oxidative, water scarcity) [130]. Al_2O_3 NPs enhanced the expression of miR397, miR399, miR395, and miR398. These miRNAs play a vital role in mediating the stress responses in plants to NP stress [144]. Chen et al. [148] reported the regulation of POD, CAT, Mn-SOD, APX, and Cu-Zn-SOD in plants under ZnO NP stress. Likewise, with melatonin application, nitro oxidative homeostasis can be achieved through the regulation of antioxidant enzymes (GR, APX, CAT, SOD) in alfalfa [149]. Previously, Huang et al. [135] reported that NO induced in rice performs a vital role to combat ZnO NP stress through the regulation of antioxidant enzymes and melatonin metabolism. The genes downregulated by ZnO NPs are mainly involved in nucleosome assembly, translation, and microtubule-based processes. Under ZnO stress, N-acetylserotonin methyltransferase and peroxidase genes are downregulated, while caffeic acid O-methyltransferase, tryptophan decarboxylase, and peroxidase genes are upregulated to alleviate the phytotoxicity induced by ZnO NPs. In *Arabidopsis,* the expressions of SLR1/IAA14, AXR3/IAA17, and AXR2/IAA7 genes were enhanced when exposed to CuO nanoparticle stress. *Arabidopsis* was exposed to CuO for 96 h, which increase the expression of these genes by 6.95, 1.48, and 2.80

times higher than the control plants, respectively [141]. These three genes are responsible for the regulation of lateral roots [150]. Salah et al. [134] investigated the expression of genes in *Oryza sativa* under ZnO NP stress. The results revealed that significant upregulation was observed in the expression of genes (APXa, APXb, CATa, CATb, CATc, SOD1, and SOD2) under different concentrations of ZnO NPs. This study proposes that these enzymes served as a fundamental defense tool to protect rice plants from oxidative stress induced by ZnO.

6 Molecular Analyses of Plants Under Nanoparticle-Induced Stress

ROS molecules provide signals for the coordination of a wide array of cellular events in the plant containing transduction and hormone perception [151]. For example, in roots, ROS can lower the sensitivity of auxin by modifying Ca+ signaling [152]. Moreover, ROS has a positive impact on ABA signaling, which performs a significant role in root development under stress conditions [153, 154]. A significant part of the current plant-NP interaction studies utilizes generally indifferent end efforts (biomass, germination, and pigment content), which give out contained guidance to grasp the interaction mechanism among plants and NPs [155]. Conversely, "omic"-based endpoints, for example, transcriptomics, proteomics, and metabolomics, can give to the point and mechanistic information on plant reactions to NP.

Reichman et al. [156] reported a transcriptome profile of changes in plants to CeO_2 NP stress. A previous study reported that *Arabidopsis* seedling, when treated with Ag NPs, enhanced the expression of IAA8, NCED3, and RD22 genes encoding auxin-inducible AUX/IAA protein, 9-cis-epoxycarotenoid dioxygenase, and ABA-mediated dehydration-responsive protein [157, 158]. Metabolomics permits the evaluation and identification of thousands of small molecules in NP-treated and NP-controlled plants. Dissimilar to other omics, metabolites (low molecular weight) are the closest connection to phenotype. Furthermore, the metabolites pool in organisms is a lot more modest than the protein and gene pool. This innovation can be utilized as a useful asset to comprehend the molecular response of plants that develop under NP stress. Babajani et al. [159] reported that Se and ZnO NPs significantly enhanced the expression of RAS (rosmarinic acid synthase) and HPPR (hydroxyphenylpyruvate reductase) genes. In plants, secondary metabolites contribute to plant communication and adaptation to the external environment (Fig. 1), though variations in plants' secondary metabolism in response to NPs are yet unclear.

Furthermore, wheat plants treated with SeNPs alter the expression of heat shock factor (HSFA4A, an anti-apoptosis agent) [160]. Results show the highest level of expression under 50 mg/L SeNP-treated plants. However, after 6 days, due to an increase in spray time, the expression of HSFA4A was reduced drastically [161]. The most important conserved signaling pathway, MAPK cascade, is

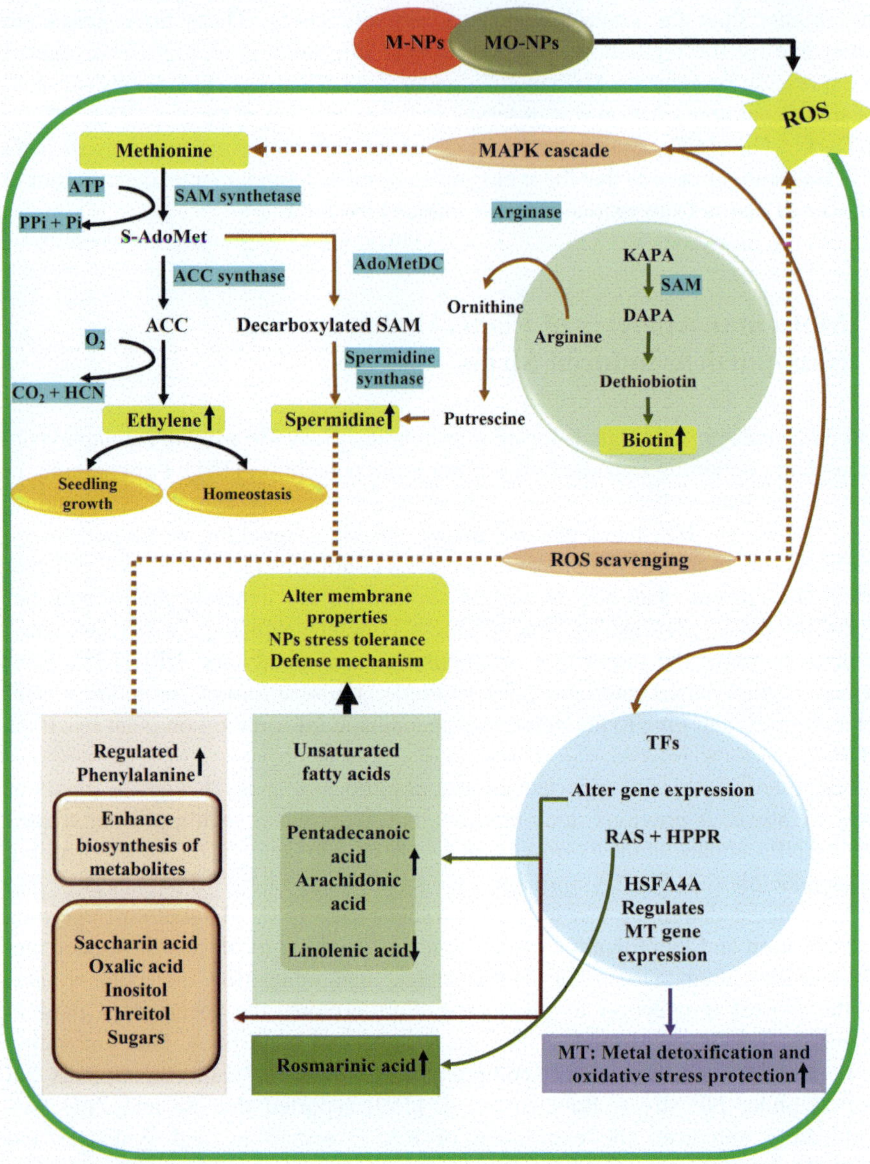

Fig. 1 Molecular mechanism of plants in response to NP stress. ROS produced in response to NP stress affects MAPK cascade, which triggers different cellular pathways. These pathways regulate the biosynthesis of secondary metabolites. Upward and downward arrows represent the increase and decrease in metabolite synthesis. Abbreviations: *ROS* reactive oxygen species, *ACC* aminocyclopropane-1-carboxylic acid, *HCN* hydrogen cyanide, *KAPA* keto-8-aminopelargonic acid, *DAPA* diaminopimelic acid, *RAS* rosmarinic acid synthase, *HPPR* hydroxyphenylpyruvate reductase, *ODC* ornithine decarboxylase, *SAM* S-adenosylmethionine

involved in the managing of various cellular reactions, such as differentiation, propagation, programmed cell death, and several responses to stress.

Here is clear molecular-based evidence suggesting HSFA4A is a substrate of the MPK3/MPK6 signaling (MAPK parts) and links to the regulation of plant responses to abiotic stress [162]. MPK3 and MPK6 regulator of several defense mechanisms and coordinator of numerous transcription factors are observed as the main contributors to control immunity of plant and cross-linking among stress stimuli, hormonal signaling, and secondary messengers, mainly Ca^{+2} and ROS [163]. ROS-triggered signaling pathways facilitate stimulation of different HSFs, transcriptionally stimulating their downstream target genes. The cross-link between MAPK and HSFA4A signaling depicts that HSFA4A act as an H_2O_2 sensor [164]. ROS-elicited signaling by SeNPs probably altered the gene expression, thus regulating growth, stress-responsive genes, metabolism, and hormonal balances.

HSFA4A enhance the cadmium (Cd) resistance in wheat (*Triticum aestivum*) and rice (*Oryza sativa*) via regulating the metallothionein gene expression [165]. Additionally, in many plant species, the genes related to photosynthesis and antioxidant machinery were changed in response to ZnO NPs [134, 166, 167]. Low molecular mass metabolites are the end product of gene expression; hence, alteration in the metabolomic profile of plant cells might be a potent approach to evaluate biological activities [168]. Metabolites produced in response to stress protect the cell from damage by altering the membrane properties and promoting plant growth under unfavorable conditions. Interaction between *Spinacia oleracea* (spinach) and cerium oxide nanoparticles (CeO_2 NPs) was examined by Zhang et al. [169] through integrating metabolic analysis. Results show that CeO_2 NPs induced metabolic reprograming that is not dose dependent. Noticeable metabolic variations occurred in roots because of their direct contact with NPs. Foliar application of CeO_2 NPs induced severe changes in the metabolic profile of leaves [170], while the amino acid decreases at a high concentration of NPs such as cycloleucine, methionine, tryptophan, l-cysteine, tyrosine, threonine, and asparagine. A decrease in glutamic acid, which is involved in the metabolism of nitrogen, may lead to a lower capacity of nitrogen assimilation through the glutamate synthase pathway. The transfer of methionine to S-adenosyl-1-methionine (SAM) regulates various important metabolite levels, for example polyamines, ethylene, phytosiderophores, and biotin [171]. Moreover, biosynthesis of secondary metabolites, chlorophyll, and cell wall formation is regulated by SAM [172]. Phenylalanine acts as a precursor for various secondary and intermediary metabolites, which perform a significant role in plant defenses to various environmental stresses [173]. During NP stress, the phenylalanine is upregulated as a response to stress. Under CeO_2 NP stress, metabolites such as ribulose-5-phosphate, allo-inositol, saccharic acid, threitol, gentiobiose, maltotriose, and oxalic acid are upregulated in a dose-dependent manner while phenylalanine and indole lactate are enhanced to a high dose of CeO_2 NPs.

Few of these metabolites, such as oxalic acid, allo-inositol, sugars, and phenylalanine could act as defense and signaling molecules [174]. Other studies revealed that exposure of *C. sativus* (cucumber plants) to silver nanoparticle (Ag NPs) stress results in upregulation of phenolic compounds (antioxidant defense systems) and phytol. Moreover, Ag NPs upregulate tricarboxylic acid intermediates (increased respiration), downregulate glycine and serine ratio (reduce photorespiration), and upregulate arachidonic acids and pentadecanoic acids while downregulating linolenic and linoleic acids (modified membrane properties) [175]. Surplus ROS is scavenged by the biosynthesized antioxidant compounds [176]. The results imply that the antioxidant defense system was stimulated by silver nanoparticles. This possibly leads to the switching of cellular energy metabolism from growth to defense, thereby increasing defense-related metabolites.

7 Conclusions and Future Perspective

Abiotic stresses negatively affect the growth and development of plants. NPs triggered different significant impacts on the physiological, morphological, and biochemical properties of plant species. NPs mitigate the negative effects of these stresses by enhancing the antioxidant enzymes (SOD, APX, and GP) and photosynthetic activities of these plants under stress. Moreover, the application of these NPs decreases MDA, EL, and H_2O_2 levels in stressed plants. These NPs translocated from root to shoot and vice versa through the phloem, which has both negative and positive effects. Under NP stress, plants produce ROS that activates different signaling pathways. These signaling pathways alter gene expression in response to stress-regulating genes related to stress. At higher concentrations, NPs may cause phytotoxicity that negatively affects the plants by damaging cell membrane, chloroplast, and electrolytic leakage. Several secondary metabolites are also produced in response to stress, which protects the plants from the adverse impact of NPs and enhances their morphological and physiological properties. Therefore, it is important to understand the complex mechanism involved in the process of stress mitigation. To overcome the negative effects of these NPs on the environment, it is important to engineer the properties of these NPs and their interaction with plants before their practical usage. The network of genes responsible for the mitigation of stress is not understood because many genes involved in the defense mechanism are unidentified until now. High-performance proteomic and transcriptomic techniques should be used to find the molecular mechanism and expression profile through biological processes. Thus, it can be concluded that metabolites produced in response to NPs can protect the plant from the negative effects of abiotic stresses by enhancing the defense mechanism.

References

1. FAO. (2009). How to feed the world in 2050, high-level expert forum. *Food and Agriculture Organization of the United Nations*, pp. 35-35.
2. Rajput, V. D., Minkina, T., Kumari, A., Singh, V. K., Verma, K. K., Mandzhieva, S., & Keswani, C. (2021). Coping with the challenges of abiotic stress in plants: New dimensions in the field application of nanoparticles. *Plants, 10*(6), 1221.

3. Francini, A., & Sebastiani, L. (2019). Abiotic stress effects on performance of horticultural crops. *Horticulturae, 5*(4), 67.
4. Acquaah, G. (2009). *Principles of plant genetics and breeding*. Wiley.
5. Calanca, P. P. (2017). Effects of abiotic stress in crop production. In *Quantification of climate variability, adaptation and mitigation for agricultural sustainability* (pp. 165–180). Springer.
6. Scrinis, G., & Lyons, K. (2007). The emerging nano-corporate paradigm: Nanotechnology and the transformation of nature, food and agri-food systems. *The International Journal of Sociology of Agriculture and Food, 15*(2), 22–44.
7. Moraru, C. I., Panchapakesan, C. P., Huang, Q., Takhistov, P., Liu, S., & Kokini, J. L. (2003). Nanotechnology: A new frontier in food science understanding the special properties of materials of nanometer size will allow food scientists to design new, healthier, tastier, and safer foods. *Nanotechnology, 57*(12), 24–29.
8. Dubchak, S., Ogar, A., Mietelski, J. W., & Turnau, K. (2010). Influence of silver and titanium nanoparticles on arbuscular mycorrhiza colonization and accumulation of radiocaesium in *Helianthus annuus*. *Spanish Journal of Agricultural Research, 8*(S1), 103–108.
9. Das, A., & Das, B. (2019). Nanotechnology a potential tool to mitigate abiotic stress in crop plants. *Abiotic and Biotic Stress in Plants*.
10. Bumbudsanpharoke, N., & Ko, S. (2015). Nano-food packaging: An overview of market, migration research, and safety regulations. *Journal of food science, 80*(5), R910–R923.
11. Gardea-Torresdey, J. L., Rico, C. M., & White, J. C. (2014). Trophic transfer, transformation, and impact of engineered nanomaterials in terrestrial environments. *Environmental Science & Technology, 48*(5), 2526–2540.
12. Schwab, F., Zhai, G., Kern, M., Turner, A., Schnoor, J. L., & Wiesner, M. R. (2016). Barriers, pathways and processes for uptake, translocation and accumulation of nanomaterials in plants–Critical review. *Nanotoxicology, 10*(3), 257–278.
13. Ma, C., White, J. C., Dhankher, O. P., & Xing, B. (2015a). Metal-based nanotoxicity and detoxification pathways in higher plants. *Environmental Science & Technology, 49*(12), 7109–7122.
14. Brain, R. A., & Cedergreen, N. (2008). Biomarkers in aquatic plants: Selection and utility. *Reviews of Environmental Contamination and Toxicology, 198*, 49–109.
15. García-Gómez, C., García, S., Obrador, A. F., González, D., Babín, M., & Fernández, M. D. (2018). Effects of aged ZnO NPs and soil type on Zn availability, accumulation and toxicity to pea and beet in a greenhouse experiment. *Ecotoxicology and Environmental Safety, 160*, 222–230.
16. Jha, S., & Pudake, R. N. (2016). Molecular mechanism of plant–nanoparticle interactions. In *Plant nanotechnology* (pp. 155–181). Springer.
17. Kole, C., Kumar, D. S., & Khodakovskaya, M. V. (2016). *Plant nanotechnology: Principles and practices*. Springer.
18. Ma, Y., Kuang, L., He, X., Bai, W., Ding, Y., Zhang, Z., & Chai, Z. (2010). Effects of rare earth oxide nanoparticles on root elongation of plants. *Chemosphere, 78*(3), 273–279.
19. Khan, M. R., Adam, V., Rizvi, T. F., Zhang, B., Ahamad, F., Jośko, I., & Mao, C. (2019). Nanoparticle–plant interactions: Two-way traffic. *Small, 15*(37), 1901794.
20. Parisi, C., Vigani, M., & Rodríguez-Cerezo, E. (2015). Agricultural nanotechnologies: What are the current possibilities. *Nano Today, 10*(2), 124–127.
21. Dan, Y., Zhang, W., Xue, R., Ma, X., Stephan, C., & Shi, H. (2015). Characterization of gold nanoparticle uptake by tomato plants using enzymatic extraction followed by single-particle inductively coupled plasma–mass spectrometry analysis. *Environmental Science & Technology, 49*(5), 3007–3014.
22. Hao, Y., Yu, F., Lv, R., Ma, C., Zhang, Z., Rui, Y., & Xing, B. (2016). Carbon nanotubes filled with different ferromagnetic alloys affect the growth and development of rice seedlings by changing the C: N ratio and plant hormones concentrations. *PLoS One, 11*(6), e0157264.
23. Rico, C. M., Lee, S. C., Rubenecia, R., Mukherjee, A., Hong, J., Peralta-Videa, J. R., & Gardea-Torresdey, J. L. (2014). Cerium oxide nanoparticles impact yield and modify

nutritional parameters in wheat (*Triticum aestivum* L.). *Journal of Agricultural and Food Chemistry, 62*(40), 9669–9675.
24. Yang, J., Cao, W., & Rui, Y. (2017). Interactions between nanoparticles and plants: Phytotoxicity and defense mechanisms. *Journal of Plant Interactions, 12*(1), 158–169.
25. Hernandez-Viezcas, J. A., Castillo-Michel, H., Andrews, J. C., Cotte, M., Rico, C., Peralta-Videa, J. R., & Gardea-Torresdey, J. L. (2013). In situ synchrotron X-ray fluorescence mapping and speciation of CeO2 and ZnO nanoparticles in soil cultivated soybean (Glycine max). *ACS Nano, 7*(2), 1415–1423.
26. Rico, C. M., Hong, J., Morales, M. I., Zhao, L., Barrios, A. C., Zhang, J. Y., & Gardea-Torresdey, J. L. (2013a). Effect of cerium oxide nanoparticles on rice: A study involving the antioxidant defense system and in vivo fluorescence imaging. *Environmental Science & Technology, 47*(11), 5635–5642.
27. Song, U., Shin, M., Lee, G., Roh, J., Kim, Y., & Lee, E. J. (2013). Functional analysis of TiO2 nanoparticle toxicity in three plant species. *Biological trace element research, 155*(1), 93–103.
28. Sheykhbaglou, R., Sedghi, M., Shishevan, M. T., & Sharifi, R. S. (2010). Effects of nano-iron oxide particles on agronomic traits of soybean. *Notulae Scientia Biologicae, 2*(2), 112–113.
29. Wu, L. J., You, Q. S., Duan, J. L., Luo, Y. X., Liu, L. J., Li, X., & Guo, X. H. (2015). Prevalence and associated factors of myopia in high-school students in Beijing. *PLoS One, 10*(3), e0120764.
30. Prasad, T. N. V. K. V., Sudhakar, P., Sreenivasulu, Y., Latha, P., Munaswamy, V., Reddy, K. R., & Pradeep, T. (2012). Effect of nanoscale zinc oxide particles on the germination, growth and yield of peanut. *Journal of Plant Nutrition, 35*(6), 905–927.
31. Yasmeen, F., Raja, N. I., Razzaq, A., & Komatsu, S. (2017). Proteomic and physiological analyses of wheat seeds exposed to copper and iron nanoparticles. *Biochimica et Biophysica Acta (BBA)-Proteins and Proteomics, 1865*(1), 28–42.
32. Thannickal, V. J., & Fanburg, B. L. (2000). Reactive oxygen species in cell signaling. *American Journal of Physiology-Lung Cellular and Molecular Physiology, 279*(6), L1005–L1028.
33. Gill, S. S., & Tuteja, N. (2010). Reactive oxygen species and antioxidant machinery in abiotic stress tolerance in crop plants. *Plant Physiology and Biochemistry, 48*(12), 909–930.
34. Meriga, B., Reddy, B. K., Rao, K. R., Reddy, L. A., & Kishor, P. K. (2004). Aluminium-induced production of oxygen radicals, lipid peroxidation and DNA damage in seedlings of rice (*Oryza sativa*). *Journal of Plant Physiology, 161*(1), 63–68.
35. Cui, D., Zhang, P., Ma, Y., He, X., Li, Y., Zhang, J., & Zhang, Z. (2014). Effect of cerium oxide nanoparticles on *Asparagus lettuce* cultured in an agar medium. *Environmental Science: Nano, 1*(5), 459–465.
36. Dimkpa, C. O., McLean, J. E., Latta, D. E., Manangón, E., Britt, D. W., Johnson, W. P., & Anderson, A. J. (2012). CuO and ZnO nanoparticles: Phytotoxicity, metal speciation, and induction of oxidative stress in sand-grown wheat. *Journal of Nanoparticle Research, 14*(9), 1–15.
37. Shaw, A. K., & Hossain, Z. (2013). Impact of nano-CuO stress on rice (*Oryza sativa* L.) seedlings. *Chemosphere, 93*(6), 906–915.
38. Lin, D., & Xing, B. (2008). Root uptake and phytotoxicity of ZnO nanoparticles. *Environmental Science & Technology, 42*(15), 5580–5585.
39. Oukarroum, A., Barhoumi, L., Pirastru, L., & Dewez, D. (2013). Silver nanoparticle toxicity effect on growth and cellular viability of the aquatic plant *Lemna gibba*. *Environmental Toxicology and Chemistry, 32*(4), 902–907.
40. Freinbichler, W., Colivicchi, M. A., Stefanini, C., Bianchi, L., Ballini, C., Misini, B., & Della Corte, L. (2011). Highly reactive oxygen species: Detection, formation, and possible functions. *Cellular and Molecular Life Sciences, 68*(12), 2067–2079.
41. Montillet, J. L., Chamnongpol, S., Rustérucci, C., Dat, J., Van De Cotte, B., Agnel, J. P., & Triantaphylidès, C. (2005). Fatty acid hydroperoxides and H2O2 in the execution of hypersensitive cell death in tobacco leaves. *Plant Physiology, 138*(3), 1516–1526.

42. Sharma, P., Jha, A. B., Dubey, R. S., & Pessarakli, M. (2012). Reactive oxygen species, oxidative damage, and antioxidative defense mechanism in plants under stressful conditions. *Journal of Botany, 2012*, 217037.
43. Halliwell, B., & Gutteridge, J. M. (2015). *Free radicals in biology and medicine*. Oxford University Press.
44. Tanou, G., Molassiotis, A., & Diamantidis, G. (2009). Induction of reactive oxygen species and necrotic death-like destruction in strawberry leaves by salinity. *Environmental and Experimental Botany, 65*(2-3), 270–281.
45. Melegari, S. P., Perreault, F., Costa, R. H. R., Popovic, R., & Matias, W. G. (2013). Evaluation of toxicity and oxidative stress induced by copper oxide nanoparticles in the green alga *Chlamydomonas reinhardtii*. *Aquatic Toxicology, 142*, 431–440.
46. Rico, C. M., Peralta-Videa, J. R., & Gardea-Torresdey, J. L. (2015). Chemistry, biochemistry of nanoparticles, and their role in antioxidant defense system in plants. In *Nanotechnology and plant sciences* (pp. 1–17). Springer.
47. Tarasenko, V. I., Garnik, E. Y., Shmakov, V. N., & Konstantinov, Y. M. (2012). Modified alternative oxidase expression results in different reactive oxygen species contents in Arabidopsis cell culture but not in whole plants. *Biologia Plantarum, 56*(4), 635–640.
48. Zhang, P., Ma, Y., Zhang, Z., He, X., Li, Y., Zhang, J., & Zhao, Y. (2015). Species-specific toxicity of ceria nanoparticles to Lactuca plants. *Nanotoxicology, 9*(1), 1–8.
49. Zhao, L., Peng, B., Hernandez-Viezcas, J. A., Rico, C., Sun, Y., Peralta-Videa, J. R., & Gardea-Torresdey, J. L. (2012b). Stress response and tolerance of *Zea mays* to CeO2 nanoparticles: Cross talk among H2O2, heat shock protein, and lipid peroxidation. *ACS Nano, 6*(11), 9615–9622.
50. Ma, Y., Zhang, P., Zhang, Z., He, X., Li, Y., Zhang, J., & Chai, Z. (2015b). Origin of the different phytotoxicity and biotransformation of cerium and lanthanum oxide nanoparticles in cucumber. *Nanotoxicology, 9*(2), 262–270.
51. Kim, S., Lee, S., & Lee, I. (2012). Alteration of phytotoxicity and oxidant stress potential by metal oxide nanoparticles in *Cucumis sativus*. *Water, Air, & Soil Pollution, 223*(5), 2799–2806.
52. Servin, A. D., Morales, M. I., Castillo-Michel, H., Hernandez-Viezcas, J. A., Munoz, B., Zhao, L., & Gardea-Torresdey, J. L. (2013). Synchrotron verification of TiO2 accumulation in cucumber fruit: A possible pathway of TiO2 nanoparticle transfer from soil into the food chain. *Environmental Science & Technology, 47*(20), 11592–11598.
53. Ma, X., Wang, Q., Rossi, L., & Zhang, W. (2016). Cerium oxide nanoparticles and bulk cerium oxide leading to different physiological and biochemical responses in *Brassica rapa*. *Environmental Science & Technology, 50*(13), 6793–6802.
54. Rico, C. M., Morales, M. I., Barrios, A. C., McCreary, R., Hong, J., Lee, W. Y., & Gardea-Torresdey, J. L. (2013b). Effect of cerium oxide nanoparticles on the quality of rice *(Oryza sativa* L.) grains. *Journal of Agricultural and Food Chemistry, 61*(47), 11278–11285.
55. Zhao, L., Peralta-Videa, J. R., Varela-Ramirez, A., Castillo-Michel, H., Li, C., Zhang, J., & Gardea-Torresdey, J. L. (2012a). Effect of surface coating and organic matter on the uptake of CeO2 NPs by corn plants grown in soil: Insight into the uptake mechanism. *Journal of Hazardous Materials, 225*, 131–138.
56. Da Costa, M. V. J., & Sharma, P. K. (2016). Effect of copper oxide nanoparticles on growth, morphology, photosynthesis, and antioxidant response in *Oryza sativa*. *Photosynthetica, 54*(1), 110–119.
57. Nair, P. M. G., & Chung, I. M. (2014). A mechanistic study on the toxic effect of copper oxide nanoparticles in soybean (*Glycine max* L.) root development and lignification of root cells. *Biological Trace Element Research, 162*(1), 342–352.
58. Nair, P. M. G., & Chung, I. M. (2015a). Study on the correlation between copper oxide nanoparticles induced growth suppression and enhanced lignification in Indian mustard (*Brassica juncea* L.). *Ecotoxicology and Environmental Safety, 113*, 302–313.

59. Zheng, L., Hong, F., Lu, S., & Liu, C. (2005). Effect of nano-TiO2 on strength of naturally aged seeds and growth of spinach. *Biological Trace Element Research, 104*(1), 83–91.
60. Linglan, M., Chao, L., Chunxiang, Q., Sitao, Y., Jie, L., Fengqing, G., & Fashui, H. (2008). Rubisco activase mRNA expression in spinach: Modulation by nanoanatase treatment. *Biological Trace Element Research, 122*(2), 168–178.
61. Yang, F., Liu, C., Gao, F., Su, M., Wu, X., Zheng, L., & Yang, P. (2007). The improvement of spinach growth by nano-anatase TiO2 treatment is related to nitrogen photoreduction. *Biological Trace Element Research, 119*(1), 77–88.
62. Changmei, L., Chaoying, Z., Junqiang, W., Guorong, W., & Mingxuan, T. (2002). Research of the effect of nanometer materials on germination and growth enhancement of *Glycine max* and its mechanism. *Soybean Science, 21*(3), 168–171.
63. Yang, F., Hong, F., You, W., Liu, C., Gao, F., Wu, C., & Yang, P. (2006). Influence of nano-anatase TiO2 on the nitrogen metabolism of growing spinach. *Biological Trace Element Research, 110*(2), 179–190.
64. López-Moreno, M. L., de la Rosa, G., Hernández-Viezcas, J. Á., Castillo-Michel, H., Botez, C. E., Peralta-Videa, J. R., & Gardea-Torresdey, J. L. (2010). Evidence of the differential biotransformation and genotoxicity of ZnO and CeO2 nanoparticles on soybean (*Glycine max*) plants. *Environmental Science & Technology, 44*(19), 7315–7320.
65. Suriyaprabha, R., Karunakaran, G., Yuvakkumar, R., Rajendran, V., & Kannan, N. (2012). Silica nanoparticles for increased silica availability in maize (*Zea mays*. L) seeds under hydroponic conditions. *Current Nanoscience, 8*(6), 902–908.
66. Siddiqui, M. H., & Al-Whaibi, M. H. (2014). Role of nano-SiO2 in germination of tomato (*Lycopersicum esculentum* seeds Mill.). *Saudi journal of biological sciences, 21*(1), 13–17.
67. Xie, Y., Li, B., Zhang, Q., Zhang, C., Lu, K., & Tao, G. (2011). Effects of nano-TiO2 on photosynthetic characteristics of *Indocalamus barbatus*. *Journal of Northeast Forestry University, 39*(3), 22–25.
68. Ramesh, M., Palanisamy, K., Babu, K., & Sharma, N. K. (2014). Effects of bulk & nano-titanium dioxide and zinc oxide on physio-morphological changes in *Triticum aestivum* Linn. *Journal of Global Biosciences, 3*(2), 415–422.
69. Iqbal, M. S., Singh, A. K., Singh, S. P., & Ansari, M. I. (2020). Nanoparticles and plant interaction with respect to stress response. In *Nanomaterials and environmental biotechnology* (pp. 1–15). Springer.
70. Maroufpour, N., Mousavi, M., Abbasi, M., & Ghorbanpour, M. (2020). Biogenic nanoparticles as novel sustainable approach for plant protection. In M. Ghorbanpour, P. Bhargava, A. Varma, & D. K. Choudhary (Eds.), *Biogenic nanoparticles and their use in agro-ecosystems* (pp. 161–172). Springer.
71. Kim, I., Viswanathan, K., Kasi, G., Thanakkasaranee, S., Sadeghi, K., & Seo, J. (2020). ZnO nanostructures in active antibacterial food packaging: Preparation methods, antimicrobial mechanisms, safety issues, future prospects, and challenges. *Food Reviews International*, 1–29.
72. Esper Neto, M., Britt, D. W., Lara, L. M., Cartwright, A., dos Santos, R. F., Inoue, T. T., & Batista, M. A. J. A. (2020). *Initial development of corn seedlings after seed priming with nanoscale synthetic zinc oxide., 10*(2), 307.
73. Rossi, L., Fedenia, L. N., Sharifan, H., Ma, X., & Lombardini, L. (2019). Effects of foliar application of zinc sulfate and zinc nanoparticles in coffee (*Coffea arabica* L.) plants. *Plant Physiology and Biochemistry, 135*, 160–166.
74. Alabdallah, N. M., & Alzahrani, H. S. (2020). The potential mitigation effect of ZnO nanoparticles on [*Abelmoschus esculentus* L. Moench] metabolism under salt stress conditions. *Saudi Journal of Biological Sciences, 27*(11), 3132–3137.
75. Pradhan, S., Patra, P., Das, S., Chandra, S., Mitra, S., Dey, K. K., & Goswami, A. (2013). Photochemical modulation of biosafe manganese nanoparticles on *Vigna radiata*: A detailed molecular, biochemical, and biophysical study. *Environmental Science & Technology, 47*(22), 13122–13131.

76. Gao, F., Hong, F., Liu, C., Zheng, L., Su, M., Wu, X., & Yang, P. (2006). Mechanism of nano-anatase TiO 2 on promoting photosynthetic carbon reaction of spinach. *Biological Trace Element Research, 111*(1), 239–253.
77. Latef, A. A. H. A., Alhmad, M. F. A., & Abdelfattah, K. E. (2017). The possible roles of priming with ZnO nanoparticles in mitigation of salinity stress in lupine (*Lupinus termis*) plants. *Journal of Plant Growth Regulation, 36*(1), 60–70.
78. Rezaei, M., & Abbasi, H. (2014). Foliar application of nanochelate and non-nanochelate of zinc on plant resistance physiological processes in cotton (*Gossipium hirsutum* L.). *Iranian. Journal of Plant Physiology, 4*(4), 1137–1144.
79. Soliman, A. S., El-feky, S. A., & Darwish, E. (2015). Alleviation of salt stress on *Moringa peregrina* using foliar application of nanofertilizers. *Journal of Horticulture and Forestry, 7*(2), 36–47.
80. Weisany, W., Sohrabi, Y., Heidari, G., Siosemardeh, A., & Ghassemi-Golezani, K. (2012). Changes in antioxidant enzymes activity and plant performance by salinity stress and zinc application in soybean (*'Glycine max'*L.). *Plant Omics, 5*(2), 60–67.
81. Mohammadi, M. H. Z., Panahirad, S., Navai, A., Bahrami, M. K., Kulak, M., & Gohari, G. (2021). Cerium oxide nanoparticles (CeO2-NPs) improve growth parameters and antioxidant defense system in Moldavian Balm (*Dracocephalum moldavica* L.) under salinity stress. *Plant Stress, 1*, 100006.
82. Rossi, L., Zhang, W., & Ma, X. (2017). Cerium oxide nanoparticles alter the salt stress tolerance of *Brassica napus* L. by modifying the formation of root apoplastic barriers. *Environmental Pollution, 229*, 132–138.
83. Dunnick, K. M., Pillai, R., Pisane, K. L., Stefaniak, A. B., Sabolsky, E. M., & Leonard, S. S. (2015). The effect of cerium oxide nanoparticle valence state on reactive oxygen species and toxicity. *Biological Trace Element Research, 166*(1), 96–107.
84. Moradbeygi, H., Jamei, R., Heidari, R., & Darvishzadeh, R. (2020). Investigating the enzymatic and non-enzymatic antioxidant defense by applying iron oxide nanoparticles in *Dracocephalum moldavica* L. plant under salinity stress. *Scientia Horticulturae, 272*, 109537.
85. Karimi, J., & Mohsenzadeh, S. (2016). Effects of silicon oxide nanoparticles on growth and physiology of wheat seedlings. *Russian Journal of Plant Physiology, 63*(1), 119–123.
86. Tubana, B. S., Babu, T., & Datnoff, L. E. (2016). A review of silicon in soils and plants and its role in US agriculture: History and future perspectives. *Soil Science, 181*(9/10), 393–411.
87. Seleiman, M. F., Refay, Y., Al-Suhaibani, N., Al-Ashkar, I., El-Hendawy, S., & Hafez, E. M. (2019). Integrative effects of rice-straw biochar and silicon on oil and seed quality, yield and physiological traits of *Helianthus annuus* L. grown under water deficit stress. *Agronomy, 9*(10), 637.
88. Parveen, N. U. S. R. A. T., & Ashraf, M. U. H. A. M. M. A. D. (2010). Role of silicon in mitigating the adverse effects of salt stress on growth and photosynthetic attributes of two maize (*Zea mays* L.) cultivars grown hydroponically. *Pakistan Journal of Botany, 42*(3), 1675–1684.
89. Mahmoud, L. M., Dutt, M., Shalan, A. M., El-Kady, M. E., El-Boray, M. S., Shabana, Y. M., & Grosser, J. W. (2020). Silicon nanoparticles mitigate oxidative stress of in vitro-derived banana (*Musa acuminata* 'Grand Nain') under simulated water deficit or salinity stress. *South African Journal of Botany, 132*, 155–163.
90. Siddiqui, M. H., Al-Whaibi, M. H., Faisal, M., & Al Sahli, A. A. (2014). Nano-silicon dioxide mitigates the adverse effects of salt stress on Cucurbita pepo L. *Environmental Toxicology and Chemistry, 33*(11), 2429–2437.
91. Saqib, M., Zörb, C., & Schubert, S. (2008). Silicon-mediated improvement in the salt resistance of wheat (*Triticum aestivum*) results from increased sodium exclusion and resistance to oxidative stress. *Functional Plant Biology, 35*(7), 633–639.
92. Dimkpa, C. O., Singh, U., Bindraban, P. S., Elmer, W. H., Gardea-Torresdey, J. L., & White, J. C. (2019). Zinc oxide nanoparticles alleviate drought-induced alterations in sorghum

performance, nutrient acquisition, and grain fortification. *Science of the Total Environment, 688*, 926–934.

93. Sun, L., Song, F., Guo, J., Zhu, X., Liu, S., Liu, F., & Li, X. (2020). Nano-ZnO-induced drought tolerance is associated with melatonin synthesis and metabolism in maize. *International Journal of Molecular Sciences, 21*(3), 782.
94. Briat, J. F., Curie, C., & Gaymard, F. (2007). Iron utilization and metabolism in plants. *Current Opinion in Plant Biology, 10*(3), 276–282.
95. Elanchezhian, R., Kumar, D., Ramesh, K., Biswas, A. K., Guhey, A., & Patra, A. K. (2017). Morpho-physiological and biochemical response of maize (*Zea mays* L.) plants fertilized with nano-iron (Fe_3O4) micronutrient. *Journal of Plant Nutrition, 40*(14), 1969–1977.
96. Zhu, H., Jia, Y., Wu, X., & Wang, H. (2009). Removal of arsenic from water by supported nano zero-valent iron on activated carbon. *Journal of Hazardous Materials, 172*(2-3), 1591–1596.
97. Adrees, M., Khan, Z. S., Ali, S., Hafeez, M., Khalid, S., Ur Rehman, M. Z., & Rizwan, M. (2020). Simultaneous mitigation of cadmium and drought stress in wheat by soil application of iron nanoparticles. *Chemosphere, 238*, 124681.
98. Ahmed, T., Noman, M., Manzoor, N., Shahid, M., Abdullah, M., Ali, L., & Li, B. (2021). Nanoparticle-based amelioration of drought stress and cadmium toxicity in rice via triggering the stress responsive genetic mechanisms and nutrient acquisition. *Ecotoxicology and Environmental Safety, 209*, 111829.
99. Pavithra, G. J., Reddy, B. R., Salimath, M., Geetha, K. N., & Shankar, A. G. (2017). Zinc oxide nano particles increases Zn uptake, translocation in rice with positive effect on growth, yield and moisture stress tolerance. *Indian Journal of Plant Physiology, 22*(3), 287–294.
100. Song, Y., Jiang, M., Zhang, H., & Li, R. (2021). Zinc Oxide Nanoparticles Alleviate Chilling Stress in Rice (*Oryza Sativa* L.) by Regulating Antioxidative System and Chilling Response Transcription Factors. *Molecules, 26*(8), 2196.
101. Venkatachalam, P., Jayaraj, M., Manikandan, R., Geetha, N., Rene, E. R., Sharma, N. C., & Sahi, S. V. (2017a). Zinc oxide nanoparticles (ZnONPs) alleviate heavy metal-induced toxicity in *Leucaena leucocephala* seedlings: A physiochemical analysis. *Plant Physiology and Biochemistry, 110*, 59–69.
102. Rajeshkumar, S., & Naik, P. (2018). Synthesis and biomedical applications of cerium oxide nanoparticles–a review. *Biotechnology Reports, 17*, 1–5.
103. Cao, Z., Rossi, L., Stowers, C., Zhang, W., Lombardini, L., & Ma, X. (2018). The impact of cerium oxide nanoparticles on the physiology of soybean (*Glycine max* (L.) Merr.) under different soil moisture conditions. *Environmental Science and Pollution Research, 25*(1), 930–939.
104. Gohari, G., Safai, F., Panahirad, S., Akbari, A., Rasouli, F., Dadpour, M. R., & Fotopoulos, V. (2020). Modified multiwall carbon nanotubes display either phytotoxic or growth promoting and stress protecting activity in *Ocimum basilicum* L. in a concentration-dependent manner. *Chemosphere, 249*, 126171.
105. Liu, R., & Lal, R. (2015). Potentials of engineered nanoparticles as fertilizers for increasing agronomic productions. *Science of the Total Environment, 514*, 131–139.
106. Wojcieszek, J., Jiménez-Lamana, J., Bierła, K., Ruzik, L., Asztemborska, M., Jarosz, M., & Szpunar, J. (2019). Uptake, translocation, size characterization and localization of cerium oxide nanoparticles in radish (*Raphanus sativus* L.). *Science of the Total Environment, 683*, 284–292.
107. Jahani, S., Saadatmand, S., Mahmoodzadeh, H., & Khavari-Nejad, R. A. (2019). Effect of foliar application of cerium oxide nanoparticles on growth, photosynthetic pigments, electrolyte leakage, compatible osmolytes and antioxidant enzymes activities of *Calendula officinalis* L. *Biologia, 74*(9), 1063–1075.
108. Jurkow, R., Sękara, A., Pokluda, R., Smoleń, S., & Kalisz, A. (2020). Biochemical response of oakleaf lettuce seedlings to different concentrations of some metal (oid) oxide nanoparticles. *Agronomy, 10*(7), 997.

109. Kurepa, J., Paunesku, T., Vogt, S., Arora, H., Rabatic, B. M., Lu, J., & Smalle, J. A. (2010). Uptake and distribution of ultrasmall anatase TiO2 Alizarin red S nanoconjugates in *Arabidopsis thaliana*. *Nano Letters, 10*(7), 2296–2302.
110. Corredor, E., Testillano, P. S., Coronado, M. J., González-Melendi, P., Fernández-Pacheco, R., Marquina, C., & Risueño, M. C. (2009). Nanoparticle penetration and transport in living pumpkin plants: In situ subcellular identification. *BMC Plant Biology, 9*(1), 1–11.
111. Miralles, P., Church, T. L., & Harris, A. T. (2012). Toxicity, uptake, and translocation of engineered nanomaterials in vascular plants. *Environmental Science & Technology, 46*(17), 9224–9239.
112. Paes, G. (2014). Fluorescent probes for exploring plant cell wall deconstruction: A review. *Molecules, 19*, 9380–9402.
113. Eichert, T., Kurtz, A., Steiner, U., & Goldbach, H. E. (2008). Size exclusion limits and lateral heterogeneity of the stomatal foliar uptake pathway for aqueous solutes and water-suspended nanoparticles. *Physiologia Plantarum, 134*(1), 151–160.
114. Kokina, I., Gerbreders, V., Sledevskis, E., & Bulanovs, A. (2013). Penetration of nanoparticles in flax (*Linum usitatissimum* L.) calli and regenerants. *Journal of Biotechnology, 165*(2), 127–132.
115. Santos, A. R., Miguel, A. S., Tomaz, L., Malhó, R., Maycock, C., Patto, M. C. V., & Oliva, A. (2010). The impact of CdSe/ZnS quantum dots in cells of *Medicago sativa* in suspension culture. *Journal of Nanobiotechnology, 8*(1), 1–14.
116. Nair, R., Varghese, S. H., Nair, B. G., Maekawa, T., Yoshida, Y., & Kumar, D. S. (2010). Nanoparticulate material delivery to plants. *Plant Science, 179*(3), 154–163.
117. Tripathi, D. K., Singh, S., Singh, V. P., Prasad, S. M., Dubey, N. K., & Chauhan, D. K. (2017). Silicon nanoparticles more effectively alleviated UV-B stress than silicon in wheat (*Triticum aestivum*) seedlings. *Plant Physiology and Biochemistry, 110*, 70–81.
118. Wang, W. N., Tarafdar, J. C., & Biswas, P. (2013). Nanoparticle synthesis and delivery by an aerosol route for watermelon plant foliar uptake. *Journal of Nanoparticle Research, 15*(1), 1–13.
119. Kaphle, A., Navya, P. N., Umapathi, A., & Daima, H. K. (2018). Nanomaterials for agriculture, food and environment: Applications, toxicity and regulation. *Environmental Chemistry Letters, 16*(1), 43–58.
120. Du, W., Sun, Y., Ji, R., Zhu, J., Wu, J., & Guo, H. (2011). TiO_2 and ZnO nanoparticles negatively affect wheat growth and soil enzyme activities in agricultural soil. *Journal of Environmental Monitoring, 13*(4), 822–828.
121. Pérez-de-Luque, A. (2017). Interaction of nanomaterials with plants: What do we need for real applications in agriculture. *Frontiers in Environmental Science, 5*, 12.
122. Ruttkay-Nedecky, B., Krystofova, O., Nejdl, L., & Adam, V. (2017). Nanoparticles based on essential metals and their phytotoxicity. *Journal of Nanobiotechnology, 15*(1), 1–19.
123. Mousavi, S. R., Galavi, M., & Ahmadvand, G. (2007). Effect of zinc and manganese foliar application on yield, quality and enrichment on potato (*Solanum tuberosum* L.). *Asian Journal of Plant Sciences, 6*(8), 1256–1260.
124. Zhu, Z. J., Wang, H., Yan, B., Zheng, H., Jiang, Y., Miranda, O. R., & Vachet, R. W. (2012). Effect of surface charge on the uptake and distribution of gold nanoparticles in four plant species. *Environmental Science & Technology, 46*(22), 12391–12398.
125. Servin, A. D., Castillo-Michel, H., Hernandez-Viezcas, J. A., Diaz, B. C., Peralta-Videa, J. R., & Gardea-Torresdey, J. L. (2012). Synchrotron micro-XRF and micro-XANES confirmation of the uptake and translocation of TiO2 nanoparticles in cucumber (*Cucumis sativus*) plants. *Environmental Science & Technology, 46*(14), 7637–7643.
126. Lv, J., Zhang, S., Luo, L., Zhang, J., Yang, K., & Christie, P. (2015). Accumulation, speciation and uptake pathway of ZnO nanoparticles in maize. *Environmental Science: Nano, 2*(1), 68–77.
127. Syu, Y. Y., Hung, J. H., Chen, J. C., & Chuang, H. W. (2014). Impacts of size and shape of silver nanoparticles on Arabidopsis plant growth and gene expression. *Plant Physiology and Biochemistry, 83*, 57–64.

128. Dong, M., Sun, R., Yang, Q., Zhang, L., Yong, Y., Fang, Y., & Xue, D. (2021). Phenotype, physiology, and gene expression of barley seedlings in response to nano zinc oxide stress. *Phyton, 90*(6), 1589.
129. Bari, R., & Jones, J. D. (2009). Role of plant hormones in plant defence responses. *Plant molecular biology, 69*(4), 473–488.
130. Landa, P., Vankova, R., Andrlova, J., Hodek, J., Marsik, P., Storchova, H., & Vanek, T. (2012). Nanoparticle-specific changes in *Arabidopsis thaliana* gene expression after exposure to ZnO, TiO2, and fullerene soot. *Journal of Hazardous Materials, 241*, 55–62.
131. Plaksenkova, I., Kokina, I., Petrova, A., Jermaļonoka, M., Gerbreders, V., & Krasovska, M. (2020). The impact of zinc oxide nanoparticles on cytotoxicity, genotoxicity, and miRNA expression in barley (*Hordeum vulgare* L.) seedlings. *The Scientific World Journal, 2020*, 6649746.
132. Hezaveh, T. A., Pourakbar, L., Rahmani, F., & Alipour, H. (2019). Interactive effects of salinity and ZnO nanoparticles on physiological and molecular parameters of rapeseed (*Brassica napus* L.). *Communications in Soil Science and Plant Analysis, 50*(6), 698–715.
133. Li, M., Ahammed, G. J., Li, C., Bao, X., Yu, J., Huang, C., & Zhou, J. (2016). Brassinosteroid ameliorates zinc oxide nanoparticles-induced oxidative stress by improving antioxidant potential and redox homeostasis in tomato seedling. *Frontiers in Plant Science, 7*, 615.
134. Salah, S. M., Yajing, G., Dongdong, C., Jie, L., Aamir, N., Qijuan, H., & Jin, H. (2015). Seed priming with polyethylene glycol regulating the physiological and molecular mechanism in rice (*Oryza sativa* L.) under nano-ZnO stress. *Scientific Reports, 5*(1), 1–14.
135. Huang, Z., Xie, W., Wang, M., Liu, X., Ashraf, U., Qin, D., & Mo, Z. (2020). Response of rice genotypes with differential nitrate reductase-dependent NO synthesis to melatonin under ZnO nanoparticles'(NPs) stress. *Chemosphere, 250*, 126337.
136. Gupta, S. D., Agarwal, A., & Pradhan, S. (2018). Phytostimulatory effect of silver nanoparticles (AgNPs) on rice seedling growth: An insight from antioxidative enzyme activities and gene expression patterns. *Ecotoxicology and Environmental Safety, 161*, 624–633.
137. Nair, P. M. G., Park, S. Y., & Choi, J. (2013). Evaluation of the effect of silver nanoparticles and silver ions using stress responsive gene expression in *Chironomus riparius*. *Chemosphere, 92*(5), 592–599.
138. Yadu, B., Chandrakar, V., Korram, J., Satnami, M. L., Kumar, M., & Keshavkant, S. (2018). Silver nanoparticle modulates gene expressions, glyoxalase system and oxidative stress markers in fluoride stressed *Cajanus cajan* L. *Journal of Hazardous Materials, 353*, 44–52.
139. Dimkpa, C. O., McLean, J. E., Martineau, N., Britt, D. W., Haverkamp, R., & Anderson, A. J. (2013). Silver nanoparticles disrupt wheat (*Triticum aestivum* L.) growth in a sand matrix. *Environmental Science & Technology, 47*(2), 1082–1090.
140. Tang, Y., He, R., Zhao, J., Nie, G., Xu, L., & Xing, B. (2016). Oxidative stress-induced toxicity of CuO nanoparticles and related toxicogenomic responses in *Arabidopsis thaliana*. *Environmental Pollution, 212*, 605–614.
141. Wang, Z., Xu, L., Zhao, J., Wang, X., White, J. C., & Xing, B. (2016a). CuO nanoparticle interaction with *Arabidopsis thaliana*: Toxicity, parent-progeny transfer, and gene expression. *Environmental Science & Technology, 50*(11), 6008–6016.
142. Hedayati, A., Hosseini, B., Palazon, J., & Maleki, R. (2020). Improved tropane alkaloid production and changes in gene expression in hairy root cultures of two Hyoscyamus species elicited by silicon dioxide nanoparticles. *Plant Physiology and Biochemistry, 155*, 416–428.
143. Mosa, K. A., El-Naggar, M., Ramamoorthy, K., Alawadhi, H., Elnaggar, A., Wartanian, S., & Hani, H. (2018). Copper nanoparticles induced genotoxicity, oxidative stress, and changes in superoxide dismutase (SOD) gene expression in cucumber (*Cucumis sativus*) plants. *Frontiers in Plant Science, 9*, 872.
144. Burklew, C. E., Ashlock, J., Winfrey, W. B., & Zhang, B. (2012). Effects of aluminum oxide nanoparticles on the growth, development, and microRNA expression of tobacco (*Nicotiana tabacum*). *PLoS One, 7*(5), e34783.

145. Ze, Y., Liu, C., Wang, L., Hong, M., & Hong, F. (2011). The regulation of TiO_2 nanoparticles on the expression of light-harvesting complex II and photosynthesis of chloroplasts of *Arabidopsis thaliana. Biological Trace Element Research, 143*(2), 1131–1141.
146. Kaveh, R., Li, Y. S., Ranjbar, S., Tehrani, R., Brueck, C. L., & Van Aken, B. (2013). Changes in *Arabidopsis thaliana* gene expression in response to silver nanoparticles and silver ions. *Environmental Science & Technology, 47*(18), 10637–10644.
147. Nair, P. M. G., & Chung, I. M. (2015b). The responses of germinating seedlings of green peas to copper oxide nanoparticles. *Biologia Plantarum, 59*(3), 591–595.
148. Chen, J., Liu, X., Wang, C., Yin, S. S., Li, X. L., Hu, W. J., & Zheng, H. L. (2015). Nitric oxide ameliorates zinc oxide nanoparticles-induced phytotoxicity in rice seedlings. *Journal of hazardous materials, 297*, 173–182.
149. Antoniou, C., Chatzimichail, G., Xenofontos, R., Pavlou, J. J., Panagiotou, E., Christou, A., & Fotopoulos, V. (2017). Melatonin systemically ameliorates drought stress-induced damage in *Medicago sativa* plants by modulating nitro-oxidative homeostasis and proline metabolism. *Journal of Pineal Research, 62*(4), e12401.
150. Mishra, B. S., Singh, M., Aggrawal, P., & Laxmi, A. (2009). Glucose and auxin signaling interaction in controlling *Arabidopsis thaliana* seedlings root growth and development. *PLoS One, 4*(2), e4502.
151. Gechev, T. S., Van Breusegem, F., Stone, J. M., Denev, I., & Laloi, C. (2006). Reactive oxygen species as signals that modulate plant stress responses and programmed cell death. *Bioessays, 28*(11), 1091–1101.
152. Jiao, Y., Sun, L., Song, Y., Wang, L., Liu, L., Zhang, L., & Hao, F. (2013). AtrbohD and AtrbohF positively regulate abscisic acid-inhibited primary root growth by affecting Ca2+ signalling and auxin response of roots in *Arabidopsis. Journal of Experimental Botany, 64*(14), 4183–4192.
153. De Smet, I., Signora, L., Beeckman, T., Inzé, D., Foyer, C. H., & Zhang, H. (2003). An abscisic acid-sensitive checkpoint in lateral root development of *Arabidopsis. The Plant Journal, 33*(3), 543–555.
154. De Smet, I., Zhang, H., Inze, D., & Beeckman, T. (2006). A novel role for abscisic acid emerges from underground. *Trends in Plant Science, 11*(9), 434–439.
155. Servin, A. D., & White, J. C. (2016). Nanotechnology in agriculture: Next steps for understanding engineered nanoparticle exposure and risk. *NanoImpact, 1*, 9–12.
156. Reichman, J. R., Rygiewicz, P. T., Johnson, M. G., Bollman, M. A., Smith, B. M., Krantz, Q. T., & Andersen, C. P. (2018). Douglas-Fir (*Pseudotsuga menziesii* (Mirb.) Franco) Transcriptome profile changes induced by diesel emissions generated with CeO2 nanoparticle fuel borne catalyst. *Environmental Science & Technology, 52*(17), 10067–10077.
157. Fujita, K., Horiuchi, H., Takato, H., Kohno, M., & Suzuki, S. (2012). Auxin-responsive grape Aux/IAA9 regulates transgenic Arabidopsis plant growth. *Molecular Biology Reports, 39*(7), 7823–7829.
158. Hao, G. P., Zhang, X. H., Wang, Y. Q., Wu, Z. Y., & Huang, C. L. (2009). Nucleotide variation in the NCED3 region of *Arabidopsis thaliana* and its association study with abscisic acid content under drought stress. *Journal of Integrative Plant Biology, 51*(2), 175–183.
159. Babajani, A., Iranbakhsh, A., Ardebili, Z. O., & Eslami, B. (2019). Differential growth, nutrition, physiology, and gene expression in *Melissa officinalis* mediated by zinc oxide and elemental selenium nanoparticles. *Environmental Science and Pollution Research, 26*(24), 24430–24444.
160. Safari, M., Ardebili, Z. O., & Iranbakhsh, A. (2018). Selenium nano-particle induced alterations in expression patterns of heat shock factor A4A (HSFA4A), and high molecular weight glutenin subunit 1Bx (Glu-1Bx) and enhanced nitrate reductase activity in wheat (*Triticum aestivum* L.). *Acta Physiologiae Plantarum, 40*(6), 1–8.
161. Plotnikov, A., Zehorai, E., Procaccia, S., & Seger, R. (2011). The MAPK cascades: Signaling components, nuclear roles and mechanisms of nuclear translocation. *Biochimica et Biophysica Acta (BBA)-Molecular Cell Research, 1813*(9), 1619–1633.

162. Pérez-Salamó, I., Papdi, C., Rigó, G., Zsigmond, L., Vilela, B., Lumbreras, V., & Szabados, L. (2014). The heat shock factor A4A confers salt tolerance and is regulated by oxidative stress and the mitogen-activated protein kinases MPK3 and MPK6. *Plant Physiology, 165*(1), 319–334.
163. Smékalová, V., Doskočilová, A., Komis, G., & Šamaj, J. (2014). Crosstalk between secondary messengers, hormones and MAPK modules during abiotic stress signalling in plants. *Biotechnology Advances, 32*(1), 2–11.
164. Miller, G. A. D., & Mittler, R. O. N. (2006). Could heat shock transcription factors function as hydrogen peroxide sensors in plants. *Annals of Botany, 98*(2), 279–288.
165. Shim, D., Hwang, J. U., Lee, J., Lee, S., Choi, Y., An, G., & Lee, Y. (2009). Orthologs of the class A4 heat shock transcription factor HsfA4a confer cadmium tolerance in wheat and rice. *The Plant Cell, 21*(12), 4031–4043.
166. Venkatachalam, P., Priyanka, N., Manikandan, K., Ganeshbabu, I., Indiraarulselvi, P., Geetha, N., & Sahi, S. V. (2017b). Enhanced plant growth promoting role of phycomolecules coated zinc oxide nanoparticles with P supplementation in cotton (*Gossypium hirsutum* L.). *Plant Physiology and Biochemistry, 110*, 118–127.
167. Wang, X., Yang, X., Chen, S., Li, Q., Wang, W., Hou, C., & Wang, S. (2016b). Zinc oxide nanoparticles affect biomass accumulation and photosynthesis in Arabidopsis. *Frontiers in Plant Science, 6*, 1243.
168. Hong, J., Yang, L., Zhang, D., & Shi, J. (2016). Plant metabolomics: An indispensable system biology tool for plant science. *International Journal of Molecular Sciences, 17*(6), 767.
169. Zhang, H., Lu, L., Zhao, X., Zhao, S., Gu, X., Du, W., & Zhao, L. (2019). Metabolomics reveals the "invisible" responses of spinach plants exposed to CeO2 nanoparticles. *Environmental Science & Technology, 53*(10), 6007–6017.
170. Forde, B. G., & Lea, P. J. (2007). Glutamate in plants: Metabolism, regulation, and signalling. *Journal of Experimental Botany, 58*(9), 2339–2358.
171. Galili, G., Amir, R., & Fernie, A. R. (2016). The regulation of essential amino acid synthesis and accumulation in plants. *Annual Review of Plant Biology, 67*, 153–178.
172. Roje, S. (2006). S-Adenosyl-L-methionine: Beyond the universal methyl group donor. *Phytochemistry, 67*(15), 1686–1698.
173. Tzin, V., & Galili, G. (2010). The biosynthetic pathways for shikimate and aromatic amino acids in *Arabidopsis thaliana*. *The Arabidopsis book/American Society of Plant Biologists, 8*, e0132.
174. Foyer, C. H., & Noctor, G. (2013). Redox signaling in plants. *Antioxid Redox Signal, 18*(16), 2087–2090.
175. Zhang, H., Du, W., Peralta-Videa, J. R., Gardea-Torresdey, J. L., White, J. C., Keller, A., & Zhao, L. (2018). Metabolomics reveals how cucumber (*Cucumis sativus*) reprograms metabolites to cope with silver ions and silver nanoparticle-induced oxidative stress. *Environmental Science & Technology, 52*(14), 8016–8026.
176. Wurzinger, B., Nukarinen, E., Nägele, T., Weckwerth, W., & Teige, M. (2018). The SnRK1 kinase as central mediator of energy signaling between different organelles. *Plant Physiology, 176*(2), 1085–1094.

Nanoelicitation: A Promising and Emerging Technology for Triggering the Sustainable In Vitro Production of Secondary Metabolites in Medicinal Plants

Rabia Javed, Buhara Yucesan, Muhammad Zia, and Ekrem Gurel

Abstract To obtain larger amounts of secondary metabolites is essentially needed by the global industrial market that is not feasible by traditional methods because these are time-consuming and result in eradication of plant stock by overexploitation. Therefore, in vitro culturing techniques are adopted to obtain the maximum quantity of secondary metabolites in a minimum time. Elicitors are key players for getting desired yields of secondary metabolites in plants. Eliciting the in vitro cell and tissue cultures is an efficient approach for the production of medicinally important plant secondary metabolites. Different parameters of optimized micropropagation protocols are exploited to be used as elicitors for further enhancement of secondary metabolites from medicinal plant species. The secondary metabolites produced by the plant cells include phenolics, flavonoids, alkaloids, terpenoids, and tannins. These metabolites are boosted under stressful conditions, whether biotic or abiotic in origin. The potential role of nanoparticles in the enhancement of secondary metabolic products in medicinal plants is a recent hot topic in the field of medicinal plant biotechnology. Nanoparticles have evolved as potent novel elicitors that significantly stimulate medicinal plant secondary metabolism. Various kinds of nanoparticles including metallic and metallic oxide nanoparticles and carbon-based nanomaterials are believed to induce abiotic stress to medicinal plants under in vitro conditions by which plant defense system is elicited, triggering biochemical as well as physiological responses, consequently producing enhanced and sustainable quantities of secondary metabolites. These industrially important bioactive metabolites are beneficial for the prevention of multiple diseases in the health-care system. Therefore, nanoelicitation should be applied as an effective tool for ameliorated stimulation and accumulation of

R. Javed (✉) · M. Zia
Department of Biotechnology, Quaid-i-Azam University, Islamabad, Pakistan

B. Yucesan
Department of Seed Science and Technology, Bolu Abant Izzet Baysal University, Bolu, Turkey

E. Gurel
Department of Biology, Faculty of Science, Bolu Abant Izzet Baysal University, Bolu, Turkey

J.-T. Chen (ed.), *Plant and Nanoparticles*,
https://doi.org/10.1007/978-981-19-2503-0_10

secondary metabolites. However, in some cases, after the efficient uptake and translocation, nanoparticles produce deleterious effects causing phytotoxicity.

Keywords Nanoparticles · Elicitation · Secondary metabolites · Medicinal plants · Biotechnology

1 Introduction

Since herbal medicine is the primary source of medical treatment in developing countries as reported by the World Health Organization (WHO), diverse and comprehensive studies on medicinally rich plants should be performed to enhance the accumulation of bioactive pharmaceutical compounds [1]. The anticancer, antidiabetic, anti-inflammatory, antibacterial, and antiviral properties of various biologically active compounds have been elucidated [2]. The extracts of different parts of medicinal plants have been extensively studied to find novel bioactive ingredients from naturally growing herbs. The wild-grown plants don't guarantee sustainable production of medicinally important compounds because of geographical and seasonal changes and certain other limitations that could be avoided by growing these plants in the customized environment of the laboratory. Moreover, the production of bioactive compounds under natural environmental conditions is scarce; hence, the plants are raised under optimized in vitro conditions to get higher quantities of such value-added products. Additionally, few of such enriched plants get endangered with time, and methods should be devised for their preservation [3]. Therefore, medicinally valuable plants are grown via tissue and cell culture techniques to obtain uniform metabolic profiles and desired amount of medicinal compounds by elicitation strategy [4].

Overexploitation, slower growth of medicinal plants, and scarce secondary metabolites production are the bottlenecks in the usage of wild plants for obtaining sustainable pharmaceutical and nutraceutical products. This is why novel approaches employing plant cell and tissue culture strategies are adopted [5]. Secondary metabolism is a defensive system that protects the plants against adverse conditions that cause internal or external stress to different plant parts. The bioactive chemical moieties produced to overcome different stresses are antioxidants. Different strategies are employed under in vitro conditions that could evoke the formation of secondary metabolites. The process of stimulating the secondary metabolites production is called elicitation, and various stimulators used for this purpose are known as elicitors that may be biotic or abiotic [6]. The abiotic triggers include environmental stresses such as light, temperature, salt, and drought stress. Recently, nanoparticles have been evolved as novel abiotic elicitors owing to their unique physicochemical and biological characteristics [7].

Secondary metabolites, also called plant specialized metabolites, are organic compounds having low molecular weight. These metabolites are synthesized when the plants are exposed to different kinds of stresses. Phenolics, terpenes, and alkaloids are the major classes of secondary metabolites. These biologically active compounds are mediators for the adaptation of plants in challenging environments [8]. They not only serve as plant defenders in biotic and abiotic stresses but also are used for the treatment of many human diseases like cancer, diabetes, inflammation, cardiovascular diseases (CVDs), neurodegenerative disorders, and several other infectious and genetic diseases [9]. Secondary metabolites are used for the manufacture of drugs against different ailments. However, many of the secondary metabolic products have been utilized in various industries like the cosmetics industry, dyes and textile industry, artificial flavoring industry, etc. [10].

This chapter provides an overview of various abiotic elicitors employed under in vitro culture conditions ensuring homogeneity as well as more productivity and reproducibility of desired secondary metabolites. The unique properties of different metals, metal oxide, and carbon-based nanoparticles, making them desirable triggers among other effective abiotic elicitors, have been defined. The proposed mechanism for interaction of nanoparticles with plants following the plant secondary metabolism elicitation, sometimes causing phytotoxicity, has been explained herein.

2 Different Abiotic Elicitors for Augmentation of Secondary Metabolites

In vitro cultures of plants are grown in the standard medium of Murashige and Skoog (MS), which is composed of 30 g/L of sucrose (carbohydrate source), 4 g/L of salts (mineral source), and 8 g/L of agar (gelling agent). Medicinal plants grown by tissue culture are exposed to changing media compositions and environmental conditions for getting increased secondary metabolites production in defense against stressful situations. Such manipulations are performed at different levels of tissue culture propagation like shoot organogenesis, root organogenesis, seed germination, callus induction, etc. [11].

Figure 1 shows a brief overview of different abiotic elicitors for secondary metabolites production. Following is the description of elicitors, with examples, that augment the production of specialized metabolites in different medicinal plants grown in vitro:

2.1 Carbohydrates

Sucrose, maltose, and fructose are the carbohydrates that are an essential source of carbon for the in vitro growth of plants. Any change in the type or concentration of carbohydrates results in increased or decreased levels of secondary metabolites formation [12].

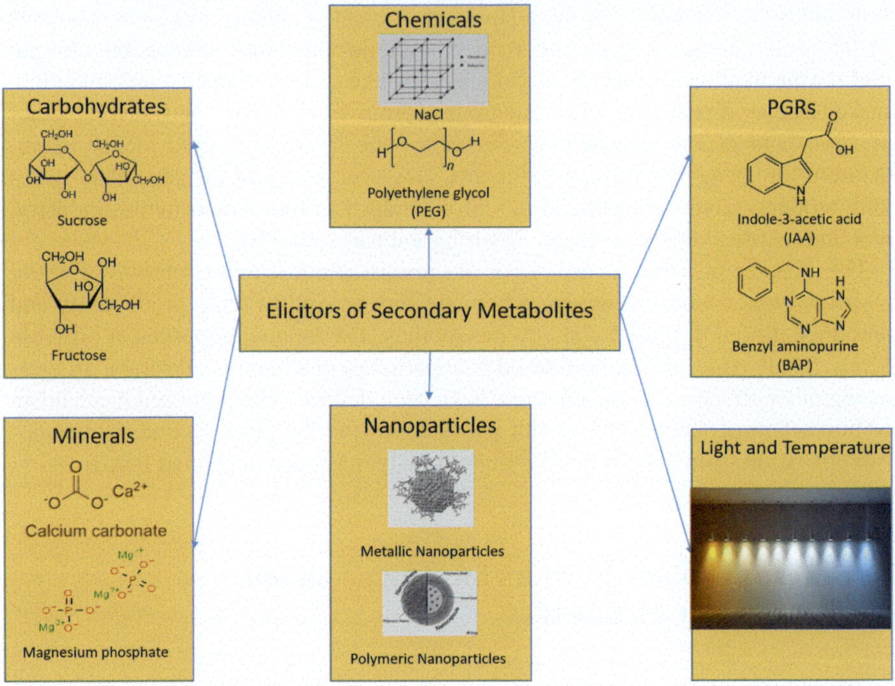

Fig. 1 Different types of abiotic elicitors for secondary metabolites production

2.2 Minerals

MS medium is composed of macronutrients (major salts) and micronutrients (minor salts) containing elements like calcium (Ca), magnesium (Mg), phosphorus (P), iron (Fe), zinc (Zn), etc., in appropriate concentrations. The alteration of the ratio of elements changes the composition of mineral salts from the MS medium producing a significant influence on the plant secondary metabolites production. There could be increased or decreased concentration or altogether removal of minerals from the media [13, 14].

2.3 Plant Growth Regulators (PGRs)

The plant growth regulators (PGRs) that include auxins, cytokinins, and gibberellins are signaling molecules involved in regulating plant growth, development, and metabolism under in vitro tissue and cell culture conditions. Examples of auxins are indole acetic acid (IAA), indole butyric acid (IBA), and α-naphthalene acetic acid (NAA), while kinetin (KIN) and 6-benzylaminopurine (BAP) are a few of the examples of cytokinins. The influence of auxins and cytokinins has been well explored regarding cell division, elongation, differentiation, shoot organogenesis, root organogenesis, embryogenesis, and secondary metabolism of different medicinal plants [15]. Gibberellins are known to produce a significant impact on plant secondary metabolites formation. Thidiazuron (TDZ) is an important cytotoxin that is known to play the functions of auxin [16]. Its role in in vitro callogenesis, embryogenesis, and organogenesis has been studied. The positive combinatorial effect of auxins and cytokinins has also been reported in the literature [17].

2.4 Light Source

Light regimes are important triggers for increasing the accumulation of secondary metabolites in different tissue culture-grown medicinal plants. Different parameters such as intensity, wavelength, and duration of light positively affect the plants in regard to their secondary metabolic activities [18].

2.5 Chemicals

The chemical manipulations produce a significant influence on plant secondary metabolites formation. The abiotic elicitors causing environmental stresses such as salt (NaCl) and drought (PEG) stress have played a key role in the augmentation of products of secondary metabolism. Such stimulating effect has been observed by many previous studies [19–21]. Nanoparticles have recently been declared potent abiotic triggers for eliciting the plant secondary metabolism.

3 The Era of Nanotechnology and Nanoparticle-Based Elicitation of Secondary Metabolites

The nanoscale particles, i.e., having 1–100 nm size, are called nanoparticles, and the field of study of nanoparticles is termed nanotechnology. Nanotechnology has been revolutionizing all sectors of life involving important implications in the fields of medicine, agriculture, pharmacy, cosmetics, electronics, engineering, dentistry, etc. [22, 23]. However, the implications of nanotechnology in the agricultural industry are still in the infancy stage. Nevertheless, for the past decade, nanoparticles have emerged as promising abiotic elicitors for the modulation of specialized secondary metabolites of economically and commercially important medicinal plants [24].

Due to the nanoparticles' smaller size and larger surface area, they have higher surface reactivity. Nanoparticles easily cross the cell membrane pores of plants and then get translocated into the cytosol where they interact with various cellular organelles, eventually entering into the nucleus. Nanoparticles interact with different metabolites and significantly affect the primary and secondary metabolism of plants. The physiology (growth and development), biochemistry (primary and secondary metabolism), and molecular biology (transcriptional changes) of plants strongly respond to the nanoparticles' uptake by the plant cells and tissues [25, 26]. The parameters like mode of synthesis, exposure time, concentration, size, and shape of nanoparticles determine the level of enhancement of secondary metabolites in medicinal plants [27]. Only a few nanoparticles have been explored till now regarding plant secondary metabolites production as described in Table 1, and many other nanoparticles remain unexplored in this regard.

3.1 Metallic Nanoparticles

Different metal nanoparticles including gold (Au), silver (Ag), copper (Cu), zinc (Zn), etc., have been employed in in vitro cultures of medicinal plants and have produced good results regarding the enhancement of secondary metabolites. Moreover, the effects of different nanoparticles of bimetallic nature such as Au-Ag, Au-Cu, and Ag-Zn in different ratios have also been reported [55].

3.2 Metallic Oxide Nanoparticles

The effects of metal oxide nanoparticles on the secondary metabolites production of different medicinal plant species have been studied by conducting different experiments. Nanoparticles such as zinc oxide (ZnO), copper oxide (CuO), aluminum oxide (Al_2O_3), titanium dioxide (TiO_2), cerium oxide (CeO_2), etc., have been used for this purpose [55].

3.3 Carbon-Based Nanomaterials

Nanomaterials including single-walled carbon nanotubes (SWCNTs), multiple-walled carbon nanotubes (MWCNTs), fullerenes, and chitosan nanoparticles have been applied to different medicinal plant species to study their impact on secondary metabolite production [56].

Table 1 Few examples of specific kinds and concentrations of nanoparticles applied under different conditions to various plants of medicinal significance for enhancing their respective secondary metabolites

Type of nanoparticles	Optimum concentration of nanoparticles	Medicinal plant species	Type of culturing and media	Effect on secondary metabolites production	Reference
Metallic nanoparticles					
Ag	45 mg L^{-1}	*Stevia rebaudiana* Bertoni	Callus culture in MS media	67% increase in stevioside content	[28]
Ag	8 mg L^{-1} 10 mg L^{-1}	*Glycyrrhiza glabra*	Seed culture in MS media	Enhancement of glycyrrhizin content	[29]
Ag	90 μg L^{-1}	*Caralluma tuberculata*	Callus culture in MS media	Increase of total phenolic and flavonoid content	[30]
Ag	2 mg L^{-1}	*Isatis constricta*	Shoot culture in MS media	Increase in indigo and tryptanthrin production	[31]
Ag	5 mg L^{-1}	*Corylus avellana* L.	Cell suspension culture in MS media	Increase in taxol content	[32]
Ag	5 mg L^{-1}	*Arabidopsis thaliana*	Seed germination in MS media	Enhancement of camalexin production	[33]
Ag	50 mg L^{-1}	*Vanilla planifolia*	Shoot culture in MS media	Enhancement of total phenolic content	[34]
Ag	25 μg mL^{-1}	*Arabidopsis thaliana*	Plantlets culture in MS media	Increase of anthocyanins production and total flavonoid content	[35]
Ag	2 mg L^{-1}	*Cucumis anguria*	Hairy root culture in MS media	Enhancement of total phenolic content and total flavonoid content	[36]
Co	5 mg L^{-1}	*Artemisia annua*	Cell suspension culture in MS media	Double increase in artemisinin content	[37]
Bimetallic nanoparticles					
Au + Ag	3:1 ratio	*Prunella vulgaris* L.	Callus culture in MS media	23% increase in total phenolic content and 4% rise in total flavonoid content	[38]

(continued)

Table 1 (continued)

Type of nanoparticles	Optimum concentration of nanoparticles	Medicinal plant species	Type of culturing and media	Effect on secondary metabolites production	Reference
Zn + Ag	19:1 ratio	*Withania somnifera*	Seed germination in potting soil	Enhancement of withanolide content	[39]
Cu + Au	N/A	*Stevia rebaudiana* Bertoni	Adventitious root culture in MS media	54% increase in phenolic and 20% increase in flavonoid content	[40]
Metallic oxide nanoparticles					
ZnO	75 mg L^{-1}	*Echinacea purpurea*	Callus culture in MS media	Enhancement of total flavonoid content	[41]
ZnO	1 mg L^{-1}	*Stevia rebaudiana* Bertoni	Shoot culture in MS media	Increased production of steviol glycosides and total phenolic and flavonoid content	[42]
ZnO	150 mg L^{-1}	*Thymus kotschyanus* *Thymus daenensis*	Callus culture in MS media	Enhancement of carvacrol and thymol contents	[43]
ZnO CuO	100 mg L^{-1} 10 mg L^{-1}	*Stevia rebaudiana* Bertoni	Callus culture in MS media	Increase in total phenolic content and total flavonoid content	[44]
ZnO CuO	2 mg L^{-1} 20 mg L^{-1}	*Stevia rebaudiana* Bertoni	Root culture in MS media	Increase in production of steviol glycosides and total phenolic content and total flavonoid content	[45]
CuO	5 mg L^{-1}	*Trigonella foenum-gaecum*	Seed germination in MS media	Rise of total phenolic content and total flavonoid content	[46]
CuO	3 mg L^{-1}	*Gymnema sylvestre*	Cell suspension in MS media	Increase in gymnemic acid, total phenolic content, and total flavonoid content	[47]

CuO	10 mg L^{-1}	*Stevia rebaudiana* Bertoni	Shoot culture in MS media	Increase in production of steviol glycosides and total phenolic content and total flavonoid content	[48]
Fe_3O_4	450 mg L^{-1} 900 mg L^{-1}	*Hyoscyamus reticulatus* L.	Hairy root culture in MS media	Increase in hyoscyamine content by 5 times	[49]
TiO_2	50 mg L^{-1} 100 mg L^{-1}	*Hypericum perforatum*	Callus culture in MS media	Increase in production of hypericin and pseudohypericin content	[50]
TiO_2	120 mg L^{-1}	*Aloe vera*	Cell suspension culture in MS media	Increase in aloin content	[51]
Al_2O_3	100 μg mL^{-1}	*Nicotiana tobaccum*	Cell suspension culture in MS media	Increase in total phenolic contents	[52]
Carbon-based nanomaterials					
Single-walled carbon nanotubes (SWCNTs)	500 mg L^{-1}	*Tanacetum parthenium* Linn.	Whole plant culture in a greenhouse	Twofold increase in parthenolide content	[53]
Multi-walled carbon nanotubes (MWCNTs)	100 μg mL^{-1}	*Satureja khuzestanica*	Callus culture in B5 basal media	12% increase in phenolic and 3% increase in flavonoid content	[54]

4 Uptake and Internalization of Nanoparticles

Entry of nanoparticles in plants takes place via three major routes, i.e., (1) soil, (2) culture medium, and (3) foliar spray. In-depth studies are required for a comprehensive understanding of the translocation and internalization of nanoparticles once these are taken up by the plant cells. Till now, the uptake of nanoparticles has been analyzed by different advanced techniques of microscopy such as transmission electron microscopy (TEM), scanning electron microscopy (SEM), and atomic force microscopy (AFM) [57]. These studies have elucidated that the nanoparticles enter into the plant cell wall using pores found in them. The cell wall pores are approximately 20 nm in size, and these pores are also reported to get stretched if nanoparticles of >20 nm are to enter them. The formation of new pores in the plant cell wall has also been documented. After crossing the cell wall, nanoparticles are engulfed by the cell membrane by the endocytosis phenomenon, from where these are penetrated toward the cytoplasm and different cellular organelles. Hence, these nanoparticles interact with different molecules of cytosol and cytoplasmic organelles, consequently disturbing the primary and secondary metabolism [58].

5 Mechanism of Triggering Behavior of Nanoparticles

The mechanism of triggering secondary metabolites by different nanoparticles is very complicated due to the involvement of multiple signaling pathways and huge cross talk between them. The origin, exposure time, dosage, way to transfer, etc., differ in the case of different nanoparticles. Similarly, different types of medicinal plants respond to external triggers differently. Hence, proposing a constant uniform model for elicitation of specialized metabolites is very difficult. However, the most probable mechanism might involve the Ca^{+2} ion influx movements in response to nanoparticles' invasion into the plant cells after being recognized by the receptors on the plasma membrane. Few oxidases like NADPH oxidase are responsible for generating reactive oxygen species (ROS), and the ROS generated due to oxidative stress result in the phosphorylation of the mitogen-activated protein kinase (MAPK) pathway [23, 59, 60]. Then the upregulation of transcription factors controlling the plant secondary metabolites production takes place [61]. Additionally, signaling compounds that control stress like salicylic acid (SA), jasmonic acid (JA), and methyl jasmonic acid (MeJA) have been reported to be involved in regulating the specialized signaling cascades of formation of biologically active ingredients under the situation of abiotic stress [62]. Figure 2 depicts the proposed mechanism of nanoelicitation of secondary metabolites.

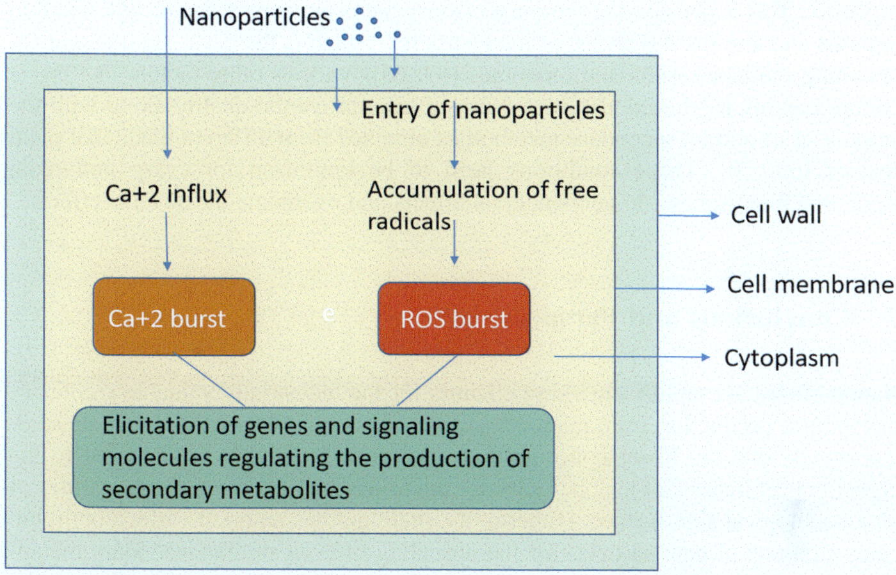

Fig. 2 Proposed mechanism of nano-based elicitation of secondary metabolites

6 Nanotoxicity

Many of the nanoelicitation studies have reported adverse effects to plants known as phytotoxicity caused by nanoparticles at higher concentrations and long-term exposures. Under specific conditions causing toxicity, plant physiology and biochemistry are badly affected leading to defective primary and secondary metabolism. The disturbance of the primary metabolism of plants has been documented, such as a decline in sugar and chlorophyll contents. These deleterious effects are significant challenges in the success of the nanoelicitation strategy that need to be avoided [63]. There are numerous reports showing nanophytotoxicity; however, the exact mechanism is still unclear. The most probable mechanism involves the overproduction of ROS. Excess of free radicals cause oxidative burst by which plant defense mechanism is distorted resulting in damage of vital organelles, like mitochondria and chloroplast, and macromolecules, like DNA, proteins, lipids, etc., finally leading to cell death [64, 65].

Finding the optimum concentration of nanoparticles at which they work as elicitors producing the maximum quantity of secondary metabolites in specialized compartments of plants is a prerequisite to preventing the harmful effects of nanotoxicity [66]. Like concentration, the size of nanoparticles is a major factor affecting the elicitation potential. If the size of nanoparticles is reduced to a minimum using capping agents, then their reactivity would get enhanced resulting in the desirable secondary metabolites production even in lesser concentration and exposure. This is because the presence of surfactants or capping agents will result in targeted and sustained delivery of nanoparticles, which is the basic requirement for obtaining the best results and outcomes [67, 68]. Various other factors like tissue culture conditions should also be optimized to improve the quality along with the amount of important secondary metabolites obtained from different medicinal plant species [69, 70]. These conditions need to be optimized for every individual medicinal plant species to get beneficial effects and overcome the adverse effects.

7 Conclusions and Perspectives

Nanoparticles are rapidly emerging elicitors for the industrially valuable secondary metabolites production from micropropagated medicinal plant parts. Previous studies suggest that the lower concentrations are more beneficial as compared to the higher concentrations of nanoparticles for the upscaling of secondary metabolites of pharmacological significance. Hitherto, the literature has reported various optimum concentrations of nanoparticles for the upscaling of even similar secondary metabolites, and this variability must be due to other involving parameters. Though the present data has the potential to open ways for upcoming systematic studies about the involvement of nanoparticles of various kinds employed as elicitors in medicinal plant biotechnology, the knowledge about the nanoparticles' interaction with plant tissues and cells and the plausible mechanism of nanoparticles' triggered secondary metabolites production is obscure and not clearly defined. In the future, more studies encompassing a combination of nanoparticles and other elicitation strategies should be employed via in vitro culturing to evaluate more cues about the mechanistic insights of the nanoparticles-based triggering of plant secondary metabolites. Furthermore, studies encompassing molecular events associated with translocation and elicitation mechanism of nanoparticles, molecular signaling pathways of secondary metabolites production, and molecular footprints of phytotoxicity caused by nanoparticles should be identified. Besides, the stimulating role of a novel class of elicitors, i.e., nanoelicitors, should be finely described by employing the different types of nanoparticles for elicitation of specialized metabolites that haven't been explored yet.

Author Contributions RJ did conceptualization, writing, and editing. All authors have read and agreed to the published version of the manuscript.

Funding: No external funding was received.

Declaration of Competing Interest: The authors declare no conflict of interest.

References

1. Wang, C. Z., Calway, T., & Yuan, C. S. (2012). Herbal medicines as adjuvants for cancer therapeutics. *The American Journal of Chinese Medicine, 40*(4), 657–669.
2. Martín Ortega, A. M., & Campos, M. R. S. (2019). Chapter 13 - Bioactive compounds as therapeutic alternatives. In M. R. S. Campos (Ed.), *Bioactive compounds* (pp. 247–264). Woodhead Publishing.
3. DiCosmo, F., & Misawa, M. (1995). Plant cell and tissue culture: Alternatives for metabolite production. *Biotechnology Advances, 13*(3), 425–453.
4. Ali, M., Abbasi, B. H., Ahmad, N., Khan, H., & Ali, G. S. (2017). Strategies to enhance biologically active-secondary metabolites in cell cultures of Artemisia–current trends. *Critical Reviews in Biotechnology, 37*, 833–851.
5. Khan, T., Abbasi, B. H., & Khan, M. A. (2019). Elicitation strategies of in-vitro cultures for the sustainable use of medicinal plants. *Proceedings of the Pakistan Academy of Sciences, 56*, 49–56.
6. Narayani, M., & Srivastava, S. (2017). Elicitation: A stimulation of stress in in vitro plant cell/ tissue cultures for enhancement of secondary metabolite production. *Phytochemistry Reviews, 16*, 1227–1252.
7. Anjum, S., Anjum, I., Hano, C., & Kausar, S. (2019). Advances in nanomaterials as novel elicitors of pharmacologically active plant specialized metabolites: Current status and future outlooks. *RSC Advances, 9*, 40404.
8. Teoh, E. S. (2016). Secondary metabolites of plants. In E. S. Teoh (Ed.), *Medicinal orchids of Asia* (pp. 59–73). Springer.
9. Akula, R., & Ravishankar, G. A. (2011). Influence of abiotic stress signals on secondary metabolites in plants. *Plant Signaling & Behavior, 6*(11), 1720–1731.
10. Hussain, M. S., Fareed, S., Saba Ansari, M., Rahman, A., Ahmad, I. Z., & Saeed, M. (2012). Current approaches toward production of secondary plant metabolites. *Journal of Pharmacy & Bioallied Sciences, 4*(1), 10–20.
11. Chandran, H., Meena, M., Barupal, T., & Sharma, K. (2020). Plant tissue culture as a perpetual source for production of industrially important bioactive compounds. *Biotechnology Reports, 26*, e00450.
12. Kumar, G. P., Subiramani, S., Govindarajan, S., Sadasivam, V., Manickam, V., Mogilicherla, K., Thiruppathi, S. K., & Narayanasamy, J. (2015). Evaluation of different carbon sources for high frequency callus culture with reduced phenolic secretion in cotton (*Gossypium hirsutum* L.) cv. SVPR-2. *Biotechnology Reports, 7*, 72–80.
13. Ahmad, M. A., Xu, D., Adeel, M., Rizwan, M., Noman, S., & Yuesuo, Y. (2021). Influence of calcium and magnesium elimination on plant biomass and secondary metabolites of *Stevia rebaudiana* Bertoni. *Biotechnology and Applied Biochemistry*.
14. Nagella, P., & Murthy, H. N. (2010). Effects of macroelements and nitrogen source on biomass accumulation and withanolide-A production from cell suspension cultures of *Withania somnifera* (L.) Dunal. *Plant Cell, Tissue and Organ Culture, 104*(1), 119–124.
15. Dobrev, P. I., Hoyerová, K., & Petrášek, J. (2017). Analytical determination of auxins and cytokinins. In T. Dandekar & M. Naseem (Eds.), *Auxins and cytokinins in plant biology: Methods and protocols* (pp. 31–39). Springer.
16. Ali, M., & Abbasi, B. H. (2014). Thidiazuron-induced changes in biomass parameters, total phenolic content, and antioxidant activity in callus cultures of *Artemisia absinthium* L. *Applied Biochemistry and Biotechnology, 172*(5), 2363–2376.
17. Javed, R., Yucesan, B., & Zia, M. (2017). Differential effects of plant growth regulators on physiology, steviol glycosides content, and antioxidant capacity in micropropagated tissues of *Stevia rebaudiana*. *Biologia, 72*, 1156–1165.
18. Fazal, H., Abbasi, B. H., Ahmad, N., Ali, S. S., Akbar, F., & Kanwal, F. (2016). Correlation of different spectral lights with biomass accumulation and production of antioxidant secondary

metabolites in callus cultures of medicinally important *Prunella vulgaris* L. *Journal of Photochemistry and Photobiology B: Biology, 159*, 1–7.
19. Javed, R., & Gurel, E. (2019). Salt stress by NaCl alters the physiology and biochemistry of tissue culture-grown *Stevia rebaudiana* Bertoni. *Turkish Journal of Agriculture and Forestry, 43*(1), 11–20.
20. Ahmad, M. A., Javed, R., Adeel, M., Rizwan, M., & Yang, Y. (2020). PEG 6000-stimulated drought stress improves the attributes of in vitro growth, steviol glycosides production, and antioxidant activities in *Stevia rebaudiana* Bertoni. *Plants, 9*, 1552.
21. Javed, R., Yücesan, B., & Gurel, E. (2018). Hydrogen peroxide-induced steviol glycosides accumulation and enhancement of antioxidant activities in leaf tissues of *Stevia rebaudiana* Bertoni. *Sugar Tech, 20*, 100–104.
22. Javed, R., Zia, M., & Naz, S. (2020). Role of capping agents in the application of nanoparticles in biomedicine and environmental remediation: Recent trends and future prospects. *Journal of Nanbiotechnology, 18*, 172.
23. Arya, S. S., Lenka, S. K., Cahill, D. M., & Rookes, J. E. (2021). Designer nanoparticles for plant cell culture systems: Mechanisms of elicitation and harnessing of specialized metabolites. *BioEssays, 43*, e2100081.
24. Javed, R., Zia, M., Yücesan, B., & Gürel, E. (2017). Abiotic stress of ZnO-PEG, ZnO-PVP, CuO-PEG and CuO-PVP nanoparticles enhance growth, sweetener compounds and antioxidant activities in shoots of *Stevia rebaudiana* Bertoni. *IET Nanobiotechnology, 11*, 898–902.
25. Riaza, M. S., Ullaha, N., Alid, H., & Nadhmane, A. (2019). *Analysis, fate, and toxicity of engineered nanomaterials in plants* (Vol. 84, p. 23). Elsevier.
26. Hatami, M., Kariman, K., & Ghorbanpour, M. (2016). Engineered nanomaterial-mediated changes in the metabolism of terrestrial plants. *Science of the Total Environment, 571*, 275–291.
27. Anjitha, K. S., Sameena, P. P., & Puthur, J. T. (2021). Functional aspects of plant secondary metabolites in metal stress tolerance and their importance in pharmacology. *Plant Stress, 2*, 100038.
28. Golkar, P., Moradi, M., & Garousi, G. A. (2018). Elicitation of stevia glycosides using salicylic acid and silver nanoparticles under callus culture. *Sugar Tech, 21*(4), 569–577.
29. Tahoori, F., Ahmad, M., Nejadsattari, T., Ofoghi, H., & Iranbakhsh, A. (2019). Qualitative and quantitative study of quercetin and glycyrrhizin in in vitro culture of liquorice (*Glycyrrhiza glabra* L.) and elicitation with $AgNO_3$. *Notulae Botanicae Horti Agrobotanici Cluj-Napoca, 47*, 143–151.
30. Ali, A., Mohammad, S., Khan, M. A., Raja, N. I., Arif, M., Kamil, A., & Mashwani, Z. U. R. (2019). Silver nanoparticles elicited in vitro callus cultures for accumulation of biomass and secondary metabolites in *Caralluma tuberculata*. *Artificial Cells, Nanomedicine, and Biotechnology, 47*, 715–724.
31. Karakas, O. (2020). Effect of silver nanoparticles on production of indole alkaloids in *Isatis constricta*. *Iranian Journal of Science and Technology, Transactions A: Science, 44*, 621–627.
32. Jamshidi, M., Ghanati, F., Rezaei, A., & Bemani, E. (2016). Change of antioxidant enzymes activity of Hazel (*Corylus avellana* L.) cells by AgNPs. *Cytotechnology, 68*, 525–530.
33. Kruszka, D., Sawikowska, A., Selvakesavan, R. K., Krajewski, P., Kachlicki, P., & Franklin, G. (2020). Silver nanoparticles affect phenolic and phytoalexin composition of *Arabidopsis thaliana*. *Science of the Total Environment, 716*, 135361.
34. Spinoso-Castillo, J., Chavez-Santoscoy, R., Bogdanchikova, N., & P'erez-Sato, J., Morales-Ramos, V., Bello-Bello, J. (2017). Nano-silver particles reduce contaminations in tissue culture but decrease regeneration rate and slows down growth and development of *Aldrovanda vesiculosa* explants. *Plant Cell, Tissue and Organ Culture, 129*, 195–207.
35. García-Sánchez, S., Bernales, I., & Cristobal, S. (2015). Transcriptome assembly, profiling and differential gene expression analysis of the halophyte *Suaeda fructicosa* provides insights into salt tolerance. *BMC Genomics, 16*, 341.
36. Chung, I. M., Rajakumar, G., & Thiruvengadam, M. (2018). Effect of silver nanoparticles on phenolic compounds production and biological activities in hairy root cultures of *Cucumis Anguria*. *Acta Biologica Hungarica, 69*, 97–109.

37. Ghasemi, B., Hosseini, R., & Nayeri, F. D. (2015). Effects of cobalt nanoparticles on artemisinin production and gene expression in *Artemisia annua*. *Turkish Journal of Botany, 39*, 769–777.
38. Fazal, H., Abbasi, B. H., Ahmad, N., & Ali, M. (2016). Elicitation of medicinally important antioxidant secondary metabolites with silver and gold nanoparticles in callus cultures of *Prunella vulgaris* L. *Applied Biochemistry and Biotechnology, 180*, 1076–1092.
39. Singh, R., Singh, D. P., Gupta, P., Jain, P., Mishra, T., Kumar, A., Dhawan, S. S., & Shirke, P. A. (2019). Nanoparticles alter the withanolide biosynthesis and carbohydrate metabolism in *Withania somnifera* (Dunal). *Industrial Crops and Products, 127*, 94–109.
40. Ghazal, B., Saif, S., Farid, K., Khan, A., Rehman, S., Reshma, A., Fazal, H., Ali, M., Ahmad, A., & Rahman, L. (2018). Stimulation of secondary metabolites by copper and gold nanoparticles in submerge adventitious root cultures of *Stevia rebaudiana* (Bert.). *IET Nanobiotechnology, 12*, 569–573.
41. Karimi, N., Behbahani, M., Dini, G., & Razmjou, A. (2018). Enhancing the secondary metabolite and anticancer activity of *Echinacea purpurea* callus extracts by treatment with biosynthesized ZnO nanoparticles. *Nanoscience and Nanotechnology, 9*, 045009.
42. Javed, R., Usman, M., Yucesan, B., Zia, M., & Gurel, E. (2017). Effect of zinc oxide (ZnO) nanoparticles on physiology and steviol glycosides production in micropropagated shoots of *Stevia rebaudiana* Bertoni. *Plant Physiology and Biochemistry, 110*, 94–99.
43. Mosavat, N., Golkar, P., Yousefifard, M., & Javed, R. (2019). Modulation of callus growth and secondary metabolites in different thymus species and *Zataria multiflora* micropropagated under ZnO nanoparticles stress. *Biotechnology and Applied Biochemistry, 66*(3), 316–322.
44. Javed, R., Yucesan, B., Zia, M., & Gurel, E. (2018). Elicitation of secondary metabolites in callus cultures of *Stevia rebaudiana* Bertoni grown under ZnO and CuO nanoparticles stress. *Sugar Tech, 20*, 194–201.
45. Ahmad, M. A., Javed, R., Adeel, M., Rizwan, M., Ao, Q., & Yang, Y. (2020). Engineered ZnO and CuO nanoparticles ameliorate morphological and biochemical response in tissue culture regenerants of Candyleaf (*Stevia rebaudiana*). *Molecules, 25*, 1356.
46. Ain, U., Haq, I. U., Abbasi, B. H., Javed, R., & Zia, M. (2017). Influence of PVP/PEG impregnated CuO NPs on physiological and biochemical characteristics of *Trigonella foenum-graecum* L. *IET Nanobiotechnology, 12*, 349–356.
47. Chung, I. M., & Thiruvengadam, M. (2019). Impact of copper oxide nanoparticles on enhancement of gymnemic acid and phenolic compounds using cell suspension culture of *Gymnema sylvestre* (Retz.) R. *British Journal of Applied Physics, 9*, 213–223.
48. Javed, R., Mohamed, A., Yucesan, B., Gurel, E., Kausar, R., & Zia, M. (2017). CuO nanoparticles significantly influence in vitro culture, steviol glycosides, and antioxidant activities of *Stevia rebaudiana* Bertoni. *Plant Cell, Tissue and Organ Culture, 131*, 611–620.
49. Moharrami, F., Hosseini, B., Shara, A., & Farjaminezhad, M. (2017). Enhanced production of hyoscyamine and scopolamine from genetically transformed root culture of *Hyoscyamus reticulatus* L. elicited by iron oxide nanoparticles. *In Vitro Cellular & Developmental Biology: Plant, 53*, 104–111.
50. Ebadollahi, R., Jafarirad, S., Kosari-Nasab, M., & Mahjouri, S. (2019). Effect of explant source, perlite nanoparticles and TiO_2/perlite nanocomposites on phytochemical composition of metabolites in callus cultures of *Hypericum perforatum*. *Scientific Reports, 9*, 12998.
51. Raei, M., Angaji, S. A., Omidi, M., & Khodayari, M. (2014). Effect of abiotic elicitors on tissue culture of *Aloe vera*. *International Journal of Biological Sciences, 5*, 74–81.
52. Poborilova, Z., Opatrilova, R., & Babula, P. (2013). Toxicity of aluminium oxide nanoparticles demonstrated using a BY-2 plant cell suspension culture model. *Environmental and Experimental Botany, 91*, 1–11.
53. Ahmadi, S., Ghorbanpour, M., Hadian, J., & SalehiArjmand, H. (2018). Impact of foliar spray of spherical nano-carbon and salicylic acid on physiological traits and parthenolide content in two feverfew cultivars (*Tanacetum parthenium* Linn. cv. *Pharmasaat* and *Jelitto*). *Journal of Medicinal Plants, 4*, 82–98.

54. Ghorbanpour, M., & Hadian, J. (2015). Multi-walled carbon nanotubes stimulate callus induction, secondary metabolites biosynthesis and antioxidant capacity in medicinal plant *Satureja khuzestanica* grown in vitro. *Carbon, 94*, 749–759.
55. Javed, R., Ahmad, M. A., Gul, A., Ahsan, T., & Cheema, M. (2021). *Chapter 7 - comparison of chemically and biologically synthesized nanoparticles for the production of secondary metabolites, and growth and development of plants, comprehensive analytical chemistry* (Vol. 94, pp. 303–329). Elsevier.
56. Husen, A., & Siddiqi, K. S. (2014). Carbon and fullerene nanomaterials in plant system. *Journal of Nanbiotechnology, 12*, 16.
57. Nair, R., Varghese, S. H., Nair, B. G., Maekawa, T., Yoshida, Y., & Kumar, D. S. (2010). Nanoparticulate material delivery to plants. *Plant Science, 179*(3), 154–163.
58. Wang, P., Lombi, E., Zhao, F. J., & Kopittke, P. M. (2016). Nanotechnology: A new opportunity in plant sciences. *Trends in Plant Science, 8*, 699–712.
59. Sosan, A., Svistunenko, D., Straltsova, D., Tsiurkina, K., Smolich, I., Lawson, T., Subramaniam, S., Golovko, V., Anderson, D., Sokolik, A., Colbeck, I., & Demidchik, V. (2016). Engineered silver nanoparticles are sensed at the plasma membrane and dramatically modify the physiology of *Arabidopsis thaliana* plants. *The Plant Journal, 85*(2), 245–257.
60. Khan, A. K., Kousar, S., Tungmunnithum, D., Hano, C., Abbasi, B. H., & Anjum, S. (2021). Nano-elicitation as an effective and emerging strategy for in vitro production of industrially important flavonoids. *Applied Sciences, 11*, 1694.
61. Phukan, U. J., Jeena, G. S., & Shukla, R. K. (2016). WRKY transcription factors: Molecular regulation and stress responses in plants. *Frontiers in Plant Science, 7*, 760.
62. Kohan-Baghkheirati, E., & Geisler-Lee, J. (2015). Gene expression, protein function and pathways of *Arabidopsis thaliana* responding to silver nanoparticles in comparison to silver ions, cold, salt, drought, and heat. *Nanomaterial, 5*, 436–467.
63. Montes, A., Bisson, M. A., Gardella, J. A., & Aga, D. S. (2017). Uptake and transformations of engineered nanomaterials: Critical responses observed in terrestrial plants and the model plant *Arabidopsis thaliana*. *Science of the Total Environment, 607-608*, 1497–1516.
64. Jamil, B., Javed, R., Qazi, A. S., & Syed, M. A. (2018). Nanomaterials: Toxicity, risk management and public perception. In M. Rai & J. Biswas (Eds.), *Nanomaterials: Ecotoxicity, safety, and public perception*. Springer.
65. Jogaiah, S., Paidi, M. K., Venugopal, K., Geetha, N., Mujtaba, M., Udikeri, S. S., & Govarthanan, M. (2021). Phytotoxicological effects of engineered nanoparticles: An emerging nanotoxicology. *Science of the Total Environment, 801*, 149809.
66. Schwab, F., Zhai, G., Kern, M., Turner, A., Schnoor, J. L., & Wiesner, M. R. (2016). Barriers, pathways and processes for uptake, translocation and accumulation of nanomaterials in plants — Critical review. *Nanotoxicology, 10*, 257–278.
67. Javed, R., Usman, M., Tabassum, S., & Zia, M. (2016). Effect of capping agents: Structural, optical and biological properties of ZnO nanoparticles. *Applied Surface Science, 386*, 319–326.
68. Javed, R., Rais, F., Fatima, H., Haq, I. U., Kaleem, M., Naz, S. S., & Ao, Q. (2020). Chitosan encapsulated ZnO nanocomposites: Fabrication, characterization, and functionalization of bio-dental approaches. *Mater Sci Eng C Mater Biol Appl, 116*, 111184.
69. Anjum, S., Komal, A., Abbasi, B. H., & Hano, C. (2021). Nanoparticles as elicitors of biologically active ingredients in plants. In A. P. Ingle (Ed.), *Nanotechnology in plant growth promotion and protection*.
70. Javed, R., & Yücesan, B. (2022). Impact of Stevia rebaudiana culturing in liquid medium: Elevation of yield and biomass, mitigation of steviol glycosides: Comparative analysis of culturing of Stevia rebaudiana in solid and liquid media. *Proceedings of the Pakistan Academy of Sciences: B. Life and Environmental Sciences, 59*(1), 69–75.

Nanomaterials as Unique Carriers in Agricultural Practices for Plant Growth and Development: A State of Current Knowledge

Amir Khan and Faheem Ahmad

Abstract In the present situation, it is critical to meet the nutritional needs of the world's rising population. Nearly one-third of crops are affected in conventional farming due to pathogens infestation. Nanotechnology can change agricultural production and improve plant growth by facilitating the cost-effective management of natural resources and other necessary inputs. Nanomaterials (NMs) in agriculture offer a once-in-a-lifetime opportunity to improve crop yield and preserve soil health. Potential NMs uptake in soil-plant systems and positive and negative effects in various crops have been observed in several studies. Because of their unique properties, NMs have received much attention in agriculture. NMs and nanotechnology improve the stability and dispersion of active ingredients, reduce residual pollution and labor costs, and maintain agricultural systems' sustainability. Many nanomaterials-based formulations have been extensively used for plant growth and development, including nano-based associated pesticides and fertilizers in the modern agricultural system. Understanding the interactions between plants and NMs opens up new paths for improving agricultural crop yield and quality. This chapter helps readers better understand the role of NMs in plant growth and development.

1 Introduction

Agriculture serves as the principal pillar of the growing economy, supplying food for a better quality of life. The scenario described above will be crucial for countries, particularly the developing world, where agricultural production is the primary source of economy and face numerous challenges in the sectors of economy, commodity reliance, poverty, and malnourishment. A considerable agricultural output is being achieved by utilizing present nanomaterials (NMs), which are aimed at effective components systems, healthy plant defense strategies, organic farming, and several other applications [1]. Recently researchers noticed that

A. Khan · F. Ahmad (✉)
Department of Botany, Aligarh Muslim University, Aligarh, Uttar Pradesh, India
e-mail: faheem.bt@amu.ac.in

J.-T. Chen (ed.), *Plant and Nanoparticles*,
https://doi.org/10.1007/978-981-19-2503-0_11

practically introducing based products lines to revolutionize advanced farming techniques. NMs are highly reactive due to high surface area-to-volume ratio and physical and chemical characteristics, delivering a concise benefit in terms of requisite alteration in response to rising supply. With the assistance of these cutting-edge materials, modern agriculture is converting into precision farming, allowing for the most outcome from the resources available. Agriculture has always been the most essential and reliable segment since it generates and supplies crude ingredients to the feed and food industry sectors. Fertilizers are required to boost crop yields, but they also reduce soil quality by disrupting soil nutrient balance [2, 3]. Pesticides, fertilizers, and antibiotics are commonly applied and adequately disintegrate. Also, pesticides cost and fertilizers are extremely huge, and this must be monitored. The use of NMs in agriculture seeks to decrease nutrient shortages to boost yields, decrease product amount lines used for crop safety [4], and mitigate production costs to maximize outcomes. It has not only fundamentally changed agriculture by implementing creative nutrients of nano-fertilizers, but it has also aided in crop protection by developing nano-pesticides, efficient irrigation systems, and rising plant efficiency in using the energy of the sun [5, 6]. The only way to meet demand is to increase productivity and income per unit of limited natural resources through efficient technological improvements.

Nanotechnology can increase agricultural production by the following ways: (1) agrochemical nano-preparations applied as pesticides and fertilizers for crop enhancement, (2) use of nano-sensors for identification of various diseases in crop protection, (3) nano-devices for genetic engineering of plants, (4) diagnosis of plant diseases, and (5) post-harvest management. NMs are materials with particle dimensions of not as much as 100 nm and have exclusive properties like size-dependent abilities, high surface-to-volume ratio, and promise optical characteristics [7]. NMs created using environmentally friendly and green techniques can improve agriculture by enhancing fertilization, plant growth regulators, and pesticides. The use of NMs in agriculture works as an alternative to agrochemicals. Noticeably, NMs increase crop yields by improving agricultural products' productivity to allow for site-targeted controlled delivery of nutrient content, confirming lesser agrochemical use.

2 Nanomaterials for Sustainable Intensification in Agriculture

In agriculture, nanotechnology offered numerous agri-techniques like nano-fertilizers, nano-pesticides, and nano-sensors, which have all demonstrated meaningful results for sustainable farming exercises. Such nano-inputs decreased the use of fertilizers or pesticides amount and delivered targeted delivery of active compounds. As a result, nontargeted organisms are unaffected by such nano-tools and the environment is protected. The use of nanotechnology in agriculture and the food industry can fundamentally change various sectors by providing modern tools for disease detection and treatment and increasing plant nutrient uptake capacity. Nano-agriculture employs nanotechnology to increase plant yield, fuel, and other drives. Nano-formulations are suggested to enhance agricultural chemical efficacy, delivery systems, plant nutrient uptake and outcome, and food quality. The nano-fertilizers, nano-pesticides, and nano-sensors, among other things, have changed traditional agricultural practices into sustainable farming. It has been reported that using NMs as an agro-based product does not affect nontargeted organisms [8]. It also improved solubilization, increased active ingredient shelf life, and governed discharging capacity. NMs are environmentally friendly for the experiment and are being utilized in field conditions. Due to agri-friendly characteristics, nano-tools are ideal for preserving ecological quality by lowering the damaging impact of synthetic chemicals.

Furthermore, research is necessary to confirm the limit for individual metal NPs in the crop system and maintain the supply of an excellent variety of concentrations. Another biocompatible NMs synthesis route for agricultural applications is green NPs synthesis. These nano-tools would also be much able to reduce the excessive use of chemical fertilizers and pesticides.

Nano-fertilizers are a relatively new agricultural advancement. Silver, iron, zinc, titanium, carbon nanotubes, molybdenum, and silica are some nano-fertilizers that are established and used in different crop frameworks. NMs are added to soil as per nano-structured fertilizers (similar to Fe, Mn, Zn, Cu, Mo NPs) or improved transport schemes to progress absorption and efficiency of fertilizer application [6]. Metallic NPs based on Fe_2O_3, ZnO, TiO_2, and Cu are used as nano-fertilizers in the soil through irrigation or foliar applications [9–11]. More prospects for using nanotechnology in agriculture exist in the sectors of plant genetic improvement [12], transfer of genes and drug particles to exact sites at cell level in plants, and nano-array-based technologies for expression of genes in plants to resolve stress, as well as the advancement of sensors and procedures for their use in smart agriculture [13]. The majority of early research for NM-based plants genetically engineered has been done in plant cell cultures. Magnetic NPs were used to ensure an effective, stable genetic transformation in cotton. Mesoporous silica NPs with lox P site integrated into chromosomal DNA was used like transporters to carry Cre recombinase in undeveloped embryos of maize. The lox P correctly recombined after introducing engineered mesoporous silica NPs in plant materials, attempting to establish an effective genetic modification [14].

Weeds are the most dangerous threat to crops because they consume nutrients that would otherwise be available to plants. Traditional methods of weed eradication, such as hand weeding, are time-consuming and labor-intensive. Several herbicides exist in the market that can destroy weed growth while causing crop damage and decreasing soil fertility. Nano-herbicides are created by utilizing nano-technological possible for active distribution of biological or chemical herbicides through nano-size or NMs established herbicide compositions. Compared to conventional herbicides, NMs-based designs might improve herbicide efficacy, solubility, and lower toxicity. Early weed management using NP-based herbicide release systems has the potential to reduce herbicide resistance, preserve active substance activity, and extend herbicide discharge over a more extended period [15]. The innovation of a particular herbicide compound encapsulated in an NP targets specific receptors found at the root of the aimed weed. The advanced NP inserts the weed's root system and is translocated to undertake its action, preventing plant root glycolysis. The focused action causes the plant to starve and thus kills it. Nano-herbicides may be a more excellent, more environmentally favorable option for weed controllers that do not leave toxic remains in the soil.

Nanotechnology applications in plant protection have impacted agriculture and enhanced yields. Metal NPs of various types like nano-formulations, nano-encapsulated active constituents, and nano-composites are reported for crop protection. Several NMs were shown to have a more significant inhibition effect against crop pathogens in the lab and the greenhouse. Nano-sensors also supplied fast and precise evidence about soil environments or pathogen recognition, allowing for a timely controller and crop safety, which aids farmers in reducing losses and improving their economic condition. Because of their recognition efficacy in small quantities, nano-sensors lessen significant crop harm by monitoring field conditions and pest attacks.

NMs are used to develop biosensors or used as "sensing materials" in crop biotechnology, agriculture, and the food industry [16]. Various nanosensors viz., plasmonic nanosensors, fluorescence resonance energy transfer-based nanosensors, carbon-based electrochemical nanosensors, and nanowire and antibody nanosensors have been used in agricultural practices. Even though the usage of nano-sensors is still in its early stages [17], some remarkable findings suggest the usage of NMs, such as apparatuses for detecting and quantifying plant metabolic flux, pesticide residues in food, bacterial and bacterial, and viral, fungal pathogens. NMs-based biosensors seem to be very encouraging because they allow for the early screening and precise quantification of virus, bacteria, and fungi in plants [18, 19]. We have summarized the agricultural applications of some metal NPs in tabular form (Table 1).

3 Nanomaterials: A New Carrier in Agricultural Development

3.1 Sources and Synthesis

Based on their origin, NMs sources can be divided into three major categories: (a) incidental NMs, which are generated as a by-product of industrial processes; (b) engineered NMs, which are produced by living beings that have specific properties required for different applications; and (c) naturally produced NMs.

Table 1 Applications of various metal nanoparticles in agriculture

S. N.	Nanoparticle	Host crop	Effect	Reference
1	TiO_2	*Triticum aestivum*	Increased total chlorophyll and carotenoids, improved stomatal conductance, and transpiration rate	[20]
2	ZnO and Si	*Mangifera indica*	Improved nutrients uptake and carbon assimilation	[21]
3	SiO_2	*Musa acuminata*	Enhanced chlorophyll content, improved photosynthesis, maintenance of K^+ and Na^+ balance	[22]
4	Fe_3O_3	*Triticum aestivum*	Enhanced photosynthesis, Fe concentration, and reduced cadmium concentration	[23]
5	Cu	*Zea mays*	Increased anthocyanin, chlorophyll, and carotenoid contents	[24]
6	CeO	*Gossypium hirsutum*	Decreased ROS levels	[25]
7	CuO	*Spinacia oleracea*	Improved photosynthesis	[26]
8	Ag	*Triticum aestivum*	Decreased anti-oxidative enzyme activity and increased POD activity	[27]
9	ZnO	*Gossypium hirsutum*	Increased photosynthetic pigments and proteins; decreased level of MDA	[28]
10	Au	*Brassica juncea*	Imbalanced ethylene and auxin production	[29]
11	MWCNT	Red spinach	Inhibition of stomatal opening or induction of stomatal closing	[30]
12	MgO	Cluster bean	Improved chlorophyll content	[31]
13	Fe	Watermelon	Enhanced photosynthesis rate and chlorophyll content	[32]
14	Si	Lupin	Enhanced germination and protein chlorophyll content	[33]

3.1.1 Incidental NMs

Usual procedures that lead to the manufacture of NMs include forest fires, volcanic eruptions, and photochemical reactions. Furthermore, detaching of skin and hair by plants and animals, which frequently occurs in nature, contributes to the composition of NMs. Natural measures like forest fires, volcanic eruptions, and dust storms have produced a large amount of nano-particulate material, significantly impacting global air quality. Similarly, human activities such as transportation, industrial operations, and charcoal burning contribute to the emergence of synthesized NMs. Throughout the universe, various types of NMs are sorted, mixed, and amended in multiple systems. In the desert and terrestrial regions, dust storms are the primary source of NMs. According to satellite images, dust storms in one province transport nano- and micro-sized crystals and pollutants thousands of kilometers away from the start. Dust storms seem to be the most significant single contributor to environmental NMs. Volcanic eruptions release a vast volume of aerosols and small elements into the atmosphere, ranging in size from micrometers to numerous nanometres. A solitary volcanic eruption can spew up to 30×106 tons of NPs into the atmosphere in ash form [34]. Grass/forest fires have been a portion of the earth's natural history for a long time and are caused mainly by lightning strikes or anthropogenic. Significant fires can have distributed ash and smoke over hundreds of square miles, increasing the amount of particulate matter, including NMs.

3.1.2 Engineered NMs

Anthropogenic actions that contribute to the development of NMs include simple combustion in vehicles, coal for power generation and fuel oil [35], chemical engineering, welding, and airplane engines [36]. Carbon and TiO_2 NPs and hydroxy-apatites are examples of NMs [37] found in various marketable cosmetics, sun-screen, toothpaste, etc. In the city, diesel and automobile exhaust are the leading causes of atmospheric nano- and micro-particles. Anthropogenic activities such as cigarette smoking and building demolition lead to the dissemination of NPs into the environment. Cigarette smoke contains a composite mix of approximately 1 lakh chemical components in NPs varying in size from 10 to 700 nm [38]. Likewise, nano- and micro-particulates smaller than 10 m are released into the atmosphere [39]. Aside from construction remains, glass, repairable fibers, and other poisonous elements from domestic resources are released as nano-sized compounds around the demolition location [39].

3.1.3 Naturally Produced NMs

NMs are found in living organisms like fungi, bacteria, algae, and viruses to plants. The understanding of nanostructures found in microorganisms is critical for future usage of these organisms in agricultural applications. A wide range of NMs derived from natural products have incredible power, light weight, transparency, and bio-compatibility, making them the best products for coatings, packaging, medicine, construction, electronics, filtration, transportation, and other applications. Given the growing concerns about environmental and sustainability, NMs derived from natural sources are gaining traction in scientific and agricultural communities. We summarized some NPs synthesized in tabular form from plants, bacteria, fungi, algae, and viruses (Table 2).

3.1.3.1 By Bacteria

Bacterial strains are broadly used as nano-factories for the production of numerous metallic NPs. It has been demonstrated that both extracellular and intracellular

Table 2 Biosynthesis of metal nanoparticles by various species of plant, bacteria, fungi, algae, and viruses

S. N.	Nanoparticles	Agricultural applications	Reference
1	Ag	Fungicidal activity against various agricultural pathogens	[40, 41]
2	CuO	Antifungal activity against *Colletotrichum gloeoesporioides*	[42]
3	ZnO	Nano-fertilizer reduces arsenic and cadmium content in rice	[43]
4	SnO_2/Pd	Nano-sensor for the detection of fungal volatile organic compounds	[44]
5	Yb_2O_3	Fluorescent sensor for detection of imazapyr herbicide	[45]
6	CuO	Biosensor for detection of *Aspergillus niger*	[46]
7	MnO	Antifungal activity against soil-borne pathogens	[47]
8	Fe/SiO_2	Nano-fertilizer to enhance plant growth and biomass	[48]
9	Ag	Inhibit the growth of *Xanthomonas axonopodis* pv. malvacearum and *Xanthomonas campestris* pv. campestris in vitro	[49]
10	MgO	Controlled bacterial wilt disease caused by *Ralstonia solanacearum*	[50]
11	Pt	Effective against *Colletotrichum acutatu* and *Cladosporium fulvum*	[51]
12	CuO	Antibacterial activity of *Ralstonia solanacearum* on Nicotiana tabacum	[52]
13	CeO_2	Disease suppression caused by *Fusarium oxysporum* on tomato	[53]
14	FeO	Nano-fertilizer for root growth of pea	[54]
15	Fe, Mg, and Zn	Increased yield and essential oil of black cumin	[55]

approaches can be used. Extracellular biosynthetic pathway happens outside of the bacterial cell using a variety of techniques, including (a) use of bacterial biomass, (b) use of bacterial culture supernatant, and (c) use of cell-free extracts. Because it does not require complex downstream processing, extracellular synthesis is favored over intracellular synthesis [56]. These NPs have found use in a variety of fields, including agriculture. Bacteria established the most significant consideration in the area of metallic NPs biosynthesis between many microorganisms. Bacteria have the unusual ability to mobilize and immobilize components, and in some cases, they can precipitate metals as small as nanometres. As a result, bacteria are referred to as bio-factories for manufacturing NMs such as silver, gold, palladium, titanium, magnetite, cadmium, and platinum. Bacterial enzymes are used in this procedure to catalyze a particular breakdown response and start producing NPs [57]. Polysaccharides, vitamins, enzymes, biodegradable polymers, and biological systems can all be used to create NPs. Extracellular secretion enzymes benefit by manufacturing many NMs ranging in size from 100 to 200 nm in pure form, free from other materials. Numerous metal NPs, including gold [58], nonmagnetic oxide [59], and ZnS [60], have been produced by various bacteria strains. The use of bacterial cells in the synthesis of NPs allows for a suitable controller of size [61]. These organisms tolerate heavy metals through various adaptations and decontamination methods and ion efflux by vigorous membrane channels. So many factors, like, alkalinity, temperature, incubation period, and substrate concentration, can influence the rate at which bacterial species synthesize NPs [61].

3.1.3.2 By Fungi

Myco-nanotechnology is a new term that refers to the production of NPs by fungi and their subsequent use. Fungi have several benefits over other microorganisms for NP synthesis, including being comparatively easy to separate, having much simpler downstream processing than bacterial fermentations, secreting huge volumes of extracellular enzymes, having an extensive range and diversity. Fungi produce more extracellular enzymes than bacteria, which has a more significant impact on NM synthesis. As opposed to bacteria, fungi can be used to make more NMs because they secrete more proteins, which subsequently increase the formation of NMs. The catalytic effect of enzymes produced by fungi during metal NP synthesis reduces salts to solid metallic NMs [62]. Fungi are generally regarded as the best source for NMs synthesis compared to other biological systems due to their ease of handling, low cost, and vast diversity.

3.1.3.3 By Plants

Plant-mediated biosynthetic pathway is a simple and low-cost method for producing NPs. Contamination makes it challenging to maintain and preserve a microbial culture. Plants could be used for this purpose to avoid the time-consuming steps of maintaining cell cultures. Plant-mediated biosynthesis is a simplified and appropriate process for making NPs on a large scale without contamination. Green NP synthesis refers to the creation of NPs from plant extract. It is currently gaining popularity due to the single-step involved in biosynthesis. As a result, it is a time-saving process with no toxicants and the availability of natural capping agents [63]. Plant material is widely existing, safe, and contains a wide range of chemical compounds. All these factors make plants preferable to other materials for NP synthesis. When compared to fungi and bacteria, phytochemicals require less time to reduce metal ions. It demonstrates that plant materials are a superior choice for the biosynthesis of NMs than bacteria and fungi. Plants are widely used in the medicinal sector for the synthesis of NPs. The choice of phytoconstituents extracts to synthesize NPs is also influenced by the source or origin of the biological matter. Plant extracts of leaves, stems, latex, roots used in green synthesis of NPs. Parts of the plant like root, stem, fruit, leaf, etc., are broadly used for green synthesis of NPs due to the high levels of phytoconstituents they yield [64]. The nature of leaf extracts and their concentration, temperature, pH, and interaction period have also influenced the rate of production and quantity of the NPs [65].

3.1.3.4 By Algae

Algae are also another significant class of living organisms that can be used in the efficient and environmentally friendly production of NMs. Heavy metals are thought to accumulate in algae, which could be used in the biologically active synthesis of metal NPs. Algae are autotrophic organisms that can thrive with only a few medium supplements. Algae cells contain various secondary metabolites and biologically dynamic composites that act as capping mediators throughout NP synthesis, transforming algal cells into a one-of-a-kind "nano-factory" for synthesizing several NPs [66]. The *Chlorella vulgaris* alga synthesizes NPs of various forms, including decahedral, icosahedral, and tetrahedral [67]. Numerous algae, including *Kappaphycu salvarezii* [68], *Fucus vesiculosus* [69], *Tetraselmisko chinensis* [70], *Chondrus crispus*, and *Spirogyra insignis* [71], were found to synthesize Ag and Au NPs. Algae is widely used for the green synthesis of several metallic and metal oxide NPs since they grow fast, are easy to handle, and their biomass growth rate is ten times faster than higher plants. To date, various algal species are studied for the green synthesis of different NPs.

3.1.3.5 By Viruses

Viruses hold excessive potential for accumulating and connecting nano-sized elements, permitting the development of organized NPs assemblies. Because of their small size, monodispersed nature, and wide range of chemical groups available for alteration, they serve as excellent support for molecular assemblage in nanoscale strategies. Because of their capacity to correlate into preferred structures with various morphologies, virus-based nanomaterials can be used as an engineering component to building smart nano-objects. Viruses are an ideal framework for the formation of nano-conjugates with noble metal NPs. Plant viruses and bacteriophages have recently gained popularity in nano-biotechnology due to their structural and chemical stabilization. The ease of manufacture, absence of toxicity, and pathogenicity in animals or humans [72] also play a key role. Viruses hold potential for gathering and linking nano-sized elements; as nanotechnology advances, such organized assemblies will interact with well-developed technologies such as lithography [73]. Viral NPs can be created by takeout viruses' genetic material and transforming them into "nano-cargoes." The virus's outer capsid protein serves a valuable purpose in synthesizing NPs by providing a susceptible surface that interrelates with metallic ions [74].

3.2 Synthesis

In recent times, there was a surge of attention in the synthesis of ecologically friendly NPs which do not generate hazardous sludge during manufacturing. It could be accomplished *via* biological synthesis methods with biotechnological practices deemed harmless and environmentally for NM production as a substitute for chemical and physical approaches. In the synthesis of NPs *via* the natural scheme, three main steps are followed: selecting a solvent medium, an eco-friendly reducing agent, and safety matter as a capping mediator to stabilize synthesized NPs [75]. Nanotechnology benefits over conventional tactics because of the accessibility of additional compounds by the biological organization for the development of NPs. The biodiversity of biological machinery has been investigated to synthesize eco-friendly NMs that can be used in various agricultural applications. Any NMs synthesis method aims to produce a material with properties that result from their characteristic length scale being in the nanometre range 1–100 nm. There are many synthesis methods reported in the literature, which are divided into two main groups, namely "Top Down" and "Bottom-Up" (Fig. 1).

3.2.1 Synthesis of NMs by Top-Down Method

The top-down synthesis produces NPs by reducing the size of a suitable starting content. Various physical and chemical treatments are used to reduce size. Top-down production approaches present flaws in the product's surface structure, which is a significant restriction since surface chemistry and additional physical characters of NPs are extremely dependent on it [76]. This method primarily employs solid and state handling of resources; it entails breaking down bulk material into minute elements *via* physical procedures like crushing, milling, and grinding. The main challenge of this procedure is the scarcity of surface structure, which

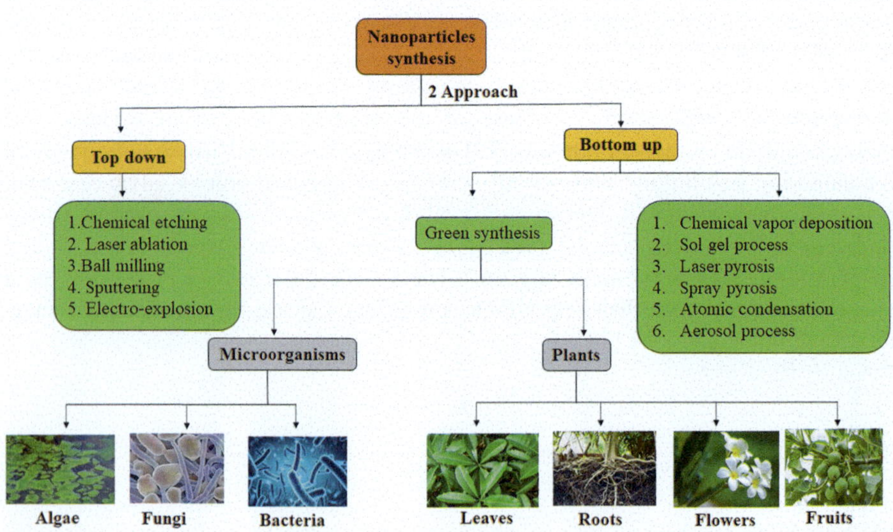

Fig. 1 Metal nanoparticles synthesis *via* the top-down and bottom-up method

influences the physical characteristics and surface chemistry of NMs. Furthermore, processed shapes suffer from significant crystallographic loss as a result of this method. Laser thinning [77], liquid exfoliation through mechanical strength [78], liquid exfoliation by oxidation [79], liquid exfoliation by ion intercalation [80], mechanical cleavage [81], selective etching [82], and ion exchange [83] are examples of top-down approaches.

3.2.2 Synthesis of NMs by Bottom-Up Method

Bottom-up approaches involve creating NPs from smaller components such as molecules and atoms that grow into nanoscopic particles using various chemical and biological methods. Bottom-up synthesis creates NPs from smaller entities, such as assembly atoms, molecules, and smaller elements. Bottom-up synthesis begins with forming nanostructured building blocks for NPs, which accumulated to yield the final part [76]. Raw materials used in these methods can be in the form of liquids, solids, or gases. NMs can be prepared molecule by molecule or atom by atom in this method to produce a large quantity. This method is more commonly used to create the majority of NMs. This method is capable of producing NMs with uniform size, shape, and distribution. It precisely controls the chemical synthesis process to avoid unwanted particle progress. This system is critical in constructing and processing NMs with improved particle dimension supply and morphology and an environmentally friendly and cost-effective approach for producing NPs. Combustion synthesis [84], gas-phase methods [85], hydrothermal synthesis [86], microwave synthesis, and sol-gel processing are just a few of the methods used to create NMs.

4 Nano-based Essential Metals

Metal NPs can be designed and synthesized through diverse functional groups, like DNA, antibodies, peptides, RNA, and prospective biocompatible polymers, such as polyethylene glycol [87]. This metal group includes NPs made of Zn, Cu, Fe, Mn, and their oxides. Zn and ZnO NPs are derived and used in many agricultural practices.

4.1 Zinc Based

Zinc is a core element of many enzymes, including alcohol dehydrogenase RNA polymerase, superoxide dismutase, and carbonic anhydrase. It also helps in chlorophyll synthesis. Zn NPs have been used as a nano-fertilizer on various crops, with positive results in optimal concentrations. Zinc NPs are metal particles that are spherical and have a large surface area. ZnO NP is also visible in agricultural sprayers as an ultraviolet ray's safeguard material [88]. ZnO contributes to protecting photosensitive pesticides in conjunction with an organic filter and is used straight for crop protection against crop protection UV radiation [89]. ZnO NPs revealed positive impacts on germination, phosphorus uptake and mobilizing enzymes, stem and root growth, and showed antifungal activities. Several studies reported that nano-Zn prevents bacterial infection [90], fungal infections [91, 92], and nematode infection [93]. Several laboratories have investigated the antagonistic activity of Zn NP against plant pathogens in the same way that Ag and Cu have [90, 92, 94].

4.2 Copper Based

Cu has long been known to inhibit fungi spore germination, but a large amount of copper is required to achieve this effect. Cu is a constituent of several plant enzymes and is also needed for plant development. According to [95] Cu NPs have antibacterial action against gram-positive and gram harmful bacteria, and are also used as a fungicide. Some researchers [96] investigated the antifungal efficacy of a Cu polymer nanocomposite against phytopathogens. Because of Cu's well-known antimicrobial properties and long record of controlling diseases in plants, nano-Cu is a rational option for plant protection. Compared to the product with cupric hydroxide, Cu NPs increase efficacy against pathogenic fungus [88]. CuO NPs were found to increase ROS (Reaction Oxygen Species) production in plants [97]. Instead, various antioxidant substances have improved in plants when treated with NP, representing that plants' protective mechanisms have been activated [97]. CuO NPs were found to reduce photosynthetic action by neutralizing PS II reaction centers [98].

4.3 Carbon Based

According to some studies, carbon-based NMs are excellent components for enhanced plant yield quality as fertilizers and products for plant protection such as pesticides and herbicides. Their connection and impacts, however, will be determined by the characteristics of the plant and NM. Carbon-based NM can boost ROS generation [99], and they can pass through different types of cells depending on their size. Carbon is an essential component of lipids, proteins, and carbohydrates. Plants use CO_2 to make food and O_2 by transforming sunlight through photosynthetic activity. A variety of carbon-based NPs, namely, single-walled carbon nanotubes (SWCNT), double-walled carbon nanotubes (DWCNT), and multi-walled carbon nanotubes (MWCNT), and fullerenes) have been tested in precision agriculture and were found to be effective in seed germination and plant growth. Carbon-based NPs have both advantages and disadvantages in terms of seed and seedling germination

[100]. According to [101], the use of SWCNT can improve the germination of barley, rice, tomato, soybean, maize, and tobacco.

Furthermore, the usage of MWCNTs aided in the uptake of water by tomato seeds after rapid germination [102]. Again, some researchers [103] found that MWCNT can trigger a stress-related gene in tomatoes, causing improved seed germination. Carbon NMs have recently been shown to have antibacterial and antifungal activities and positive effects on plant growth [104].

4.4 Manganese Based

Mn is considered as a micronutrient required for plant growth. It is essential for both direct and indirect oxygenic photosynthesis. Plant nutritional disorders are the most severe consequences of manganese deficiency. [105] also demonstrated that MgO NPs efficiently inhibited *R. solanacearum*, which caused bacterial wilt in tobacco. Physical damage to cell membranes and ROS accumulation were proposed as mechanisms. [106] synthesized magnesium hydroxide NPs and compared their antimicrobial effects to marketable pesticides such as Kocide 3000, which contains copper hydroxide. *Pseudomonas syringae*, *Xanthomonas alfalfae*, and *E. coli* are all inhibited by magnesium hydroxide NPs. [107] created MgO NPs and investigated their antibacterial properties. On tomatoes, researchers confirmed that MgO NPs increased systemic resistance against *Ralstonia solanacearum*. They also discovered that MgO-treated roots generate ROS rapidly, upregulation of PR1, ethylene, jasmonic acid, and systemic resistance-associated genes [50].

4.5 Titanium Based

Photochemically active TiO_2 NPs have antimicrobial activities; they have agricultural significance as nano-pesticides. TiO_2 NPs also revealed an excellent correlation with plant enzymatic activity, promoting crop growth when exposed to sunlight; they enhance photosynthetic action. [108] discovered that photocatalytic TiO_2 NPs have antibacterial activity against *Xanthomonas perforans* (a pathogen that causes spot disease in tomato). Nano-photocatalytic TiO_2 actions may contribute to its antifungal action. [108] synthesized TiO_2/Zn NPs to prevent *Xanthomonas* sp. that cause bacterial leaf spot on rose. Researchers [109] reported that nano-TiO_2 showed high antifungal activity. TiO_2 NP exposure improved chlorophyll content and biomass by activating antioxidant enzyme, after lower hydrogen peroxide and malondialdehyde levels, increased generation of soluble sugars and proline, thereby sustaining osmotic balance [110]. TiO_2 can increase plant hydration by enhancing the action of the nitrate reductase (NR) enzyme, which increases osmolyte gathering. Increased NR enzyme activity results in the production of nitric oxide, which induces the synthesis of glycine betaine and proline [111]. TiO_2 NPs had enzymatic as well as nonenzymatic stress defense schemes in plants. Another study on medicinal herbs [112] discovered that providing salicylic acid, methyl jasmonate, and TiO_2 NPs alleviated drought stress. The foliar application of TiO_2 NP resulted in improved plant growth, increased fruit yield, and chlorophyll content in tomatoes [113].

4.6 Silver Based

Because of their historically known antimicrobial action, Ag NPs were investigated for managing plant diseases. Ag NPs have received much consideration as a potential nano-pesticide in agriculture due to their broad spectral range of antibacterial efficacy. Furthermore, Ag NPs are found to be effective against nematodes, a common soil-borne pathogen. Several metabolites found in plants or bacteria act as reducing and capping mediators during the fabrication of Ag NPs. [114] green-synthesized Ag NPs from turnip leaf extract demonstrated antifungal potential against some wood-degrading fungi, including *Chaetomium globosum, Phanerochaete sordida, Gloeophyllum abietinum,* and *G. trabeum*. Plant pathogens like *F. culmorum, Phythium ultimum, R. solani, Biploaris sorokiniana, B. cinerea, Colletotrichum gloeosporioides, Magnaporthe grisea, Scalerotinia sclerotiorum,* and *Magnaporthe grisea* are reported to be controlled by Ag NPs [115, 116]. [115] studied the impact of biologically synthesized Ag NPs against *Candida albicans, Trichoderma* sp., and *Phoma glomerata*. Plant growth was enhanced using Ag NP of 200–800 nm size [117], whereas Ag NP of 35–40 nm positively influenced several crops' growth [117]. According to recent research, when Ag NP is combined with different composites, diverse influence on plants is reported [118]. Ag NP has also been evaluated as fungicidal activity, and their significant impact was observed [119]. However, their usage in agriculture is still debatable because Ag NP is known to release silver ions as it ages, and they can also affect biomass accumulation [120].

4.7 Silicon Based

Silicon (Si) is the second most common element in the earth after oxygen and is regarded as a nonessential component for plants; if present, plants can benefit adequately. Si NPs interact with plants directly or indirectly, causing morphological and physiological alterations that deliver stress tolerance. They promote the growth of plants, increase biomass, physiology, and anatomy, alter tissue diversity, stimulate defense schemes, and aid in acclimatization to environmental stresses [121]. Si NPs exhibited anti-stress properties against drought stress in *Crataegus* sp. at several concentrations; the diverse responses in seedlings depend upon concentrations. Among these effects are increased photosynthetic capacity, membrane electrolyte leakage, water content, more levels of proline, carotenoids, and chlorophylls [122]. Furthermore, under salinity stress, SiO_2 NPs increased water use efficiency, transpiration proportion, stomatal conductance, and decreased chlorophyll degradation [123]. Silica [124] also helped the expansion of diseased plants' stress resistance ability.

4.8 Other Metal Based

Iron (Fe) NP represents an emerging generation of ecological remediation machinery that can deliver cost-effective resolutions to some problematic matters. Iron oxide (Fe_2O_3) NPs could be used in place of Fe fertilizers in agriculture. [125] investigated efficacy of Fe_2O_3 NPs as fertilizer to replace traditional Fe fertilizers. In another study, Fe_2O_3 NPs after foliar spray had a significant impact on the yield of *Vigna unguiculata*, Fe content in leaf, stability of plasma membrane, and chlorophyll content [126]. According to [127], Fe NP for wheat seed treatment can enhance germination frequency and shoot and root length. Lower concentrations of Fe_2O_3 NP have helped plants and improved germination [128]. Some researchers [129] observed an increase in germination and root length percentage in rice seeds treated with silica and molybdenum (Mo) NP. Researchers [53] recently designed a greenhouse assay to examine cerium oxide (CeO_2) NP-mediated Fusarium wilt suppression in tomatoes. Experiments in the field and on the soil with wheat and rice revealed that the use of CeO_2 NP reduced grain quality [130]. Molybdenum (Mo) NP is a significant component of the nitrogen fixation scheme in plants. Researchers [131] found that applying Mo NP to chickpea seedlings increased growth by two 2-3 times. According to researchers [132], gold (Au) NP used for seed treatment of maize positively affects germination and increases seed quality parameters. [133] produced nano-Au from the extract of lady's finger and demonstrated its antifungal activity against *Candida albicans, Aspergillus niger, A. flavus,* and *Puccinia graminis*. In vitro data presented by [134] showed that nano-sulfur inhibits *Venturia inaequalis* and *Fusarium solani*. Without affecting photosynthetic activity at low concentrations, cadmium oxide (CdO) NPs enhanced amino acid production [135].

5 Mechanism of Nanomaterial Uptake, Translocation, and Action

Plants are exposed to NPs through two routes: foliar and root exposure. The cuticle is thought to be primarily a physical border against NPs entering, since the waxy cuticle protects the leaves of higher plants from water loss and uncontrolled exchange of other solutes [136]. The bioavailability and toxicity of NPs are determined by a series of bio/geo-transformations in soil. Subsequently interacting with plant roots, NPs translocate to the aerial side and collect in cellular or subcellular organelles. The initial step in accumulation is the adsorption of NPs from the soil through roots. The size of NP is directly related to its absorption as it is an important factor that permits it to enter *via* cell wall pore spaces or stomata. Small NPs have been observed to pierce plant roots *via* capillary forces, osmotic pressure, or directly by root epidermal cells [137]. Epidermal cells of the root are semipermeable and comprise minor pores, limiting the passage of large NPs. Nano-pores aid foliar entry in leaves, which facilitate NP uptake and transport within leaves. Aquaporins have been proposed as NMs transporters inside the cell [138]; however, their minor pore diameter, varying between 2.8 and 3.4 A° [139], marks them unlikely as NP penetration frequencies [140]. NMs can also enter cells *via* plasmodesmata, which are particular structures transporting materials between cells [141]. The uptake of NP by plants is influenced by numerous factors related to NP nature, plant physiology, and NM interaction with the environment. The properties of NP will significantly impact their behavior and, as a result, whether or not plants will absorb them. Size appears to be among the most significant barriers to penetration into plant tissues. Some reports have been on maximum dimensions that allow NP to move and accumulate inside cells, with 40–50 nm as a size exclusion limit [142]. Furthermore, the type of NP and its chemical composition impact uptake [138], and morphology is determinant in several conditions [143]. The functionalization and coating of NM surfaces can substantially modify and affect NM absorption and accumulation properties by plants [144]. Furthermore, the occurrence of other organisms like fungi and bacteria affects the uptake of NPs by plants, particularly if those microorganisms form symbiotic relationships with plants, as mycorrhizal fungi do [145]. Prospective strategies must be developed for tracking NMs inside plants; additionally, more critical data is required to measure uptake and translocation of NPs within plants and as discharge in the environment. The uptake and dispersal of TiO2 NPs studied in rice plants and found that NPs transported long distances *via* the vascular scheme. Photosynthate, sugars, and macromolecules have conventionally been able to transport downward to shoots and roots by phloem system [146]. Overall, long-distance liquid transportation in higher plants happens *via* the vascular system, consisting of the xylem and phloem. Flow direction in the xylem system is from bottom to top (from root to shoot), whereas flow direction in the phloem system is from top to bottom (from shoot to root). The whole plant's vascular scheme is noncirculatory, representing substances moving downward in phloem that do not return to their original locations *via* xylem [146]. Once inside the plant, NPs can move through two types of pathways: apoplast and symplast. Apoplastic transportation occurs in the outer plasma membrane *via* extracellular places, cell walls of neighboring cells, and xylem vessels [147]. In contrast, symplastic transport occurs within the cytoplasm of adjacent cells *via* specific structures known as plasmodesmata [141] and sieve plates. The apoplastic process is essential for radial mobility within plant tissues because it enables NMs to attain the central cylinder of root and vascular tissues. They can move upwards to the aerial part [148]. NPs can move through the xylem to the aerial part of the central cylinder by subsequent transpiration stream [148, 149]. Significant symplastic transport is also possible, utilizing sieve tube components in phloem and permitting dispersion to nonphotosynthetic tissues and organs [143]. In the scenario of foliar spray, NMs must pass through the cuticle barrier, either *via* lipophilic or hydrophilic way [150, 151]. Since the diameter of cuticular pores is approximated to be about 2 nm [151], the stomatal pathway has seemed to be the best possible path for NPs penetration, with the size limit of 10 nm or greater [152]. The movement of NMs within plants is critical because it can indicate which plant parts they could attain and wherever they may accumulate. For instance, NPs transported primarily *via* xylem rather than phloem, most possibly move from root to shoot and leaves rather than downwards, so applied to roots to ensure good distribution in the plant. On the opposite, foliar spraying must be used if NPs show significant translocation by phloem. Furthermore, NMs trying to move along phloem are likely to acquire plant organs that act as sinks. Though translocation is not always limited to a single cell type, lateral movement of NMs among xylem and phloem is possible. The features and nature of NMs and plant types had an important effect on translocation and gathering in plant tissues. For instance, variances in translocation and accumulation of the same NP were observed in diverse plant species, so even small alterations in analogous NMs lead to altered outcomes within the same plant [153].

Nanoscience is a new scientific innovation platform that entails the progress of strategies to various low-cost applications and is helpful to improve the growth and development of plants. In this regard, numerous research shows that the use of NMs had a positive effect on germination and plant growth. Still, fundamental mechanisms by which NMs can stimulate germination remain unknown. The application of nano-SiO_2 and -TiO_2 encourages the germination of seeds [154]. Research has shown that NMs can pierce seed coats and improve water absorption and consumption, regulate enzymatic scheme as a result, and increase germination and seedling growth [155]. NMs like ZnO, TiO_2, MWCNTs, FeO, Zn, Fe, CuO, and hydroxy-fullerenes also shown to enhance the growth and development of crops while improving crop quality in a variety of crops such as mustard, peanut, potato, tomato, spinach, onion, wheat, soybean, and mung bean [156, 157]. Although the exact mechanism underlying the improvement of plant growth is unknown, it might be clarified that NMs absorb more nutrients and water, which supports the vigor of root systems through improved enzymatic action [156]. TiO_2 NMs promote plant growth by enhancing photosynthesis and nitrogen (N_2) metabolism [158]. Plant contact with NMs caused excitation of genes associated with water channel protein and for better cell growth by regulating cell cycle; these impacts of NMs are reflected in the form of enhanced seed germination and plant growth [159]. Treated plants with NMs are more resistant to abiotic stresses, and these treated plants have higher photosynthesis rate, transpiration activity, water use efficiency, chlorophyll (Chl), proline content, stomatal conductance, and high concentrations of carbonic anhydrase action [123, 160]. NPs could mitigate damaging photosynthesis, which is caused by *Ultraviolet*-B (UV-B) radiations. NPs also improve photosynthesis by preventing oxidative stress, increasing Chl synthesis, Rubisco activity, energy transformation, and light absorbance [161, 162]. Plants are protected against different abiotic

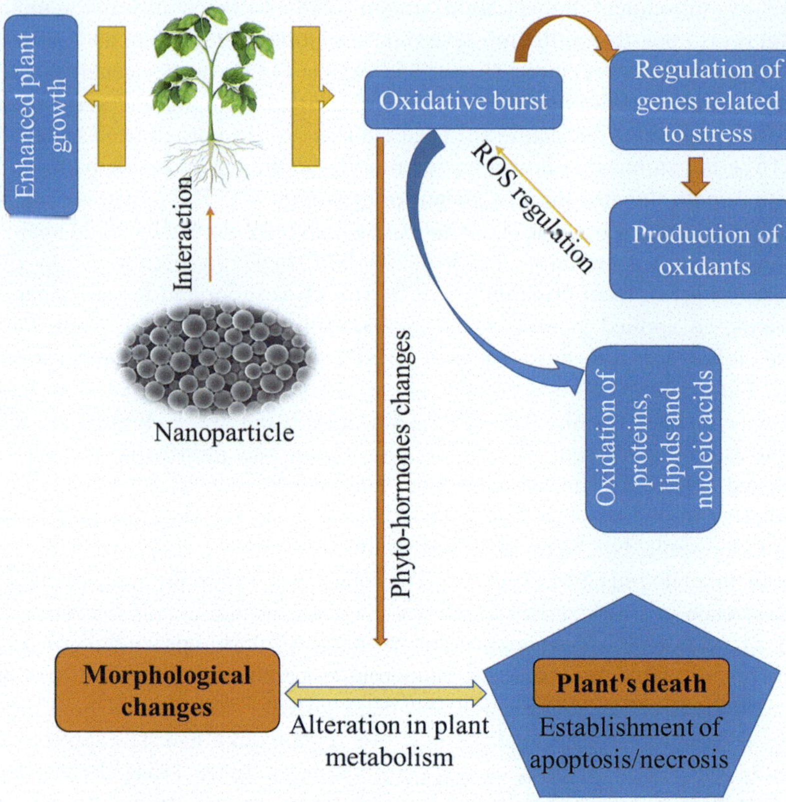

Fig. 2 Interaction, translocation, and mechanism induced by nanoparticles in plants

stresses by NMs, which stimulate antioxidant enzymes' actions and gather free amino acids, nutrients, and osmolytes. Mesoporous Si NPs enhance photosynthesis by interacting with chloroplasts, resultant in enhanced chlorophyll content, total protein, and seed germination. Interaction, translocation, and mechanism induced by NPs in plants are given in (Fig. 2).

6 Nanomaterials Interaction and Physiochemical Response of Plants

With the development of nanotechnology, more effective and contaminant-free nano-formulations for sustainable farming are being developed regularly. The uptake of NMs within plants is heavily influenced by the chemical properties, size and functional groups, and coating type. Interaction and uptake of NMs cause molecular deviations that affect plant physiology [163]. Adsorption on the root surface, integration into the cell wall, and cell uptake are all potential interactions of NPs with plant roots [164]. Furthermore, knowledge of the interaction of NMs with plants, whether negative or positive, is mandatory for the controlled delivery of bioactive substances. The potential of NMs to pierce tough coating of seeds and permit water import determines increased growth and vigor. The NPs transported to various parts of the plant and interact with cellular mechanisms, stimulating the growth of plants. NMs can essentially be applied to either root or vegetable parts of plants, preferentially leaves. NPs can be passively taken up at shooting surfaces *via* natural plant openings like stomata and hydathodes. [165]. NPs availability may be influenced by symbiotic relationships between organisms, soil organic matter, and mucilage and exudates. To better understand the dynamics of NPs-plant connections, plant anatomical, and physiological characteristics must be measured. Damages and injuries in plants' aerial and hypogeal parts also serve as feasible for NPs internalization [166]. In the root, rhizodermis lateral root may allow easy entry of NMs, particularly close root tip, whereas the upper portions are impermeable due to the presence of suberin [167]. Root mucilage and exudates, e.g., which are generally excreted in the rhizosphere, play two roles: firstly, they promote NP adhesion to the root's surface and may enhance NP internalization proportion. Secondly, these jelly components also stimulate NP absorption and accumulation [168]. The rate of NPs accumulation by roots of plant is influenced by NPs' properties and ecological factors.

NPs gain entry plants through a variety of ways, most common of which are roots and leaves. Different NPs have been shown to encourage the germination of seeds, development, and growth [169]. The mechanism by which NPs application increase germination of seeds remains unknown. Treatments with NPs increase seed absorption and moisture holding, which enhance the germination of seeds [170]. NPs have been shown to benefit crop plants in the following ways: improve metabolites compounds [171], enhanced root and shoot measurement [172], increased production of fruits, and significantly increased seedlings and vegetative biomass of several crops. Likewise, the impact of NPs on various biochemical parameters such as improved N_2 efficacy and enhanced photosynthetic activity in some chief crops, including soybeans [173], peanuts [174]. NPs are also well-recognized for increasing nutrient consumption and resistance to plants against several diseases and abiotic stresses [175]. NPs can influence plant growth and development by altering a few physiological processes in plants. Numerous studies show that foliar application of metal NPs significantly increases chlorophyll content, allowing them to absorb extra light energy and improve photosynthesis. SiO_2 NP treatments significantly increased photosynthesis rate due to higher action of carbonic anhydrase and photosynthetic pigment synthesis [176]. Many research indicates that NPs caused toxicity above specific concentrations, and plant toxicity evaluated their effect on germination percentage and biomass production [177]. Zn has been used as a cofactor in some enzymes, including catalase & superoxide dismutase, and protects plant cells from oxidative damage [143]. We have summarized some physiological responses of various metal nanoparticles on different host crops in tabular form (Table 3).

Table 3: Physiological response of various metal nanoparticles on different host crops.

S.N.	Nanoparticles	Source of organisms	Reference
By plants			
1	Au	*Pimenta dioica*	[178]
2	ZnO	*Acalypha fruticosa*	[179]
3	Cu	*Orobanche aegyptiaca*	[180]
4	Pt	*Nigella sativa*	[181]
5	TiO_2	Lemon peel extract	[182]
6	Fe_2O_3	*Medicago sativa*	[183]
7	Zn	*Lycopersicum esculentum*	[184]
By bacteria			
1	Ag	*Bacillus subtilis*	[185]
2	Au	*Staphylococcus epidermidis*	[186]
3	ZnO	*Bacillus subtillis*	[187]
4	Pt and Pd	*Desulfovibrio vulgaris*	[188]
5	Pd	*Pseudomonas putida*	[189]
6	Mg	*Magnetotactic bacteria*	[190]
By fungi			
1	Fe_2O_3	*Aspergillus niger*	[191]
2	Ag	*Aspergillus terreus*	[192]
3	Cu	*Aspergillus niger*	[193]
4	ZnO	*Aspergillus terreus*	[194]
5	TiO_2	*Aspergillus flavus*	[195]
6	Fe_3O_4	Verticillium sp.	[196]
By algae			
1	Ag, Au	*Turbinaria conoides*	[197]
2	ZnO	*Sargassum muticum*	[198]
4	Fe	Chlorococcum sp.	[199]
5	CuO	*Bifurcaria bifurcata*	[200]
6	Fe_3O_4	*Sargassum muticum*	[201]
By virus			
1	SiO_2, CdS, PbS, Fe_2O_3	Tobacco mosaic virus	[202]
4	TiO_2	M13 virus	[203]
5	Nano-carriers	Potato virus X	[204]
6	Nano-assemblies	Cucumber mosaic virus	[205]

7 Toxicological Impact and Health Hazards in Agriculture

Diverse research has also found that the use of metals and metal oxides in the synthesis of nano-fertilizers and nano-pesticides had adverse and toxic effects on plants and the environment. Metal, metal oxide, and synthetic polymers have been discovered to be nonbiocompatible, nondegradable, and harmful at various concentrations, raising serious concerns about the utilization of nanotechnology in agriculture. NMs persist in the environment, and their concentration rises as a result of their nano-size, according to researchers. Excess amounts of NMs are toxic to people, nontargeted lifeforms, and affect climate. NPs accumulation in plants can modify physiological activities. In specific scenarios, reductions in photosystem quantum yield and transpiration were also detected [206]. According to a series of studies, NPs can affect crops by minimizing germination of seeds, reducing shoot and root length, varying photosynthesis, inducing oxidative stress, antioxidants, and balancing the nutritional substance of eatable crops and yield [206, 207]. Deposition of NPs in plant tissues may also harm protein, lipid, and nucleic acid content *via* hydroxyl radicals [208]. NMs promote plant growth and productivity while protecting against biotic and abiotic stresses.

On the other hand, NMs cause cytotoxicity and genotoxicity in plants [209]. It prominently diminished biochemical and physiological activities [210], growth [211], and compact nutritive worth of crops [212]. Toxic effects of NMs on plants are primarily determined by the size, concentration, and chemistry of the NMs and the chemical properties of subcellular where NMs deposited [213]. NPs larger than cell wall pore size stick to epithelial root cells caused mechanical damages of cells [214], blocking pores, and reducing hydraulic conductivity, resulting in decreased water uptake and nutrient acquisition capability [215]. The direct interaction of NPs with the cell surface and cellular membranes induces mechanical interruption and impacts the integrity and role of the cell membrane and walls [216]. NP accumulation leads to the decline of the PSII reaction center, modification of O_2 evolving complex, downregulation of electron transport and chlorophyll composition [217], a smaller proportion of thylakoids, lower transpiration, stomatal conductance, CO_2 absorption, and photosynthetic pigments [98]. Understanding NM toxicity in crops is still in its early stages but critical for developing innovative nanotools and functions. With the rapid evolution of nanotechnology, there is apprehension about the accumulation of NMs and their potential entrance into the food chain [218]. Conventional foods have various NMs, but the use of many engineered NMs in water, agriculture, and food may pose hazards for human service, usage, the atmosphere, or all of them. Furthermore, a category of NPs are found to be toxic to plants by retarding germination and root elongation. Phytotoxicity of NPs is connected with the discharge of lethal elements from NPs, generating radicals via NPs interaction with plant or environment.

8 Concluding Remarks and Future Directions

Considering the significant challenges we will be facing, mainly due to a growing worldwide population and climate change, the application of NMs in agriculture can potentially contribute as a unique carrier in agricultural practices. According to data gathered, the influence of NPs differs from plant to plant and is dependent on the mechanism of application, size, morphology, and concentrations. Nanotechnology seems to have the ability to change pest management technologies and provide solutions for agricultural applications. A piece of complete knowledge about properties of NMs like morphology, functional groups, and size serves as a beneficial preliminary fact for selecting appropriate NMs. Agricultural nanotechnology is encouraging options for the emerging quality output, which are being discovered. A few specific areas of agricultural nanotechnology research may require additional attention shortly: (1) New eco-friendly and reliable delivery methods for specific food/feed substances, plant nutrients, and so on, (2) Nanotechnology-related (bio)-sensors play a vital role in controlling pests and agricultural food products, (3) The characterizations of NMs should be closely reviewed, and (4) Nano-toxicity is critical with fertilizers; ideal dose estimation should also be inspected.

References

1. Tarafdar, J. C., Sharma, S., & Raliya, R. (2013). Nanotechnology, interdisciplinary science of applications. *African Journal of Biotechnology, 12*, 219–226.
2. Dethier, J. J., & Effenberger, A. (2011). Agriculture and development: A brief review of the literature. In *World Bank policy research working paper N° 5553*. World Bank.
3. Solanki, P., Bhargava, A., Chhipa, H., Jain, N., & Panwar, J. (2015). Nano-fertilizers and their smart delivery system. In M. Rai (Ed.), *Nanotechnologies in food and agriculture* (pp. 81–101). Springer.
4. Usman, M., Farooq, M., Wakeel, A., Nawaz, A., Cheema, S. A., Rehman, H., Ashraf, I., & Sanaullah, M. (2020). Nanotechnology in agriculture: Current status, challenges and future opportunities. *Science of the Total Environment, 721*, 137778.
5. Ditta, A., Arshad, M., & Ibrahim, M. (2015). Nanoparticles in sustainable agricultural crop production, applications and perspectives. In *Nanotechnology and plant sciences* (pp. 55–75). Springer International Publishing.
6. Liu, R., & Lal, R. (2015). Potentials of engineered nanoparticles as fertilizers for increasing agronomic productions. *Science of the Total Environment, 514*, 131–139.
7. Dutschk, V., Karapantsios, T., Liggieri, L., McMillan, N., Miller, R., & Starov, V. M. (2014). Smart and green interfaces: From single bubbles/drops to industrial environmental and biomedical applications. *Advances in Colloid and Interface Science, 209*, 109–126.
8. Chhipa, H. (2017). Nanofertilizers and nanopesticides for agriculture. *Environmental Chemistry Letters, 15*(1), 15–22.
9. Dimkpa, C. O., & Bindraban, P. S. (2017). Nanofertilizers: New products for the industry? *Journal of Agricultural and Food Chemistry, 66*, 6462–6473.
10. Rameshaiah, G. N., Pallavi, J., & Shabnam, S. (2015). Nano fertilizers and nano sensors–an attempt for developing smart agriculture. *International Journal of Engineering Research and General Science, 3*, 314–320.
11. Verma, S. K., Das, A. K., Patel, M. K., Shah, A., Kumar, V., & Gantait, S. (2018). Engineered nanomaterials for plant growth and development: A perspective analysis. *Science of the Total Environment, 630*, 1413–1435.
12. Kuzma, J. (2007). Moving forward responsibly: Oversight for the nanotechnology-biology interface. *Journal of Nanoparticle Research, 9*, 165–182.
13. Day, W. (2005). Engineering precision into variable biological systems. *Annals of Applied Biology, 146*, 155–162.
14. Valenstein, J. S., Lin, V. S. Y., Lyznik, L. A., Martin-Ortigosa, S., Wang, K., Peterson, D. J., et al. (2013). Mesoporous silica nanoparticle-mediated intracellular Cre protein delivery for maize genome editing via loxP site excision. *Plant Physiology, 164*, 537–547.
15. Manjunatha, S. B., Biradar, D. P., & Aladakatti, Y. R. (2016). Nanotechnology and its applications in agriculture: A review. *Journal of Farm Sciences, 29*(1), 1–13.

16. Chaudhry, N., Dwivedi, S., Chaudhry, V., Singh, A., Saquib, Q., Azam, A., et al. (2018). Bio-inspired nanomaterials in agriculture and food: Current status, foreseen applications and challenges. *Microbial Pathogenesis, 123*, 196–200.
17. Rai, V., Acharya, S., & Dey, N. (2012). Implications of nanobiosensors in agriculture. *Journal of Biomaterials and Nanobiotechnology, 03*, 315–324.
18. Duhan, J. S., Kumar, R., Kumar, N., Kaur, P., Nehra, K., & Duhan, S. (2017). Nanotechnology: The new perspective in precision agriculture. *Biotechnology Reports, 15*, 11–23.
19. Yao, K. S., Li, S. J., Tzeng, K. C., Cheng, T. C., Chang, C. Y., Chiu, C. Y., et al. (2009). Fluorescence silica nanoprobe as a biomarker for rapid detection of plant pathogens. *Advanced Materials Research, 79–82*, 513–516.
20. Faraji, J., & Sepehri, A. (2020). Exogenous nitric oxide improves the protective effects of TiO_2 nanoparticles on growth, antioxidant system, and photosynthetic performance of wheat seedlings under drought stress. *The Journal of Soil Science and Plant Nutrition, 20*, 703–714.
21. Elsheery, N. I., Helaly, M. N., El-Hoseiny, H. M., & Alam-Eldein, S. M. (2020). Zinc oxide and silicone nanoparticles to improve the resistance mechanism and annual productivity of salt-stressed mango trees. *Agronomy, 10*, 558.
22. Mahmoud, L. M., Dutt, M., Shalan, A. M., El-Kady, M. E., et al. (2020). Silicon nanoparticles mitigate oxidative stress of in vitroderived banana (*Musa acuminata* 'Grand Nain') under simulated water deficit or salinity stress. *South African Journal of Botany, 132*, 155–163.
23. Adrees, M., Khan, Z. S., Ali, S., Hafeez, M., Khalid, S., Ur Rehman, M. Z., et al. (2020). Simultaneous mitigation of cadmium and drought stress in wheat by soil application of iron nanoparticles. *Chemosphere, 238*, 124681.
24. Van-Nguyen, D., Nguyen, H. M., Le, N. T., Nguyen, K. H., Le, H. M., et al. (2021). Copper nanoparticle application enhances plant growth and grain yield in maize under drought stress conditions. *Journal of Plant Growth Regulation, 41*, 364–375. https://doi.org/10.1007/s00344-021-10301-w
25. An, J., Hu, P., Li, F., Wu, H., Shen, Y., White, J. C., et al. (2020). Emerging investigator series: Molecular mechanisms of plant salinity stress tolerance improvement by seed priming with cerium oxide nanoparticles. *Environmental Science Nano, 7*, 2214–2228.
26. Wang, Y., Lin, Y., Xu, Y., Yin, Y., Guo, H., & Du, W. (2019). Divergence in response of lettuce (var. *ramosa* Hort.) to copper oxide nanoparticles/microparticles as potential agricultural fertilizer. *Environmental Pollutants and Bioavailability, 31*, 80–84.
27. Mohamed, A. K. S., Qayyum, M. F., Abdel-Hadi, A. M., Rehman, R. A., Ali, S., & Rizwan, M. (2017). Interactive effect of salinity and silver nanoparticles on photosynthetic and biochemical parameters of wheat. *Archives of Agronomy and Soil Science, 63*, 1736–1747.
28. Venkatachalam, P., Priyanka, N., Manikandan, K., Ganeshbabu, I., et al. (2017). Enhanced plant growth promoting role of phycomolecules coated zinc oxide nanoparticles with P supplementation in cotton (*Gossypium hirsutum* L.). *Plant Physiology and Biochemistry, 110*, 118–127.
29. Arora, S., Sharma, P., Kumar, S., Nayan, R., Khanna, P. K., & Zaidi, M. G. H. (2012). Gold nanoparticle induced enhancement in growth and seed yield of *Brassica juncea*. *Plant Growth Regulation, 66*(3), 303–310.
30. Begum, P., & Fugetsu, B. (2012). Phytotoxicity of multi-walled carbon nanotubes on red spinach (*Amaranthus tricolor* L.) and the role of ascorbic acid as an antioxidant. *The Journal of Hazardous Materials, 243*, 212–222.
31. Raliya, R., Tarafdar, J., Singh, S., Gautam, R., Choudhary, K., Maurino, V. G., & Saharan, V. (2014). MgO nanoparticles biosynthesis and its effect on chlorophyll contents in the leaves of clusterbean (*Cyamopsis tetragonoloba* L.). *Advanced Science, Engineering and Medicine, 6*, 538–545.
32. Wang, W. N., Tarafdar, J. C., & Biswas, P. (2013). Nanoparticle synthesis and delivery by an aerosol route for watermelon plant foliar uptake. *Journal of Nanoparticle Research, 15*, 1417.

33. Sun, D., Hussain, H. I., Yi, Z., Rookes, J. E., Kong, L., & Cahill, D. M. (2016). Mesoporous silica nanoparticles enhance seedling growth and photosynthesis in wheat and lupin. *Chemosphere, 152*, 81–91.
34. Taylor, D. A. (2002). Dust in the wind. *Environmental Health Perspectives, 110*, A80.
35. Linak, W. P., Miller, C. A., & Wendt, J. O. L. J. (2000). Comparison of particle size distributions and elemental partitioning from the combustion of pulverized coal and residual fuel oil air waste manage. *Journal of the Air & Waste Management Association, 50*, 1532–1544.
36. Rogers, F., Arnott, P., Zielinska, B., Sagebiel, J., Kelly, K. E., Wagner, D., Lighty, J. S., & Sarofim, A. F. J. (2005). Real-time measurements of jet aircraft engine exhaust. *Air & Waste Management Association, 55*, 583–593.
37. Sadat-Shojai, M., Atai, M., Nodehi, A., & Khanlar, L. N. (2010). Hydroxyapatite nanorods as novel fillers for improving the properties of dental adhesives: Synthesis and application. *Dental Materials, 26*, 471–482.
38. Cao, G. (2004). *Synthesis, properties and applications*. World Scientific.
39. Stefani, D., Wardman, D., & Lambert, T. J. (2005). The implosion of the Calgary General Hospital: Ambient air quality issues. *Air & Waste Management Association, 55*, 52–59.
40. Paulkumar, K., Gnanajobitha, G., Vanaja, M., Rajeshkumar, S., Malarkodi, C., Pandian, K., & Annadurai, G. (2014). *Piper nigrum* leaf and stem assisted green synthesis of silver nanoparticles and evaluation of its antibacterial activity against agricultural plant pathogens. *The Scientific World Journal, 2014*, 829894.
41. Mishra, S., & Singh, H. B. (2015). Biosynthesized silver nanoparticles as a nano weapon against phytopathogens: Exploring their scope and potential in agriculture. *Applied Microbiology and Biotechnology, 99*, 1097–1107.
42. Oussou-Azo, A., Nakama, T., Nakamura, M., Futagami, T., & Vestergaard, M. (2020). Antifungal potential of nanostructured crystalline copper and its oxide forms. *Nanomaterials (Basel), 10*, 1003.
43. Ma, X., Sharifan, H., Dou, F., & Sun, W. (2020). Simultaneous reduction of arsenic (As) and cadmium (Cd) accumulation in rice by zinc oxide nanoparticles. *Chemical Engineering Journal, 384*, 123802.
44. Shokrzadeh, L., Mohammadi, P., Mahmoudian, M. R., Basirun, W. J., & Bahreini, M. (2020). L-glycine-assisted synthesis of SnO_2/Pd nanoparticles and their application in detection of biodeteriorating fungi. *Materials Chemistry and Physics, 240*, 122172.
45. Kumar, S., Sachdeva, S., Chaudhary, S., & Chaudhary, G. R. (2020). Assessing the potential application of bio-compatibly tuned nanosensor of Yb_2O_3 for selective detection of imazapyr in real samples. *Colloids and Surfaces A: Physicochemical and Engineering Aspects, 593*, 124612.
46. Etefagh, R., Azhir, E., & Shahtahmasebi, N. (2013). Synthesis of CuO nanoparticles and fabrication of nanostructural layer biosensors for detecting *Aspergillus niger* fungi. *Scientia Iranica, 20*, 1055–1058.
47. Chen, J., Wu, L., Lu, M., Lu, S., Li, Z., & Ding, W. (2020). Comparative study on the fungicidal activity of metallic MgO nanoparticles and macroscale MgO against soil borne fungal phytopathogens. *Frontiers in Microbiology, 11*, 365.
48. Disfani, M. N., Mikhak, A., Kassaee, M. Z., & Maghari, A. (2017). Effects of nano Fe/SiO_2 fertilizers on germination and growth of barley and maize. *Archives of Agronomy and Soil Science, 63*, 817–826.
49. Vanti, G. L., Nargund, V. B., Basavesha, K. N., Vanarchi, R., Kurjogi, M., Mulla, S. I., Tubaki, S., & Patil, R. R. (2019). Synthesis of *Gossypium hirsutum*-derived silver nanoparticles and their antibacterial efficacy against plant pathogens. *Applied Organometallic Chemistry, 33*, 4630.
50. Imada, K., Sakai, S., Kajihara, H., Tanaka, S., & Ito, S. (2016). Magnesium oxide nanoparticles induce systemic resistance in tomato against bacterial wilt disease. *Plant Pathology, 65*, 551–560.

51. Velmurugan, P., Shim, J., Kim, K., & Oh, B. T. (2016). Prunus × yedoensis tree gum mediated synthesis of platinum nanoparticles with antifungal activity against phytopathogens. *Materials Letters, 174*, 61–65.
52. Chen, J., Mao, S., Xu, Z., & Ding, W. (2019). Various antibacterial mechanisms of biosynthesized copper oxide nanoparticles against soilborne *Ralstonia solanacearum*. *RSC Advances, 9*(7), 3788–3799.
53. Adisa, I. O., Reddy-Pullagurala, V. L., Rawat, S., Hernandez-Viezcas, J. A., Dimkpa, C. O., Elmer, W. H., et al. (2018). Role of cerium compounds in Fusarium wilt suppression and growth enhancement in tomato (*Solanum lycopersicum*). *Journal of Agricultural and Food Chemistry, 66*(24), 5959–5970.
54. Palchoudhury, S., Jungjohann, K. L., Weerasena, L., et al. (2018). Enhanced legume root growth with pre-soaking in α-Fe_2O_3 nanoparticle fertilizer. *RSC Advances, 8*(43), 24075–24083.
55. Rezaei-Chiyaneh, E., Rahimi, S., Rahimi, A., Hadi, H., & Mahdavikia, H. (2018). Response of seed yield and essential oil of black cumin (*Nigella sativa* L.) affected as foliar spraying of nano-fertilizers. *Journal of Medicinal Plants and By-products, 7*(1), 33–40.
56. Correa-Llantén, D. N., Munoz-Ibacache, S. A., Castro, M. E., Munoz, P. A., & Blamey, J. M. (2013). Gold nanoparticles synthesized by *Geobacillus* sp. strain ID17 a thermophilic bacterium isolated from Deception Island, Antarctica. *Microbial Cell Factories, 12*, 75.
57. Bali, R., Razak, N., Lumb, A., & Harris, A. T. (2006). The synthesis of metallic nanoparticles inside live plants. In *International conference on nanoscience and nanotechnology, ICONN'06*. IEEE.
58. Nair, B., & Pradeep, T. (2002). Coalescence of nanoclusters and formation of submicron crystallites assisted by *Lactobacillus* strains. *Crystal Growth & Design, 2*, 293–298.
59. Jha, A. K., & Prasad, K. (2010). Ferroelectric $BaTiO_3$ nanoparticles: Biosynthesis and characterization. *Colloids and Surfaces B: Biointerfaces, 75*, 330–334.
60. Labrenz, M., Druschel, G. K., Thomsen-Ebert, T., Gilbert, B., et al. (2000). Formation of sphalerite (ZnS) deposits in natural biofilms of sulfate-reducing bacteria. *Science, 290*, 1744–1747.
61. Singh, C. R., Kathiresan, K., & Anandhan, S. (2015). A review on marine based nanoparticles and their potential applications. *African Journal of Biotechnology, 14*(18), 1525–1532.
62. Sweeney, R. Y., Mao, C., Gao, X., Burt, J. L., Belcher, A. M., Georgiou, G., & Iverson, B. L. (2004). Bacterial biosynthesis of cadmium sulfide nanocrystals. *Chemistry & Biology, 11*(11), 1553–1559.
63. Saxena, J., Sharma, M. M., Gupta, S., & Singh, A. (2014). Emerging role of fungi in nanoparticle synthesis and their applications. *The World Journal of Pharmaceutical Sciences, 3*, 1586–1613.
64. Iravani, S. (2011). Green synthesis of metal nanoparticles using plants. *Green Chemistry, 13*, 2638–2650.
65. Dwivedi, A. D., & Gopal, K. (2010). Biosynthesis of silver and gold nanoparticles using *Chenopodium album* leaf extract. *Colloids and Surfaces A: Physicochemical and Engineering Aspects, 369*, 27–33.
66. Fawcett, D., Verduin, J. J., Shah, M., Sharma, S. B., & Poinern, G. E. J. (2017). A review of current research into the biogenic synthesis of metal and metal oxide nanoparticles via marine algae and seagrasses. *Journal of Nanoscience, 2017*, 8013850.
67. Luangpipat, T., Beattie, I., Chisti, Y., & Haverkamp, R. (2011). Gold nanoparticles produced in a microalga. *Journal of Nanoparticle Research, 13*, 6439–6445.
68. Rajasulochana, P., Dhamotharan, R., Murugakoothan, P., Subbiah, M., & Krishnamoorthy, P. (2011). Biosynthesis and characterization of gold nanoparticles using the alga *Kappaphycus alvarezii*. *International Journal of Nanoscience, 9*, 511–516.
69. Mata, Y., Torres, E., Blázquez, M. L., Ballester, A., González, F., & Muñoz, J. (2008). Gold (III) biosorption and bioreduction with the brown alga *Fucus vesiculosus*. *Journal of Hazardous Materials, 166*, 612–618.

70. Senapati, S., Syed, A., Moeez, S., Kumar, A., & Ahmad, A. (2012). Intracellular synthesis of gold nanoparticles using alga *Tetraselmis kochinensis*. *Materials Letters, 79*, 116–118.
71. Castro, L., Blázquez, M. L., Muñoz, J., González, F., & Ballester, A. (2013). Biological synthesis of metallic nanoparticles using algae. *IET Nanobiotechnology, 7*, 109–116.
72. Liu, L., Cañizares, M. C., Monger, W., Perrin, Y., Tsakiris, E., Porta, C., Shariat, N., Nicholson, L., & Lomonossoff, G. P. (2005). Cowpea mosaic virus-based systems for the production of antigens and antibodies in plants. *Vaccine, 23*, 1788–1792.
73. Blum, A. S., Soto, C. M., Wilson, C. D., Brower, T. L., Pollack, S. K., Schull, T. L., Chatterji, A., Lin, T., Johnson, J. E., Amsinck, C., Franzon, P., Shashidhar, R., & Ratna, B. R. (2005). An engineered virus as a scaffold for three-dimensional self-assembly on the nanoscale. *Small, 1*, 702–706.
74. Makarov, V., Love, A. J., Sinitsyna, O., Makarova, S., Yaminsky, I., Taliansky, M., et al. (2014). "Green" nanotechnologies: Synthesis of metal nanoparticles using plants. *Acta Naturae, 6*, 35–44.
75. Singh, M., Manikandan, S., & Kumaraguru, A. K. (2011). Nanoparticles: A new technology with wide applications. *Research Journal of Nanoscience and Nanotechnology, 1*(1), 1–11.
76. Thakkar, K. N., Mhatre, S. S., & Parikh, R. Y. (2010). Biological synthesis of metallic nanoparticles. *Nanomedicine: Nanotechnology, Biology and Medicine, 6*, 257–262.
77. Castellanos-Gomez, A., Barkelid, M., Goossens, A. M., Calado, V. E., et al. (2012). Laser-thinning of MoS_2: On demand generation of a single-layer semiconductor. *Nano Letters, 12*(6), 3187–3192.
78. Ciesielski, A., & Samorì, P. (2014). Graphene via sonication assisted liquid-phase exfoliation. *Chemical Society Reviews, 43*, 381–398.
79. Li, D., Müller, M. B., Gilje, S., Kaner, R. B., & Wallace, G. G. (2008). Processable aqueous dispersions of graphene nanosheets. *Nature Nanotechnology, 3*, 101–105.
80. Zheng, J., Zhang, H., Dong, S., Liu, Y., Tai, C., Nai, H., et al. (2014). High yield exfoliation of two-dimensional chalcogenides using sodium naphthalenide. *Nature Communications, 5*, 2995.
81. Yi, M., & Shen, Z. (2015). A review on mechanical exfoliation for the scalable production of graphene. *Journal of Materials Chemistry A, 3*, 11700–11715.
82. Anasori, B., Xie, Y., Beidaghi, M., Lu, J., et al. (2015). Two-dimensional, ordered, double transition metals carbides (MXenes). *ACS Nano, 9*(10), 9507–9516.
83. Liang, J., Renzhi, M., Iyi, N., Ebina, Y., Takada, K., & Sasaki, T. (2010). Topochemical synthesis, anion exchange, and exfoliation of Co-Ni layered double hydroxides: A route to positively charged Co-Ni hydroxide nanosheets with tunable composition. *Chemistry of Materials, 22*(2), 371–378.
84. Nagaveni, K., Hedge, M. S., Ravishankar, N., Subbanna, G. N., & Madras, G. (2004). Synthesis and structure of nanocrystal line TiO_2 with lower band gap showing high photocatalytic activity. *Langmuir, 20*, 2900–2907.
85. Wang, W. N., Lenggoro, I. W., Terashi, Y., Kim, T. O., & Okuyama, K. (2005). One-step synthesis of titanium oxide nanoparticles by spray pyrolysis of organic precursors. *Materials Science and Engineering: B, 123*, 194–202.
86. Yin, H., Wada, Y., Kitamura, T., Kambe, S., Murasawa, S., Mori, H., Sakata, T., & Yanagida, S. (2001). Hydrothermal synthesis of nanosized anatase and rutile TiO_2 using amorphous phase TiO_2. *Journal of Materials Chemistry, 11*, 1694–1703.
87. Fan, G., Dundas, C. M., Zhang, C., Lynd, N. A., & Keitz, B. K. (2018). Sequence dependent peptide surface functionalization of metal–organic frameworks. *ACS Applied Materials & Interfaces, 10*, 18601–18609.
88. Gogos, A., Knauer, K., & Bucheli, T. D. (2012). Nanomaterials in plant protection and fertilization: Current state, foreseen applications, and research priorities. *Journal of Agricultural and Food Chemistry, 60*, 9781–9792.

89. Ishaque, M., Schnabel, G., & Anspaugh, D. D. *Agrochemical formulation comprising a pesticide, an organic UV photo protective filter and coated metal oxide nanoparticles.* In: Patentscope WO/2009/153231; 2009
90. Graham, J. H., Johnson, E. G., Myers, M. E., Young, M., Rajasekaran, P., et al. (2016). Potential of nano-formulated zinc oxide for control of citrus canker on grapefruit trees. *Plant Disease, 100*(12), 2442–2447.
91. Sardella, D., Gatt, R., & Valdramidis, V. (2017). *Physiological effects and mode of action of ZnO nanoparticles against postharvest fungal contaminants.* Paper presented at the 109th Annual Meeting of the American Phytopathological Society, San Antonio, TX.
92. Wani, A. H., & Shah, M. A. (2012). A unique and profound effect of MgO and ZnO nanoparticles on some plant pathogenic fungi. *Journal of Applied Pharmaceutical Science, 2*(3), 40–44.
93. Kaushik, H., & Dutta, P. (2017). *Chemical synthesis of zinc oxide nanoparticle: Its application for antimicrobial activity and plant health management.* Paper presented at the 109th Annual Meeting of the American Phytopathological Society, San Antonio, TX.
94. Azam, A., Ahmed, A. S., Oves, M., Khan, M. S., Habib, S. S., & Memic, A. (2012). Antimicrobial activity of metal oxide nanoparticles against Gram-positive and Gram-negative bacteria: A comparative study. *International Journal of Nanomedicine, 7*, 6003–6009.
95. Esteban-Tejeda, L., Malpartida, F., Esteban-Cubillo, A., Pecharromán, C., & Moya, J. (2009). Antibacterial and antifungal activity of a soda-lime glass containing copper nanoparticles. *Nanotechnology, 20*, 505701.
96. Cioffi, N., Torsi, L., Ditaranto, N., Tantillo, G., et al. (2005). Copper nanoparticle/polymer composites with antifungal and bacteriostatic properties. *Chemistry of Materials, 17*, 5255–5262.
97. Shaw, A. K., Ghosh, S., Kalaji, H. M., Bosa, K., Brestic, M., Zivcak, M., et al. (2014). Nano-CuO stress induced modulation of antioxidative defense and photosynthetic performance of syrian barley (*Hordeum vulgare* L.). *Environmental and Experimental Botany, 102*, 37–47.
98. Da Costa, M. V. J., & Sharma, P. K. (2016). Effect of copper oxide nanoparticles on growth, morphology, photosynthesis, and antioxidant response in *Oryza sativa*. *Photosynthetica, 54*, 110–119.
99. Zaytseva, O., & Neumann, G. (2016). Carbon nanomaterials: Production, impact on plant development, agricultural and environmental applications. *Chemical and Biological Technologies in Agriculture, 3*, 17.
100. Vithanage, M., Herath, I., Almaroai, Y. A., Rajapaksha, A. U., et al. (2017). Effects of carbon nanotube and biochar on bioavailability of Pb, Cu and Sb in multi-metal contaminated soil. *Environmental Geochemistry and Health, 39*(6), 1409–1420.
101. Milewska-Hendel, A., Gawecki, R., Zubko, M., Stróż, D., & Kurczyńska, E. (2016). Diverse influence of nanoparticles on plant growth with a particular emphasis on crop plants. *Acta Agrobotanica, 69*(4), 1694.
102. Aslani, F., Bagheri, S., Muhd Julkapli, N., Juraimi, A. S., Hashemi, F. S. G., & Baghdadi, A. (2014). Effects of engineered nanomaterials on plants growth: An overview. *The Scientific World Journal, 2014*, 641759.
103. Hossain, Z., Mustafa, G., & Komatsu, S. (2015). Plant responses to nanoparticle stress. *International Journal of Molecular Sciences, 16*(11), 26644–26653.
104. Wang, X., Liu, X., Han, H. Y., Gu, X. X., Chen, K., & Lu, D. L. (2012). Multi-walled carbon nanotubes can enhance root elongation of wheat (*Triticum aestivum*) plants. *Journal of Nanoparticle Research, 14*(6), 841–850.
105. Cai, L., Chen, J., Liu, Z., Wang, H., Yang, H., & Ding, W. (2018). Magnesium oxide nanoparticles: Effective agricultural antibacterial agent against *Ralstonia solanacearum*. *Frontiers in Microbiology, 9*, 790.
106. Huang, Z., Rajasekaran, P., Ozcan, A., & Santra, S. (2018). Antimicrobial magnesium hydroxide nanoparticles as an alternative to Cu biocide for crop protection. *Journal of Agricultural and Food Chemistry, 66*(33), 8679–8686.

107. Leung, Y. H., Ng, A. M., Xu, X., Shen, Z., Gethings, L. A., Wong, M. T., et al. (2014). Mechanisms of antibacterial activity of MgO: Non-ROS mediated toxicity of MgO nanoparticles towards *Escherichia coli*. *Small, 10*, 1171–1183.
108. Paret, M. L., Vallad, G. E., Averett, D. R., Jones, J. B., & Olson, S. M. (2013). Photocatalysis: Effect of light-activated nanoscale formulations of TiO_2 on *Xanthomonas perforans* and control of bacterial spot of tomato. *Phytopathology, 103*(3), 228–236.
109. Boxi, S. S., Mukherjee, K., & Paria, S. (2016). Ag doped hollow TiO_2 nanoparticles as an effective green fungicide against *Fusarium solani* and *Venturia inaequalis* phytopathogens. *Nanotechnology, 27*, 085103.
110. Abdel Latef, A. A. H., Srivastava, A. K., El-Sadek, M. S. A., Kordrostami, M., & Tran, L. S. P. (2018). Titanium dioxide nanoparticles improve growth and enhance tolerance of broad bean plants under saline soil conditions. *Land Degradation & Development, 29*, 1065–1073.
111. Khan, M. N., Al-Solami, M. A., Basahi, R. A., Siddiqui, M. H., Al-Huqail, A. A., Abbas, Z. K., et al. (2020). Nitric oxide is involved in nano-titanium dioxide-induced activation of antioxidant defense system and accumulation of osmolytes under water-deficit stress in *Vicia faba* L. *Ecotoxicology and Environmental Safety, 190*, 110152.
112. Karamiana, R., Ghasemloub, F., & Amirib, H. (2020). Physiological evaluation of drought stress tolerance and recovery in *Verbascum sinuatum* plants treated with methyl jasmonate, salicylic acid and titanium dioxide nanoparticles. *Plant Biosystems, 154*(3), 277–287.
113. Raliya, R., Nair, R., Chavalmane, S., Wang, W. N., & Biswas, P. (2015). Mechanistic evaluation of translocation and physiological impact of titanium dioxide and zinc oxide nanoparticles on the tomato (*Solanum lycopersicum* L.) plant. *Metallomics, 7*, 1584–1594.
114. Narayanan, K. B., & Park, H. H. (2014). Antifungal activity of silver nanoparticles synthesized using turnip leaf extract (*Brassica rapa* L.) against wood rotting pathogens. *The European Journal of Plant Pathology, 140*(2), 185–192.
115. Gajbhiye, M., Kesharwani, J., Ingle, A., Gade, A., & Rai, M. (2009). Fungus-mediated synthesis of silver nanoparticles and their activity against pathogenic fungi in combination with fluconazole. *Nanomedicine: Nanotechnology, Biology and Medicine, 5*, 382–386.
116. Gopal, M., Gogoi, R., Srivastava, C., Kumar, R., et al. (2011). Nanotechnology and its application in plant protection. In *Plant pathology in India: Vision 2030*. Indian Pathology Society.
117. Jasim, B., Thomas, R., Mathew, J., & Radhakrishnan, E. K. (2016). Plant growth and diosgenin enhancement effect of silver nanoparticles in Fenugreek (*Trigonella foenum-graecum* L.). *Saudi Pharmaceutical Journal, 25*, 443–447.
118. Tripathi, D. K., Singh, S., Singh, S., Srivastava, P. K., Singh, V. P., Singh, S., et al. (2017). Nitric oxide alleviates silver nanoparticles (Ag NPs)-induced phytotoxicity in *Pisum sativum* seedlings. *Plant Physiology and Biochemistry, 110*, 167–177.
119. Jo, Y. K., Kim, B. H., & Jung, G. (2009). Antifungal activity of silver ions and nanoparticles on phytopathogenic fungi. *Plant Disease, 93*, 1037–1043.
120. Liu, J., & Hurt, R. H. (2010). Ion release kinetics and particle persistence in aqueous nano-silver colloids. *Environmental Science & Technology, 44*, 2169–2175.
121. Babajani, A., Iranbakhsh, A., Ardebili, Z. O., & Eslami, B. (2019). Differential growth, nutrition, physiology, and gene expression in *Melissa officinalis* mediated by zinc oxide and elemental selenium nanoparticles. *Environmental Science and Pollution Research, 26*, 24430–24444.
122. Ashkavand, P., Tabari, M., Zarafshar, M., Tomášková, I., & Struve, D. (2015). Effect of SiO_2 nanoparticles on drought resistance in hawthorn seedlings. *Forest Research Papers, 76*, 350–359.
123. Haghighi, M., & Pessarakli, M. (2013). Influence of silicon and nano-silicon on salinity tolerance of cherry tomatoes (*Solanum lycopersicum* L.) at early growth stage. *Scientia Horticulturae, 161*, 111–117.

124. Kanto, T., Miyoshi, A., Ogawa, T., Maekawa, K., & Aino, M. (2004). Suppressive effect of potassium silicate on powdery mildew of strawberry in hydroponics. *Journal of General Plant Pathology, 70*, 207–211.
125. Rui, M., Ma, C., Hao, Y., Guo, J., Rui, Y., et al. (2016). Iron oxide nanoparticles as a potential iron fertilizer for peanut (*Arachis hypogaea*). *Frontiers in Plant Science, 7*, 815.
126. Delfani, M., Firouzabadi, M. B., Farrokhi, N., & Makarian, H. (2014). Some physiological responses of black-eyed pea to iron and magnesium nanofertilizers. *Communications in Soil Science and Plant Analysis, 45*, 530–540.
127. Alam, M. J., Sultana, F., & Iqbal, M. T. (2015). Potential of iron nanoparticles to increase germination and growth of wheat seedling. *Journal of Nanoscience with Advanced Technology, 1*(3), 14–20.
128. Li, J., Hu, J., Ma, C., Wang, Y., Wu, C., Huang, J., et al. (2016). Uptake, translocation and physiological effects of magnetic iron oxide (g-Fe_2O_3) nanoparticles in corn (*Zea mays* L.). *Chemosphere, 159*, 326–334.
129. Adhikari, T., Kundu, S., & Rao, A. S. (2013). Impact of SiO_2 and Mo nano particles on seed germination of rice (*Oryza sativa* L.). *International Journal of Agriculture Food Science Technology, 4*(8), 809–816.
130. Du, W., Gardea-Torresdey, J. L., Ji, R., Yin, Y., Zhu, J., Peralta-Videa, J. R., et al. (2015). Physiological and biochemical changes imposed by CeO_2 nanoparticles on wheat: A life cycle field study. *Environmental Science & Technology, 49*, 11884–11893.
131. Taran, N. Y., Gonchar, O. M., Lopatko, K. G., Batsmanova, L. M., Patyka, M. V., & Volkogon, M. V. (2014). The effect of colloidal solution of molybdenum nanoparticles on the microbial composition in rhizosphere of *Cicer arietinum* L. *Nanoscale Research Letters, 9*, 289.
132. Mahakham, W., Theerakulpisut, P., Maensiri, S., Phumying, S., & Sarmah, A. K. (2016). Environmentally benign synthesis of phytochemicals-capped gold nanoparticles as nanopriming agent for promoting maize seed germination. *Science of the Total Environment, 573*, 1089–1102.
133. Jayaseelan, C., Ramkumar, R., Rahuman, A. A., & Perumal, P. (2013). Green synthesis of gold nanoparticles using seed aqueous extract of *Abelmoschus esculentus* and its antifungal activity. *Industrial Crops and Products, 45*, 423–429.
134. Rao, K. J., & Paria, S. (2013). Use of sulfur nanoparticles as a green pesticide on *Fusarium solani* and *Venturia inaequalis* phytopathogens. *RSC Advances, 3*(26), 10471–10478.
135. Vecerova, K., Vecera, Z., Docekal, B., Oravec, M., Pompeiano, A., Tˇríska, J., et al. (2016). Changes of primary and secondary metabolites in barley plants exposed to CdO nanoparticles. *Environmental Pollution, 218*, 207–218.
136. Pollard, M., Beisson, F., Li, Y. H., & Ohlrogge, J. B. (2008). Building lipid barriers: Biosynthesis of cutin and suberin. *Trends in Plant Science, 13*, 236–246.
137. Du, W., Sun, Y., Ji, R., Zhu, J., Wu, J., & Guo, H. (2011). TiO_2 and ZnO nanoparticles negatively affect wheat growth and soil enzyme activities in agricultural soil. *Journal of Environmental Monitoring, 13*(4), 822–828.
138. Rico, C. M., Majumdar, S., Duarte-Gardea, M., Peralta-Videa, J. R., & Gardea-Torresdey, J. L. (2011). Interaction of nanoparticles with edible plants and their possible implications in the food chain. *Journal of Agricultural and Food Chemistry, 59*, 3485–3498.
139. Wu, B., & Beitz, E. (2007). Aquaporins with selectivity for unconventional permeants. *Cellular and Molecular Life Sciences, 64*, 2413–2421.
140. Schwab, F., Zhai, G., Kern, M., Turner, A., Schnoor, J. L., & Wiesner, M. R. (2015). Barriers, pathways and processes for uptake, translocation and accumulation of nanomaterials in plants-critical review. *Nanotoxicology, 10*, 257–278.
141. Roberts, A. G., & Oparka, K. J. (2003). Plasmodesmata and the control of symplastic transport. *Plant, Cell & Environment, 26*, 103–124.
142. Taylor, A. F., Rylott, E. L., Anderson, C. W., & Bruce, N. C. (2014). Investigating the toxicity, uptake, nanoparticle formation and genetic response of plants to gold. *PLoS One, 9*, e93793.

143. Raliya, R., Franke, C., Chavalmane, S., Nair, R., Reed, N., & Biswas, P. (2016). Quantitative understanding of nanoparticle uptake in watermelon plants. *Frontiers in Plant Science, 7*, 1288.
144. Judy, J. D., Unrine, J. M., Rao, W., Wirick, S., & Bertsch, P. M. (2012). Bioavailability of gold nanomaterials to plants: Importance of particle size and surface coating. *Environmental Science & Technology, 46*, 8467–8474.
145. Wang, F., Liu, X., Shi, Z., Tong, R., Adams, C. A., & Shi, X. (2016). Arbuscular mycorrhizae alleviate negative effects of zinc oxide nanoparticle and zinc accumulation in maize plants-A soil microcosm experiment. *Chemosphere, 147*, 88–97.
146. Lough, T. J., & Lucas, W. J. (2006). Integrative plant biology: Role of phloem long-distance macromolecular trafficking. *Annual Review of Plant Biology, 57*, 203–232.
147. Sattelmacher, B. (2001). The apoplast and its significance for plant mineral nutrition. *New Phytologist, 149*, 167–192.
148. Sun, D., Hussain, H. I., Yi, Z., Siegele, R., Cresswell, T., Kong, L., et al. (2014). Uptake and cellular distribution, in four plant species, of fluorescently labelled mesoporous silica nanoparticles. *Plant Cell Reports, 33*, 1389–1402.
149. Robards, A. W., & Robb, M. E. (1972). Uptake and binding of uranyl ions by barley roots. *Science, 178*, 980–982.
150. Schönherr, J. (2002). A mechanistic analysis of penetration of glyphosate salts across astomatous cuticular membranes. *Pest Management Science, 58*, 343–351.
151. Eichert, T., & Goldbach, H. E. (2008). Equivalent pore radii of hydrophilic foliar uptake routes in stomatous and astomatous leaf surfaces–further evidence for a stomatal pathway. *Physiologia Plantarum, 132*, 491–502.
152. Eichert, T., Kurtz, A., Steiner, U., & Goldbach, H. E. (2008). Size exclusion limits and lateral heterogeneity of the stomatal foliar uptake pathway for aqueous solutes and water-suspended nanoparticles. *Physiologia Plantarum, 134*, 151–160.
153. Zhu, Z. J., Wang, H., Yan, B., Zheng, H., Jiang, Y., Miranda, O. R., et al. (2012). Effect of surface charge on the uptake and distribution of gold nanoparticles in four plant species. *Environmental Science & Technology, 46*, 12391–12398.
154. Manjaiah, K. M., Mukhopadhyay, R., Paul, R., Datta, S. C., Kumararaja, P., & Sarkar, B. (2019). Clay minerals and zeolites for environmentally sustainable agriculture. In M. Mariano, S. Binoy, & L. Alessio (Eds.), *Modified clay and zeolite nanocomposite materials* (pp. 309–329). Elsevier.
155. Banerjee, J., & Kole, C. (2016). Plant nanotechnology: An overview on concepts, strategies, and tools. In C. Kole, D. Kumar, & M. Khodakovskaya (Eds.), *Plant nanotechnology* (pp. 1–14). Springer.
156. Shojaei, T. R., Salleh, M. A. M., Tabatabaei, M., Mobli, H., Aghbashlo, M., Rashid, S. A., & Tan, T. (2019). Applications of nanotechnology and carbon nanoparticles in agriculture. In A. R. Suraya, N. I. R. O. Raja, & Z. H. Mohd (Eds.), *Synthesis, technology and applications of carbon nanomaterials* (pp. 247–277). Elsevier.
157. Shalaby, T. A., Bayoumi, Y., Abdalla, N., Taha, H., Alshaal, T., Shehata, S., Amer, M., Domokos-Szabolcsy, É., & El-Ramady, H. (2016). Nanoparticles, soils, plants and sustainable agriculture. In R. Shivendu, D. Nandita, & L. Eric (Eds.), *Nanoscience in food and agriculture 1* (pp. 283–312). Springer.
158. Yang, F., Hong, F., You, W., Liu, C., Gao, F., Wu, C., & Yang, P. (2006). Influences of nanoanatase TiO2 on the nitrogen metabolism of growing spinach. *Biological Trace Element Research, 110*, 179–190.
159. Khodakovskaya, M. V., de Silva, K., Biris, A. S., Dervishi, E., & Villagarcia, H. (2012). Carbon nanotubes induce growth enhancement of tobacco cells. *ACS Nano, 6*, 2128–2135.
160. Siddiqui, M. H., Al-Whaibi, M. H., Faisal, M., & Alsahli, A. A. (2014). Nano-silicon dioxide mitigates the adverse effects of salt stress on *Cucurbita pepo* L. *Environmental Toxicology and Chemistry, 33*, 2429–2437.

161. Hong, F. S., Liu, C., Zheng, L., Wang, X. F., Wu, K., Song, W. P., Lv, S. P., Tao, Y., & Zhao, G. W. (2005). Formation of complexes of rubisco-rubisco activase from La^{3+}, Ce^{3+} treatment spinach. *Science China Series B Chemistry, 48*, 67–74.
162. Gao, F. Q., Hong, F. S., Liu, C., Zheng, L., Su, M. Y., Wu, X., Yang, F., Wu, C., & Yang, P. (2006). Mechanism of nanoanatase TiO_2 on promoting photosynthetic carbon reaction of spinach: Inducing complex of Rubisco-Rubisco activase. *Biological Trace Element Research, 11*, 239–254.
163. Jin, Y., Fan, X., Li, X., Zhang, Z., Sun, L., Fu, Z., et al. (2017). Distinct physiological and molecular responses in *Arabidopsis thaliana* exposed to aluminum oxide nanoparticles and ionic aluminum. *Environmental Pollution, 228*, 517–527.
164. Nowack, B., & Bucheli, T. (2007). Occurrence, behavior and effects of nanoparticles in the environment. *Environmental Pollution, 150*, 5–22.
165. Kurepa, J., Paunesku, T., Vogt, S., Arora, H., Rabatic, B. M., Lu, J., et al. (2010). Uptake and distribution of ultrasmall anatase TiO2 alizarin red s nanoconjugates in *Arabidopsis thaliana*. *Nano Letters, 10*, 2296–2302.
166. Al-Salim, N., Barraclough, E., Burgess, E., Clothier, B., Deurer, M., Green, S., et al. (2011). Quantum dot transport in soil, plants, and insects. *Science of the Total Environment, 409*, 3237–3248.
167. Chichiricco, G., & Poma, A. (2015). Penetration and toxicity of nanomaterials in higher plants. *Nanomaterials, 5*, 851–873.
168. Avellan, A., Schwab, F., Masion, A., Chaurand, P., Borschneck, D., Vidal, V., et al. (2017). Nanoparticle uptake in plants: Gold nanomaterial localized in roots of *Arabidopsis thaliana* by X-ray computed nanotomography and hyperspectral imaging. *Environmental Science & Technology, 51*, 8682–8691.
169. Hatami, M. (2017). Stimulatory and inhibitory effects of nanoparticulates on seed germination and seedling vigor indices. In M. Ghorbanpour, K. Manika, & A. Varma (Eds.), *Nanoscience and plant–soil systems* (Soil biology) (Vol. 48, pp. 357–385). Springer.
170. Khodakovskaya, M., Dervishi, E., Mahmood, M., et al. (2009). Carbon nanotubes are able to penetrate plant seed coat and dramatically affect seed germination and plant growth. *ACS Nano, 3*, 3221–3227.
171. Kole, C., Kole, P., Randunu, K. M., Choudhary, P., Podila, R., & Ke, P. C. (2013). Nanobiotechnology can boost crop production and quality: First evidence from increased plant biomass, fruit yield and phytomedicine content in bitter melon (*Momordica charantia*). *BMC Biotechnology, 13*(1), 37.
172. Hafeez, A., Razzaq, A., Mahmood, T., & Jhanzab, H. M. (2015). Potential of copper nanoparticles to increase growth and yield of wheat. *Journal of Nanoscience with Advanced Technology, 1*(1), 6–11.
173. Ngo, Q. B., Dao, T. H., Nguyen, H. C., & Tran, T. V. (2014). Effects of nano crystalline powders (Fe, Co and Cu) on the germination, growth, crop yield and product quality of soybean (Vietnamese species DT-51). *Advances in Natural Sciences: Nanoscience and Nanotechnology, 5*, 1–7.
174. Prasad, T. N. V. K. V., Sudhakar, P., Sreenivasulu, Y., et al. (2012). Effect of nanoscale zinc oxide particles on the germination, growth and yield of peanut. *Journal of Plant Nutrition, 35*, 905–927.
175. Wang, P., Lombi, E., Zhao, F. J., & Kopittke, P. M. (2016). Nanotechnology: A new opportunity in plant sciences. *Trends in Plant Science, 21*(8), 699–712.
176. Siddiqui, M. H., & Al-Whaibi, M. H. (2014). Role of nano-SiO_2 in germination of tomato (*Lycopersicum esculentum* seeds Mill.). *Saudi Journal of Biological Sciences, 21*, 13–17.
177. Aken Van, B. (2015). Gene expression changes in plants and microorganisms exposed to nanomaterials. *Current Opinion in Biotechnology, 33*, 206–219.
178. Kharey, P., Dutta, S. B., Gorey, A., et al. (2020). *Pimenta dioica* mediated biosynthesis of gold nanoparticles and evaluation of its potential for theranostic applications. *ChemistrySelect, 5*, 7901–7908.

179. Vijayakumar, S., Arulmozhi, P., Kumar, N., et al. (2020). *Acalypha fruticose* L. leaf extract mediated synthesis of ZnO nanoparticles: Characterization and antimicrobial activities. *Materials Today: Proceedings, 23*, 73–80.
180. Akhter, G., Khan, A., Ali, S. G., et al. (2020). Antibacterial and nematicidal properties of biosynthesized Cu nanoparticles using extract of holoparasitic plant. *SN Applied Sciences, 2*, 1268.
181. Aygün, A., Özdemir, S., Gülcan, M., et al. (2020). Synthesis and characterization of Reishi mushroom-mediated green synthesis of silver nanoparticles for the biochemical applications. *Journal of Pharmaceutical and Biomedical Analysis, 178*, 112970.
182. Nabi, G., Ain, Q. U., Tahir, M. B., et al. (2020). Green synthesis of TiO_2 nanoparticles using lemon peel extract: Their optical and photocatalytic properties. *International Journal of Environmental Analytical Chemistry, 102*, 434–442. https://doi.org/10.1080/03067319.2020.1722816
183. Herrera-Becerra, R., Zorrilla, C., Rius, J. L., & Ascencio, J. A. (2008). Electron microscopy characterization of biosynthesized iron oxide nanoparticles. *Applied Physics A: Materials Science and Processing, 91*, 241–246.
184. Sutradhar, P., & Saha, M. (2016). Green synthesis of zinc oxide nanoparticles using tomato *(Lycopersicon esculentum)* extract and its photovoltaic application. *Journal of Experimental Nanoscience, 11*(5), 314–327.
185. Jampílek, J., & Kráľová, K. (2015). Application of nanotechnology in agriculture and food industry, its prospects and risks. *Ecological Chemistry and Engineering S, 22*(3), 321–361.
186. Patil, M. P., Kang, M. J., Niyonizigiye, I., Singh, A., Kim, J. O., Seo, Y. B., & Kim, G. D. (2019). Do Extracellular synthesis of gold nanoparticles using the marine bacterium *Paracoccus haeundaensis* BC74171T and evaluation of their antioxidant activity and antiproliferative effect on normal and cancer cell lines. *Colloids and Surfaces B: Biointerfaces, 183*, 110455.
187. Dhandapani, P., Prakash, A. A., AlSalhi, M. S., Maruthamuthu, S., Devanesan, S., & Rajasekar, A. (2020). Ureolytic bacteria mediated synthesis of hairy ZnO nanostructure as photocatalyst for decolorization of dyes. *Materials Chemistry and Physics, 243*, 122619.
188. Martins, M., Mourato, C., Sanches, S., Noronha, J. P., Crespo, M. T. B., & Pereira, I. A. C. (2017). Biogenic platinum and palladium nanoparticles as new catalysts for the removal of pharmaceutical compounds. *Water Research, 108*, 160–168.
189. Gericke, M., & Pinches, A. (2006). Biological synthesis of metal nanoparticles. *Hydrometallurgy, 83*, 132–140.
190. Xie, J., Chen, K., & Chen, X. (2009). Production, modification and bio-applications of magnetic nanoparticles gestated by magnetotactic bacteria. *Nano Research, 2*(4), 261–278.
191. Kumari, R. M., Kumar, V., Kumar, M., Agrawal, A., Pareek, N., & Nimesh, S. (2020). Extracellular biosynthesis of silver nanoparticles using *Aspergillus terreus*: Evaluation of its antibacterial and anticancer potential. *Materials Today: Proceedings*. https://doi.org/10.1016/j.matpr.2020.04.494
192. Manjunath, H. M., & Joshi, C. G. (2019). Characterization, antioxidant and antimicrobial activity of silver nanoparticles synthesized using marine endophytic fungus-*Cladosporium cladosporioides*. *Process Biochemistry, 82*, 199–204.
193. Zhang, H., Zhou, H., Bai, J., Li, Y., Yang, J., Ma, Q., & Qu, Y. (2019). Biosynthesis of selenium nanoparticles mediated by fungus *Mariannaea* sp. HJ and their characterization. *Colloids and Surfaces A: Physicochemical and Engineering Aspects, 571*, 9–16.
194. Raliya, R., & Tarafdar, J. C. (2014). Biosynthesis and characterization of zinc, magnesium and titanium nanoparticles: An eco-friendly approach. *International Nano Letters, 4*, 93.
195. Raliya, R., Biswas, P., & Tarafdar, J. C. (2015). TiO_2 nanoparticle biosynthesis and its physiological effect on mung bean (*Vigna radiata* L.). *Biotechnology Reports, 5*, 22–26.
196. Moghaddam, A. B., Namvar, F., Moniri, M., Tahir, P. M., Azizi, A., & Mohamad, R. (2015). Nanoparticles biosynthesized by fungi and yeast: A review of their preparation, properties, and medical applications. *Molecules, 20*, 16540–16565.

197. Vijayan, S. R., Santhiyagu, P., Singamuthu, M., Ahila, N. K., Jayaraman, R., & Ethiraj, K. (2014). Synthesis and characterization of silver and gold nanoparticles using aqueous extract of seaweed, *Turbinaria conoides*, and their anti-microfouling activity. *The Scientific World Journal, 2014*, 938272. https://doi.org/10.1155/2014/938272
198. Azizi, S., Ahmad, M. B., Namvar, F., & Mohamad, R. (2014). Green biosynthesis and characterization of zinc oxide nanoparticles using brown marine macroalga *Sargassum muticum* aqueous extract. *Materials Letters, 116*, 275–277.
199. Subramaniyam, V., Subashchandrabose, S. R., Thavamani, P., Megharaj, M., Chen, Z., & Naidu, R. (2015). *Chlorococcum* sp. MM11-a novel phyco-nanofactory for the synthesis of iron nanoparticles. *Journal of Applied Phycology, 27*, 1861–1869.
200. Abboud, Y., Saffaj, T., Chagraoui, A., El Bouari, A., Brouzi, K., Tanane, O., & Ihssane, B. (2014). Biosynthesis, characterization and antimicrobial activity of copper oxide nanoparticles (CONPs) produced using brown alga extract (*Bifurcaria bifurcata*). *Applied Nanoscience, 4*, 571–576.
201. Mahdavi, M., Namvar, F., Ahmad, M. B., & Mohamad, R. (2013). Green biosynthesis and characterization of magnetic iron oxide (Fe_3O_4) nanoparticles using seaweed (*Sargassum muticum*) aqueous extract. *Molecules, 18*, 5954–5964.
202. Shenton, W., Douglas, T., Young, M., Stubbs, G., & Mann, S. (1999). Inorganic-organic nanotube composites from template mineralization of tobacco mosaic virus. *Advanced Materials, 11*, 253–256.
203. Chen, P. Y., Dang, X., Klug, M. T., Courchesne, N. M. D., et al. (2015). M13 virus-enabled synthesis of titanium dioxide nanowires for tunable mesoporous semiconducting networks. *Chemistry of Materials, 27*, 1531–1540.
204. Le, D. H., Lee, K. L., Shukla, S., Commandeur, U., & Steinmetz, N. F. (2017). Potato virus X, a filamentous plant viral nanoparticle for doxorubicin delivery in cancer therapy. *Nanoscale, 9*, 2348–2357.
205. Zeng, Q., Wen, H., Wen, Q., Chen, X., et al. (2013). Cucumber mosaic virus as drug delivery vehicle for doxorubicin. *Biomaterials, 34*, 4632–4642.
206. Rajput, V., Minkina, T., Fedorenko, A., Sushkova, S., Mandzhieva, S., Lysenko, V., Duplii, N., Fedorenko, G., Dvadnenko, K., & Ghazaryan, K. (2018). Toxicity of copper oxide nanoparticles on spring barley (*Hordeum sativum distichum*). *Science of the Total Environment, 645*, 1103–1113.
207. Gao, X., Avellan, A., Laughton, S., Vaidya, R., Rodrigues, S. M., Casman, E. A., & Lowry, G. V. (2018). CuO nanoparticle dissolution and toxicity to wheat (*Triticum aestivum*) in rhizosphere soil. *Environmental Science & Technology, 52*, 2888–2897.
208. Halliwell, B., & Gutteridge, J. M. C. (1985). The importance of free radicals and catalytic metal ions in human diseases. *Molecular Aspects of Medicine, 8*, 89–193.
209. Ghosh, M., Bandyopadhyay, M., & Mukherjee, A. (2015). Genotoxicity of titanium dioxide (TiO_2) nanoparticles at two trophies levels: Plant and human lymphocytes. *Chemosphere, 81*, 1253–1262.
210. Gunjan, B., Zaidi, M. G. H., & Sandeep, A. (2014). Impact of gold nanoparticles on physiological and biochemical characteristics of *Brassica juncea*. *Journal of Plant Biochemistry & Physiology, 2*, 133.
211. Begum, P., Ikhtiari, R., Fugetsu, B., Matsuoka, M., Akasaka, T., & Watari, F. (2012). Phytotoxicity of multi-walled carbon nanotubes assessed by selected plant species in the seedling stage. *Applied Surface Science, 262*, 120–124.
212. Peralta-Videa, J. R., Hernandez-Viezcas, J. A., Zhao, L., Diaz, B. C., Ge, Y., Priester, J. H., Holden, P. A., & Gardea-Torresdey, J. L. (2014). Cerium dioxide and zinc oxide nanoparticles alter the nutritional value of soil cultivated soybean plants. *Plant Physiology and Biochemistry, 80*, 128–135.
213. Dietz, K. J., & Herth, S. (2011). Plant nanotoxicology. *Trends in Plant Science, 16*, 582–589.

214. Miralles, P., Church, T. L., & Harris, A. T. (2012). Toxicity, uptake, and translocation of engineered nanomaterials in vascular plants. *Environmental Science & Technology, 46*, 9224–9239.
215. Martínez-Fernández, D., Vítková, M., Bernal, M. P., & Komárek, M. (2015). Effects of nano-maghemite on trace element accumulation and drought response of *Helianthus annuus* L. in a contaminated mine soil. *Water, Air, & Soil Pollution, 226*, 101.
216. Jang, H., Pell, L. E., Korgel, B. A., & English, D. S. (2003). Photoluminescence quenching of silicon nanoparticles in phospholipid vesicle bilayers. *Journal of Photochemistry and Photobiology A: Chemistry, 158*, 111–117.
217. Barhoumi, L., Oukarroum, A., Taher, L. B., Smiri, L. S., Abdelmelek, H., & Dewez, D. (2015). Effects of superparamagnetic iron oxide nanoparticles on photosynthesis and growth of the aquatic plant *Lemna gibba*. *Archives of Environmental Contamination and Toxicology, 68*, 510–520.
218. Priester, J. H., et al. (2012). Soybean susceptibility to manufactured nanomaterials with evidence for food quality and soil fertility interruption. *Proceedings of the National Academy of Sciences of the United States of America, 109*(37), E2451–E2456.

Nanotechnologies and Sustainable Agriculture for Food and Nutraceutical Production: An Update

Pragya Tiwari

Abstract The recent era has seen the emergence of nanotechnology as a promising tool in modern agriculture, projected to have a major impact in boosting food and nutraceutical production, shortly. The recent advance in nanotechnology has made a substantial contribution towards the transformation of traditional practices in agriculture, subject to the development of nanosensors, nanofertilizers, smart food packaging methods, aiming at boosting agriculture output. The development of promising and novel nanomaterials for enhancing crop growth and yield, quality of food, and monitoring amidst fluctuating climatic conditions calls for the efficient utilization/application of nanotechnologies as well as addressing the associated challenges to a certain level. The application of nanotechnology in agriculture highlights novel avenues in the development of diagnostics for plant diseases, nanosensors in crop protection, animal health, and poultry, detection of food quality/contamination, application of nanofertilizers among other promising techniques for boosting agricultural productivity. The nanotechnologies are based on nanoparticle-mediated DNA/gene transfer in plants, aiming at the generation of improved plant varieties, increased shelf-life, food processing/packaging leading to technologies in biomass to biofuel production. Considering the emerging popularity and advances in the field of nanotechnologies, how these techniques aim to expedite and revolutionize smart agriculture in the present era forms the underlying theme of the chapter. The multi-faceted applications of nanotechnological methods in food and nutraceutical production, the emerging prospects, associated limitations, and the solutions and the future directions/outcomes are further discussed.

Keywords Active packaging · Food-quality · Nano-enabled food · Nanopesticides · Nutraceuticals · Sustainable agriculture · Toxicity assessment

---

P. Tiwari (✉)
Department of Biotechnology, Yeungnam University, Gyeongsan, Gyeongbuk, Republic of Korea

J.-T. Chen (ed.), *Plant and Nanoparticles*,
https://doi.org/10.1007/978-981-19-2503-0_12

1 Sustainable Agriculture in the Era of Nanotechnology

In the present era, the field of nanotechnology has been witnessed as a prospective way to boost crop productivity, leading to agricultural sustainability. The recent advances in the development of nanotechnologies and their applications in diverse areas have been prospective in impacting the agri-food sector, and reducing limitations in agricultural practices, aiming towards better food security for the rising global population. Statistics have suggested the rise in the world's population to approximately nine billion by 2050, thus necessitating the increase in global food output to feed the rising population [1]. Agriculture forms the backbone of the developing economies and agri-food production is of key importance. The limited land areas for crop production as well as pharmaceuticals and biofuels production for energy and healthcare, and dwindling natural resources account for some major concerns.

Nanotechnological advancements hold great potential to revolutionize agriculture and allied sectors, including fisheries and aquaculture. Nano-based agriculture employs nanosized particles to increase crop and livestock productivity, primarily targeting farming systems [2, 3]. The nanotechnology-based techniques have only been recently used for promoting plant growth, plant disease detection, food quality/trait improvement, towards targeting increased food production by "sustainable intensification" [4–6]. Agri-food nanotechnology highlights multidisciplinary nature, with diverse applications in biotechnology, hydroponics, livestock, nanotoxicology, and food sectors, with agriculture comprising of a recent sector for nanotechnological applications. The nanoscale materials are 1–100 nm approximately and one dimension nanoscale (layers, e.g. thin films, graphene), two-dimension nanoscale (nanotube), and three dimensions (quantum dots) are employed in different areas of agriculture [1]. Furthermore, the engineered nanomaterial comprises of specific composition and is produced to have specific properties, different from those of their conventional counterparts. The nanoparticles of biological origin (e.g. lipoproteins, carotenoid lycopene, ferritin, viruses) exhibit diverse biological roles, are reproducible and biocompatible, making them ideal candidates in biomedical applications [1]. With the development of new food packaging products, pathogen detection by nanosensors, nanoparticles-based water purification methods, there has been considerable improvement in food industries [7, 8]. Figure 1 diagrammatic representation of the multi-faceted application of nanotechnology in the environment and sustainable agriculture.

In food industries, nanobiotechnology has made great strides in monitoring food quality through nanosensors to detect the presence of fungus/insects, genetic modification of golden rice (http://www.azonano.com/, [9]), use of nanocapsules for genetic manipulation towards the generation of novel plant varieties [10], increasing fruit yield and medicinal content [11] among other significant ones. The use of smart sensors in precision farming show prospects in improving agricultural productivity providing better management, time frame, and input reduction. Moreover, technologies comprising geographic systems, remote sensors, and satellite monitoring are

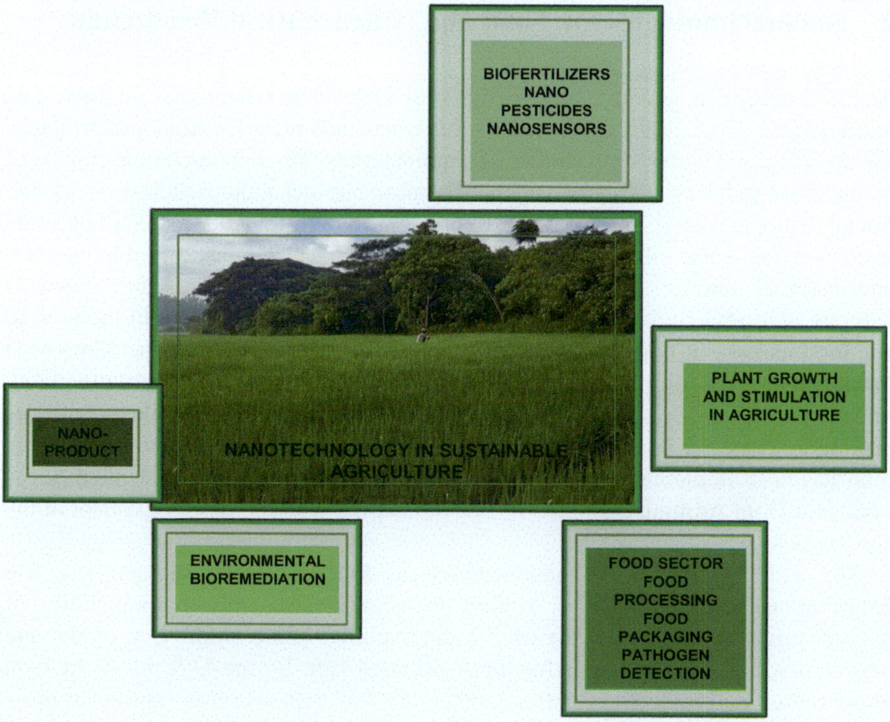

Fig. 1 Diagrammatic representation of the multi-faceted application of nanotechnology in the environment and sustainable agriculture

helpful in the detection of abiotic stress and plant pests [1]. The signature initiative "Nanotechnology for Sensors and Sensors for Nanotechnology: Improving and Protecting Health, Safety, and the Environment" is launched by the National Nanotechnology Initiative and nanodevices show potential in fast detection of chemicals, pathogens, diseases leading to faster actions [12].

The increasing recognition of the potential applications of nanotechnology in agriculture and food systems has opened new avenues suggesting that novel nanomaterials co-integrated with engineering approaches would be innovative towards nanotechnological advances in agriculture: design/development of new nanodevices and materials [13]. Nanotechnological interventions in agriculture offer prospects improving food quality/yield, monitoring/control of diseases, enhanced nutrient uptake/utilization by plants, development of safe/effective new-generation pesticides, addressing soil/water contamination, and increasing shelf-life of flowers/vegetables through the application of nanoparticles [14]. An account of the existing and upcoming potential of nanotechnological interventions in agriculture has been discussed, the recent/future trends, the need to address the associated challenges, and prospects/scenarios form the key theme.

2 Nanotechnologies for Food and Nutraceutical Production

Nanotechnology is gaining increased momentum in diverse areas in food and nutraceutical production, attributed to the recent advances in high performance, low toxicity, and wider applicability of nanomaterials. The nanomaterials employed in the food industry comprise organic (natural product nanoparticles), inorganic (metal nanoparticles), and combined forms [15]. While gold nanoparticles are studied as sensors, silver nanoparticles have broader applications due to their antimicrobial activity. Natural product-based nanoparticles are used/designed as delivery platforms and in food industries, as food components [15]. In the present era, food nanotechnology has been widely used in different sections, including food packaging, preservation/processing, and food additives, ensuring the safety/quality of food items. However, the potential risks associated with the use of nanoparticles need to be carefully examined/addressed. The guidelines on nanoparticles usage in food has been implemented by the European Commission (EC) and United States Food and Drug Administration (U.S. FDA), keeping with the risk assessment of the nanoparticle size.

The wider application of nanomaterials as food additives, supplements, and preservatives among others is subject to good design and reproducibility of nanomaterials. Additionally, the use of nanoparticles (Zinc oxide, iron oxide, and copper oxide) in animal feeds has been declared safe by the U.S. FDA. In food preservation, edible coating with nanomaterials has demonstrated significant prospects: freshly coated fruits/vegetables remained in good condition during storage [15]. Furthermore, the broad use of nanoclay as the food packaging material is due to its thermal, and mechanical properties and less cost. Gabr et al. [16] discussed that 3% nanoclay loaded nanocomposites remarkably improved the toughness and transition temperature of glass by approx. 6 °C [16]. Another study suggested that nano-edible films (<100 μm) may be used to increase mechanical properties, such as moisture/gas barrier, and enhance the shelf life of fresh food items [17]. The application of TiO_2 nanoparticles in several fields can be attributed to its anti-proliferative and biocidal functions [18], including novel antimicrobials [19]. In food industries, nanoparticles are used to increase nano-nutraceuticals and health supplements bioavailability, towards the improvement of stability, taste, and food texture [20]. In addition, the antimicrobial properties of nanoparticles in food packaging maintain the safety of food for consumption and increase shelf life. In food processing, the use of nanocapsules is gaining momentum, as additives, nutritional supplements, nano-sized ingredients, in food [20, 21]. Although nanotechnological interventions in food and nutraceutical production are emerging rapidly, there is a growing concern among people regarding the adverse effects of nanoparticles, leading to limited utilization. In the present time, it is necessary to devise guidelines/promote public awareness facilitating the management of potential risks, associated with the application of nanoparticles.

3 Sustainable and Novel Nanomaterials

The use of nanomaterials in food industries has gained considerable interest, with new/novel functions of nanoparticles opening new avenues to improve the performance of traditional products, through diverse sectors in pharmaceutical, agricultural, and industrial fields [22, 23]. The nanomaterials used in food industries are naturally synthesized or engineered [24]. The casein micelles in milk (300–400 nm) produced by farm animals matches with nanoscale dimensions while the engineered nanoparticles, e.g. nano-salt are developed for reducing dietary intake [25], have been developed. The encapsulated nanomaterial as a carrier for minerals and vitamins offers the most promising approach in food industries. Moreover, studies have reported the nanomaterial-mediated delivery of food supplements, and beverages, keeping the taste in its original state [26].

3.1 Biosynthesized Nanomaterials

The biological systems namely, microbes, plants, enzymes have shown good prospects as biosynthetic platforms for nanomaterials and show distinct advantages: during biosynthesis, biological systems act as stabilizing, reducing, and capping agents, minimizing the use of hazardous substances, the biocompatible/less toxic nature of synthesized nanomaterials and biosynthesis at ambient conditions reduce the use of chemicals and energy resources [15]. Studies have shown that lipids and proteins (having functional carboxyl and amide groups) are efficiently adsorbed on the surface of nanoparticles, e.g. the SDS-PAGE of the proteins (extracellular) stabilized the silver nanoparticles during its biosynthesis [27]. Furthermore, chemical synthesis is performed at high temperatures, extreme pH, and high pressure [28]. In addition, the functions of biosynthesized nanomaterials are easily verified by the FT-IR technique and induce many biological processes [15]. Table 1 shows the key representatives of bio-based synthesis of nanoparticles and their socioeconomic relevance.

3.2 Bioinspired Nanomaterials

The bioinspired method aims at nanomaterials with novel functions (having similar morphologies and functions to a biological substance) and is widely employed in the biomedical sector [41] and other fields [42, 43]. The biosynthesis methods rely on biological systems (e.g. plant extracts) for the synthesis of less toxic and green nanomaterials, one such key example is Gecko-inspired adhesives—the adhesive system is based on nanoscale fibers and uses van der Waals forces to produce adhesion [44]. The bioinspired nanomaterials dynamically adjust their

Table 1 Key representatives of bio-based synthesis of nanoparticles and their socioeconomic relevance

Biological species	Nanoparticles	Agricultural applications	Reference
Fungal species			
Fusarium solani	Silver	Antifungal activity	Ingle et al. [29]
Aspergillus niger	Gold	Mosquito larvae control	Bhambure et al. [30]
Rhizopus stolonifer	Gold	–	Binupriya et al. [31]
Algal species			
Chlamydomonas reinhardtii	Silver	–	Rao and Gautam [32]
Chlorella vulgaris	Gold	Antipathogen activity	Annamalai and Nallamuthu [33]
Enteromorpha flexuosa	Silver	Antimicrobial activity	Yousefzadi et al. [34]
Bacterial species (as probiotics)			
Staphylococcus aureus	Silver	Antimicrobial activity	Nanda and Saravanan [35]
Lactobacillus. casei sp.	Silver	Antimicrobial activity	Singh et al. [36]
Actinobacter spp.	Gold	Nanofertilizer	Bharde et al. [37]
Non-probiotic species			
Escherichia coli K12	Gold	Bioremediation	Srivastava et al. [38]
Acinetobacter sp. GWRVA25	Gold	Antioxidant activity	Nadhe et al. [39]
Klebsiella aerogenes	Cadmium sulfide	Antibacterial activity	Holmes et al. [40]

physicochemical properties about changes in environmental conditions. However, challenges exist in the control of nanomaterials at high resolution, comparable to natural biological systems [15]. The application of bioinspired nanomaterials in agriculture/food is limited, with some significant studies on artificial bio-inspired devices for pathogen/chemical detection, environment sensing, and delivery of pesticides. Although being less explored, some interesting studies on bioinspired nanomaterials focusing on agriculture and food sectors are the following: Liang et al. [45] discussed that bioinspired mussel avermectin NPs have strong adhesion properties to crop foliage, decreasing the pesticide contamination and loss of soil. The nanostructure showed potential storage stability, high retention of avermectin, and release sustainably. Another key study discussed the application of bioinspired nanomaterial in environmental sensing which showed high compatibility and sensitivity to flow velocity and direction [46], highlighting significant application. It is imperative to witness the development of bioinspired nanomaterials which completely mimic such biofunctions, attributed to inputs from low-cost methods and large-scale production.

4 Application of Nanomaterials in Food Industries

The development of innovative nanotechnologies has made substantial contributions to the food industry. The application of nanoparticles in the food sector, specifically in food storage, packaging, and development of novel products has witnessed a remarkable increase. The use of nanoparticles is aimed at improving the taste and flavor of food, stability, the bioavailability of nutraceuticals, among others [20, 47]. In addition, antimicrobial properties of nanoparticles may be included in food packaging to ensure food safety and increase shelf life. Furthermore, the application of nanoparticles leads to nano-formulated fertilizers, additives, pesticides for agri-food applications [21]. Table 2 provides comprehensive overview of the nanotechnological interventions in food and nutraceutical production.

Table 2 Nanotechnological interventions in food and nutraceutical production

Nanoparticles	Purpose	Research outcome	Reference
ZnO_2 quantum dots	Detection of pesticides	R&D	Sahoo et al. [48]
Chitosan-based coatings	Edible coating	Experimental testing	Shi et al. [49]
Fluorescent nanosensors	Detection of toxins	R&D	Sun et al. [50]
Nanosilver	Nanosilver food containers	Approved	–
TiO_2 NPs	Color additives	Exempted from certification	Code of Federal Regulations CFR [51]
CuO_2 NPs	Dietary supplement	Approved (animal feed)	U.S. FDA [52]
Ag-silica NPs	Preservative	Approved	U.S. FDA [52]
SiO_2 NPs	Vegetable/fruit marking	Exempted from certification	Code of Federal Regulations CFR [51]
Electrochemical biosensors	Detection of L-alanine in food	–	Sertova [53]
Electrochemical biosensors	Food contamination	–	Sertova [53]
PVF nanofibers	Assessment of food safety (food allergen/pathogens)	Approved	Asadnia et al. [46]
Carbon quantum dots	Analysis of nutritional content of food	Approved	Wang et al. [54]
Mussel avermectin NPs	Nanocarrier-controlled release of avermectin	Approved	Liang et al. [45]
Graphene NPs	Detection of food pathogens	Approved	Mannoor et al. [55]
TiO_2 NPs	Enzyme as reducing agents	–	Ahmad et al. [56]

4.1 Food Processing

Nanoparticles are of great importance in food processing and are used for improving flow properties, flavor, nutritional content, shelf life, and others. The nanotechnological interventions lead to the development of food products (rich in health contents), namely lower in sugar, fats, and salt content to reduce food-associated diseases [21]. In addition, many micro- and nano-sized assemblies of nanoparticles were designed and developed for food ingredients encapsulation, nutritional supplements, and use as a functional food [57]. The nanoemulsion and nano-encapsulation methods are used widely in food processing, with the engineered nanoparticles showing good potential. Furthermore, inorganic oxide chemicals such as TiO_2 and MgO (approved by U.S. FDA), are used as a carrier of food flavor, food additives, and anti-caking agents [15]. A key example highlights the use of TiO_2 as food additives namely in candies, cake icing, puddings, etc. In addition, nanoparticles of iron oxide, zinc oxide, and copper oxide are recognized as safe (GRAS) by U.S. FDA in animal feed as dietary supplements. Some recent studies showed that encapsulated vitamins were delivered into the bloodstream with high efficiency, while few discussed the use of nanoparticles emulsions in ice cream, which improves the uniformity and texture [21].

4.2 Preservatives

Some nanoparticles, like TiO_2 and ZnO_2, demonstrate antimicrobial activity, therefore show good prospects in food preservation (with improved functional properties). A study by Bryaskova et al. discussed the synthesis and application of silver nanoparticles (AgNPs) as effective antibacterials against food-borne *Pseudomonas aeruginosa*, *E. coli*, and *Bacillus subtilis* [58]. In addition to the pharmaceutical application, the study showed that promising AgNPs may be used for food packaging and preservation [59]. In another study, silver nanoparticles were synthesized (using *Serratia* sp. culture) and they showed potential antifungal activity against the wheat pathogen [60]. ZnO_2 nanoparticles synthesized from *Catharanthus roseus* showed effective antibacterial activity against *E. coli*, *P. aeruginosa*, and *B. thuringiensis* [61]. Furthermore, metal nanoparticles are mainly employed to prevent food contamination, while the different sizes and high surface area are some properties of nanoparticles that lead to diverse applications in the food industry. Biosensors find relevance in food pathogen detection/food contamination, thereby evaluation of food quality. Similarly, other nanoparticles such as silver, gold, and titanium oxide are used commercially to control food-borne pathogens [59].

4.3 Color Additives

Nanotechnological interventions in the food sector provide novel avenues for manufacturers to monitor the color of the products in a sophisticated manner. However, the nanotechnology-based color additives must be approved by FDA for commercial use. Color additives are defined as a pigment, dye, or substance that is synthesized naturally or chemically, that adds color to food, cosmetics, or the human body [62]. A much different range of color additives was synthesized and studied for their prospects in the food industry. TiO_2 is approved (not exceeding 1% w/w) as a food additive [49]. TiO_2 in combination with SiO_2/Al_2O_3 has also been approved and used as color additives in food items. Nano-sized additives/ingredients in functional food comprise antimicrobials, antioxidants, and preservatives for increased absorption and taste [20]. Moreover, beta-carotene, phytosterols, and lycopene represent nutraceuticals incorporated in the carriers and used in food to reduce cholesterol levels [63].

4.4 Nutritional Dietary Supplements

The successful application of nanotechnology to improve food products and their functional attributes has been documented. The nanotechnological interventions in the food industry aim at healthier food ingredients/components as well as sustainability, exemplified by diverse food ranges with safe intake, good taste, and affordability. The new range of food products is characterized by low-calorie content, prolonged digestion, and the bioavailability of fortified food [64]. Several ongoing types of research are focused on the development of functional food, aiming at efficient delivery of nutrients, in keeping with the requirements in the body. The development of nanocapsules and their inclusion in food for nutrient delivery is an emerging aspect including the application of nanoparticles to enhance nutrient absorption in food. Nanoencapsulation of nanoparticles for packaging aims at forming nanocapsules for targeted delivery of medicines (https://just-food.nridigital.com/). In the food industry food packaging with nanoparticles shows better efficiency in the delivery of nutrients to cells. Moreover, nanoencapsulation of vitamins for increased wellness has witnessed a tremendous upsurge; the commercial market for vitamins has increased from £417 m in 2013, predicted to reach £477 m by 2023 (https://just-food.nridigital.com/). The additional advantages of vitamins encapsulation include masking of bitter flavor (during consumption) and increased bioavailability of the vitamins. However, for the general use of nanoparticles, it is important to understand the general impact on human health (before use) and promote public awareness.

4.5 Polymer Production

Nanocomposites with nanoparticles aid in improving the property of the polymer and provide diverse chemical functions therefore used for high barrier properties development [65]. The use of nanocomposites maintains the freshness of the food product and restricts microbial contamination for a certain time. In carbonated beverages, these act as barriers for gas leakage (carbon dioxide) from the bottles [65]. Another key example is Nanoclay, a nanocomposite (with nanoparticles) that acts as gas barriers and comprise phyllosilicates, the naturally occurring aluminum silicates, and classified as exfoliated nanocomposites and intercalated nanocomposites [66]. While the exfoliated nanocomposites are layers of clay dispersed in a polymer matrix, the intercalated nanocomposites are defined as multilayer structures. Another example is the single-walled nanotubes combined with SiO_2 nanoparticles, forming effective gas barriers [67]. Some of the commercially marketed Nanoclays are Durethan, Aegis, and Imperm [67]. Nanoencapsulation-based Nanocoatings (nanolaminates), are used in the food industry to coat vegetables, fruits, meat products, and baked items. In addition, polymers combined with metals have antimicrobial properties, for instance, nanomagnesium oxide [68].

5 Food Contact Packaging

There have been wider implications of food packaging with nanoparticles, in the current scenario. With the increasing application of nanoparticles in food packaging, concerns regarding the safety of consumption are still to achieve public approval/acceptance. However, amidst popularity and highlighted challenges, nanoparticles-mediated food applications are gaining widespread recognition. Some of the popular and widespread applications of nanoparticles in food packaging comprise the following:

5.1 Detection of Pathogens in Food

Food products are likely to get contaminated by food-borne pathogens, leading to food wastage and the risk of food-associated diseases. Several applications of nanobiosensors in food analysis suggest monitoring food and animal feed for the presence of pathogens and toxin contaminants. These are also effective in the detection of mycotoxins, toxic metabolites produced by fungi, and require selective methods for detection [53]. Ochratoxin A (an abundant mycotoxin) contaminates coffee, animal feeds, wines, etc., and employing biosensors for its detection has been an efficient way to assess food quality/suitability for consumption. Recent advances in the development of nanostructured immunosensors (antigen bound to gold nanoparticles) efficiently measure low levels of mycotoxins in food components [69]. A significant example of nanosensors for the detection of food-borne pathogens is employing poly-(dimethylsiloxane) (PDMS) chips (with specific antibodies) for immunosensing of enterotoxin B of *Staphylococcus* sp. [70]. Moreover, lysozyme-coated polystyrene nanoparticles demonstrated bactericidal function against *Listeria monocytogenes* [71], showing potent activity for the control of food-borne pathogens. The UV-activated TiO_2 nanoparticles were dispersed through EVOH films and exhibited photo-activated biocidal properties against *Vibrio parahaemolyticus*, *L. monocytogenes*, and others [72].

6 Nanotechnology and Nutraceutical Production

Nanotechnological interventions in the nutraceutical industry are gaining increased momentum, enhancing the bioavailability, solubility, and biological functions of the encapsulated food components [73]. The functional ingredients (nutraceuticals) are the key components of many industrial products, displaying multiple properties; however, these are utilized via different delivery systems. Moreover, the key features of the delivery products should be protection, compatibility, transportation, and regulated release. The primary delivery systems which are studied include association colloids and nanoemulsions [74]. Nanoemulsions offer distinct benefits including high kinetic stability and the ability to include lipophilic/hydrophilic compounds into emulsion state [74]. Particularly, the nano-sized iron materials in the deliveries composed of nutraceuticals improve the bioavailability of the food products. Furthermore, the bioavailability of vitamins and other compounds like carotenoids and curcumin is low and may be increased by using nanoformulations [73]. In addition, nanotechnology-based systems are employed to increase the delivery of bioactive compounds and their bioavailability. While the nanoparticles comprise food macromolecules namely, lipids, polysaccharides, and surfactants, composite nanoparticles are synthesized by different combinations of these macromolecules [75]. The multiple properties of bioactive food constituents (stability and efficiency) may be improved by food-grade nanoparticles [76]. Moreover, a combination of pharmaceuticals and nutrition, known as nanonutraceuticals includes the production of dietary supplements (herbal products, bioactive constituents, etc.) employing the nanoformulation method [15]. Nanotechnology in nutraceutical delivery offers several advantages—nanofibers, nanotubes, nanosheets, etc., are delivered through microemulsions, liposomes, nanosensors, microgels, and other methods [77]. However, it is pertinent to understand the benefits and potential toxicity of nanocarriers in food components. Recent advances in fabrication and design improvement of multiple food-grade nanoparticles have contributed to the development of efficient delivery systems. For example, the bioavailability of the food components may be improved by colloidal delivery-based foods, leading to increased absorption and increased bioactivity, in addition to multiple health advantages [15].

7 Nanotechnology in Agriculture: Nanoparticles in Plant Growth Promotion and Stress Tolerance

Nanotechnology constitutes an important tool to promote sustainable agriculture, projected to be a defining platform in generating socioeconomic returns. The application of nanotechnological interventions in agriculture aims to decrease pesticide usage and provide eco-friendly agrochemicals and novel delivery methods for crop production. The beneficial applications of nanotechnology in agriculture include nanodevices for plant genetic engineering, nanoformulations to reduce pesticide usage/crop improvement, diagnosis of plant diseases, nanosensors in crop protection, and animal health [1]. The enlightening field of nanotechnology holds a good potential to impact and accelerate biomass for the technological production of biofuels. However, assessment and a deep understanding of the adverse/toxic effects of nanomaterials and generating public awareness are necessary to explore completely the nanotechnological interventions in agriculture.

7.1 Nanofertilizers

Different types of fertilizers and their development aim to boost nutrient availability for plants, thereby aiming to boost crop productivity. The Food and Agriculture Organization (FAO) suggested that the use of mineral fertilizers reached 48.5 megatonnes phosphate (P_2O_5), 38.7 megatonnes potash (K_2O), and 110 megatonnes of nitrogen (N) (2016), in contrast to 2002 [78]. Although the use of fertilizers aims to increase crop productivity, extensive usage threatens the environment, leading to adverse implications in terms of pollution of groundwater, poor soil quality, greenhouse emissions, and hazard to human health [79, 80]. Considering the emerging trend, the application of nanofertilizers has made a substantial contribution by maintaining ecological balance and increasing crop productivity. Nanofertilizers are also known as smart carriers for the nutrient composition accessible to the plant parts, leading to better crop yield and fitness [81]. Encapsulation of macro- and micronutrients within the nanoparticles may be directly delivered through nanoemulsions [82, 83]. The application of nanofertilizers projects distinct advantages in terms of less production cost, multiple nutrient accessibility to plants, increasing plant fitness and yield, and reducing the usage of conventional pesticides. Furthermore, nanofertilizers are efficient in multiple nutrient deliveries, prevent loss of nutrients and targeted delivery to the plant parts, and thereby prevent nutrient losses [84]. An interesting example shows that chitosan-based substances as one of the most effective nano-agrochemicals for efficient nutrient delivery in agriculture [85]. In addition to the above, nano-enhanced fertilizers comprise nanomaterials comprising of silicon NPs, zeolites, TiO_2 nanoparticles among others [81], and assist plants in nutrient uptake. Table 3 provides a tabular representation of key nanomaterial-based fertilizers and prospects in sustainable agriculture.

Table 3 Commercially available nanomaterials used as fertilizers and their beneficial role in sustainable agriculture

Nano-based fertilizers	Plant species	Biological effect on plant growth	Reference
Chitosan nanoparticles	Seedlings of coffee	Promotes plant height and leaf area/number	Ha et al. [86]
Hydroxyapatite NPs-urea	*Oryza sativa*	Delays nitrogen release	Kottegoda et al. [87]
Nanoparticles of Cu and Zn metals	*Ocimum basilicum*	Enhanced chlorophyll and carotenoid leaves content	Abbasifar et al. [88]
Nano-chitosan	*Triticum aestivum*	Promotes shoot and root length, crop index	Aziz et al. [89]
Fe_2O_3 nanocomposites (zeolites)	–	Slow iron release, promotes plant growth	Jahangirian et al. [90]
ZnO nanoparticles	*T. aestivum*	In drought stress, increase Zn uptake for plant	Dimkpa et al. [91]
Urea-based nanofertilizers	*O. sativa*	High nitrogen recovery for plant and high yield	Naseem et al. [92]
Nanostructured alumina	–	Protection of seeds against insect pests	Belhamel et al. [93]
Silver nanoparticles	*Punica granatum*	Control of bacterial blight disease	Sherkhane et al. [94]
Copper nanoparticles	Multiple crops	Control of fungal pathogens	Pariona et al. [95]
Metsulfuron-based pectin nanoparticles	–	Control of *Chenopodium album* (weeds)	Kumar et al. [96]
Nanoencapsulated essential oil of (*Satureja hortensis* L.)	–	Control of different herbs	Taban et al. [97]
Graphene oxide nanobiosensors	–	Assessment of nitrate in water and soil	Ali et al. [98]
Quantum dots	–	Pesticide detection	Kołataj et al. [99]
Titanium nanoparticles	*Nicotiana benthamiana*	Control of plant virus diseases	Hao et al. [100]

7.2 *Nanopesticides*

Pesticides are employed in agriculture to promote plant health and productivity. The development and use of nanopesticides in agriculture offer promise to address the existing problems with routine pesticides [101]. Nanopesticides comprise multiple products; consisting of organic polymers, surfactants, and metal nanoparticles as nanoformulations [1]. The techniques adopted to improve solubility for nanoagrochemicals employed the microencapsulation method to enhance aqueous dispersion in the controlled release of hydrophobic pesticides [102]. The initial studies showed that the insecticide ethiprole was encapsulated by nanospheres comprising of poly-lactic acid and polycaprolactone. The study showed a controlled release of agrochemicals and enhanced penetration in the plant [103]. Another study performed in *Spodoptera littoralis* (cotton leafworm) showed that nanoparticle toxicity was similar to the formulations used commercially to control the leafworm [104]. Nanomaterial application as additives (for active constituents) is demonstrated by a good example: Imidacloprid formulations, synthesized from aliphatic diacids and polyethylene glycol (via encapsulation), were effective in the management of pests causing plant diseases. Furthermore, the commercial formulation, as well as the above-mentioned formulation, was evaluated against soybean pests and the result suggested better efficacy of Imidacloprid formulations, as compared to commercial ones as well as higher plant yield [105, 106].

Several nanomaterials namely, gold nanoparticles, polymeric nanoparticles, and iron oxide nanoparticles, are widely used as pesticides. The various properties/functions of nanoparticles including formulations, potential functions, and characteristics were studied for plant disease management [107]. Another important application of nanoparticles was in the management of insect pests. The pest *Helicoverpa armigera* was managed through employing nanotechnological interventions [108]. *Tinospora cordifolia* extract in a combination of silver nanoparticles was effective against *Pediculus humanus* and *Anopheles subpictus* and *Culex-quinque fasciatus* (larvae), demonstrating significant larvicidal activity [109]. Several examples showed the promising effects of nanofertilizers in plant disease management. Nanopesticides include different sub-classes, namely nanofungicides, nanoinsecticides, nanobactericides, nanoinsecticides, and nanoweedicindes, addressing the various purposes of agricultural applications [110, 111]. While nanoencapsulation of pesticides is beneficial to improve the active ingredient, nanocarriers (safe, biocompatible, and environment friendly) improve the solubility of bioactive constituents [112], improving agricultural practices to a considerable extent. In addition, formulations of different constituents namely, lipid NP, polymeric NP, nanogels, carbon substances, and nanoemulsions, have been used as effective nanopesticides in agriculture [113, 114].

7.3 Nanosensors for Smart Agriculture

In agricultural application, nanosensors are effective in the detection of different pesticides, pathogens, fertilizers, and pH of the soil, and their regulated application may promote crop production [115]. The use of "smart sensors" for precision farming offers an attractive opportunity with reduced inputs, efficient management of fertilization, and proper time use. Furthermore, NP-based delivery systems in agriculture aim at efficient utilization of natural resources namely nutrients, water, and chemical for better farming practices. In recent times, precision farming practices are proving significant in smart agriculture and comprise remote-sensing systems, satellite-positioning systems, and geographic systems for the detection of pests and environmental stresses. Other significant applications of nanosensors include the detection of environmental pollution by nano-smart dust (tiny wireless sensors) [116], quality of agricultural products by nanobarcodes [117], and regulation of plant hormones through nanotechnology for understanding plant root adaptation in soil [118]. Considering the emerging importance, nanosensors have the potential to impact the agri-food sector and the environment with the signature initiative "Nanotechnology for Sensors and Sensors for Nanotechnology: Improving and Protecting Health, Safety, and the Environment" put forth by the National Nanotechnology Initiative.

8 Current Status, Prospects, and Challenges

8.1 The Need for Legalization, Public Awareness, and Acceptance

The present era has witnessed the significant and emerging contributions of nanotechnological interventions in the agri-food interface as well as in the environment; however, there is a fundamental requirement in understanding the existing and potential implications. With associated challenges in nanotechnology applications, there is a growing need for regulation and legislation in the socioeconomic interface as well as generating public awareness. The main procedure/guidelines for legislation include: Guidelines/suggestions by academic bodies (organizations/individuals), Suggestions taken by the government and suggested by the respective head (legislation or head of country) which is further passed or approved by the country and enforced as law, and the respective legislation mostly includes the general/broad aspects for industry [15]. Furthermore, when the law is enforced by regulators, a detailed/specific guideline is implemented and usually pertains to a broader scope of the industry. For the application of nanomaterials, certain standard protocols are followed, which require an understanding of different properties of nanoparticles and their implications (toxic/adverse effects). In 2003, government agencies implemented guidelines for legislation of nanotechnology in the food industry; according to the European Union, Registration, Evaluation, Authorisation, and Restriction of Chemicals (REACH) was undertaken as key regulation, led by Food and Drug Administration (FDA, USA).

8.2 Toxicity/Risk Assessment

The potential of nanotechnological interventions in the agri-food sector is immense (food components, livestock feed, food packaging, and smart nano-systems); however, there is little awareness about the safety of these NPs in food and nutraceutical production [1]. With the unavailability of clear information, the uncertainty concerning the nature and potential associated risks of NPs remains. The current application of NPs in food additives/ingredients comprise of inclusion of NPs as food component or in food packaging. Public opinion needs to be taken into account regarding the development of NP-based food products and their wider implications [119]. A global concern comprises assessing the benefits and risks of the silver NPs application as an antibacterial agent in the food and healthcare sector. Moreover, the absence of proper data and regulation procedures continues to affect global procedures regarding the commercial application of NPs [47].

8.3 Prospects/Directions

Nanotechnology continues to make great strides in different areas of food, livestock, and agriculture, impacting human lives and generating socioeconomic returns. The prospects of nanotechnological interventions are immense, with nanotechnology-based precision farming aiming to maximize crop productivity through efficient utilization of pesticides, water, and fertilizers by plants. Recent innovations in product development (NP-mediated food packaging, and processing) continue to benefit the food industry and farmers by generating new resources for food and nutraceutical production. In addition, the development/application of nanosensors for monitoring soil quality, plant health, and pathogen detection and nanoencapsulation for efficient delivery of agrochemicals, nanoemulsions (antimicrobials) in food quality assessment highlights distinct advantages in promoting agri-food sectors for the betterment of the global population. While maximizing the benefits of nanotechnological interventions in multiple socioeconomic applications, it is equally important to generate public awareness/implement guidelines for understanding the potential benefits and addressing the associated challenges with NPs. The future of nanotechnology in promoting sustainable agriculture looks promising as far as measures to safeguard the social and environmental well-being are addressed.

Acknowledgments The author acknowledges her institution for encouragement and support.

Conflict of Interest No conflict of interest was declared.

References

1. Sekhon, B. S. (2014). Nanotechnology in agri-food production: An overview. *Nanotechnology, Science and Applications, 2014*(7), 31–53.
2. Batsmanova, L. M., Gonchar, L. M., Taran, N. Y., & Okanenko, A. A. (2013). Using a colloidal solution of metal nanoparticles as micronutrient fertiliser for cereals. In *Proceedings of the International Conference on Nanomaterials: Applications and Properties* (September 16–21, 2013, Crimea, Ukraine) Retrieved April 18, 2014, from http://nap.sumdu.edu.ua/index.php/nap/nap2013/paper/view/1097/504

3. Scott, N., & Chen, H. (2002). Nanoscale science and engineering for agriculture and food systems. In *National Planning Workshop* (November 18–19, 2002, Washington, DC) Retrieved April 18, 2014, from http://www.nseafs.cornell.edu/web.roadmap.pdf
4. Gruère, G., Narrod, C., & Abbott, L. (2011). Agriculture, food, and water nanotechnologies for the poor: Opportunities and constraints. *IFPRI Policy Brief, 19*, 1. Retrieved May 6, 2014, from http://www.ifpri.org/sites/default/files/publications/bp019.pdf
5. Frewer, L. J., Norde, W., Fischer, A. R. H., & Kampers, F. W. H. (Eds.). (2011). *Nanotechnology in the agri-food sector: Implications for the future*. Wiley-VCH.
6. Sonkaria, S., Ahn, S. H., & Khare, V. (2012). Nanotechnology and its impact on food and nutrition: A review. *Recent Patents on Food, Nutrition & Agriculture, 4*(1), 8–18.
7. Senturk, A., Yalcın, B., & Otles, S. (2013). Nanotechnology as a food perspective. *Journal of Nanomaterials & Molecular Nanotechnology, 2*, 6.
8. Boom, R. M. (2011). Nanotechnology in food production. In L. J. Frewer, W. Norde, A. R. H. Fischer, & F. W. H. Kampers (Eds.), *Nanotechnology in the agri-food sector: Implications for the future* (pp. 39–58). Wiley-VCH.
9. AZoNano.com. (2003). *Nanofibers to be used in drug delivery, gene therapy, crop engineering and environmental monitoring*. AZoM.com Pty. Ltd. Updated June 11, 2013. Retrieved April 19, 2014, from http://www.azonano.com/article.aspx?ArticleID=114
10. Torney, F., Trewyn, B. G., Lin, V. S., & Wang, K. (2007). Mesoporous silica nanoparticles deliver DNA and chemicals into plants. *Nature Nanotechnology, 2*(5), 295–300.
11. Kole, C., Kole, P., Randunu, K. M., et al. (2013). Nanobiotechnology can boost crop production and quality: First evidence from increased plant biomass, fruit yield and phytomedicine content in bitter melon (Momordica charantia). *BMC Biotechnology, 13*, 37.
12. Farrell, D., Hoover, M., Chen, H., & Friedersdorf, L. (2013). *Overview of resources and support for nanotechnology for sensors and sensors for nanotechnology: Improving and protecting health, safety, and the environment*. US National Nanotechnology Initiative. Retrieved April 19, 2014, from http://nano.gov/sites/default/files/pub_resource/nsi_nanosensors_resources_for_web.pdf
13. Scott, N., & Chen, H. (2013). Nanoscale science and engineering for agriculture and food systems. *Industrial Biotechnology, 9*, 17–18.
14. Mukhopadhyay, S. S. (2014). Nanotechnology in agriculture: Prospects and constraints. *Nanotechnology, Science and Applications, 2014*(7), 63–71.
15. He, X., Deng, H., & Hwang, H.-m. (2019). The current application of nanotechnology in food and agriculture. *Journal of Food and Drug Analysis, 27*(2019), 1–21.
16. Gabr, M. H., Okumura, W., Ueda, H., Kuriyama, W., Uzawa, K., & Kimpara, I. (2015). Mechanical and thermal properties of carbon fiber/polypropylene composite filled with nano-clay. *Composites Part B: Engineering, 69*, 94–100.
17. Flores-Lopez, M. L., Cerqueira, M. A., de Rodríguez, D. J., & Vicente, A. A. (2016). Perspectives on utilization of edible coatings and nano-laminate coatings for extension of postharvest storage of fruits and vegetables. *Food Engineering Reviews, 8*, 292–305.
18. Blake, D. M., Maness, P.-C., Huang, Z., Wolfrum, E. J., Jacoby, W. A., & Huang, J. (1999). Application of the photocatalytic chemistry of titanium dioxide to disinfection and the killing of cancer cells. *Separation and Purification Reviews, 28*(1), 50.
19. Duncan, T. V. (2011). Applications of nanotechnology in food packaging and food safety: Barrier materials, antimicrobials and sensors. *Journal of Colloid and Interface Science, 363*(1), 1 24.
20. Momin, J. K., Jayakumar, C., & Prajapati, J. B. (2013). Potential of nanotechnology in functional foods. *Emirates Journal of Food and Agriculture, 25*(1), 10 19.
21. Berekaa, M. M. (2015). Nanotechnology in food industry; Advances in food processing, packaging and food safety. *International Journal of Current Microbiology and Applied Sciences, 4*(5), 345–357.

22. Landsiedel, R., Ma-Hock, L., Kroll, A., Hahn, D., Schnekenburger, J., Wiench, K., & Wohlleben, W. (2010). Testing metal-oxide nanomaterials for human safety. *Advanced Materials, 22*(24), 2601–2627.
23. Lue, J. T. (2007). *Physical properties of nanomaterials. Encyclopedia of nanoscience and nanotechnology* (Vol. X, pp. 1–46). American Scientific Publishers.
24. Magnuson, B. A., Jonaitis, T. S., & Card, J. W. (2011). A brief review of the occurrence, use, and safety of food-related nanomaterials. *Journal of Food Science, 76*(6), R126–R133.
25. Rasouli, F., & Zhang, W. (2006). *Nanoscale materials*. U.S. Patent US20060286239 A1.
26. Bumbudsanpharoke, N., & Ko, S. (2015). Nano-food packaging: An overview of market, migration research, and safety regulations. *Journal of Food Science, 80*(5), R910–R923.
27. Jain, N., Bhargava, A., Majumdar, S., Tarafdar, J., & Panwar, J. (2011). Extracellular biosynthesis and characterization of silver nanoparticles using *Aspergillus flavus* NJP08: A mechanism perspective. *Nanoscale, 3*, 635–641.
28. He, X., Aker, W. G., Pelaez, M., Lin, Y., Dionysiou, D. D., & Hwang, H.-m. (2016). Assessment of nitrogenefluorine-codoped TiO2 under visible light for degradation of BPA: Implication for field remediation. *Journal of Photochemistry and Photobiology, A: Chemistry, 314*, 81–92.
29. Ingle, A., Rai, M., Gade, A., & Bawaskar, M. (2009). *Fusarium solani*: A novel biological agent for the extracellular synthesis of silver nanoparticles. *Journal of Nanoparticle Research, 11*, 2079–2085.
30. Bhambure, R., Bule, M., Shaligram, N., Kamat, M., & Singhal, R. (2009). Extracellular biosynthesis of gold nanoparticles using *Aspergillus niger*—Its characterization and stability. *Chemical Engineering & Technology: Industrial Chemistry, Plant Equipment, Process Engineering, Biotechnology, 32*(7), 1036–1041.
31. Binupriya, A., Sathishkumar, M., & Yun, S.-I. (2010). Biocrystallization of silver and gold ions by inactive cell filtrate of *Rhizopus stolonifer*. *Colloids and Surfaces, B: Biointerfaces, 79*, 531–534.
32. Rao, D., & Gautam, P. (2014). A facile one-pot synthesis of gold nanoparticles by *Chlamydomonas reinhardtii*. *Asian Journal of Microbiology Biotechnology & Environmental Sciences, 16*, 633–639.
33. Annamalai, J., & Nallamuthu, T. (2015). Characterization of biosynthesized gold nanoparticles from aqueous extract of *Chlorella vulgaris* and their anti-pathogenic properties. *Applied Nanoscience, 5*(5), 603–607.
34. Yousefzadi, M., Rahimi, Z., & Ghafori, V. (2014). The green synthesis, characterization and antimicrobial activities of silver nanoparticles synthesized from green alga *Enteromorpha flexuosa* (wulfen). J Agardh. *Materials Letters, 137*, 1–4.
35. Nanda, A., & Saravanan, M. (2009). Biosynthesis of silver nanoparticles from *Staphylococcus aureus* and its antimicrobial activity against MRSA and MRSE. *Nanomedicine: Nanotechnology, Biology and Medicine, 5*, 452–456.
36. Singh, S., Singh, B., Yadav, S., & Gupta, A. (2014). Applications of nanotechnology in agricultural and their role in disease management. *Research Journal of Nanoscience and Nanotechnology, 5*, 1–5.
37. Bharde, A., Kulkarni, A., Rao, M., Prabhune, A., & Sastry, M. (2007). Bacterial enzyme mediated biosynthesis of gold nanoparticles. *Journal of Nanoscience and Nanotechnology, 7*(12), 4369–4377.
38. Srivastava, S. K., Yamada, R., Ogino, C., & Kondo, A. (2013). Biogenic synthesis and characterization of gold nanoparticles by *Escherichia coli* K12 and its heterogeneous catalysis in degradation of 4-nitrophenol. *Nanoscale Research Letters, 8*, 70.
39. Nadhe, S. B., Wadhwani, S. A., Singh, R., & Chopade, B. A. (2020). Green synthesis of AuNPs by *Acinetobacter* sp. GWRVA25: Optimization, characterization, and its antioxidant activity. *Frontiers in Chemistry, 8*, 474.

40. Holmes, J. D., Smith, P. R., Evans-Gowing, R., Richardson, D. J., Russell, D. A., & Sodeau, J. R. (1995). Energy-dispersive X-ray analysis of the extracellular cadmium sulfide crystallites of Klebsiella aerogenes. *Archives of Microbiology, 163*(2), 143–147.
41. Yoo, J. W., Irvine, D. J., Discher, D. E., & Mitragotri, S. (2011). Bio-inspired, bioengineered and biomimetic drug delivery carriers. *Nature Reviews. Drug Discovery, 10*, 521.
42. Zong, L., Li, M., & Li, C. (2017). Bioinspired coupling of inorganic layered nanomaterials with marine polysaccharides for efficient aqueous exfoliation and smart actuating hybrids. *Advanced Materials, 29*, 1604691.
43. Feng, Y., Zhu, W., Guo, W., & Jiang, L. (2017). Bioinspired energy conversion in nanofluidics: A paradigm of material evolution. *Advanced Materials, 29*, 1702773.
44. Autumn, K., Sitti, M., Liang, Y. A., Peattie, A. M., Hansen, W. R., Sponberg, S., et al. (2002). Evidence for van der Waals adhesion in gecko setae. *Proceedings of the National Academy of Sciences of the United States of America, 99*, 12252–12256.
45. Liang, J., Yu, M., Guo, L., Cui, B., Zhao, X., Sun, C., et al. (2018). Bioinspired development of P (SteMAA) eavermectin nanoparticles with high affinity for foliage to enhance folia retention. *Journal of Agricultural and Food Chemistry, 66*, 6578–6584.
46. Asadnia, M., Kottapalli, A. G. P., Karavitaki, K. D., Warkiani, M. E., Miao, J., Corey, D. P., et al. (2016). From biological cilia to artificial flow sensors: Biomimetic soft polymer nanosensors with high sensing performance. *Scientific Reports, 6*, 32955.
47. Chaudhry, Q., Scotter, M., Blackburn, J., et al. (2008). Applications and implications of nanotechnologies for the food sector. *Food Additives & Contaminants. Part A, Chemistry, Analysis, Control, Exposure & Risk Assessment, 25*(3), 241–258.
48. Sahoo, D., Mandal, A., Mitra, T., Chakraborty, K., Bardhan, M., & Dasgupta, A. K. (2018). Nanosensing of pesticides by zinc oxide quantum dot: An optical and electrochemical approach for the detection of pesticides in water. *Journal of Agricultural and Food Chemistry, 66*(2), 414–423.
49. Shi, H., Magaye, R., Castranova, V., & Zhao, J. (2013). Titanium dioxide nanoparticles: A review of current toxicological data. *Particle and Fibre Toxicology, 10*, 15. https://doi.org/10.1186/1743-8977-10-15. PMID: 23587290; PMCID: PMC3637140.
50. Sun, J. C., Xiao, T., Shi, X., Li, X., Zhao, Q., Li, D., & Chen, J. (2018). Development of a selective fluorescence nanosensor based on molecularly imprinted-quantum dot optosensing materials for saxitoxin detection in shellfish samples. *Sensors and Actuators B: Chemical, 258*, 408–414. https://doi.org/10.1016/j.snb.2017.11.143
51. Code of Federal Regulations (CFR). (2018). *Electronic code of federal regulations. Title 21: Food and drugs. PART 73—LISTING OF COLOR ADDITIVES EXEMPT FROM CERTIFICATION.* The United States office of the federal register (OFR) and the United States, Government Publishing Office. Retrieved from https://www.ecfr.gov/cgi-bin/text-idx?SID=79a76b1d7e7a98ae9459d88005ab7058&mc=true&node=pt21.1.73&rgn=div5Aili
52. U.S. FDA. (2015). *Color additive status list.* United States Food & Drug Administration. Retrieved August 7, 2018.
53. Sertova, N. M. (2015). Application of nanotechnology in detection of mycotoxins and in agricultural sector. *Journal of Central European Agriculture, 16*(2), 117–130.
54. Wang, H., Liu, S., Song, Y., Zhu, B.-W., & Tan, M. (2019). Universal existence of fluorescent carbon dots in beer and assessment of their potential toxicity. *Nanotoxicology, 13*(2), 160–173. https://doi.org/10.1080/17435390.2018.1530394
55. Mannoor, M. S., Tao, H., Clayton, J. D., Sengupta, A., Kaplan, D. L., Naik, R. R., et al. (2012). Graphene-based wireless bacteria detection on tooth enamel. *Nature Communications, 3*.
56. Ahmad, R., Mohsin, M., Ahmad, T., & Sardar, M. (2015). Alpha amylase assisted synthesis of TiO_2 nanoparticles: Structural characterization and application as antibacterial agents. *Journal of Hazardous Materials, 283*, 171–177. https://doi.org/10.1016/j.jhazmat.2014.08.073
57. Augustin, M. A., & Hemar, Y. (2009). Nano- and micro-structured assemblies for encapsulation of food ingredients. *Chemical Society Reviews, 38*, 902–912.

58. Morais, M. G. D., Martins, V. G., Steffens, D., Pranke, P., & Costal, J. A. V. D. (2014). Biological applications of nanobiotechnology. *Journal of Nanoscience and Nanotechnology, 14*, 1007–1017.
59. Prakash, J., Vignesh, K., Anusuya, T., Kalaivani, T., Ramachandran, C., Sudha Rani, R., Rub, M., Khan, I., Elahi, F., Deog-Hwan, O., & Devanand, V. G. (2019). Application of nanoparticles in food preservation and food processing. *Journal of Food Hygiene and Safety, 34*(4), 317–324. p ISSN 1229-1153, e ISSN 2465-9223.
60. Mishra, A., Kumari, M., Pandey, S., Chaudhry, V., Gupta, K. C., & Nautiyal, C. S. (2014). Biocatalytic and antimicrobial activities of gold nanoparticles synthesized by *Trichoderma* sp. *Bioresource Technology, 166*, 235–242.
61. Bhumi, G., & Savithramma, N. (2014). Biological synthesis of zinc oxide nanoparticles from *Catharanthus roseus* (l.) G. Don. leaf extract and validation for antibacterial activity. *International Journal of Drug Development and Research, 6*, 208–214.
62. Barrows, J. N., et al. (2003). Color additives: FDA's regulatory process and historical perspectives. *Food Safety Magazine*. FFDCA § 201(t), 21 U.S.C. § 321(t); see also 21 C.F. R. § 70.3(g). Note that a colorant added to a food packaging material or other food contact substance is regulated as a food additive, not as a color additive. Retrieved from http://www.fda.gov/ForIndustry/ColorAdditives/RegulatoryProcessHistoricalPerspectives/default.htm
63. Mozafari, M. R., Flanagan, J., & Matia-Merino, L. (2006). Recent trends in the lipid based nanoencapsulation of antioxidants and their role in foods. *Journal of Science and Food Agriculture, 86*, 2038–2045.
64. McClements, D. J. (2020). Advances in nanoparticle and microparticle delivery systems for increasing the dispersibility, stability, and bioactivity of phytochemicals. *Biotechnology Advances, 38*, S0734-9750(18)30136-8.
65. Pandey, S., Zaidib, M. G. H., & Gururani, S. K. (2013). Recent developments in clay-polymer nano composites. *Scientific Journal of Review, 2*(11), 296–328.
66. Davis, D., Guo, X., Musavi, L., Lin, C.-S., Chen, S.-H., & Wu, V. C. H. (2013). Gold nanoparticle-modified carbon electrode biosensor for the detection of *Listeria monocytogenes*. *Industrial Biotechnology, 9*(1), 31–36.
67. Flanagan, J., & Singh, H. (2006). Microemulsions: A potential delivery system for bioactives in food. *Critical Reviews in Food Science and Nutrition, 46*(3), 221–237.
68. Pradhan, N., Singh, S., Ojha, N., Shrivastava, A., Barla, A., Rai, V., & Bose, S. (2015). Facets of nanotechnology as seen in food processing, packaging, and preservation industry. *BioMed Research International, 2015*, 365672. https://doi.org/10.1155/2015/365672. 17 pages.
69. Bonel, L., Vidal, J., Duato, P., & Castillo, J. (2010). Ochratoxin A nanostructured electrochemical immunosensors based on polyclonal antibodies and gold nanoparticles coupled to the antigen. *Analytical Methods, 2*, 335–341.
70. Dong, Y., Phillips, K. S., & Cheng, Q. (2006). Immunosensing of *Staphylococcus* exterotoxin B (SEB) in milk with PDMS microfluidic systems using reinforced supported bilayer membranes (r-SBMs). *Lab on a Chip, 6*(675), 681.
71. Yang, H., Qu, L., Lin, Y., Sun, Y., & Jiang, X. (2007). Detection of *Listeria monocytogenes* in biofilms using immune-nanoparticles. *Journal of Biomedical Nanotechnology, 3*(131), 138.
72. Kim, B., Kim, D., Cho, D., & Cho, S. (2003). Bactericidal effect of TiO2 photocatalyst on selected food-borne pathogenic bacteria. *Chemosphere, 52*, 277–281.
73. Yadav, S. K. (2017). Tissue science & engineering realizing the potential of nanotechnology for agriculture and food technology. *Journal of Tissue Science & Engineering, 8*, 8–11.
74. Chauhan, H., & Prasad, D. (2017). Nanofood materials: Characteristics and evaluations. In S. Sen & Y. Pathak (Eds.), *Nanotechnology in nutraceuticals: Production to consumption*. Taylor and Francis Group, CRC Press. ISBN 9781498721882.
75. Nile, S. H., Baskar, V., Selvaraj, D., Nile, A., Xiao, J., & Kai, G. (2020). Nanotechnologies in food science: Applications, recent trends, and future perspectives. *Nano-Micro Letters, 12*, 45. 1–34.

76. Hildeliza, Q. B., Chanona-pe, J., Jose, L. S. M., Gutie, G. F., & Jimene, A. (2010). Nanoencapsulation: A new trend in food engineering processing. *Food Engineering Reviews, 2*, 39–50.
77. Cushen, M., Kerry, J., Morris, M., Cruz-Romero, M., & Cummins, E. (2012). Nanotechnologies in the food industry-recent developments, risks and regulation. *Trends in Food Science and Technology, 24*, 30–46.
78. FAO. (2018). *World food and agriculture: Statistical pocketbook*. Author.
79. Li, D., & Wu, Z. (2008). Impact of chemical fertilizers application on soil ecological environment. *Journal of Applied Ecology, 19*, 1158–1165.
80. Sharma, N., & Singhvi, R. (2017). Effects of chemical fertilizers and pesticides on human health and environment: A review. *International Journal of Agriculture Environment and Biotechnology, 10*, 675–680.
81. Singh, H., Sharma, A., Bhardwaj, S. K., Arya, S. K., Bhardwaj, N., & Khatri, M. (2021). Recent advances in the applications of nanoagrochemicals for sustainable agricultural development. *Environmental Science: Processes & Impacts, 2021*(23), 213–239.
82. Ghormade, V., Deshpande, M. V., & Paknikar, K. M. (2011). Perspectives for nano-biotechnology enabled protection and nutrition of plants. *Biotechnology Advances, 29*, 792–803.
83. França, D., Messa, L. L., Souza, C. F., & Faez, R. (2019). *Polymers for agri-food applications* (pp. 29–44). Springer.
84. Kah, M., Kookana, R. S., Gogos, A., & Bucheli, T. D. (2018). A critical evaluation of nanopesticides and nanofertilizers against their conventional analogues. *Nature Nanotechnology, 13*, 677–684.
85. Maluin, F. N., & Hussein, M. Z. (2020). Chitosan-based agro-nanochemicals as a sustainable alternative in crop protection. *Molecules, 25*, 1611.
86. Ha, N. M. C., Nguyen, T. H., Wang, S.-L., & Nguyen, A. D. (2019). Preparation of NPK nanofertilizer based on chitosan nanoparticles and its effect on biophysical characteristics and growth of coffee in green house. *Research on Chemical Intermediates, 45*, 51–63.
87. Kottegoda, N., Sandaruwan, C., Priyadarshana, G., Siriwardhana, A., Rathnayake, U. A., Berugoda Arachchige, D. M., Kumarasinghe, A. R., Dahanayake, D., Karunaratne, V., & Amaratunga, G. A. (2017). Urea-hydroxyapatite nanohybrids for slow release of nitrogen. *ACS Nano, 11*, 1214–1221.
88. Abbasifar, A., Shahrabadi, F., & ValizadehKaji, B. (2020). Effects of green synthesized zinc and copper nano-fertilizers on the morphological and biochemical attributes of basil plant. *Journal of Plant Nutrition, 43*, 1104–1118.
89. Aziz, H. M. A., Hasaneen, M. N., & Omer, A. M. (2016). Nano chitosan-NPK fertilizer enhances the growth and productivity of wheat plants grown in sandy soil. *Spanish Journal of Agricultural Research, 14*, 17.
90. Jahangirian, H., Rafiee-Moghaddam, R., Jahangirian, N., Nikpey, B., Jahangirian, S., Bassous, N., Saleh, B., Kalantari, K., & Webster, T. J. (2020). Green synthesis of zeolite/Fe2O3 nanocomposites: Toxicity & cell proliferation assays and application as a smart iron nanofertilizer. *International Journal of Nanomedicine, 15*, 1005.
91. Dimkpa, C. O., Andrews, J., Sanabria, J., Bindraban, P. S., Singh, U., Elmer, W. H., Gardea-Torresdey, J. L., & White, J. C. (2020). Interactive effects of drought, organic fertilizer, and zinc oxide nanoscale and bulk particles on wheat performance and grain nutrient accumulation. *Science of the Total Environment, 722*, 137808.
92. Naseem, F., Zhi, Y., Farrukh, M. A., Hussain, F., & Yin, Z. (2020). Mesoporous ZnAl2Si10O24 nanofertilizers enable high yield of *Oryza sativa* L. *Scientific Reports, 10*, 1–11.
93. Belhamel, C., Boulekbache-Makhlouf, L., Bedini, S., Tani, C., Lombardi, T., Giannotti, P., Madani, K., Belhamel, K., & Conti, B. (2020). Nanostructured alumina as seed protectant against three stored-product insect pests. *Journal of Stored Products Research, 87*, 101607.

94. Sherkhane, A., Suryawanshi, H., Mundada, P., & Shinde, B. (2018). Control of bacterial blight disease of pomegranate using silver nanoparticles. *Journal of Nanoscience and Nanotechnology, 9*, 1–5.
95. Pariona, N., Mtz-Enriquez, A. I., Sanchez-Rangel, D., Carrion, G., Paraguay-Delgado, F., & Rosas-Saito, G. (2019). Green-synthesized copper nanoparticles as a potential antifungal against plant pathogens. *RSC Advances, 9*, 18835–18843.
96. Kumar, S., Bhanjana, G., Sharma, A., Dilbaghi, N., Sidhu, M., & Kim, K. H. (2017). Development of nanoformulation approaches for the control of weeds. *Science of the Total Environment, 586*, 1272–1278.
97. Taban, A., Saharkhiz, M. J., & Khorram, M. (2020). Formulation and assessment of nano encapsulated bioherbicides based on biopolymers and essential oil. *Industrial Crops and Products, 149*, 112348.
98. Ali, M. A., Jiang, H., Mahal, N. K., Weber, R. J., Kumar, R., Castellano, M. J., & Dong, L. (2017). Microfluidic impedimetric sensor for soil nitrate detection using graphene oxide and conductive nanofibers enabled sensing interface. *Sensors and Actuators, B: Chemical, 239*, 1289–1299.
99. Kołataj, K., Krajczewski, J., & Kudelski, A. (2019). *Environmental nanotechnology* (pp. 255–287). Springer.
100. Hao, Y., Yuan, W., Ma, C., White, J. C., Zhang, Z., Adeel, M., Zhou, T., Rui, Y., & Xing, B. (2018). Engineered nanomaterials suppress Turnip mosaic virus infection in tobacco (*Nicotiana benthamiana*). *Environmental Science: Nano, 5*, 1685–1693.
101. Sasson, Y., Levy-Ruso, G., Toledano, O., & Ishaaya, I. (2007). Nanosuspensions: Emerging novel agrochemical formulations. In I. Ishaaya, R. Nauen, & A. R. Horowitz (Eds.), *Insecticides design using advanced technologies* (pp. 1–39). Springer.
102. Perlatti, B., de Souza Bergo, P. L., da Silva, M. F., et al. (2013). Polymeric nanoparticle-based insecticides: A controlled release purpose for agrochemicals, insecticides. In S. Trdan (Ed.), *Insecticides: Development of safer and more effective technologies* (pp. 523–550). InTech.
103. Boehm, A. L., Martinon, I., Zerrouk, R., Rump, E., & Fessi, H. (2003). Nanoprecipitation technique for the encapsulation of agrochemical active ingredients. *Journal of Microencapsulation, 20*(4), 433–441.
104. Elek, N., Hoffman, R., Raviv, U., Resh, R., Ishaaya, I., & Magdassi, S. (2010). Novaluron nanoparticles: Formation and potential use in controlling agricultural insect pests. *Colloids and Surfaces A: Physicochemical and Engineering Aspects, 372*(1–3), 66–72.
105. Adak, T., Kumar, J., Shakil, N. A., & Walia, S. (2012a). Development of controlled release formulations of imidacloprid employing novel nano-ranged amphiphilic polymers. *Journal of Environmental Science and Health, Part B, Pesticides, Food Contaminants, and Agricultural Wastes, 47*(3), 217–225.
106. Adak, T., Kumar, J., Dey, D., Shakil, N. A., & Walia, S. (2012b). Residue and bio-efficacy evaluation of controlled release formulations of imidacloprid against pests in soybean (*Glycine max*). *Journal of Environmental Science and Health, Part B, Pesticides, Food Contaminants, and Agricultural Wastes, 47*(3), 226–231.
107. Al-Samarrai, A. M. (2012). Nanoparticles as alternative to pesticides in management plant diseases-a review. *International Journal of Scientific and Research Publications, 2*(4), 1–4.
108. Vinutha, J. S., Bhagat, D., & Bakthavatsalam, N. (2013). Nanotechnology in the management of polyphagous pest Helicoverpa armigera. *Journal of Academia and Industrial Research, 1* (10), 606–608.
109. Jayaseelan, C., Rahuman, A. A., Rajakumar, G., et al. (2011). Synthesis of pediculocidal and larvicidal silver nanoparticles by leaf extract from heartleaf moonseed plant, *Tinospora cordifolia* Miers. *Parasitology Research, 109*(1), 185–194.
110. Baker, S., Satish, S., Prasad, N., & Chouhan, R. S. (2019). *Industrial applications of nanomaterials* (pp. 341–363). Elsevier.
111. Yadav, A. S., & Srivastava, D. (2015). Application of nanotechnology in weed management: A review. *Research & Reviews: Journal of Crop Science and Technology, 4*, 21–23.

112. Jordan, W. (2010). *Nanotechnology and pesticides*. Pesticide Program Dialogue Committee.
113. Usman, M., Farooq, M., Wakeel, A., Nawaz, A., Cheema, S. A., Rehman, H., Ashraf, I., & Sanaullah, M. (2020). Nanotechnology in agriculture: Current status, challenges and future opportunities. *Science of the Total Environment, 721*, 137778.
114. Kah, M., & Hofmann, T. (2014). Nanopesticide research: Current trends and future priorities. *Environment International, 63*, 224–235.
115. Rai, V., Acharya, S., & Dey, N. (2012). Implications of nanobiosensors in agriculture. *Journal of Biomaterials and Nanobiotechnology, 3*, 315–324.
116. Mousavi, S. R., & Rezaei, M. (2011). Nanotechnology in agriculture and food production. *Journal of Applied Environmental and Biological Sciences, 1*(10), 414–419.
117. Li, Y., Cu, Y. T., & Luo, D. (2005). Multiplexed detection of pathogen DNA with DNA-based fluorescence nanobarcodes. *Nature Biotechnology, 23*(7), 885–889.
118. McLamore, E. S., Diggs, A., Calvo Marzal, P., et al. (2010). Non-invasive quantification of endogenous root auxin transport using an integrated flux microsensor technique. *The Plant Journal, 63*(6), 1004–1016.
119. López-Vázquez, E., Brunner, T. A., & Siegrist, M. (2012). Perceived risks and benefits of nanotechnology applied to the food and packaging sector in México. *British Food Journal, 114*(2), 197–205.

Green Synthesis of Plant-Assisted Manganese-Based Nanoparticles and Their Various Applications

Canh Minh Vu, Suresh Ghotekar, Nguyen Minh Viet, Harshal Dabhane, Rajeshwari Oza, and Arpita Roy

Abstract Nanotechnology is now one of the most advanced concepts globally, which has remarkable properties and implications in every discipline of modern science and engineering. The need for biocompatible materials for diverse uses in varied fields such as health, medicine, and water treatment has attracted greater attention to this topic in recent years. In recent years, the green fabrication ovdf inorganic nanoparticles (NPs) has been intensively explored. On the other hand, manganese has received less focus as a high-performance metal in a variety of disciplines, including medicine, biosensors, biomedicine, catalysis, electrochemistry, electronics, photoelectronics, and water treatment. Manganese and manganese oxides (Mn oxides/based) have diverse structures, including MnO, MnO_2, Mn_2O_3, Mn_3O_4, and Mn_5O_8, and can be employed in a myriad of implementations. Mn-based NPs have a lot of potential for long-term nanotechnology. This chapter focuses on Mn-based NPs' green synthesis, uses, and prospects. Various techniques of green production of Mn-based NPs have been researched and described, namely

C. M. Vu
Advanced Institue of Science and Technology, The University of Da Nang, Da Nang, Vietnam

S. Ghotekar (✉)
Department of Chemistry, Smt. Devkiba Mohansinhji Chauhan College of Commerce and Science, Silvassa 396 230, University of Mumbai, Dadra and Nagar Haveli (UT), Mumbai, India

N. M. Viet
VNU Key Laboratory of Advanced Materials for Green Growth, Faculty of Chemistry, University of Science, Vietnam National University, Hanoi, Vietnam

H. Dabhane
Department of Chemistry, G.M.D Arts, B.W Commerce and Science College, Sinnar 422 103, Savitribai Phule Pune University, Pune, Maharashtra, India

R. Oza
Department of Chemistry, S.N. Arts, D.J.M. Commerce and B.N.S. Science College, Sangamner 422 605, Savitribai Phule Pune University, Pune, Maharashtra, India

A. Roy
Department of Biotechnology, School of Engineering & Technology, Sharda University, Greater Noida, India

J.-T. Chen (ed.), *Plant and Nanoparticles*,
https://doi.org/10.1007/978-981-19-2503-0_13

fabrication using plant extract. Different implications of green-produced Mn-based NPs have also been discussed. Moreover, the future direction of plant extract-mediated eco-benevolent fabrication of Mn-based NPs and their efficient uses is discussed carefully.

Keywords Green synthesis · Plant extracts · Microorganism · Mn-based NPs · Applications

1 Introduction

Nowadays, modern nanotechnology research is gaining popularity due to its revolutionary and promising results in a variety of fields [1–3]. Advanced nanotechnology is evolving as a unique sector of research concerned with the fabrication of multifunctional nanoparticles (NPs) and/or nanomaterials for use in a variety of disciplines including biomedicine, catalysis, cosmetics, electrochemistry, electronics, energy science, food technology, healthcare, mechanics, membrane modification, optical devices, pharmaceutics, sensors, space industry, textile industry, and water treatment, due to their highly multipurpose, modular, and efficient features [4–10].

Due to the high surface area of NPs, they could reveal interesting chemical and physical characteristics in their bulk [11, 12]. NPs can be produced by various species, including actinomycetes, algae, bacteria, fungus, plants, vitamins, and yeasts, and they significantly alter the characteristics of metals [13, 14]. Diverse research and review articles [20–35] have been published on the green synthesis of copper [15], gold [16], silver [17], platinum [18], palladium [19], and other metal oxides [20–35].

Nanobiotechnology is a concept used to describe the connection between nanotechnology and biology [36]. Modern nanobiotechnology is a new and intriguing branch that involves a diverse study sector, including biology, chemistry, engineering, medicine, material science, and physics [21–26]. Because biological entities may access more components to synthesize NPs, nanotechnology has much more advantages than the other conventional protocol. Bionanomaterials should be created using the rich biodiversity of related biological species [37].

To synthesize NPs, two methodologies are frequently used: bottom-up and top-down (Fig. 1) [38]. Bulk substances are normally broken down to NPs in the top-down technique, whereas atoms are assembled to NPs in the bottom-up strategy. The bottom-up technique is frequently applied for eco-benign and chemical fabrication of NPs. Green NPs synthesis emerged from nanobiotechnology [33], and eco-benevolent NPs are now the primary focus of nanotechnology exploration [21–25]. Green fabrication of NPs is gaining popularity as a non-noxious, ecologically benign, clean, affordable, and almost novel protocol that can be carried out at ambient temperature [30–35]. The preparation of biocompatible NPs, which are the updated practical technique of integrating material science and bioengineering, can be viewed as an alternative to the green synthesis of NPs [33]. As a result,

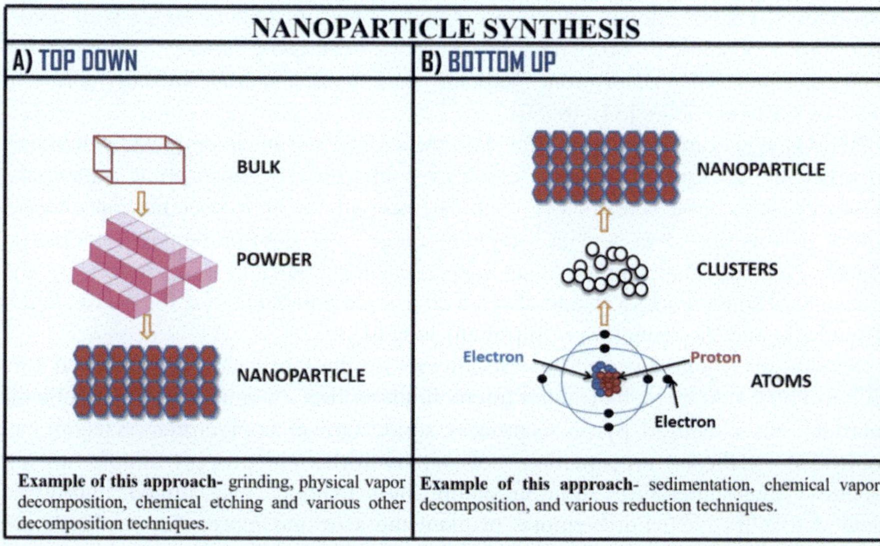

Fig. 1 Diverse protocols for the fabrication of NPs (Reproduced from Ref. 38)

eco-friendly production of NPs with controlled topology and size utilizing genetic engineering approaches, plant extracts, and other eco-friendly procedures will be a significant step forward in modern nanobiotechnology [39, 40].

In particular, manganese has been overlooked despite its intriguing and essential characteristics. In this chapter, an overview of the eco-friendly synthesis, uses, and future direction of Mn-based NPs is under-considered critically.

2 Mn and Mn-Based NPs

Mn-oxides, for instance, MnO, MnO_2, Mn_2O_3, Mn_3O_4, and Mn_5O_8, have piqued curiosity among the various 3d transition metal-oxides due to their diverse structural and elemental variations [41]. Mn-based oxide NPs have a lot of scope for long-term nanotechnology [34]. Because of their favourable chemical and chemical features, Mn-based oxides can be employed in batteries, catalysts, drug delivery, magnetic materials, molecular sieves, optoelectronics, and solar cells [32, 34, 42]. Furthermore, Mn-oxides are generally less harmful than other chemicals, such as different chalcogenides, and they are also cost-effective, have high specific capacitance, and are environmentally compatible [42–46]. The structural versatility of Mn-based NPs comprises a myriad of physicochemical features. Diverse methods have been employed to create a variety of Mn-based nanostructures, including nanobelts, nanorods, nanosheets, nanotubes, nanowires, and nanofibers [46, 47].

Manganese dioxide (MnO_2) is the foremost important material, and numerous scientists have focused on its effectiveness and the materials' electromagnetic capabilities. Under ambient conditions, MnO_2 is the most stable oxide with favorable physicochemical properties. Biosensors, catalysis, energy storage, ion exchange, molecular adsorption, medicine, and supercapacitors are only a few of the uses for MnO_2 NPs [34, 48–50].

Mn_3O_4 is a combined oxide that has shown interest in diverse uses, including adsorbents, anode materials, catalysts, microwave absorption materials, supercapacitors, and sensors [51, 52]. Because of its high specific capacitance, ability to function over a wide potential range, and environmental friendliness, Mn_3O_4 NPs are also employed as supercapacitors. Mn_3O_4 is crystalline in the usual spinel framework [53], and also a well-known potent catalyst for the oxidation of methane and the reduction of nitrobenzene [54].

Several divergent approaches were utilized for the fabrication of Mn-based NPs such as solvothermal, self-reacting microemulsion, wet chemical, and photochemical route, reflux method, pyrolysis process, sonochemical, coprecipitation technique, controlled synthesis, precipitation, sol-gel, hydrothermal process simple stirring process, low-temperature solution combustion method, and forced hydrolysis method [32, 34, 55]. Furthermore, because the size and morphology of NPs are connected, a fabrication approach that allows for monodispersity, size, and morphology control is an important field of research.

2.1 Green Synthesis of Mn and Mn-Based NPs

Researchers are now implementing the concepts of green chemistry approaches for the creation of NPs because they are non-noxious and eco-benign, and this eco-benign approach is known as biosynthesis/green synthesis. As a result, green approaches for manufacturing Mn-based NPs are considered from an environmental standpoint, as a special reagent does not need to be reduced and/or stabilized, and its fabrication can be performed under room temperature [32, 34]. Bacteria, fungi, biopolymers, raw bio-materials, and plant extracts are used to create Mn-based NPs in the biological synthesis of Mn-based NPs [32, 34]. However, optimizing the size and morphology of eco-benignly produced Mn-based NPs and their uses are still two significant obstacles in advanced nanobiotechnology [33].

Furthermore, biosynthesis of nanotechnology means manufacturing Mn-based NPs without employing perilous substances that cause deleterious residues. Traditional approaches can indeed produce large-scale Mn-based NPs with expected size and morphology. Nevertheless, again, these protocols necessitate expensive, outdated, and complicated methodologies. Contrary to traditional protocols, green synthesis approaches have several merits, such as simple, swift, cost-effective, clean, safer, and low-waste formation [34].

2.2 *Eco-Benevolent Production of Mn and Mn-Based NPs Employing Diverse Plant Extract*

Environmentally gracious fabrication of Mn-based NPs has been demonstrated in a variety of ways. Utilizing diverse plant extract, reduction, and/or stabilization Mn ion into Mn-based NPs are the most basic, affordable, and eco-benevolent green chemistry techniques [32, 34]. Biocompatibility, medical application, and scalability are all benefits of green fabrication using plant extracts [33]. When utilizing plant extracts to synthesize NPs, the chosen plant extract is merely blended with the aqua solution of Mn salt, and the reaction takes only a few minutes to finish. The Mn ion reduction is connected to flavonoids, polysaccharides, polyphenols, terpenoids, and tannins; all found in the plant extract [31–35]. Multiple plant extracts were employed in the production of Mn-based NPs so far.

Amatya et al. (2021) revealed the biosynthesis of Mn NPs using *Brassica oleraceae* leaves extract as a natural fuel, with a mean diameter of Mn NPs estimated to be around 10.70 nm [56]. Bio-fabrication of Mn NPs employing a leaves extract of *Fumaria officinalis* was described by Li et al. (2021), and SEM microphotographs are displayed in Fig. 2. In this experiment, the reaction mixture was continuously stirred for 1 h on a magnetic stirrer at 60 °C [59]. Jayandran et al. (2015) demonstrated the production of Mn NPs from manganese acetate using methanolic lemon extract as a reductant. During the reaction, the pH was controlled in the range of 3 to 4, and the reaction temperature was kept at 60 °C [60].

Khan et al. (2020) employed an aqueous leaves extract of *Abutilon indicum* to produce MnO NPs with a mean diameter of 80 ± 0.5 nm. This eco-benevolent production of MnO NPs has occurred at 55 °C [62]. The schematic layout for the

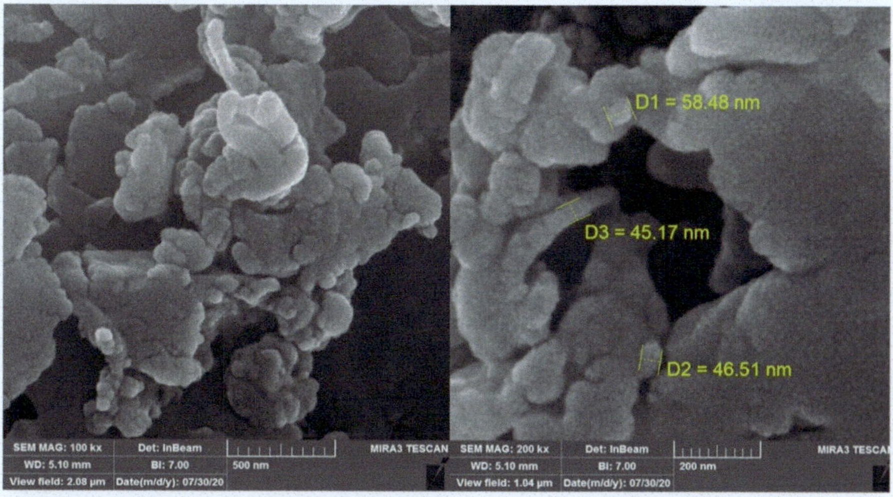

Fig. 2 SEM images of the *Fumaria officinalis* leaves extract-mediated Mn NPs (Reproduced from Ref. 59)

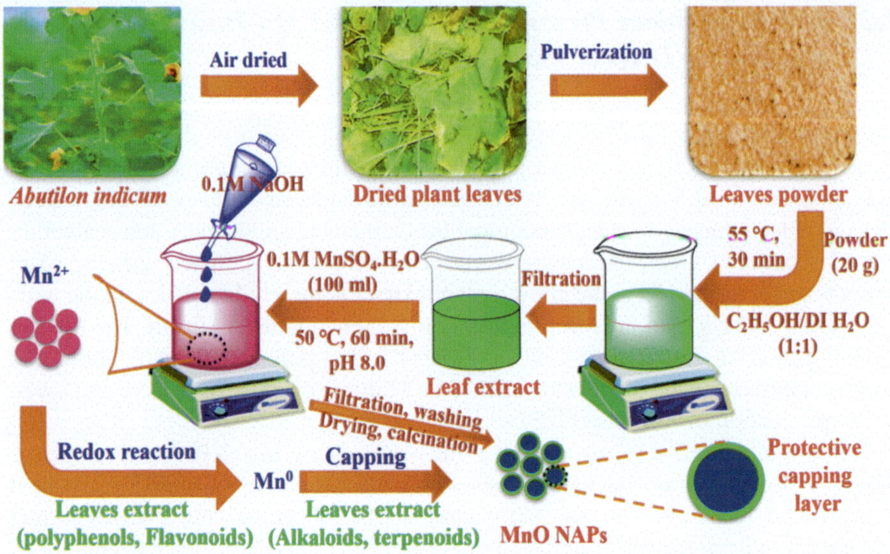

Fig. 3 Schematic presentation of eco-friendly fabrication of MnO NPs. (Reproduced from Ref. 62)

eco-benevolent fabrication of MnO NPs is presented in Fig. 3. Souri et al. (2018) used RSM to optimize the eco-benign production of MnO NPs employing extract from *Dittrichia graveolens*. The influence of the metal salt concentration, plant extract, pH, and time on the fabrication of MnO NPs was optimized. The plant extract to metal salt concentration was discovered to be the most effective metric. The mean size of MnO NPs in optimum conditions was around 38 nm [63]. Kumar et al. (2017) used *Syzygium aromaticum* aqueous extract as a natural fuel to create MnO NPs. According to XRD analysis, as-synthesized MnO NPs estimated to be 1.8 and 2.5 nm in size [64].

Ullah et al. (2020) applied an aqueous leaves extract of *Bryophyllum pinnatum* to produce MnO_2 NPs with a mean NPs size of 4–18 nm [66]. Facile green fabrication of MnO_2 NPs employing an aqueous leaves extract of *Euphorbia heterophylla* was described by Dewi and Yulizar (2020) with a median diameter of 56.68 nm. In this experiment, the reaction mixture was stirred for 1 h at 80 °C and calcinated for 2 h at 500 °C [68]. In other studies, MnO_2 NPs were prepared using *Gardenia resinifera, Kalopanax pictus, Orange, Phyllanthus amarus, Artemisia dracunculus, Origanum vulgare, Rosmarinus officinalis, Sapindus mukorossi,* and *Vernonia amygdalina* leaves extract [69–75]. Moreover, Hoseinpour et al. (2018) have revealed the eco-benevolent production of MnO_2 NPs employing leaf extract of *Yucca Gloriosa* [76]. The fabrication of MnO_2 NPs was revealed by the XRD study. The Debye-Scherer formula was also applied to get the median size of 80 nm.

Furthermore, Mn_3O_4 NPs were synthesized using *Aspalathus linearis, Azadirachta indica, Phoenix dactylifera,* and *Simarouba glauca* leaf extracts at varying temperatures. The bio-fabrication of Mn_3O_4 NPs was revealed by diverse characterization techniques [78–81].

Green fabrication of Mn-based NPs employing diverse medicinal plants and their structural properties are presented in Table 1.

3 Application of Green Synthesized Mn-Based NPs

3.1 *Antimicrobial Agent*

The antimicrobial efficacy of NPs has been attributed to their potential to produce highly reactive oxygen species on their surfaces, associated with fungal and bacterial death [33]. The majority of works emphasize the use of Mn NPs in catalytic and electrical characteristics, whereas the antimicrobial capabilities of Mn NPs are rarely addressed [60]. The disc diffusion procedure was applied to assess the bactericidal properties of biogenically produced Mn NPs for *S. aureus, B. subtilis, E. coli,* and *S. bacillus,* and the zone of inhibition was compared with chloramphenicol [66]. Mn NPs outperformed chloramphenicol in bactericidal activity against *S. aureus* and had approximately identical efficacy for *E. coli* [63]. The antifungal properties of Mn NPs were also revealed using the diffusion approach against four fungal pathogens: *T. simii, A. niger, C. lunata,* and *C. albicans* [60].

To conclude, as mentioned in Table 2, a few investigations on the antimicrobial efficacy of biosynthesized Mn-based NPs use diverse biological sources on different microbes.

3.2 *Photocatalytic Agent*

The active oxygen species are usually produced by forming electron-hole pairs between the conduction and valence bands of NPs. On the other side, active oxygen species are accountable for the breakdown of dyes into less toxic materials [30–33]. The textile and paper industries drain many carcinogenic substances, environmental contaminants, and nondegradable colors. Because of their efficiency in degrading dyes, photocatalytic techniques have gained a lot of interest in previous decades [30–33]. The ability of MnO_2 NPs to degrade Safranin O (SO) and Congo red (CR) was investigated in one study [70]. When comparing chemically fabricated MnO_2 NPs to biogenically produced MnO_2 NPs, a time-dependent study of CR degradation revealed that biogenically produced MnO_2 NPs degraded more rapidly. Biogenically and chemically fabricated MnO_2 NPs had similar dye degradation performance to degrade SO dye. The effectiveness variation between biogenically and chemically fabricated MnO_2 NPs could be attributed to changes in size [70].

Moreover, MnO NPs were also employed to degrade Rhodamine B and light green dyes [63]. In 17 and 22 min, respectively, Rhodamine B and light green were

Table 1 Environmentally benevolent fabrication of Mn-based NPs employing diverse medicinal plant extracts with their size and topology

Mn-based NPs	Name of the plant	Plant part	Shape	Size (nm)	Applications	References
Mn	*Brassica oleraceae*	Leaves	Spherical	10.70	Antibacterial activity	[56]
	Cinnamomum verum	Bark	Spherical	50–100	Antibacterial and photocatalytic activity	[57]
	Ctenolepis garcinii	–	Elliptical	57–69	–	[58]
	Fumaria officinalis	Leaves	Spherical	50.05	Anticancer activity	[59]
	Lemon	Fruits	Spherical and eclipse	50	Antimicrobial activity	[60]
	Ziziphora clinopodioides	Leaves	Spherical	48.10	Antioxidant, antifungal, antibacterial, cytotoxicity, and wound healing activities	[61]
MnO	*Abutilon indicum*	Leaves	Spherical	80 ± 0.5	Antibacterial, cytotoxicity, and photocatalytic activity	[62]
	Dittrichia graveolens	–	Spherical	38	Photocatalytic activity	[63]
	Syzygium aromaticum	Flower	–	1.8	Electrochemical sensing of p-nitrophenol	[64]
MnO_2	*Aloe vera*	Leaves	Spherical	–	Antibacterial activity	[65]
	Bryophyllum pinnatum	Leaf	Spherical	4–18	–	[66]
	Datura stramonium	Leaf	–	–	Antimicrobial activity	[67]
	Euphorbia heterophylla	Leaf	Irregular	56.68	–	[68]
	Gardenia resinifera	Leaf	Spherical	17–35	Antibacterial activity	[69]
	Kalopanax pictus	Leaf	Spherical	18.6	Photocatalytic activity	[70]
	Orange	Fruits and peels	Nonrectangular	2–50	Electrochemical properties	[71]

	Phyllanthus amarus	–	Rod	40–50	–	[72]
	Artemisia dracunculus	–	Spherical	2.9	Used as a catalyst for biofuel production	[73]
	Origanum vulgare	–	Spherical	3.4	Used as a catalyst for biofuel production	[73]
	Rosmarinus officinalis	–	Spherical	3.1	Used as a catalyst for biofuel production	[73]
	Sapindus mukorossi	–	Needle	10	Oxidative polymerization	[74]
	Vernonia amygdalina	Leaf	Ball-shaped flower	20–22	–	[75]
	Yucca gloriosa	Leaf	Spherical	80	Photocatalytic activity	[76]
	Yucca gloriosa	Leaf	Spherical	32		[77]
Mn_2O_3	*Sapindus mukorossi*	–	Spherical	28.7–63.1	Oxidative polymerization	[74]
Mn_3O_4	*Aspalathus linearis*	Leaves	Quasi-spherical	17–18	–	[78]
	Azadirachta indica	Leaf	Spherical	18.2	Used as a chemical sensor	[79]
	Phoenix dactylifera	–	Spherical	20.4–24.2	Electrochemical application	[80]
	Sapindus mukorossi	–	Cubic	21.6–65.4	Oxidative polymerization	[74]
	Simarouba glauca	Leaf	Spherical	15	Photocatalytic activity	[81]

Table 2 Antimicrobial efficacy of Mn-based NPs against various pathogens

Type of Mn-based NPs	Name of the plants	Tested microbes	References
Mn	*Brassica oleraceae*	*S. aureus, S. typhi,* and *E. coli*	[56]
	Cinnamomum verum	*S. aureus* and *E. coli*	[57]
	Lemon	*S. aureus, B. subtilis, E. coli, S. bacillus, C. albicans, C. lunata, A. niger,* and *T. simii*	[60]
	Ziziphora clinopodioides	*S. Pneumonia, S. aureus, S. typhimurium, B. subtilis, E. coli, P. aeruginosa, C. guilliermondii, C. glabrata, C. krusei,* and *C. albicans*	[61]
MnO	*Abutilon indicum*	*E. coli, B. bronchiseptica, S. aureus, and B. subtilis*	[62]
MnO_2	*Aloe vera*	*E. coli, S. mutans, and S. aureus*	[65]
	Datura stramonium	*S. aureus, S. mutans, S. typhi, P. vulgaris,* and *E. coli*	[67]
	Gardenia resinifera	*S. aureus, P. aeruginosa,* and *S. arcescens*	[69]

Table 3 Degradation of different dyes employing Mn-based NPs

Type of Mn-based NPs	Dye	Light source	Efficiency (%)	Time	References
Mn	Congo red	UV irradiation	78.5	60 min	[57]
MnO	Methylene blue	UV lamp Sunlight	97–99	180 min	[62]
	Light green and Rhodamine B	UV irradiation	–	20 min	[63]
MnO_2	Congo red and safranin O	UV irradiation	68.7	8 min	[70]
	Acid orange	UV irradiation	33	20 min	[76]
Mn_3O_4	Malachite green	Dark condition	78	30 min	[81]

solely degraded [63]. Green manufactured MnO_2 NPs were also used in another investigation to decolorize acid orange dye and demonstrated encouraging results for the breakdown of this organic pollutant [76].

Table 3 offers a few more reports of diverse dye degradation employing Mn-based NPs as a nanocatalyst.

4 Future Perspectives

As stated previously, there really are numerous techniques for green NP synthesis, but only a few strategies for Mn-based NPs production were used, and hence there is a broad scope to utilize diverse medicinal plants and microbial biomass for the biosynthesis of Mn-based NPs. Because different structures for Mn-oxides (MnO, MnO_2, Mn_2O_3, Mn_3O_4, and Mn_5O_8) exist, as well as the variety of biological entities that have been employed in bio-fabrication, such as actinomycetes, algae, bacteria, fungi, plants, and yeasts, more research into the eco-benign production of Mn-based NPs could be interesting. Furthermore, looking into the possibilities of manufacturing metallic nanomaterials (such as CuO-MnO, Au-MnO_2, ZnO-MnO, and so on) could enhance the uses of Mn-based NPs. Plants having significant antioxidant capabilities could be investigated for the green production of metallic nanocomposites, as one example. Also, the implementation of bio-waste materials like eggshells, biopolymers, starch, honey, gum, and cellulose are needed to be explored for the bio-inspired fabrication of Mn-based NPs.

5 Conclusion

Nanobiotechnology is growing in appeal as a new branch of nanotechnology. The need for biocompatible materials for diverse applications in diverse sectors such as biosensors, textile, and food industries, health, medicine, water treatment, and so on has drawn greater attention to this topic in recent years. On the contrary, developing an eco-benign strategy has become a future imperative need for the sector. As a result, more study into green NPs synthesis could be very fruitful. Many investigations on the eco-benign preparation of metallic NPs have been conducted so far. Multifunctional metallic NPs, such as copper, gold, platinum, iron, palladium, silver, and zinc, have also been proposed for a variety of uses. Despite their impressive capabilities, not enough studies on Mn-based NPs have been done.

Green syntheses of Mn-based NPs employing diverse medicinal plant extracts are discussed in this study. Each method's morphology and size of biogenically produced Mn-based NPs are compared. Different potential uses of biologically produced Mn-based NPs have also been highlighted. Mn-based NPs made with green chemistry can be utilized in electronic, biological, biomedical, environmental, medicinal, and other implementations, including tissue engineering.

References

1. Vanlalveni, C., Lallianrawna, S., Biswas, A., Selvaraj, M., Changmai, B., & Rokhum, S. L. (2021). Green synthesis of silver nanoparticles using plant extracts and their antimicrobial activities: A review of recent literature. *RSC Advances, 11*(5), 2804–2837.

2. Javed, B., Ikram, M., Farooq, F., Sultana, T., & Raja, N. I. (2021). Biogenesis of silver nanoparticles to treat cancer, diabetes, and microbial infections: A mechanistic overview. *Applied Microbiology and Biotechnology*, 1–15.
3. Dabhane, H., Ghotekar, S. K., Tambade, P. J., Pansambal, S., Ananda Murthy, H. C., Oza, R., & Medhane, V. (2021). Cow urine mediated green synthesis of nanomaterial and their applications: A state-of-the-art review. *Journal of Water and Environmental Nanotechnology, 6*(1), 81–91.
4. Gawande, M. B., Goswami, A., Felpin, F. X., Asefa, T., Huang, X., Silva, R., Zou, X., Zboril, R., & Varma, R. S. (2016). Cu and Cu-based nanoparticles: Synthesis and applications in catalysis. *Chemical Reviews, 116*(6), 3722–3811.
5. Dabhane, H., Ghotekar, S., Tambade, P., Pansambal, S., Murthy, H. A., Oza, R., & Medhane, V. (2021). A review on environmentally benevolent synthesis of CdS nanoparticle and their applications. *Environmental Chemistry and Ecotoxicology, 3*, 209–219.
6. Khan, I., Saeed, K., & Khan, I. (2019). Nanoparticles: Properties, applications and toxicities. *Arabian Journal of Chemistry, 12*(7), 908–931.
7. Ghotekar, S., Pagar, K., Pansambal, S., Murthy, H. A., & Oza, R. (2020). A review on eco-friendly synthesis of $BiVO_4$ nanoparticle and its eclectic applications. *Advanced Journal of Science and Engineering, 1*, 106–112.
8. Sekoai, P. T., Ouma, C. N. M., Du Preez, S. P., Modisha, P., Engelbrecht, N., Bessarabov, D. G., & Ghimire, A. (2019). Application of nanoparticles in biofuels: An overview. *Fuel, 237*, 380–397.
9. Korde, P., Ghotekar, S., Pagar, T., Pansambal, S., Oza, R., & Mane, D. (2020). Plant extract assisted eco-benevolent synthesis of selenium nanoparticles-a review on plant parts involved, characterization and their recent applications. *Journal of Chemical Reviews, 2*(3), 157–168.
10. Polshettiwar, V., & Varma, R. S. (2010). Green chemistry by nano-catalysis. *Green Chemistry, 12*(5), 743–754.
11. Polshettiwar, V., Cha, D., Zhang, X., & Basset, J. M. (2010). High-surface-area silica nanospheres (KCC-1) with a fibrous morphology. *Angewandte Chemie International Edition, 49*(50), 9652–9656.
12. Virkutyte, J., & Varma, R. S. (2011). Green synthesis of metal nanoparticles: Biodegradable polymers and enzymes in stabilization and surface functionalization. *Chemical Science, 2*(5), 837–846.
13. Singh, P., Kim, Y. J., Zhang, D., & Yang, D. C. (2016). Biological synthesis of nanoparticles from plants and microorganisms. *Trends in Biotechnology, 34*(7), 588–599.
14. Ghotekar, S., Pagar, T., Pansambal, S., & Oza, R. (2020). A review on green synthesis of sulfur nanoparticles via plant extract, characterization and its applications. *Advanced Journal of Chemistry-Section B*, 128–143.
15. Al-Hakkani, M. F. (2020). Biogenic copper nanoparticles and their applications: A review. *SN Applied Sciences, 2*(3), 1–20.
16. Santhoshkumar, J., Rajeshkumar, S., & Kumar, S. V. (2017). Phyto-assisted synthesis, characterization and applications of gold nanoparticles–a review. *Biochemistry and Biophysics Reports, 11*, 46–57.
17. Abdelghany, T. M., Al-Rajhi, A. M., Al Abboud, M. A., Alawlaqi, M. M., Magdah, A. G., Helmy, E. A., & Mabrouk, A. S. (2018). Recent advances in green synthesis of silver nanoparticles and their applications: About future directions. A review. *BioNanoScience, 8*(1), 5–16.
18. Jameel, M. S., Aziz, A. A., & Dheyab, M. A. (2020). Green synthesis: Proposed mechanism and factors influencing the synthesis of platinum nanoparticles. *Green Processing and Synthesis, 9*(1), 386–398.
19. Vishnukumar, P., Vivekanandhan, S., & Muthuramkumar, S. (2017). Plant-mediated biogenic synthesis of palladium nanoparticles: Recent trends and emerging opportunities. *ChemBioEng Reviews, 4*(1), 18–36.

20. Ghotekar, S. (2019). A review on plant extract mediated biogenic synthesis of CdO nanoparticles and their recent applications. *Asian Journal of Green Chemistry, 3*(2), 187–200.
21. Pagar, T., Ghotekar, S., Pagar, K., Pansambal, S., & Oza, R. (2019). A review on bio-synthesized Co_3O_4 nanoparticles using plant extracts and their diverse applications. *Journal of Chemical Reviews, 1*(4), 260–270.
22. Bandeira, M., Giovanela, M., Roesch-Ely, M., Devine, D. M., & da Silva Crespo, J. (2020). Green synthesis of zinc oxide nanoparticles: A review of the synthesis methodology and mechanism of formation. *Sustainable Chemistry and Pharmacy, 15*, 100223.
23. Nikam, A., Pagar, T., Ghotekar, S., Pagar, K., & Pansambal, S. (2019). A review on plant extract mediated green synthesis of zirconia nanoparticles and their miscellaneous applications. *Journal of Chemical Reviews, 1*(3), 154–163.
24. Pagar, T., Ghotekar, S., Pansambal, S., Oza, R., & Marasini, B. P. (2020). Facile plant extract mediated eco-benevolent synthesis and recent applications of CaO-NPs: A state-of-the-art review. *Journal of Chemical Reviews, 2*(3), 201–210.
25. Matussin, S., Harunsani, M. H., Tan, A. L., & Khan, M. M. (2020). Plant-extract-mediated SnO_2 nanoparticles: Synthesis and applications. *ACS Sustainable Chemistry & Engineering, 8*(8), 3040–3054.
26. Ghotekar, S. (2019). Plant extract mediated biosynthesis of Al_2O_3 nanoparticles-a review on plant parts involved, characterization and applications. *Nanochemistry Research, 4*(2), 163–169.
27. Ghotekar, S., Dabhane, H., Pansambal, S., Oza, R., Tambade, P., & Medhane, V. (2020). A review on biomimetic synthesis of Ag_2O nanoparticles using plant extract, characterization and its recent applications. *Advanced Journal of Chemistry-Section B, 2*(3), 102–111.
28. Ghotekar, S., Pansambal, S., Bilal, M., Pingale, S. S., & Oza, R. (2021). Environmentally friendly synthesis of Cr_2O_3 nanoparticles: Characterization, applications and future perspective—a review. *Case Studies in Chemical and Environmental Engineering, 3*, 100089.
29. Bouafia, A., & Laouini, S. E. (2021). Plant-mediated synthesis of iron oxide nanoparticles and evaluation of the antimicrobial activity: A review. *Mini-Reviews in Organic Chemistry, 18*(6), 725–734.
30. Dabhane, H., Ghotekar, S., Tambade, P., & Medhane, V. (2020). Plant mediated green synthesis of lanthanum oxide (La_2O_3) nanoparticles: A review. *Asian Journal of Nanosciences and Materials, 3*(4), 291–299.
31. Dabhane, H., Ghotekar, S., Tambade, P., Pansambal, S., Oza, R., & Medhane, V. (2021). MgO nanoparticles: Synthesis, characterization, and applications as a catalyst for organic transformations. *European Journal of Chemistry, 12*(1), 86–108.
32. Hoseinpour, V., & Ghaemi, N. (2018). Green synthesis of manganese nanoparticles: Applications and future perspective–a review. *Journal of Photochemistry and Photobiology B: Biology, 189*, 234–243.
33. Cuong, H. N., Pansambal, S., Ghotekar, S., Oza, R., Hai, N. T. T., Viet, N. M., & Nguyen, V. H. (2022). New frontiers in the plant extract mediated biosynthesis of copper oxide (CuO) nanoparticles and their potential applications: A review. *Environmental Research, 203*, 111858.
34. Pagar, T., Ghotekar, S., Pagar, K., Pansambal, S., & Oza, R. (2021). Phytogenic synthesis of manganese dioxide nanoparticles using plant extracts and their biological application. In *Handbook of Greener Synthesis of Nanomaterials and Compounds* (pp. 209–218). Elsevier.
35. Abinaya, S., Kavitha, H. P., Prakash, M., & Muthukrishnaraj, A. (2021). Green synthesis of magnesium oxide nanoparticles and its applications: A review. *Sustainable Chemistry and Pharmacy, 19*, 100368.
36. Lowe, C. R. (2000). Nanobiotechnology: The fabrication and applications of chemical and biological nanostructures. *Current Opinion in Structural Biology, 10*(4), 428–434.
37. Narayanan, K. B., & Sakthivel, N. (2011). Green synthesis of biogenic metal nanoparticles by terrestrial and aquatic phototrophic and heterotrophic eukaryotes and biocompatible agents. *Advances in Colloid and Interface Science, 169*(2), 59–79.

38. Singh, K. R., Nayak, V., Sarkar, T., & Singh, R. P. (2020). Cerium oxide nanoparticles: Properties, biosynthesis and biomedical application. *RSC Advances, 10*(45), 27194–27214.
39. Dikshit, P. K., Kumar, J., Das, A. K., Sadhu, S., Sharma, S., Singh, S., & Kim, B. S. (2021). Green synthesis of metallic nanoparticles: Applications and limitations. *Catalysts, 11*(8), 902.
40. Salem, S. S., & Fouda, A. (2021). Green synthesis of metallic nanoparticles and their prospective biotechnological applications: An overview. *Biological Trace Element Research, 199*(1), 344–370.
41. Prasad, A. S. (2017). Green synthesis of nanocrystalline manganese (II, III) oxide. *Materials Science in Semiconductor Processing, 71*, 342–347.
42. Hu, H., Cheng, H., Liu, Z., & Yu, Y. (2015). Facile synthesis of carbon spheres with uniformly dispersed MnO nanoparticles for lithium ion battery anode. *Electrochimica Acta, 152*, 44–52.
43. Liu, X., Chen, C., Zhao, Y., & Jia, B. (2013). A review on the synthesis of manganese oxide nanomaterials and their applications on lithium-ion batteries. *Journal of Nanomaterials, 2013*, 736375.
44. Ghosh, D., Bhandari, S., & Khastgir, D. (2016). Synthesis of MnO_2 nanoparticles and their effective utilization as UV protectors for outdoor high voltage polymeric insulators used in power transmission lines. *Physical Chemistry Chemical Physics, 18*(48), 32876–32890.
45. Wang, S., Xing, Y., Xu, H., & Zhang, S. (2014). MnO nanoparticles interdispersed in 3D porous carbon framework for high performance lithium-ion batteries. *ACS Applied Materials & Interfaces, 6*(15), 12713–12718.
46. Jankovský, O., Sedmidubský, D., Šimek, P., Sofer, Z., Ulbrich, P., & Bartůněk, V. (2015). Synthesis of MnO, Mn_2O_3 and Mn_3O_4 nanocrystal clusters by thermal decomposition of manganese glycerolate. *Ceramics International, 41*(1), 595–601.
47. Fei, J. B., Cui, Y., Yan, X. H., Qi, W., Yang, Y., Wang, K. W., He, Q., & Li, J. B. (2008). Controlled preparation of MnO_2 hierarchical hollow nanostructures and their application in water treatment. *Advanced Materials, 20*(3), 452–456.
48. Raj, B. G. S., Asiri, A. M., Qusti, A. H., Wu, J. J., & Anandan, S. (2014). Sonochemically synthesized MnO_2 nanoparticles as electrode material for supercapacitors. *Ultrasonics Sonochemistry, 21*(6), 1933–1938.
49. Yamaguchi, Y., Aono, R., Hayashi, E., Kamata, K., & Hara, M. (2020). Template-free synthesis of mesoporous β-MnO_2 nanoparticles: Structure, formation mechanism, and catalytic properties. *ACS Applied Materials & Interfaces, 12*(32), 36004–36013.
50. Sankar, S., Inamdar, A. I., Im, H., Lee, S., & Kim, D. Y. (2018). Template-free rapid sonochemical synthesis of spherical α-MnO_2 nanoparticles for high-energy supercapacitor electrode. *Ceramics International, 44*(14), 17514–17521.
51. Dubal, D. P., Dhawale, D. S., Salunkhe, R. R., Fulari, V. J., & Lokhande, C. D. (2010). Chemical synthesis and characterization of Mn_3O_4 thin films for supercapacitor application. *Journal of Alloys and Compounds, 497*(1–2), 166–170.
52. Bai, Z., Sun, B., Fan, N., Ju, Z., Li, M., Xu, L., & Qian, Y. (2012). Branched mesoporous Mn_3O_4 nanorods: Facile synthesis and catalysis in the degradation of methylene blue. *Chemistry–a. European Journal, 18*(17), 5319–5324.
53. He, X., Wang, Z. H., Geng, D. Y., & Zhang, Z. D. (2011). Structure and magnetic properties of S-doped Mn_3O_4/S composited nanoparticles and Mn3O4 nanoparticles. *Journal of Materials Science & Technology, 27*(6), 503–506.
54. Djerdj, I., Arcon, D., Jagličić, Z., & Niederberger, M. (2007). Nonaqueous synthesis of manganese oxide nanoparticles, structural characterization, and magnetic properties. *The Journal of Physical Chemistry C, 111*(9), 3614–3623.
55. Sukhdev, A., Challa, M., Narayani, L., Manjunatha, A. S., Deepthi, P. R., Angadi, J. V., … Pasha, M. (2020). Synthesis, phase transformation, and morphology of hausmannite Mn_3O_4 nanoparticles: Photocatalytic and antibacterial investigations. *Heliyon, 6*(1), e03245.
56. Amatya, S. P., & Shrestha, S. (2021). Biosynthesis of manganese nanoparticles (MnNPs) from brassica oleraceae (cabbage leaves) and its antibacterial activity. *Asian Journal of Chemical Sciences*, 1–11.

57. Kamran, U., Bhatti, H. N., Iqbal, M., Jamil, S., & Zahid, M. (2019). Biogenic synthesis, characterization and investigation of photocatalytic and antimicrobial activity of manganese nanoparticles synthesized from Cinnamomum verum bark extract. *Journal of Molecular Structure, 1179*, 532–539.
58. Paul, J. J. P., Sakunthala, M., & Udhaya, I. (2017). Green synthesis of manganese nanoparticles using the aqueous extract of Ctenolepis garcini (Burm. F.) CB Clarke. *International Journal of Botany Studies, 2*, 71–75.
59. Li, C., Zhang, Y., Li, M., Zhang, H., Zhu, Z., & Xue, Y. (2021). Fumaria officinalis-assisted synthesis of manganese nanoparticles as an anti-human gastric cancer agent. *Arabian Journal of Chemistry, 14*(10), 103309.
60. Jayandran, M., Haneefa, M. M., & Balasubramanian, V. (2015). Green synthesis and characterization of manganese nanoparticles using natural plant extracts and its evaluation of antimicrobial activity. *Journal of Applied Pharmaceutical Science, 5*(12), 105–110.
61. Mahdavi, B., Paydarfard, S., Zangeneh, M. M., Goorani, S., Seydi, N., & Zangeneh, A. (2020). Assessment of antioxidant, cytotoxicity, antibacterial, antifungal, and cutaneous wound healing activities of green synthesized manganese nanoparticles using Ziziphora clinopodioides lam leaves under in vitro and in vivo condition. *Applied Organometallic Chemistry, 34*(1), e5248.
62. Khan, S. A., Shahid, S., Shahid, B., Fatima, U., & Abbasi, S. A. (2020). Green synthesis of MnO nanoparticles using abutilon indicum leaf extract for biological, photocatalytic, and adsorption activities. *Biomolecules, 10*(5), 785.
63. Souri, M., Hoseinpour, V., Shakeri, A., & Ghaemi, N. (2018). Optimisation of green synthesis of MnO nanoparticles via utilising response surface methodology. *IET Nanobiotechnology, 12*(6), 822–827.
64. Kumar, V., Singh, K., Panwar, S., & Mehta, S. K. (2017). Green synthesis of manganese oxide nanoparticles for the electrochemical sensing of p-nitrophenol. *International Nano Letters, 7*(2), 123–131.
65. Joshi, N. C., Siddiqui, F., Salman, M., & Singh, A. (2020). Antibacterial activity, characterizations, and biological synthesis of manganese oxide nanoparticles using the extract of Aloe vera. *Asian Pacific Journal of Health Sciences, 7*(3), 27–29.
66. Ullah, A. A., Haque, M. M., Akter, M., Hossain, A., Tamanna, A. N., Hosen, M. M., Kibria, A. F., Khan, M., & Khan, M. A. (2020). Green synthesis of Bryophyllum pinnatum aqueous leaf extract mediated bio-molecule capped dilute ferromagnetic α-MnO_2 nanoparticles. *Materials Research Express, 7*(1), 015088.
67. Joshi, N. C., Joshi, E., & Singh, A. (2020). Biological synthesis, Characterisations and antimicrobial activities of manganese dioxide (MnO_2) nanoparticles. *Research Journal of Pharmacy and Technology, 13*(1), 135–140.
68. Dewi, N. O. M., & Yulizar, Y. (2020). Euphorbia heterophylla L. leaf extract-mediated synthesis of MnO_2 nanoparticles and its characterization. *Materials Today: Proceedings, 22*, 199–204.
69. Manjula, R., Thenmozhi, M., Thilagavathi, S., Srinivasan, R., & Kathirvel, A. (2020). Green synthesis and characterization of manganese oxide nanoparticles from Gardenia resinifera leaves. *Materials Today: Proceedings, 26*, 3559–3563.
70. Moon, S. A., Salunke, B. K., Alkotaini, B., Sathiyamoorthi, E., & Kim, B. S. (2015). Biological synthesis of manganese dioxide nanoparticles by Kalopanax pictus plant extract. *IET Nanobiotechnology, 9*(4), 220–225.
71. Abuzeid, H. M., Elsherif, S. A., Ghany, N. A. A., & Hashem, A. M. (2019). Facile, cost-effective and eco-friendly green synthesis method of MnO_2 as storage electrode materials for supercapacitors. *Journal of Energy Storage, 21*, 156–162.
72. Prasad, K. S., & Patra, A. (2017). Green synthesis of MnO_2 nanorods using Phyllanthus amarus plant extract and their fluorescence studies. *Green Processing and Synthesis, 6*(6), 549–554.
73. Stegarescu, A., Lung, I., Leoştean, C., Kacso, I., Opriş, O., Lazăr, M. D., Copolovici, L., Guţoiu, S., Stan, M., Popa, A., Pană, O., Porav, A. S., & Soran, M. L. (2019). Green synthesis,

characterization and test of MnO_2 nanoparticles as catalyst in biofuel production from grape residue and seeds oil. *Waste and Biomass Valorization*, 1–11.
74. Jassal, V., Shanker, U., Gahlot, S., Kaith, B. S., Iqubal, M. A., & Samuel, P. (2016). Sapindus mukorossi mediated green synthesis of some manganese oxide nanoparticles interaction with aromatic amines. *Applied Physics A, 122*(4), 271.
75. Dessie, Y., Tadesse, S., & Eswaramoorthy, R. (2020). Physicochemical parameter influences and their optimization on the biosynthesis of MnO_2 nanoparticles using Vernonia amygdalina leaf extract. *Arabian Journal of Chemistry, 13*(8), 6472–6492.
76. Hoseinpour, V., Souri, M., & Ghaemi, N. (2018). Green synthesis, characterisation, and photocatalytic activity of manganese dioxide nanoparticles. *Micro & Nano Letters, 13*(11), 1560–1563.
77. Souri, M., Hoseinpour, V., Ghaemi, N., & Shakeri, A. (2019). Procedure optimization for green synthesis of manganese dioxide nanoparticles by Yucca gloriosa leaf extract. *International Nano Letters, 9*(1), 73–81.
78. Diallo, A., Tandjigora, N., Ndiaye, S., Jan, T., Ahmad, I., & Maaza, M. (2021). Green synthesis of single phase hausmannite Mn_3O_4 nanoparticles via Aspalathus linearis natural extract. *SN Applied Sciences, 3*(5), 1–11.
79. Sharma, J. K., Srivastava, P., Ameen, S., Akhtar, M. S., Singh, G., & Yadava, S. (2016). Azadirachta indica plant-assisted green synthesis of Mn_3O_4 nanoparticles: Excellent thermal catalytic performance and chemical sensing behavior. *Journal of Colloid and Interface Science, 472*, 220–228.
80. Sackey, J., Akbari, M., Morad, R., Bashir, A. K. H., Ndiaye, N. M., Matinise, N., & Maaza, M. (2021). Molecular dynamics and bio-synthesis of phoenix dactylifera mediated Mn_3O_4 nanoparticles: Electrochemical application. *Journal of Alloys and Compounds, 854*, 156987.
81. Nair Sreekala, G., Abdullakutty, F., & Beena, B. (2019). Green synthesis, characterization, and photo catalytic degradation efficiency of Trimanganese tetroxide nanoparticle. *International Journal of Nano Dimension, 10*(4), 400–409.

Biogenic Synthesis of Lead-Based Nanoparticles and Their Recent Applications

Khanderao Pagar, Suresh Ghotekar, Onkar Pardeshi, Shreyas Pansambal, Sachin Pawar, Jigna Machhi, and Balasaheb Pagar

Abstract Lead nanoparticles (Pb NPs) are a type of metallic NPs employed in diverse uses, including sensors, ceramics, glasses, pigments, batteries, and solar cells. The production of harmful chemicals and noxious contaminants is a major issue in the chemical synthesis of Pb-based NPs. Many research investigations on the eco-benign fabrication of Pb-based NPs employing microbial biomass and plant extracts without creating toxic waste have been performed to deal with these problems. Plants could be particularly useful for studying the biosynthesis of Pb-based NPs among green sources. The green synthesis of Pb-based NPs like Pb NPs, PbO NPs, and PbS NPs using diverse plant extracts and microbes in the absence of harmful capping agents has been discussed. The current advancement and future direction in the eco-benevolent production of Pb-based NPs are discussed in this chapter. Furthermore, the biosynthesized Pb-based NPs' uses have been highlighted.

K. Pagar · S. Pawar
Department of Chemistry, S.S.R College of Arts, Commerce and Science College, Savitribai Phule Pune University, Silvassa, India

S. Ghotekar (✉)
Department of Chemistry, Smt. Devkiba Mohansinhji Chauhan College of Commerce and Science, University of Mumbai, Silvassa, India

O. Pardeshi
Department of Electronics, KKHA Arts, SMGL Commerce and SPHJ Science College, Savitribai Phule Pune University, Chandwad, Maharashtra, India

S. Pansambal
Department of Chemistry, Shri Saibaba College, Savitribai Phule Pune University, Shirdi, Maharashtra, India

J. Machhi
Department of Chemistry, Government Science College Songadh, Tapi, Gujarat, India

B. Pagar
Department of Chemistry, S.V.K.T Arts, Commerce and Science College, Savitribai Phule Pune University, Nashik, Maharashtra, India

J.-T. Chen (ed.), *Plant and Nanoparticles*,
https://doi.org/10.1007/978-981-19-2503-0_14

Keywords Green synthesis · Plant extracts · Microorganisms · Pb-based nanoparticles · Applications

1 Introduction

Modern nanoscience has recently gotten a lot of popularity because of its revolutionary and future implications in diverse industries [1–5]. Nanotechnology is an interdisciplinary branch of research that covers the development, analysis, and application of nanomaterials [6–9]. Nanoparticle fabrication has a large surface area-to-volume ratio, which is responsible for important uses in food technology, optical devices, cosmetics, defense, textile industry, electrochemistry, pharmaceutics, space industry, optics, mechanics, sensors, water treatment and purification, catalysis, and electronics [10–25].

Nowadays, modern green nanotechnology has risen to prominence because it uses plant extracts and a variety of bio-materials to create a safer, cleaner, ecologically sound, and environmentally friendly alternative to traditional physiochemical approaches [26, 27]. The use of plant resources for NP synthesis does not require complicated methods. Green chemistry is a long-term replacement for traditional reduction processes that require natural chemicals with dual reducing and capping properties [28, 29]. Green synthesis approaches have advantages over physical and chemical synthesis because they are highly cost-effective and environmentally friendly for synthesizing NPs with high yield [30, 31]. In addition, plant extracts have frequently been used as abundant sources of phytomolecules, facilitating the creation of stable NPs that are free of harmful chemicals and have robust medicinal properties [32].

Lead is a p-block element typically applied in industrial uses and is generally detrimental to the environment and human health; it is especially fatal when consumed orally or inhaled [33, 34]. Lead-containing substances can have a variety of negative consequences, including oxidative stress [35], genetic toxicity [36], and neurological impacts [37]. Pb NPs perform an essential impact in electronic sciences, particularly sensors. Pb-NPs have been produced employing a variety of approaches, including UV-light reduction, reverse micelles, tetrazolium-based ionic liquids, and reduction of lead salts [38].

Due to its superior electrical, mechanical, and optical characteristics, lead oxide nanoparticles (PbO NPs) are commonly used in some fields, including gas sensors, colors, ceramics, glass, batteries, and semiconductors [39–43]. In addition, it is commonly applied in road construction, construction skeleton, and shipbuilding due to its antibacterial, rustproof, and anti-algae features [44, 45]. PbO is a semiconducting material with two separate crystal structures: tetragonal (litharge) and orthorhombic (massicot) [46, 47]. Tetragonal crystals have a red color with the α-PbO form demonstrated to be stable at lower temperatures, whereas orthorhombic crystals have a yellow color with the β-PbO form, that is, at higher temperatures. Also, PbO is a viable photovoltaic material [48] mostly used as an optically active semiconductor [49, 50] with band gaps of 1.92 and 2.7 eV. To date, PbO NPs have been fabricated using a variety of approaches, including solvothermal method [51], chemical deposition [52], microwave radiation [53], sol-gel pyrolysis [54], thermal decomposition [49], aerogel decomposition [55], and sonochemical method [56]. Moreover, the lead chalcogenide of PbS possesses unique electrical and optical properties. Therefore, they have a variety of photonic applications in sensing, solar cells, light-emitting diodes, and photo-detectors for telecommunications [57–60].

Unfortunately, as mentioned above, these approaches have some pitfalls. Physical fabrication strategies require high vacuum and energy, whereas chemical approaches are connected with noxious and pernicious waste production. Alternatively, green and eco-benign approaches have been developed to mitigate the burden of hazardous waste and energy usage [3–5]. For producing multifunctional metal, metal oxide, and metal sulfide NPs, the interaction between therapeutic plants and metal-based NPs has been regarded as a promising field of research. Although the chemical synthesis of Pb-based NPs has been effectively described, there are only a few reports on the biosynthesis of Pb-based NPs in the literature.

Thus, the focus of this chapter is to offer a report on the eco-friendly synthesis of Pb-based NPs and their recent application.

2 Green Fabrication of Pb-Based NPs

Nowadays, green nanotechnology synthesis involves the production of NPs and/or nanomaterials without the use of hazardous chemicals that produce harmful waste materials. In other words, eco-benevolent synthesis is an environmentally sustainable, simple, and less expensive method of producing nanoparticles that is not harmful to human health or the environment. Indeed, existing conventional methods can produce NPs in large amounts with precise size and topology. Regardless, these methods necessitate costly protocols, as well as sophisticated, time-consuming, hazardous, and out-of-date protocols [1]. On another side, the green approaches have some advantages, including a simple manufacturing technique, a quick and simple economic model, and minimal waste creation. Furthermore, bio-materials, fungi, microbes, and plant extracts are employed in the biogenic fabrication of Pb-based NPs [1–3]. Green fabrication of Pb-based NPs employing diverse microbes and medicinal plant extracts is presented in Table 1.

2.1 Synthesis of Pb NPs from Microbial Biomass and Plant Extracts

In the green synthesis process of NPs, biomolecules found in plant extracts and microbial biomass can serve as both bio-reductant and bio-stabilizers. Various aqua-soluble plant metabolites and coenzymes are among the bio-reductants and

Table 1 Environmentally gracious fabrication of Pb-based NPs employing diverse microbes and medicinal plant extracts with structural properties

NPs	Name of entities (plants/microbes)	Parts	Characterization techniques	Shape	Size (nm)	Ref.
Pb	*Avivennia marina*	Leaves	UV-vis, XRD, TEM, FTIR	–	15–25	[61]
	Cocos nucifera	–	UV-vis, XRD, TEM,	Spherical	47	[38]
	Aspergillus sp.	Biomass	SEM, TEM	–	5–20	[62]
	Jatropha curcas	Latex	UV-vis, XRD, TEM,	Spherical	5–17.5	[63]
	Serratia plymuthica	Biomass	UV-vis, XRD, TEM, DLS	–	92.93	[64]
PbO	*Sageretia thea*	Leaf	UV-vis, XRD, TEM, FTIR, HR-TEM	Quasi-spherical	27	[65]
	Averrhoa bilimbi	–	XRD, FTIR, SEM	Nonuniform	–	[66]
	Datura sternum	Leaf	UV-vis, XRD,	–	60	[67]
	Eucalyptus globulus	Leaves	UV-vis, FTIR, SEM, HR-TEM, XRD, PL, EDS	–	34.61	[68]
PbS	*Aspergillus flavus*	Biomass	UV-vis, FTIR, SEM, TEM, EDX, XRD	–	35–100	[69]
	Aspergillus sp.	Biomass	UV-vis, XRD, PSA, TEM	Spherical	10–15	[70]
	Desulfotomaculum sp.	Biomass	XRD, TEM	Spheroidal	13	[71]
	Rhodosporidium diobovatum	Yeast	UV-vis, TEM, XRD, EDS	Spherical	2–5	[72]
	Torulopsis sp.	Biomass	UV-vis, XRD, XPS, TEM	–	–	[73]

bio-stabilizers involved. Bio-materials, fungi, microbes, and plant extracts are utilized in the biological production of Pb NPs. The biogenic synthesis of Pb NPs has been successfully achieved using extracts from various plant species.

Elango et al. [38] revealed the green fabrication of spherical-shaped Pb NPs employing *Cocos nucifera* extract as a bio-reductant as well as a bio-stabilizer with a mean diameter of Pb NPs was estimated to be around 47 nm [38]. Green synthesis of Pb NPs employing a leaf extract of *Avivennia marina* was described by Shankar et al. [61]. This experiment stirred the reaction mixture on a magnetic stirrer at 60 °C [61]. Pavani et al. [62] reported the fabrication of Pb NPs from lead acetate utilizing *Aspergillus* sp. biomass as a natural fuel [62]. Facile biosynthesis of spherical-shaped Pb NPs employing an aqueous latex extract of *Jatropha curcas* was described by Joglekar et al. [63] with a median size of 5–17.5 nm [63]. Ramadan et al. [64] applied biomass of *Serratia plymuthica* to produce Pb NPs that had a mean NPs diameter of 92.93 nm [64].

2.2 Synthesis of PbO NPs from Plant Extracts

Implementing plant extracts in the fabrication of NPs is an effective way to adopt a green chemical strategy. Herein, Khalil et al. [65] employed an aqueous leaves extract of *Sageretia thea* to synthesize PbO NPs with a mean diameter of 27 nm. This eco-benevolent production of PbO NPs has occurred at 60 °C [65]. The schematic layout for the eco-benevolent synthesis of PbO NPs is presented in Fig. 1. Also, Hamid et al. [67] described the green production of PbO NPs

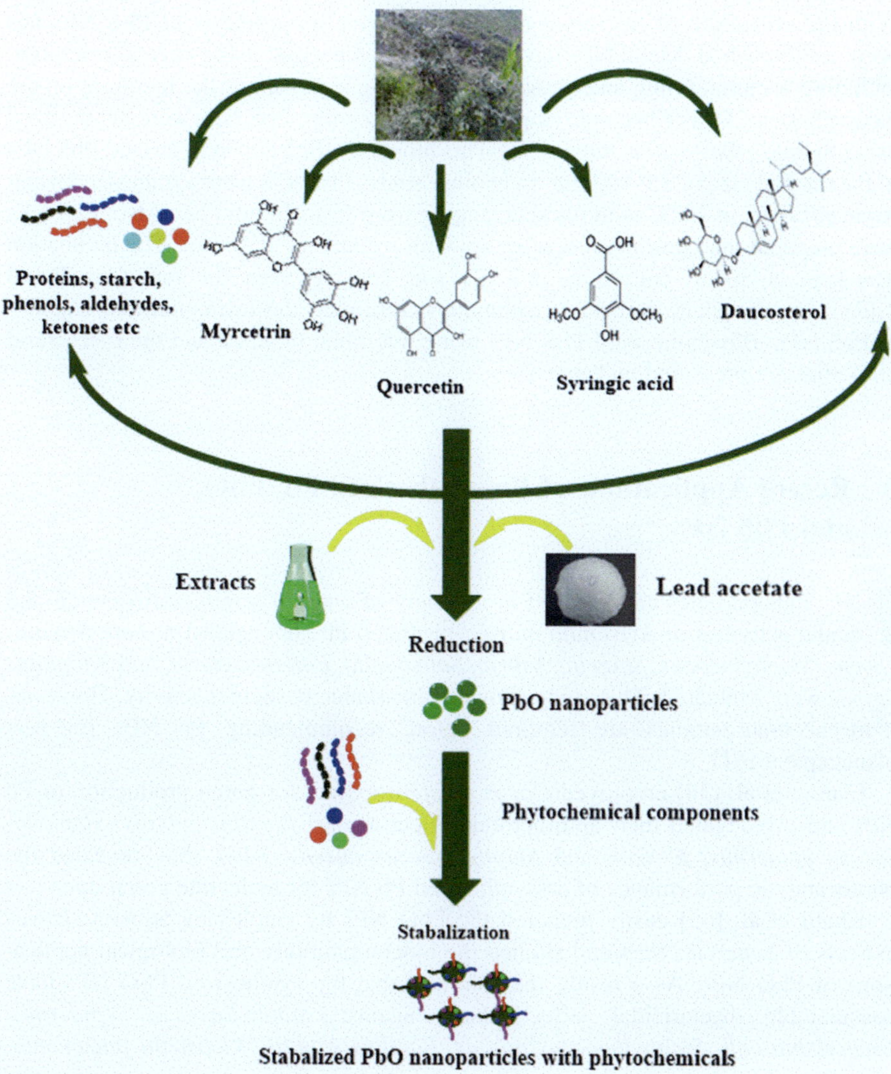

Fig. 1 Schematic presentation of eco-friendly fabrication of PbO NPs. (Reproduced from ref. [65])

employing leaf extract of *Datura sternum*. The average diameter of PbO NPs in optimum conditions was around 60 nm [67]. Moreover, Tailor et al. [68] used *Eucalyptus globulus* aqueous leaf broth as a natural fuel to create PbO NPs. As-synthesized PbO NPs estimated 34.61 nm in size, according to XRD analysis [68].

2.3 Green Synthesis of PbS NPs from Microbial Biomass

With the utilization of microbial biomass, effective biosynthesis of PbS NPs has been accomplished. Metal NPs have long been synthesized using several microbes, including bacteria, fungi, and yeast. The following summarizes the literature on the biosynthesis of PbS NPs employing microbes. Recently, PbS NPs were synthesized using biomass *Aspergillus flavus* [69]. This biomass efficiently synthesized PbS NPs of the median size of 35–100 nm. In another study, PbS NPs were synthesized using *Aspergillus* sp. in 25 °C with the size ranging from 10 to 15 nm [70]. Also, PbS NPs were prepared via *Desulfotomaculum* sp. with a diameter of 13 nm. The incubation was kept at 30 °C for 48 h, and pH was kept between the range of 5 and 9 [71]. Diverse characterization techniques confirmed the formation and stability of PbS NPs. Biosynthesis of PbS NPs using microbial biomass and their structural properties are presented in Table 1.

3 Recent Applications of Biosynthesized Pb, PbO, and PbS NPs

Pb-NPs nanoparticles can be used in a variety of ways. Therein, antibacterial and pesticidal activities of *Avivennia marina*-mediated Pb NPs against *E. coil*, *Streptococcus*, *Staphylococus*, *Shigella*, *Vibrio*, *Salmonella*, *Enterobacteria*, and *Sitophilus oryzae* were evaluated. This study showed remarkable pesticidal activity. However, more accurate studies are required before recommending Pb NPs for pest management [61].

Elango et al. [38] employed *Cocos nucifera* extract for green production of Pb NPs and investigated their antimicrobial effect against *Bacillus subtilis*, *Staphylococcus epidermis*, *E. coli*, and *Staphylococcus aureus*. Also, they revealed the photocatalytic performance of as-synthesized Pb NPs for malachite green dye.

Khalil et al. [65] easily prepared the PbO NPs by employing aqueous leaves extracts of *Sageretia thea* and studied the biocompatibility and biological applications of PbO NPs. As a result, they showed that the synthesized PbO NPs have considerable bactericidal effectiveness against *Staphylococcus epidermis*, *Escherichia coli*, *Staphylococcus aureus*, *Bacillus subtilis*, *Klebsiella pneumonia*, and *Pseudomonas aeruginosa*. They also reported the antioxidant, enzyme

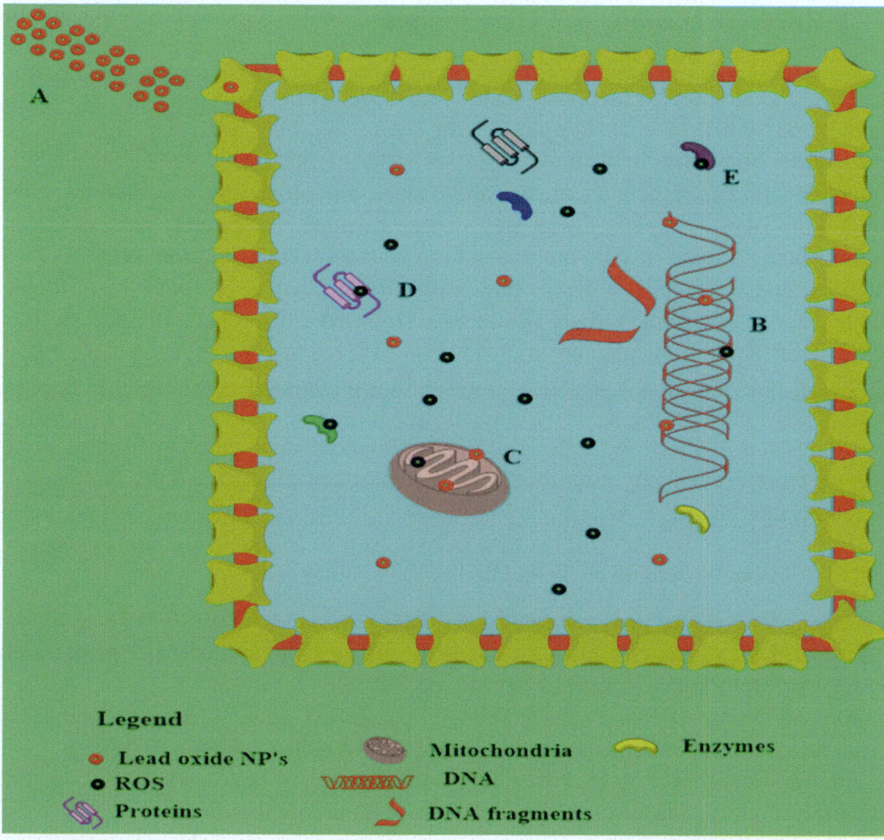

Fig. 2 Schematic presentation of the cytotoxic nature for biosynthesized PbO NPs. (Reproduced from ref. [65])

inhibition, and antileishmanial activities of PbO NPs. Furthermore, the authors investigated the MTT cytotoxicity study of PbO NPs and depicted their schematic mechanism in Fig. 2.

Moreover, Tailor et al. [68] investigated the eco-benign synthesis of PbO NPs utilizing leaf extract of *Eucalyptus globulus* and reported their bactericidal effect. The synthesized PbO NPs exhibit a significant bactericidal effect against *E. coli* and *Staphylococcus aureus*.

Priyanka et al. [69] revealed the biosynthesis of PbS NPs employing biomass of *Aspergillus flavus* was in the diameter of 35–100 nm. Therefore, these as-synthesized PbS NPs were effectively used to detect arsenic in the aqueous medium.

4 Future Directions and Conclusion

According to a report on the most recent breakthroughs in this field, these NPs were synthesized with low-cost equipment employing microbes and plant extracts in green, easy, nontoxic, and quick processes. One of the safest, most sustainable, and most efficient green chemistry approaches are producing metal-based NPs employing plant extracts. Surprisingly, no harmful compounds are used or manufactured. Pb-based NPs are valuable in sensors, ceramics, glasses, pigments, batteries, and solar cells, among other applications. Only a few parts of different plants have been employed to synthesize Pb-based NPs. Plant extracts' active phytochemical/biomolecules are used to create Pb-based NPs. These plant phytochemicals not only speed up the reaction with a predictable output, but they also entirely prevent polluting the environment. In addition, microbes were employed in the production of Pb-based NPs. The most prominent implications of these green NPs structures formed by microbial biomass and plant extracts are the removal of hazardous chemicals.

Despite current breakthroughs in the biosynthesis of MNPs from plant sources, a couple of obstacles remain to be handled in the coming time:

- Only a few plants and microbes have been explored for the biosynthesis of Pb-based NPs. There is an opportunity to utilize diverse medicinal plants and microbes for the biosynthesis of Pb-based NPs.
- Implementation of bio-waste materials like eggshell, starch, gum, and cellulose are needed to be examined for the bio-inspired synthesis of Pb-based NPs.
- For the mechanism of Pb-based NPs formation, it is necessary to clarify the biomolecules available in the plant extract and microbial biomass.
- A study of the effects of time, temperature, concentration, and pH on the synthesis of Pb-based NPs is required.
- Biosynthesized Pb-based NPs can be explored by advanced characterization techniques.

Therefore, we believe that sustainable approaches have a significant role in the economic and industrial synthesis of Pb-based NPs. Furthermore, among the green chemistry technologies available, plants- and microbes-mediated synthesis of Pb-based NPs appear to be the most effective and ideal for cheap and scalable production.

References

1. Jadoun, S., Arif, R., Jangid, N. K., & Meena, R. K. (2021). Green synthesis of nanoparticles using plant extracts: A review. *Environmental Chemistry Letters, 19*(1), 355–374.
2. Vanlalveni, C., Lallianrawna, S., Biswas, A., Selvaraj, M., Changmai, B., & Rokhum, S. L. (2021). Green synthesis of silver nanoparticles using plant extracts and their antimicrobial activities: A review of recent literature. *RSC Advances, 11*(5), 2804–2837.

3. Cuong, H. N., Pansambal, S., Ghotekar, S., Oza, R., Hai, N. T. T., Viet, N. M., & Nguyen, V. H. (2022). New frontiers in the plant extract mediated biosynthesis of copper oxide (CuO) nanoparticles and their potential applications: A review. *Environmental Research, 203*, 111858.
4. Pandit, C., Roy, A., Ghotekar, S., Khusro, A., Islam, M. N., Emran, T. B., Lam, S. E., Khandaker, M. U., & Bradley, D. A. (2022). Biological agents for synthesis of nanoparticles and their applications. *Journal of King Saud University-Science, 34*, 101869.
5. Ghotekar, S., Pansambal, S., Bilal, M., Pingale, S. S., & Oza, R. (2021). Environmentally friendly synthesis of Cr2O3 nanoparticles: Characterization, applications and future perspective—a review. *Case Studies in Chemical and Environmental Engineering, 3*, 100089.
6. Dikshit, P. K., Kumar, J., Das, A. K., Sadhu, S., Sharma, S., Singh, S., Gupta, P. K., & Kim, B. S. (2021). Green synthesis of metallic nanoparticles: Applications and limitations. *Catalysts, 11*(8), 902.
7. Salem, S. S., & Fouda, A. (2021). Green synthesis of metallic nanoparticles and their prospective biotechnological applications: An overview. *Biological Trace Element Research, 199*(1), 344–370.
8. Dabhane, H., Ghotekar, S., Tambade, P., Pansambal, S., Murthy, H. A., Oza, R., & Medhane, V. (2021). A review on environmentally benevolent synthesis of CdS nanoparticle and their applications. *Environmental Chemistry and Ecotoxicology, 3*, 209–219.
9. Dabhane, H., Ghotekar, S., Tambade, P., Pansambal, S., Oza, R., & Medhane, V. (2021). MgO nanoparticles: Synthesis, characterization, and applications as a catalyst for organic transformations. *European Journal of Chemistry, 12*(1), 86–108.
10. Verma, R., Pathak, S., Srivastava, A. K., Prawer, S., & Tomljenovic-Hanic, S. (2021). ZnO nanomaterials: Green synthesis, toxicity evaluation and new insights in biomedical applications. *Journal of Alloys and Compounds, 876*, 160175.
11. Ghotekar, S. (2019). A review on plant extract mediated biogenic synthesis of CdO nanoparticles and their recent applications. *Asian Journal of Green Chemistry, 3*(2), 187–200.
12. Nikam, A., Pagar, T., Ghotekar, S., Pagar, K., & Pansambal, S. (2019). A review on plant extract mediated green synthesis of zirconia nanoparticles and their miscellaneous applications. *Journal of Chemical Reviews, 1*(3), 154–163.
13. Pagar, T., Ghotekar, S., Pagar, K., Pansambal, S., & Oza, R. (2019). A review on bio-synthesized Co3O4 nanoparticles using plant extracts and their diverse applications. *Journal of Chemical Reviews, 1*(4), 260–270.
14. Korde, P., Ghotekar, S., Pagar, T., Pansambal, S., Oza, R., & Mane, D. (2020). Plant extract assisted eco-benevolent synthesis of selenium nanoparticles-a review on plant parts involved, characterization and their recent applications. *Journal of Chemical Reviews, 2*(3), 157–168.
15. Paiva-Santos, A. C., Herdade, A. M., Guerra, C., Peixoto, D., Pereira-Silva, M., Zeinali, M., Mascarenhas-Melo, F., Paranhos, A., & Veiga, F. (2021). Plant-mediated green synthesis of metal-based nanoparticles for dermopharmaceutical and cosmetic applications. *International Journal of Pharmaceutics, 597*, 120311.
16. Dabhane, H., Ghotekar, S., Zate, M., Kute, S., Jadhav, G., & Medhane, V. (2022). Green synthesis of MgO nanoparticles using aqueous leaf extract of Ajwain (Trachyspermum ammi) and evaluation of their catalytic and biological activities. *Inorganic Chemistry Communications, 138*, 109270.
17. Kalia, R., Chauhan, A., Verma, R., Mansi, K., Batoo, K. M., Kumar, R., Hussain, S., Ghotekar, S., & Ijaz, M. F. (2022). Photocatalytic degradation properties of Li-Cr ions substituted CoFe2O4 nanoparticles for wastewater treatment application. *Physica Status Solidi A, 219*, 2100539.
18. Kelele, K. G., Tadesse, A., Desalegn, T., Ghotekar, S., Balachandran, R., & Murthy, H. C. A. (2021). Synthesis and characterizations of metal ions doped barium strontium titanate (BST) nanomaterials for photocatalytic and electrical applications: A mini review. *International Journal of Materials Research, 112*(8), 665–677.

19. Ghotekar, S., Pagar, K., Pansambal, S., Murthy, H. A., & Oza, R. (2020). A review on eco-friendly synthesis of BiVO4 nanoparticle and its eclectic applications. *Advanced Journal of Science and Engineering, 1*(4), 106–112.
20. Dabhane, H., Ghotekar, S., Tambade, P., & Medhane, V. (2020). Plant mediated green synthesis of lanthanum oxide (La2O3) nanoparticles: A review. *Asian Journal of Nanosciences and Materials, 3*(4), 291–299.
21. Ndwandwe, B. K., Malinga, S. P., Kayitesi, E., & Dlamini, B. C. (2021). Advances in green synthesis of selenium nanoparticles and their application in food packaging. *International Journal of Food Science & Technology, 56*(6), 2640–2650.
22. Ghotekar, S., Pagar, T., Pansambal, S., & Oza, R. (2020). A review on green synthesis of sulfur nanoparticles via plant extract, characterization and its applications. *Advanced Journal of Chemistry, Section B: Natural Products and Medical Chemistry, 2*, 128–143.
23. Ghotekar, S., Dabhane, H., Pansambal, S., Oza, R., Tambade, P., & Medhane, V. (2020). A review on biomimetic synthesis of Ag2O nanoparticles using plant extract, characterization and its recent applications. *Advanced Journal of Chemistry, Section B: Natural Products and Medical Chemistry, 2*(3), 102–111.
24. Pagar, T., Ghotekar, S., Pansambal, S., Oza, R., & Marasini, B. P. (2020). Facile plant extract mediated eco-benevolent synthesis and recent applications of CaO-NPs: A state-of-the-art review. *Journal of Chemical Reviews, 2*(3), 201–210.
25. Ghotekar, S. (2019). Plant extract mediated biosynthesis of Al2O3 nanoparticles-a review on plant parts involved, characterization and applications. *Nanochemistry Research, 4*(2), 163–169.
26. Tran, T. V., Nguyen, D. T. C., Kumar, P. S., Din, A. T. M., Jalil, A. A., & Vo, D. V. N. (2022). Green synthesis of ZrO2 nanoparticles and nanocomposites for biomedical and environmental applications: A review. *Environmental Chemistry Letters, 20*, 1–23.
27. Dabhane, H., Ghotekar, S. K., Tambade, P. J., Pansambal, S., Ananda Murthy, H. C., Oza, R., & Medhane, V. (2021). Cow urine mediated green synthesis of nanomaterial and their applications: A state-of-the-art review. *Journal of Water and Environmental Nanotechnology, 6*(1), 81–91.
28. Prakash, M., Kavitha, H. P., Abinaya, S., Vennila, J. P., & Lohita, D. (2022). Green synthesis of bismuth based nanoparticles and its applications-A review. *Sustainable Chemistry and Pharmacy, 25*, 100547.
29. Pagar, T., Ghotekar, S., Pagar, K., Pansambal, S., & Oza, R. (2021). *Phytogenic synthesis of manganese dioxide nanoparticles using plant extracts and their biological application. Handbook of greener synthesis of nanomaterials and compounds* (pp. 209–218). Elsevier.
30. Gur, T., Meydan, I., Seckin, H., Bekmezci, M., & Sen, F. (2022). Green synthesis, characterization and bioactivity of biogenic zinc oxide nanoparticles. *Environmental Research, 204*, 111897.
31. Ghotekar, S., Pagar, K., Pansambal, S., Murthy, H. A., & Oza, R. (2021). Biosynthesis of silver sulfide nanoparticle and its applications. In *Handbook of greener synthesis of nanomaterials and compounds* (pp. 191–200). Elsevier.
32. Hano, C., & Abbasi, B. H. (2022). Plant-based green synthesis of nanoparticles: Production, characterization and applications. *Biomolecules, 12*(1), 31.
33. Martinez-Haro, M., Green, A. J., & Mateo, R. (2011). Effects of lead exposure on oxidative stress biomarkers and plasma biochemistry in waterbirds in the field. *Environmental Research, 111*(4), 530–538.
34. Kordas, K., Roy, A., Vahter, M., Ravenscroft, J., Mañay, N., Peregalli, F., Martínez, G., & Queirolo, E. I. (2018). Multiple-metal exposure, diet, and oxidative stress in Uruguayan school children. *Environmental Research, 166*, 507–515.
35. Tandon, S. K., Singh, S., Prasad, S., Srivastava, S., & Siddiqui, M. K. J. (2002). Reversal of lead-induced oxidative stress by chelating agent, antioxidant, or their combination in the rat. *Environmental Research, 90*(1), 61–66.

36. Kaehler, T. (1994). Nanotechnology: Basic concepts and definitions. *Clinical Chemistry, 40*(9), 1797–1797.
37. Miri, A., Sarani, M., Hashemzadeh, A., Mardani, Z., & Darroudi, M. (2018). Biosynthesis and cytotoxic activity of lead oxide nanoparticles. *Green Chemistry Letters and Reviews, 11*(4), 567–572.
38. Elango, G., & Roopan, S. M. (2015). Green synthesis, spectroscopic investigation and photocatalytic activity of lead nanoparticles. *Spectrochimica Acta Part A: Molecular and Biomolecular Spectroscopy, 139*, 367–373.
39. Dumbaugh, W. H., & Lapp, J. C. (1992). Heavy-metal oxide glasses. *Journal of the American Ceramic Society, 75*(9), 2315–2326.
40. Jaffe, B., Roth, R. S., & Marzullo, S. (1955). Properties of piezoelectric ceramics in the solid-solution series lead titanate-lead zirconate-lead oxide: Tin oxide and lead titanate-lead hafnate. *Journal of Research of the National Bureau of Standards, 55*(5), 239–254.
41. Senvaitiene, J., Smirnova, J., Beganskiene, A., & Kareiva, A. (2007). XRD and FTIR characterisation of lead oxide-based pigments and glazes. *Acta Chimica Slovenica, 54*(1), 185.
42. Šljukić, B., Banks, C. E., Crossley, A., & Compton, R. G. (2007). Lead (IV) oxide–graphite composite electrodes: Application to sensing of ammonia, nitrite and phenols. *Analytica Chimica Acta, 587*(2), 240–246.
43. Sonmez, M. S., & Kumar, R. V. (2009). Leaching of waste battery paste components. Part 1: Lead citrate synthesis from PbO and PbO2. *Hydrometallurgy, 95*(1–2), 53–60.
44. Blair, T. L. (1998). Lead oxide technology—Past, present, and future. *Journal of Power Sources, 73*(1), 47–55.
45. Ragg, M. (1933). The protective action of lead pigments against rust. *Transactions of the Electrochemical Society, 64*(1), 59.
46. Klein, C. A. (1968). Bandgap dependence and related features of radiation ionization energies in semiconductors. *Journal of Applied Physics, 39*(4), 2029–2038.
47. Eya, D. D. O. (2006). Influence of thermal annealing on the structural and optical properties of PbO thin films prepared by chemical bath deposition technique. *Pacific Journal of Science and Technology, 7*(2), 114–119.
48. Darwish, A. A. A., El-Zaidia, E. F. M., El-Nahass, M. M., Hanafy, T. A., & Al-Zubaidi, A. A. (2014). Dielectric and electrical conductivity studies of bulk lead (II) oxide (PbO). *Journal of Alloys and Compounds, 589*, 393–398.
49. Salavati-Niasari, M., Mohandes, F., & Davar, F. (2009). Preparation of PbO nanocrystals via decomposition of lead oxalate. *Polyhedron, 28*(11), 2263–2267.
50. Bangi, U. K., Han, W., Yoo, B., & Park, H. H. (2013). Effects of successive additions of two capping ligands on the structural properties of PbO nanoparticles. *Journal of Nanoparticle Research, 15*(11), 1–8.
51. Gao, P., Liu, Y., Bu, X., Hu, M., Dai, Y., Gao, X., & Lei, L. (2013). Solvothermal synthesis of α-PbO from lead dioxide and its electrochemical performance as a positive electrode material. *Journal of Power Sources, 242*, 299–304.
52. Mythili, N., & Arulmozhi, K. T. (2014). Characterization studies on the chemically synthesized α and β phase PbO nanoparticles. *International Journal of Scientific and Engineering Research, 5*(1), 412–416.
53. Li, S., Yang, W., Chen, M., Gao, J., Kang, J., & Qi, Y. (2005). Preparation of PbO nanoparticles by microwave irradiation and their application to Pb (II)-selective electrode based on cellulose acetate. *Materials Chemistry and Physics, 90*(2–3), 262–269.
54. Karami, H., & Ghamooshi-Ramandi, M. (2013). Synthesis of sub-micro and nanometer sized lead oxide by sol-gel pyrrolysis method and its application as cathode and anode of lead-acid batteries. *International Journal of Electrochemical Science, 8*(7553), e7564.
55. Lyons, S. W., Xiong, Y., Ward, T. L., Kodas, T. T., & Pratsinis, S. E. (1992). Role of particle evaporation during synthesis of lead oxide by aerosol decomposition. *Journal of Materials Research, 7*(12), 3333–3341.

56. Ghasemi, S., Mousavi, M. F., Shamsipur, M., & Karami, H. (2008). Sonochemical-assisted synthesis of nano-structured lead dioxide. *Ultrasonics Sonochemistry, 15*(4), 448–455.
57. Karami, H., Ghasemi, M., & Matini, S. (2013). Synthesis, characterization and application of lead sulfide nanostructures as ammonia gas sensing agent. *International Journal of Electrochemical Science, 8*(10), 11661–11679.
58. Koleilat, G. I., Levina, L., Shukla, H., Myrskog, S. H., Hinds, S., Pattantyus-Abraham, A. G., & Sargent, E. H. (2008). Efficient, stable infrared photovoltaics based on solution-cast colloidal quantum dots. *ACS Nano, 2*(5), 833–840.
59. Konstantatos, G., Huang, C., Levina, L., Lu, Z., & Sargent, E. H. (2005). Efficient infrared electroluminescent devices using solution-processed colloidal quantum dots. *Advanced Functional Materials, 15*(11), 1865–1869.
60. Konstantatos, G., Howard, I., Fischer, A., Hoogland, S., Clifford, J., Klem, E., Levina, L., & Sargent, E. H. (2006). Ultrasensitive solution-cast quantum dot photodetectors. *Nature, 442*(7099), 180–183.
61. Sankar, M. V., & Abideen, S. (2015). Pesticidal effect of green synthesized silver and lead nanoparticles using Avicennia marina against grain storage pest Sitophilus oryzae. *International Journal of Nanomaterials and Biostructures, 5*(3), 32–39.
62. Pavani, K. V., Kumar, N. S., & Sangameswaran, B. B. (2012). Synthesis of lead nanoparticles by Aspergillus species. *Polish Journal of Microbiology, 61*(1), 61–63.
63. Joglekar, S., Kodam, K., Dhaygude, M., & Hudlikar, M. (2011). Novel route for rapid biosynthesis of lead nanoparticles using aqueous extract of Jatropha curcas L. latex. *Materials Letters, 65*(19–20), 3170–3172.
64. Ramadan, O. A., Sabry, A. A., Kesht, A. T., & Amer, A. A. (2017). Biosynthesis and characterization of lead sulfide nanoparticles using wastewater bacteria. *Biochemistry Letters, 13*(1), 64–84.
65. Khalil, A. T., Ovais, M., Ullah, I., Ali, M., Jan, S. A., Shinwari, Z. K., & Maaza, M. (2020). Bioinspired synthesis of pure massicot phase lead oxide nanoparticles and assessment of their biocompatibility, cytotoxicity and in-vitro biological properties. *Arabian Journal of Chemistry, 13*(1), 916–931.
66. Sutjaritvorakul, T., & Chutipaijit, S. (2020). Biological synthesis and characterization of lead oxide nanoparticles using Averrhoa bilimbi Linn. aqueous extract. *AIP Conference Proceedings, 2279*(1), 130001.
67. Hamid, A., Khan, M., Hayat, A., Raza, J., Zada, A., Ullah, A., Raziq, F., Li, T., & Hussain, F. (2020). Probing the physio-chemical appraisal of green synthesized PbO nanoparticles in PbO-PVC nanocomposite polymer membranes. *Spectrochimica Acta Part A: Molecular and Biomolecular Spectroscopy, 235*, 118303.
68. Tailor, G., & Lawal, A. M. (2021). Phytochemical screening; green synthesis, characterization and biological significance of lead oxide nanoparticles from Eucalyptus globulus Labill. (leaves). *Nanotechnology for Environmental Engineering, 6*(3), 1–11.
69. Priyanka, U., Akshay, G. K. M., Elisha, M. G., Surya, T. B., Nitish, N., & Raj, M. B. (2017). Biologically synthesized PbS nanoparticles for the detection of arsenic in water. *International Biodeterioration & Biodegradation, 119*, 78–86.
70. Kaur, P., Jain, P., Kumar, A., & Thakur, R. (2014). Biogenesis of PbS nanocrystals by using rhizosphere fungus i.e., Aspergillus sp. isolated from the rhizosphere of chickpea. *BioNanoScience, 4*(2), 189–194.
71. Gong, J., Zhang, Z., Bai, H., & Yang, G. (2007). Microbiological synthesis of nanophase PbS by Desulfotomaculum sp. *Science in China Series E: Technological Sciences, 50*(3), 302–307.
72. Seshadri, S., Saranya, K., & Kowshik, M. (2011). Green synthesis of lead sulfide nanoparticles by the lead resistant marine yeast, Rhodosporidium diobovatum. *Biotechnology Progress, 27*(5), 1464–1469.
73. Kowshik, M., Vogel, W., Urban, J., Kulkarni, S. K., & Paknikar, K. M. (2002). Microbial synthesis of semiconductor PbS nanocrystallites. *Advanced Materials, 14*(11), 815–818.

Nanofertilizers and Nanopesticides for Crop Growth

Nam Nghiep Tran, Tu Nguyen Quang Le, Hue Quoc Pho, Tung Thanh Tran, and Volker Hessel

Abstract In the last decades, advanced applications of nanotechnology in agriculture have gained good momentum with many methods being developed to widespread the production and application of nanofertilizers and nanopesticides for plants. Nanotechnology develops new types of nanopesticides and nanofertilizers to enhance crop productivity while reducing the advert effects on the surrounding environment. Nanopesticides can protect plants against phytopathogens, while nanofertilizers stimulate plant growth and ensure large-scale food production all over the world. In this chapter, popular nanofertilizers and nanopesticides and their applications on plants in practice were comprehensively introduced.

1 Nanopesticides

Nanopesticides are belonging to an emerging field in the modern agriculture sector, which uses nanotechnology to protect crops by providing novel nanomaterials to enhance the effectiveness of active ingredients, in addition to their formulation and delivery. Up to date, nanopesticides are classified into four categories (Fig. 1) that significantly contribute to agricultural development and sustainability all over the world.

N. N. Tran
School of Chemical Engineering and Advanced Materials, The University of Adelaide, Adelaide, SA, Australia

Department of Chemical Engineering, Can Tho University, Can Tho, Vietnam

T. N. Q. Le · T. T. Tran
School of Chemical Engineering and Advanced Materials, The University of Adelaide, Adelaide, SA, Australia

H. Q. Pho · V. Hessel (✉)
School of Chemical Engineering and Advanced Materials, The University of Adelaide, Adelaide, SA, Australia

School of Engineering, University of Warwick, Coventry, London, UK
e-mail: volker.hessel@adelaide.edu.au

J.-T. Chen (ed.), *Plant and Nanoparticles*,
https://doi.org/10.1007/978-981-19-2503-0_15

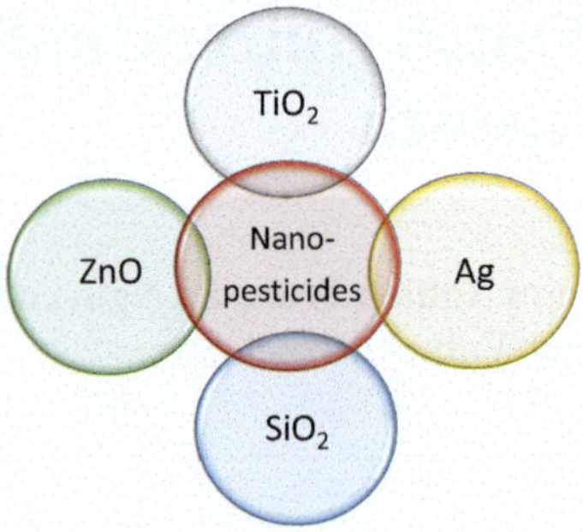

Fig. 1 Popular types of nanomaterials as nanopesticides

Table 1 The list of SiO_2 nanoparticles as nanopesticides

Name of insects	Types of plants	Concentrations	Reference
Sitophilus oryzae	Rice	0.5–2 g/kg	[7]
Plutella xylostella	Cabbage	0.125–1 mg/cm^2	[3]
Mythimna separata	Wheat	5% in soil	[8]
Rhyzopertha dominica	Wheat, barley	50–300 mg/kg	[9]
Liriomyza trifolii	Faba bean	100–500 mg/L	[10]
Spodoptera littoralis	Castor	250–1000 ppm	[5]
Tuta absoluta	Tomato	150–600 ppm	[11]
True spider	Sugar beet	15–75 g/4200 m^2	[12]
Sitophilus granaries	Wheat	1–3 g/kg	[13]
Coccinella spp.	Sugar beet	15–75 g/4200 m^2	[12]

1.1 Silica Nanoparticles

Among nanomaterials, silica (SiO_2) nanoparticles (NPs) have attracted remarkable attention due to their unique properties, such as water solubility compared to traditional pesticides [1, 2]. Hydrophilic properties can increase bioactivity and coverage uniformity [1]. These nanomaterials can be used in a small volume and quickly penetrate cells. Therefore, the use of SiO_2 can prevent or slow the resistance of target pests. The pesticide activity of SiO_2 is believed to penetrate the cuticle directly [3–5] or adsorbs through the cuticular layers [5]. Another study also showed that SiO_2 NPs might indirectly affect pests by inactivating the digestive tract [6]. Some studies on the direct effects of SiO_2 NPs on various pests have been carried out on the laboratory scale, as given in Table 1. In fields, SiO_2 NPs showed dose-dependent performance of pesticide effects on a chewing insect (moth: *Spodoptera littoralis*) [12], a piercing-sucking insect (aphid: *Aphis craccivora*) [10], and an internal feeder (leaf-mining fly: *Liriomyza trifolii*) [10].

To inactivate the development of agricultural insect pests, silica NPs (3–5 nm) can be modified on the surface. Another study showed that silica in the form of a thin film on seeds was used to inactivate the fungal development and trigger cereal germination [14]. In addition, silica NPs have fewer side effects on plants because they have no impact on photosynthesis of leaves or respiration in several horticultural plants [15]. Importantly, silica NPs do not have an impact on gene expression in insect trachea [3]. For this reason, silica NPs were approved as nanobiopesticide. In addition, World Health Organization (WHO) approved amorphous silica as a nanobiopesticide, which is safe for humans [16].

Apart from acting themselves as nanopesticides, surface-modified silica nanoparticles have also been widely used to load conventional pesticides. For example, insecticide chlorfenapyr was loaded with dispersible SNPs to enhance biological efficacy compared to conventional chlorfenapyr [17]. Moreover, porous hallow surface-modified silica nanoparticles have been used to protect the short half-life active ingredient (avermectin) from degradation under UV exposure, which might induce a slower release (approximately 30 days) [18].

1.2 Silver Nanoparticles

Pests are one of the most threatening issues in agriculture that leads to significant crop loss [19]. In addition, over the past several decades, synthetic pesticides have been globally over-used, which causes negative impacts on the environment [20]. Similar to other nanoparticles, silver nanoparticles act as an effective agent in pest control because of their antibacterial and antifungal activities, low toxicity, high surface area, crystallographic structure, and adaptability to various substrates [20]. In addition, it is non-toxic and safe for the environment. In comparison with synthetic fungicides, silver NPs can minimize toxicity to human beings, and reduce pest resistance as well as environmental degradation [21, 22]. Many fabrication methods have been studied to produce green silver NPs [23]. However, in the green synthesis of silver NPs, the biological agents used for their synthesis can be microbes [24] or plants [25] and flavonoids [26] to be non-toxic to the plants. Apart from that, the antifungal activity of the silver nanoparticles has also been discovered and proved to be used in treating the fungal pathogens of the plants. For example, silver nanoparticles were revealed to inactivate the development of fungal phytopathogen *Raffaelea*—the destructive disease oak wilt [21]. Kim et al. have reported that silver NPs are able to inhibit fungal growth at various concentrations [21]. Sap-lam et al. reported that under UV irradiation, silver nanoparticles are active in controlling pests such as mosquito larvae [27].

1.3 Titanium Dioxide Nanoparticles

TiO_2 NPs are low toxicity nanopesticides for humans approved by the American Food and Drug Administration (FDA) to be widely used in agricultural practice [28]. The antimicrobial activity of this type of nanopesticide is primarily attributed to a mechanism that which TiO_2 NPs generate reactive oxygen species (ROS) under

UV irradiation such as hydroxyl radicals ($^{\bullet}OH$), superoxide anions ($O^{\bullet}{}_2{}^-$), and hydrogen peroxide (H_2O_2) molecules [29]. As a result, the cell membrane of microorganisms can be damaged by these ROS groups [29]. Finally, these microorganisms will be suppressed.

Recently, it was revealed that TiO_2 and TiO_2 doped with zinc and silver were investigated to inhibit bacterial spot disease in tomatoes [22]. Additionally, TiO_2 NPs can inactivate *Rhipicephalus* (*Boophilus*) *microplus* larvae as well as adults of *Haemaphysalis bispinosa* [30, 31]. It is revealed that TiO_2 NPs can suppress the growth of *Hypocrea lixii* (white rot) and *Nucor circibelloides* (brown rot) fungi on various types of wood such as Scots pine, silver fir, walnut, wild cherry, sessile oak, beech, and ash [31, 32]. At low concentrations, TiO_2 can slow the development of foliar and pod diseases of cowpea [33]. Also, Mathew et al. indicated that the growth of the mung bean plant was triggered when exposed to TiO_2 nanoparticles [34]. Another study also reported that different strands of bacteria (*Aeromonas hydrophila*, *Escherichia coli*, *Staphylococcus aureus*, *Proteus mirabilis*, and *Pseudomonas aeruginosa* were also actively inhibited by TiO_2 NPs [35, 36].

1.4 Zinc Oxide Nanoparticles

Like other nanopesticides, ZnO NPs as a pesticide have attracted great attention in agricultural applications due to their potential to inhibit harmful microorganisms. ZnO NPs are belonging to a type of photocatalyst that significantly influences chemical compounds and biological systems [36]. Various studies have demonstrated that the fundamental mechanism of ZnO NPs to kill bacteria and fungi is attributed to ROS generation under UV irradiation [36]. Therefore, when exposed to UV light, a variety of ROS groups such as hydroxyl radicals ($^{\bullet}OH$), superoxide anions ($O^{\bullet}{}_2{}^-$), and hydrogen peroxide (H_2O_2) is generated from ZnO NPs. Ultimately, these ROS groups damage the cell membrane, leading to the cell death of pests. The explanation for this is that these ROS groups are involved in the decomposition of lipids, DNA, and proteins in the cell of pests [37]. Based on this mechanism, ZnO NPs have been broadly applied in various applications, one of which is to control the phytopathogenic pest caused by various fungi and bacteria like *fungus-like B. cinerea* and *P. expansum* [37], were gray and blue mold on table grapes and rotting of stored apples and pears can be observed. In addition, ZnO also exhibits strong antifungal activity against fungus-caused diseases such as *Fusarium oxysporum*, *Alternaria alternata*, *Mucor plumbeus*, and *Rhizopus stolonifer* [37].

2 Nanofertilizers

Fertilizers are popularly used to amend soils in agricultural productions to increase crop yields. However, the use of conventional fertilizers causes the barrenness of soil due to their persistence in soil for a relatively long time [38]. This problem leads to the lack of essential nutrients for plants which can be overcome by the application of nanofertilizers. Nanofertilizers offer the best solution for crop and plant cultivations since they can be released on-demand, are plant and site-specific, efficient, and easily soluble [39]. Recently, the development of bio nanofertilizers has been considered a game-changer since it provides the farmers with an environmentally friendly and cost-effective solution for farming [40].

Table 2 shows the advantages of nanofertilizers in comparison to conventional technologies.

In general, four popular types of nanofertilizers are widely applied for farming including zeolites, nanocomposites, super-absorbent fertilizers, and carbon nanotubes.

Table 2 Comparison of nanotechnology-based formulations and conventional fertilizers applications [41]

Criteria	Nanofertilizers	Conventional fertilizers
Mineral micronutrient solubility and dispersion	The nanostructure can improve the solubility and dispersion of mineral micronutrients. It can also enable the solubility of fertilizers in the soil while reducing both absorption and fixation processes. Thus, the bioavailability of soil can be improved.	The larger particle size and significantly low solubility of conventional fertilizers caused less bioavailability to soils and plants
The efficiency of nutrient consumption	Enhance fertilizer efficiency and the availability to adsorb nutrients from soils while saving the fertilizer used.	Efficiency is reduced due to the unavailable bulk composite for roots resources.
Active control of the release of nutrients	Encapsulated nanofertilizers using semipermeable membranes can enhance the release pattern and rate of water.	Conventional fertilizers often cause an excess release that leads to the generation of toxic compounds which might contaminate water and break up the existing ecological balance in the soil.
Nutrient release duration	The release duration is more effective with the application of nanostructured fertilizer.	Uptaking only by plants at the application, the rest is transformed into insoluble salts or leaching.
Nutrients loss rate	Prevent nutrients loss	The loss of nutrients is usually high due to runoff, leaching, and washout.

2.1 Zeolites

Zeolites are derived from aluminosilicate minerals which are widely used as adsorbents thanks to their microscopic and crystalline hydrated properties [42]. Depending on the ability to exchange ions, to retain or release water, zeolites are classified into various types such as clinoptilolite, stilbite, chabazite, analcime, and natrolite, etc. [43].

Zeolites are not only used as an additive to soils but also be applied as a carrier for various nutrient distribution or a regulator for nutrient mineral fertilizers. They can also be applied in combination with potassium and phosphorus compounds to control the release of these nutrients in agriculture and horticulture [44]. In particular, adding a specific loading of zeolites to nitrogen and potassium-containing fertilizers can improve the efficiency of those fertilizers to soil amendment and plant nutrient uptakes [45].

2.2 Nanocomposites

Nanocomposites are defined as the multiphase matrixes of silicates that can be mixed with nanoparticles to modify the function of a specific material [46]. Nanocomposites are nanoscale reinforcing components integrated into a continuous phase matrix, allowing them to have a relatively high aspect ratio and surface-area-to-volume ratio [47]. In recent years, nanocomposites have received greater interest, especially for nano-based agri-products development, since they are effective at even very low concentrations, have low environmental impacts, and are amenable to experimentation under field conditions [48]. One of the most important applications of nanocomposites is to develop nanofertilizers with special properties that can be utilized on demand the farmers. For example, ammonium-loaded clinoptilolite can be combined with phosphorite to create a very good carrier to control the release of nitrogen and phosphorus in soil [49]. It was reported that clinoptilolite loaded with the combination of ammonium, potassium, and phosphorite can effectively control the slow release of elements such as calcium, phosphorus, potassium, and nitrogen [50].

2.3 Super-Absorbent Fertilizers (SAF)

Super adsorbent fertilizers are the copolymers of NPK fertilizers and super-absorbent polymers created through the links of hydrogen bonds [51]. The water capturing and holding potential of this type of nanofertilizer can be adjusted by changing the hydrophilic functional groups that are presented in their molecular structure [52]. Experiment show that the development of super-absorbent nitrogen fertilizers could contribute to the control of nitrogen and water release in soils, offering a solution to enhance the water and fertilizer use efficiency by the crops [53].

Experiments have also been performed on wheat to compare the effects of chemical fertilizers and nanofertilizers (incorporated and coated with nanoparticles) in terms of yield and nutrient contents. Results showed that both protein content and productivity of wheat could be improved if nanoparticles are applied. Thus, nanoparticles can be applied to control the release of nutrients to achieve higher yield and nutritional value for wheat [39].

2.4 *Carbon Nano Tubes (CNT)*

Carbon nano tubes are a special type of nanofertilizer which possess not only nano structure but also allotropic forms of carbon, allowing them to be widely used in agricultural productions. Similar to the other nanofertilizers, CNT has also been widely applied for the control of fertilizer release, especially for seeds germination. Results showed that there is a twofold increase in both germination rate and seeding biomass of the experimental plants in comparison to the control ones. Additionally, further investigation of the seeds has also confirmed the presence of CNT inside their shells which enhances the ameliorating effects on both germination and growth [54].

On the other hand, the application of CNT can also improve the water absorption ability and retention capacity of seeds thanks to the formation of new pores within the seed shells due to the penetration of CNT [55].

3 Examples of Nanopesticides and Nanofertilizers Synthetic Processes

3.1 *Nanopesticides*

There is a wide range of nanomaterials that can be used in nanopesticides and their role will be either as active ingredients or as carriers. This section will cover examples of the synthetic process of the most commonly used nanoparticles in pathogen and pest control.

3.1.1 Polymeric Nanoparticles

Polymeric nanoparticles used in nanopesticides originate from natural sources such as chitosan, alginate, collagen, and gelatine and they can either be used as active ingredients or nanocarriers. Chitosan nanoparticles for application in resistance against downy mildew of pearl millet were synthesized based on the ionic gelation process of low molecular weight chitosan and a polyanion such as TPP

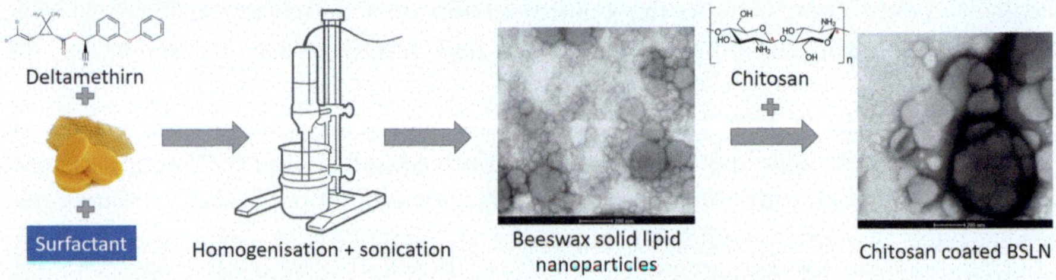

Fig. 2 Chitosan-coated beeswax solid nanoparticles for the loading of deltamethrin. (With kind permission from Wiley [58])

[56]. Harpin$_{pss}$ loaded chitosan nanoparticles were prepared by mixing harpin$_{pss}$ and chitosan nanoparticles at the ratio of 0.1:1 for enhancing disease resistance in tomatoes [57]. Nguyen et al. studied beeswax solid lipid nanoparticles coated with chitosan (CH-BSLNs) for photoprotection of deltamethrin [58]. BSLNs were first synthesized using a combination of homogenization and sonication at high temperatures and followed by the coating of chitosan solution under mechanic stirring (Fig. 2) [58]. Amphiphilic nanopolymers used for carrying pesticides were synthesized using PEG with different molecular weights as hydrophilic head and aliphatic di-acids and aromatic di-esters as linker hydrophobic moiety and were then used to encapsulate pesticides such as carbofuran [59], imidacloprid [60], thiamethoxam [61], and β-cyfluthrin [62, 63].

3.1.2 Metal and Metal Oxide NPs

Metal and metal oxide nanoparticles are the most common form of active ingredients in nanopesticide and there are a vast number of commercial suppliers of these nanomaterials. Additionally, metal and metal oxide nanoparticles can have their surface further modified with elements (e.g. N) using advanced technologies such as plasma-assisted synthesis to enhance their activity and this approach was also discussed in other literature [31]. However, due to the rising trend of green chemistry, more studies have focused on the eco-friendly, non-toxic preparation of metal and metal oxides nanoparticles (Fig. 3). The preparation of silver nanoparticles from an aqueous solution of silver nitrate via green chemistry for nanopesticide can be performed using a wide range of microorganisms or extracts such as an aqueous extract of endophytic non-pathogenic *Alternaria solani* [64], *Chaetomium globosum* [65], *Fusarium oxysporum* [66], extract solutions of carrageenan seaweeds (*Hypnea musciformis* and *Spyridia hypnoides*), and agar seaweeds (*Gracilaria corticata* and *Gracilaria edulis*) [67]. The preparation of zinc oxide nanoparticles for reducing Cercospora leaf spot disease using 32 various plant aqueous extracts was investigated by Farahat [68]. The extracts of plants were obtained using a simple method and were used as reducing agents for the reaction with $Zn(NO_3)_2$ solution [68]. Similarly, copper oxide nanoparticles were synthesized using papaya leaf extract and were investigated for bactericidal activity against *Ralstonia solanacearum* [69].

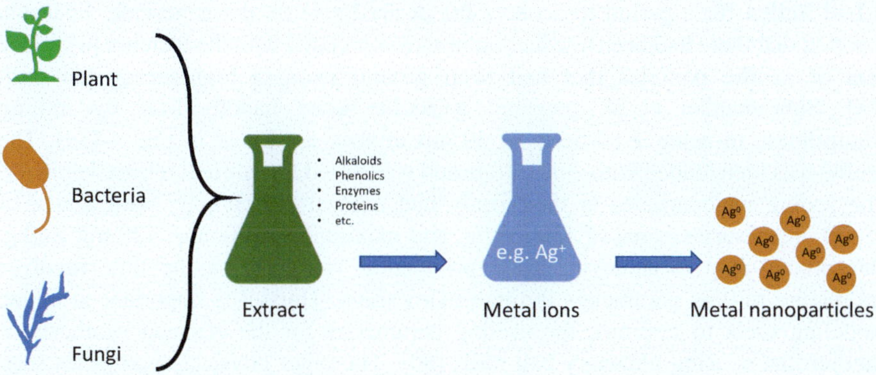

Fig. 3 Schematic illustration of preparing Ag nanoparticles via biosynthesis

3.1.3 Other Nanomaterials

Mesoporous silica nanoparticles (MSN) have attracted lots of attention due to their potential for pesticide delivery. Sol-gel is among the most common methods used to prepare MSN from TEOS and surfactant (e.g. CTAB). MSN can be directly used as a pesticide [7] or as a carrier to carry active ingredients such as validamycin [70], cyantraniliprole [71], and copper ions [72]. Cao et al. studied the combination of an in situ carbon dots (CD) fabrication method and a selective-etching strategy to prepare luminescent double-shelled hollow MSN, without fluorophore, via combining as pesticide delivery carrier [73].

Carbon-based nanomaterials can also be used as carriers for agrochemicals. Wei et al. prepared porous carbon from silkworm excrement via thermal treatment and then the porous carbon was modified using various Lewis acid metal ions via ultrasonic-assisted immersion method [74]. Salak et al. prepared a hybrid material by polymerizing citric acid onto the surface of oxidized carbon nanotubes and used the hybrid material to encapsulate pesticides such as zineb and mancozeb [75]. Garrido-herrera et al. prepared alginate-bentonite-activated carbon formulations using the gelling properties of alginate in the presence of divalent cations to study the controlled release of isoproturon, imidacloprid, and cyromazine [76].

3.2 Nanofertilizers

3.2.1 Physical Synthesis

Physical synthesis methods generally provide a basic approach to mass-producing nanofertilizers with ease of operation and short duration [77]. Among those, mechanical grinding/milling is the most suitable technique to meet the demand of an economically feasible large-scale production. Generally, an adsorbent or subtract is ball milled for a period to achieve the desired size on the nanoscale. Sharmila reported that there had been a reduction in size associated with the increase in surface area of zeolite particles that had been ground using a high-energy ball mill [78]. Subramanian et al. prepared a zeolite-based nanofertilizer by milling clinoptilolite to achieve particles of 30 nm in size, as shown in Fig. 4 [79]. The zeolite was then loaded with zinc sulfate and was tested for nutrient release behavior. The zeolite nanoparticles in their study had cubical morphology with a size of 25–30 nm, a surface area of 1300 m^2/g, and maximum sorption of 429 mg Zn/kg. One crucial step to achieving good production yield and saving time involves optimizing milling parameters through many trials. Thus, it is important to utilize modeling tools to optimize the milling parameters for the efficient synthesis of nanofertilizers using planetary ball mills [80]. The other physical approaches such as laser ablation or sputtering have been utilized for nanomaterials without functionalization or surface modification which is only suitable for industrial purposes [81]. Nevertheless, the physical synthesis exhibited limitations in source-precursor incompatibility, dispersed particle size and surface properties (defects and imperfections), high level of impurities and equipment maintenance cost.

3.2.2 Chemical Synthesis

The chemical synthesis for the preparation of nanoparticles is the most common approach. Table 3 illustrated the examples of chemical synthesis processes for typical nanofertilizers. By choosing appropriate methods and precursors, these nanoparticles could further be tailored made into nanofertilizers. The level of impurities, morphology, and composition of the prepared nanomaterials depends on the chosen reaction and can have better control compared with physical methods.

Different synthesis procedures are corresponding to different nutrients for the best results. For macronutrients (i.e. N, P, K, Ca, S, and Mg), the most suitable approach is to prepare nanocarrier (adsorbent or coating) or nanocomposite with release

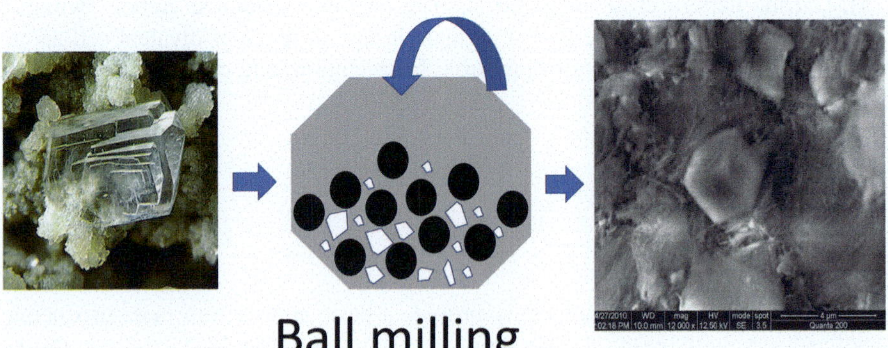

Fig. 4 Preparation of zeolite nanoparticles via ball milling. (With kind permission from Diva Enterprises Pvt. Ltd. [79])

Table 3 Examples of chemical synthesis process for common nanofertilizers

Nanofertilizer	Method
Chitosan-NPK	Polymerization and impregnation [82–85]
CNT-NPK	Room temperature preparation of CNTs [86, 87]
Hydroxyapatite	Precipitation/neutralization methods [88, 89]
N-Zeolite	Impregnation and liquid hydrothermal [90, 91]
Fe NPs	Precipitation [92, 93]
Zn NPs	Polyol method [94]
Mg NPs	Impregnation [95]
Mn NPs	
Cu NPs	
B NPs	

behavior matching the crop's nutrient requirements. Chitosan nanoparticles for loading of NPK were synthesized using the two-step process which included dissolving of chitosan in methacrylic acid solution and polymerization [82]. The NPK fertilizer was then loaded with different concentrations by dissolving in chitosan nanoparticles dispersion. This procedure was simple and easy to perform and thus it is adapted in many similar studies [83–85]. One example of a major nutrient nanofertilizer that does not require a carrier is hydroxyapatite (HA) nanoparticles ($Ca_{10}(PO_4)_6(OH)_2$). HA NPs can be synthesized by the neutralization method [89] or precipitation [88] where the Ca^{2+} ion reacts with H_3PO_4 at the ratio of Ca/P is 10:6 to form a white solid.

Micronutrients including manganese (Mn), copper (Cu), zinc (Zn), iron (Fe), molybdenum (Mo), boron (B), and chloride (Cl) are required in a small amount to optimize crop growth and thus micro nanofertilizers can be delivered directly in their nanoscale morphology or through a nanocarrier. The most common and simple method to prepare micro-nanofertilizer of metal elements is precipitation using a base (such as NaOH). A wide range of oxide nanoparticles (such as CuO, MnO_x, ZnO, and FeO_x) can be obtained using this approach. Soliman et al. prepared ZnO by adding dropwise NaOH solution into zinc acetate solution under stirring resulting in spherical nanoparticles of 10–15 nm in size and they also prepared Fe_3O_4 nanoparticles with the size of 10–12 nm by co-precipitating a solution of Fe^{2+} and Fe^{3+} with aqueous ammonia [92]. Liu et al. prepared various oxides for the study on the germination of lettuce using a similar procedure [93]. In their study, drops of sodium hydroxide were added to solutions containing sulfate salt of the corresponding metal. The CMC was added to the metal precursor solution as a stabilizer for nanoparticles. Other approaches, such as the impregnation of metal ions in nanocarriers, require preparing nanocarriers using previously described methods. In a study on the influence on the growth of coffee, Wang et al. prepared a nanofertilizer consisting of Zn and B on chitosan nanoparticles [95]. Those chitosan-based nanoparticles were first synthesized using TPP ionic gelation and $ZnSO_4$ and H_3BO_3 solution was later dissolved in the chitosan nanoparticles dispersion for the loading of nutrients.

3.2.3 Green/Biological Synthesis

The new trend in green chemistry is shifting the synthesis process of nanomaterials into eco-friendly, safer, and non-toxic approaches. Thus, biosynthesis which is the synthesis of nanoparticles using plant extracts or microorganisms has been receiving booming attention. Shah et al. have reported that compounds or molecules such as phenolics, alkaloids, proteins, and enzymes are responsible for reducing precursors to form nanoparticles [96]. The general procedure for preparing nanoparticles using plants extracts is mixing these compounds with the precursors for the reaction to take place. Chaudhuri et al. reported that mixing and incubating a mixture of zinc acetate dihydrate, the leaf extract of *Calotropis gigantea* and sodium hydroxide at appropriate conditions can result in zinc oxide nanoparticles (average size of 11 nm) that can affect the tree seedling growth [97]. Chahar et al. also reported the use of the root extract of *Plumbago zeylanica* as a reducing agent for the biosynthesis of silver nanoparticles [98]. Singular and bimetal iron and manganese nanofertilizers were obtained from mixing a fermented broth of an endophytic *Paenibacillus polymyxa* bacterium with a solution of $FeCl_3$ and $MnSO_4$ and incubating at 45 °C in the absence of light for a study on germination and plant development of maize [99]. *Bacillus licheniformis* with its ability to produce gluconic acid to enhance phosphorus solubilization from unavailable sources has been used in preparing nanohydroxyapatite for application as fertilizer [100]. Nevertheless, the microbes used in the biosynthesis of nanomaterials must have resistance toward the prepared materials or they will be killed within a few minutes before finishing their job [101, 102].

4 Potential Applications of Nanopesticides and Nanofertilizers in Agriculture

Nanotechnology has gained intense attention over the past decade for potential applications in the agriculture sector. Particularly, the development of nanoagrochemicals in the form of nanopesticides and nanofertilizers for sustainable agriculture with multiple targeting for crop improvement, nutrition, and food safety while minimizing or eliminating any potential hazards [103]. These nanotechnology-based approaches can increase agriculture production in many ways including (1) formulations of nanoscale agrochemicals for use as pesticides and fertilizers for the growth of crops; (2) nanoparticles mediated gene or DNA transfer in plants for use of insect-resistant crop varieties, food processing, and preservation; (3) reduce nitrogen loss caused by leaching and emissions, and soil microorganisms; (4) avoiding concerns of the occupational health of farmers, of breeding, nutrition and animal health; and (5) Improving postharvest handling and reducing losses of biomass-to-fuel production [104]. We will consider how nanoagrochemicals could help revolutionize farming in terms of plant growth promotion and protection. Figure 5 illustrates commonly used nanoparticles as

Fig. 5 Schematic represents the potential application of nanopesticides and nanofertilizers toward sustainable agriculture [105]

nanopesticides (left part) for killing fungi, insects, herbivore, bacterial, and nanofertilizers as the most promising for soil or foliar applications (right part).

4.1 Potential Applications of Nanopesticides

The use of agrochemicals is crucial to modern agriculture, with a large number of pesticides consumed on crops each year, which was estimated at 2.5 million tons annually, and harm caused by pesticide use reaches $100 billion globally [106]. Some reasons for that are included: (1) the high toxicity and non-biodegradable properties of pesticides; and (2) the residual pesticides in soil, water resources, and crops that negatively affect human and animal health [107]. Thus, nanoscale-based antimicrobials are developed and recently added to the fight against fungal pathogens, insect pests, and weeds, replacing toxic elements like heavy metals. The major benefits of nanoscale particles in the form of nanopesticides are including improved solubility of ingredients, better stability of the formulation, slow release of the active ingredient, and fast mobility caused by smaller particle size. In this regard, nanopesticides play a dual role in both controlled deliveries of pesticides and achieving greater effects while lower chemical doses. Several types of nanopesticides can be formulated in the form of active ingredients, that are either manufactured nanomaterials from metal nanoparticles (e.g. silver and copper) [108–112] to metal oxide nanoparticles (e.g. ZnO, CuO, Mn_2O, TiO_2) [113–115], SiO_2-based nanoparticles [116, 117]. The development of nanopesticides aims to increase the efficiency and durability of a pesticide, meanwhile to decrease the number of active ingredients contained. By shrinking the size of individual nanopesticide droplets, the number of toxins sprayed on agricultural fields could be significantly reduced. As smaller droplets have a higher total specific surface area, which allows greater contact with crop pests. These tiny particles can be engineered for example, with a physical shell called a capsule which offers longer-lasting protection than conventional pesticides. However, that shell can alter the physical properties of pesticides, such as solubility in water. As a result, a reduction in the amount of pesticide needed to assure crop protection may be achieved in several ways such as by improved apparent solubility, controlled release, targeted delivery, enhanced bioavailability, and increased leaf adhesion [118]. Literature reported that in comparison with conventional once similar toxicity or increased pesticidal toxicity [119–121] or similar toxicity at lower concentrations [122, 123] could be archived for nanopesticide formulations.

4.1.1 Development of Nanoscale Materials Used as Nanofungicides, Nanobactericides, Nanoinsecticides, and Nanoherbicides

Nanofungicide is used to describe any fungicidal formulations that intentionally include entities at the nano scale (up to 100 nm), which are biointerfaces nanocide materials being used as a new environment-friendly antimicrobial against various fungal pathogenic organisms of plants [124, 125]. Nanofungicides include a wide range of products which are including different types of nanopesticides such as organic ingredients, solid nanoparticles (silica, diatoms, alumina, etc.), titanium dioxide, polymer-based inorganic silica-based nanoparticles [116, 117], or nanoemulsions and nanoclays in various forms (e.g., particles, micelles) [126, 127]. According to literature, metallic nanoparticles are effective nanocides against plant fungal pathogenic organisms. For example, the silver nanoparticle has been reported recently against phytopathogen *Colletotrichum gloeosporioides* [128, 129]. Other nanoparticles such as Fe, Cu, Si, Al, Zn, ZnO, TiO_2, CeO_2, Al_2O_3, and CNTs have been explored to have some adverse effects on plant growth apart from the antimicrobial properties [130–132]. They also affect the growth of useful soil bacteria, such as *Pseudomonas putida KT2440* [133]. Among the various inorganic nanoparticle-based antimicrobial agents, silver has been extensively investigated due to it having several advantages over other nanoparticles such as copper, zinc, gold, ZnO, Al_2O_3, and TiO_2. In particular, the green biosynthesized Ag nanoparticles revealed strong antifungal activity against *Bipolaris sorokiniana* and effectively controlled its infection in wheat plants [134]. Ocsoy et al. fabricated DNA grafted silver nanoparticles decoratedon graphene oxide which decreased the cultured activity of *Xanthomonas* perforans [135], Thi bacterium usually make a 10–50% reduction in yields of tomato by causing bacterial spots. In the other work, Cromwell et al. fabricated Ag nanoparticles that exhibited efficiency against nematodes, a common soil-borne organism. It was reported that a concentration of 150 mg/L of Ag nanoparticles agent reduced the nematodes by 82% and 92% at Day 2 and Day 4, respectively [136].

The increase in antimicrobial activity effect and fungal resistance and the reduction in the usage of pesticides may be achieved via the combination of multiple active ingredients. For example, the solar radiation activity exhibited in inorganic nanoparticles (i.e. Ag, ZnO, CuO, MgO, S) combined with biopolymer showed positive effects in antimicrobial activity and also reduced the fungicide residue when compared with the same yet separated nanoparticles [137]. The motivation in developing new green nanocide capable of synergizing the antimicrobial activity and subsequently photocatalytic degrading pesticide residue for controlling pathogens in a proficient and eco-friendly way has to lead to the development of low-cost but stable nanoparticles with high sensitivity toward pathogen fungi [104, 138]. For an instant, the mixing of several bio-based chemicals formed a nanobiocide which was reported to eliminate *Magnaporthe grisea* fungus (rice blast disease) [139]. Metal oxides with optical characterizations can also exhibit excellent antimicrobial activities. When TiO_2 nanoparticles are exposed to a suitable light wavelength in the presence of oxygen, excited electrons moved from the valance band to the conduction band and then are directly transferred onto the oxygen molecules to form superoxide radicals which have an antibacterial potential and noticeable photocatalytic properties against the pathogens. At a similar concentration of 500–800 mg/L, the ZnO/TiO_2 nanoparticles demonstrated a better performance in reducing the bacterial spots caused by *X. perforans* compared to crops treated with copper and untreated controls in a greenhouse condition [140]. Similarly, cerium oxide-based nanoparticles had been tested for their ability to counter Fusarium wilt in tomato plants under a greenhouse setting. The concentration of nanoparticles applied to the plants via foliar and root pathways were 50 and 250 mg/L with a corresponding reduction in the severity of infection of 50% and 57% [141].

The antifungal characteristics of carbon-based nanomaterials also enable them to be used as efficient fungicides which could be utilized with a significant role in an efficient remediation method for the decontamination of soil. Six kinds of carbon nanomaterials [activated carbon (AC), graphene oxide (GO), reduced graphene oxide (rGO), single-walled carbon nanotubes (SWCNTs), multi-walled carbon nanotubes (MWCNTs), and fullerenes] with concentrations between 62.5 and 500 mg/L have been studied for their antifungal activity against *Fusarium graminearum* and *Fusaium poae*. Researchers found that SWCNTs, MWCNTs, GO, and rGO showed excellent antifungal activities at the concentration of 500 mg/L and the carbon nanomaterials also displayed the induction of plasmolysis and inhibition of water uptake [142]. Carbon dots were fabricated in another work for the modification of Ag nanoparticles to develop a colorimetric sensor to regulate phoxin in fruit samples and the environment. The prepared nanomaterial with high selectivity and good recovery values displayed excellent sensitivity and could detect phoxim at a concentration as low as 0.04 μM [143]. Carbon nanomaterials with various dimensions also increased the rate of germination and water uptake of treated rice seeds compared to control samples. However, in a study on the effect of MNWCTs on zucchini plants, not only did the plants show an insignificant response to the exposure of carbon nanoparticles in terms of root elongation but there was a decrease in the biomass at a later stage of growing [144].

Although displaying superior properties compared to conventional pesticides, there is still a long way to fully utilize nanotechnology in plant pathology. To achieve this, future studies will need to focus on developing a new disease management plan via investigating the physiology and interaction of plants and pathogens, disease infection process, and diagnosis to develop new nanopesticides formulations that are less harmful to the environment [118].

4.2 Potential Applications of Nanofertilizers

The use of nanotechnology for fertilizers is still in its infancy but is already adopted for sustainable agriculture applications. Nanofertilizer can be classified as nanoscale fertilizers, nanoscale additives, and nanoscale coatings [145] are the most promising engineered materials in the agriculture sector that are being tested, either for soil or foliar applications.

Low nutrient use efficiency (NUE) in fertilizer is the major challenge for sustainable agriculture. The current NUE of nitrogen remains below 50% with the loss caused by volatilization and leaching regardless of many attempts of research activities to enhance conventional products [146, 147] and the leaching of nutrients causes severe impacts on the environment such as eutrophication [148]. Hence, nanofertilizers containing nitrogen are expected to be one of many possible solutions for increasing NUE and reducing nutrient loss via enhancing N delivery to plants and matching the nutrient need. In general, nanotechnology will increase the NUE of fertilizers by two main routes: improving the delivery of poorly bioavailable nutrients (e.g. P, Zn) and/or reducing losses of nutrients to the surrounding environment (e.g. NO_3^-) [45]. In addition, alternative growth promoters such as TiO_2 or CNTs are also being investigated [149, 150]. Nutrients delivered in nanofertilizers are immobilized and/or encapsulated into a particular nanocarrier whose release behavior can be triggered by three different factors: (1) degradation of synthetic polymeric materials, that allow the release of nutrients and fixation into the soil; (2) biological factors that degrade the biodegradable coating (i.e. bacteria, fungi, and other microorganisms); and (3) the chemical stimulus (e.g. moisture, pH variation, soil type, solubilization) [147].

Directly focusing on the development of high NUE products for macronutrients (i.e. N, P, K, Ca, Mg, and S) should be prioritized as it is the strategic entrance to the fertilizer market of nanofertilizer. These nutrients are commonly used in high volume but they are low NUE and more environmentally impactful. The cost of fertilizer per unit area of the crop would significantly drop if nanofertilizer could allow the farmer to apply less kg per area, fewer applications per season but with the same or higher NUE than conventional fertilizers [151]. Based on the nutrient in use and the role of nanofertilizer, nanofertilizers can be classified into three different categories (1) macronutrient nanofertilizer, (2) micronutrient nanofertilizer, and (3) nanocarriers for macronutrients. Nanofertilizers have advantages over their conventional counterparts in terms of lowering the release rate of nutrients but increasing the stability without causing changes in chemical speciation [152]. Some noticeable examples of macronutrient nanofertilizers are (1) increasing the bioavailability of P for plants by fixation of P in rock phosphate nanoparticles, and (2) supporting regulation of N, P, and K for plants via the use of modified zeolites with high specific surface area [153]. Promisingly, compared to soybean (*Glycine max* L.) treated with conventional fertilizer, the use of phosphatic nanofertilizers has improved the growth rate and seed yield by 32% and 20%, respectively [154]. SiO_2 and TiO_2 nanoparticles have been reported to enhance the activity of nitrate reductase in soybeans and intensified plant absorption capacity, resulting in the use of water and fertilizer being more efficient [155].

Nanofertilizers can be a solution to deal with problems caused by abiotic stress. For example, they can help protect plants from the attack of herbivores or infection by a pathogen which are important factors in designing the productivity of the crops. Microorganisms can also be integrated into nanofertilizers to form nanobiofertilizers which can further enhance the nutrient of soil. Studies have also been conducted to investigate the benefits of SiO_2 nanoparticles in combination with biofertilizers. Results show that SiO_2 nanobiofertilizers could significantly improve crop yield [156]. The nanoscale of ZnO, MnO_2, and FexOy have also been prepared via a microwave-assisted hydrothermal method (20–60 nm) and employed as a fertilizer for squash [157]. It was reported that both vegetative growth and photosynthetic pigments characteristics could be enhanced if MnO_2 nanoparticles can be applied to the plants. In addition, the application of FexOy nanoparticles can also increase the quality of squash in terms of nutritional values, lipids, protein, and other organic matters. In another study, de França Bettencourt et al. [99] reported the application of both FeOx and MnOx nanoparticles (NPs) which were rapidly synthesized through a simple process employing bacteria supernatant. It is expected that the bimetallic MnOx/FeOx NPs can be applied as micronutrient nanofertilizers and have positive effects on plant growth such as fresh weight, root growth, and germination rates [99]. The application of N-P-K nanoparticles as foliar fertilizers (applying liquid fertilizer directly to the leaves) in potato farms (approximately 50% rate) resulted in better production yield and product quality in comparison to similar applications to soils. Thus, it can be recommended that foliar application of nanofertilizers might reduce the environmental impacts while increasing economic benefits thanks to the lower application rates [158]. Abdel-Aziz et al. [159] investigated the utilization of nanochitosan and carbon nanotubes (CNTs), either as a single application or loaded with N-P-K, for French beans. It was reported that those foliar nanofertilizers have positive effects on both plants (antioxidant system, yield, and growing rate) and seeds (biochemical content). In particular, harvesting days, in comparison to control and seed priming, could be reduced if foliar treatment is applied (only 80 days vs. 110 days). Interestingly, chitosan nanoparticles have better foliar performance in terms of yield and growth rate in comparison to CNTs. Several studies also reported that zinc nanofertilizers can enhance plant growth (shoot and root system) and increase the leaves' chlorophyll content. The utilization of zinc nanofertilizers could significantly increase the productivity of peanut farming. These nanofertilizers also improve the seed production of vegetables [160]. Similarly, CNTs-based fertilizers were reported to reduce the days to germination while also promote the development of plant root systems in rice seedlings [161].

On the other hand, contaminated soils can also be treated by the application of nanoparticles through catalytic degradation and mineralization of organic pollutants. The fundamental of this process is based on the application of different oxidants for pollutant oxidations. For instance, Fe_xO_y nanoparticles (together with chelating agents) could be combined with modified Fenton oxidation to treat pyrene-contaminated soils [162]. In this process, chelating agents play a vital role since they can adjust the oxidation efficiency. Heavy metals can also be removed or treated by the application of hydroxyapatite thanks to its adsorption ability. Particle size (~40 nm), applied for the treatment of both contaminated soils and sediments, has a significant effect on the immobilization of metal compounds since it can further reduce the metal exchangeable fractions, allowing the removal of heavy metals in the water stoma [163, 164]. 2D nanostructured clays have recently been applied as a nutrient carrier thanks to their neutrality. Those nanoparticles possess both cations and anions that interact and neutralize each other [165]. The ability to accommodate both inorganic and organic irons together with its ionic characterization make clays an excellent carrier for controlled and slow nutrients release fertilizers.

Despite aiding in sustainable crop production, the limitations of nanofertilizers should be carefully considered before commercializing. The limitations and adverb effects of using nanofertilizers for farming mainly arise due to the absence of rigorous monitoring and lack of knowledge in agric-ecotoxicity. Existing problems and challenges of nanoparticles in terms of biological effects need to be addressed before those fertilizers could be widely applied to plants.

5 Conclusion and Outlook

The efficiency of nitrogen use in farming can be increased when applying nanofertilizers [147]. Yet, besides those pros, cons have been reported on shortcomings and adverse effects [166]. Nanofertilizers are not different from other emerging technologies in that they only have proven performance on a laboratory scale [167]. The right way of dosage is essential, for example, the detrimental effect of foliar fertilizers have been reported [168]. There is also an indication that they are sensitive to details of the application, such as the time, season, and climate. Technical and economic issues remain for the application of nanoparticles [166].

Future research has to provide a solution to elucidate the mechanism of how nitrogen molecules (i.e. NO_3^-, NH_4^+) interact in the soil and fertilizer uptake of the plant for the benefit of fewer nitrogen losses to the environment. We need to understand better the transformation of nitrogen fertilizers in the soil and how they interact with proteins and metabolites [166]. To support this, we need to improve the nanoformulation approaches.

The nanomaterials and nanoformulations need to be as well characterized as their conventional fertilizer counterparts, and field studies are essential [167]. Nanofertilizers will not be used as a standalone concept, but rather need to be complemented by other nutrition and soil strategies such as biofertilizers and biostimulants, and the entire concept of precision agriculture with, e.g., multispectral imaging of drones. There is a need for sustainability evaluation, and life cycle assessment has a prime role as well as techno-economic analysis. Field grassland trials are needed here to supply robust data. A reconsideration of the whole supply chain is needed, and that finally counts for the entire business model. Nanofertilizers and nanoparticles will not be introduced to the market without consultation and the acceptability of the designated end-users, which are the farmers.

Acknowledgments The authors acknowledge support from the ERC Synergy Grant Surface-COnfined fast modulated Plasma for process and Energy intensification (SCOPE) from the European Commission with Grant No. 810182.

References

1. Thabet, A. F., Boraei, H. A., Galal, O. A., El-Samahy, M. F., Mousa, K. M., Zhang, Y. Z., et al. (2021). Silica nanoparticles as pesticide against insects of different feeding types and their non-target attraction of predators. *Scientific Reports, 11*(1), 1–13.
2. Torabi, Z., & MohammadiNafchi, A. (2013). The effects of SiO2 nanoparticles on mechanical and physicochemical properties of potato starch films. *Journal of Chemical Health Risks, 3*(1), 33.
3. Shoaib, A., Elabasy, A., Waqas, M., Lin, L., Cheng, X., Zhang, Q., et al. (2018). Entomotoxic effect of silicon dioxide nanoparticles on Plutella xylostella (L.)(Lepidoptera: Plutellidae) under laboratory conditions. *Toxicological and Environmental Chemistry, 100*(1), 80–91.
4. Rastogi, A., Tripathi, D. K., Yadav, S., Chauhan, D. K., Živčák, M., Ghorbanpour, M., et al. (2019). Application of silicon nanoparticles in agriculture. *3 Biotech, 9*(3), 1–11.
5. Ayoub, H. A., Khairy, M., Rashwan, F. A., & Abdel-Hafez, H. F. (2017). Synthesis and characterization of silica nanostructures for cotton leaf worm control. *Journal of Nanostructure in Chemistry, 7*(2), 91–100.
6. Galal, O., & El-Samahy, M. (2012). Genetical effects of using silica nanoparticles as biopesticide on Drosophila melanogaster. *Egyptian Journal of Genetics and Cytology, 41*(1), 87–106.
7. Debnath, N., Das, S., Seth, D., Chandra, R., Bhattacharya, S. C., & Goswami, A. (2011). Entomotoxic effect of silica nanoparticles against Sitophilus oryzae (L.). *Journal of Pesticide Science, 84*(1), 99–105.
8. Mousa, K., Elsharkawy, M., Khodeir, I., El-Dakhakhni, T., & Youssef, A. (2014). Growth perturbation, abnormalities and mortality of oriental armyworm Mythimna separata (Walker) (Lepidoptera: Noctuidae) caused by silica nanoparticles and Bacillus thuringiensis toxin. *Egyptian Journal of Biological Pest Control, 24*(2), 347.
9. Ziaee, M., & Ganji, Z. (2016). Insecticidal efficacy of silica nanoparticles against Rhyzopertha dominica F. and Tribolium confusum Jacquelin du Val. *Journal of Plant Protection Research, 56*, 250.
10. El-Samahy, M., & Galal, O. A. (2012). Evaluation of silica nanoparticles as a new approach to control faba bean (Vicia faba L.) insects and its genotoxic effect on M2 plants. *Egyptian Journal of Agricultural Research, 90*(2), 869–888.
11. Fouad, H. A., El-Gepaly, H. M. K. H., & Fouad, O. A. (2016). Nanosilica and jasmonic acid as alternative methods for control Tuta absoluta (Meyrick) in tomato crop under field conditions. *Archives of Phytopathology and Plant Protection, 49*(13–14), 362–370.
12. El-Samahy, M., Khafagy, I., & El-Ghobary, A. (2015). Efficiency of silica nanoparticles, two bioinsecticides, peppermint extract and insecticide in controlling cotton leafworm, spodoptera littoralis Boisd. and their effects on some associated natural enemies in sugar beet fields. *Journal of Plant Protection and Pathology, 6*(9), 1221–1230.
13. Rouhani, M., Samih, M. A., Zarabi, M., Beiki, K., Gorji, M., & Aminizadeh, M. R. (2019). Synthesis and entomotoxicity assay of zinc and silica nanoparticles against Sitophilus granarius (Coleoptera: Curculionidae). *Journal of Plant Protection Research, 59*, 26–31.

14. Abdelrhim, A. S., Mazrou, Y. S. A., Nehela, Y., Atallah, O. O., El-Ashmony, R. M., & Dawood, M. F. A. (2021). Silicon dioxide nanoparticles induce innate immune responses and activate antioxidant machinery in wheat against Rhizoctonia solani. *Plants (Basel), 10*(12), 2758.
15. Shabbaj, I. I., Madany, M., Tammar, A., Balkhyour, M. A., & AbdElgawad, H. (2021). Silicon dioxide nanoparticles orchestrate carbon and nitrogen metabolism in pea seedlings to cope with broomrape infection. *Environmental Science: Nano, 8*, 1960.
16. Deka, B., Babu, A., Baruah, C., & Barthakur, M. (2021). Nanopesticides: A systematic review of their prospects with special reference to tea pest management. *Frontiers in Nutrition, 8*, 686131.
17. Song, M.-R., Cui, S.-M., Gao, F., Liu, Y.-R., Fan, C.-L., Lei, T.-Q., et al. (2012). Dispersible silica nanoparticles as carrier for enhanced bioactivity of chlorfenapyr. *Journal of Pesticide Science, 37*(3), 258–260.
18. Hayles, J., Johnson, L., Worthley, C., & Losic, D. (2017). 5 - Nanopesticides: A review of current research and perspectives. In A. M. Grumezescu (Ed.), *New pesticides and soil sensors* (pp. 193–225). Academic Press.
19. Paini, D. R., Sheppard, A. W., Cook, D. C., De Barro, P. J., Worner, S. P., & Thomas, M. B. (2016). Global threat to agriculture from invasive species. *Proceedings of the National Academy of Sciences, 113*(27), 7575–7579.
20. Sharma, A., Kumar, V., Shahzad, B., Tanveer, M., Sidhu, G. P. S., Handa, N., et al. (2019). Worldwide pesticide usage and its impacts on ecosystem. *SN Applied Sciences, 1*(11), 1446.
21. Kim, S.-W., Kim, K.-S., Lamsal, K., Kim, Y.-J., Kim, S.-B., Jung, M.-Y., et al. (2009). An in vitro study of the antifungal effect of silver nanoparticles on oak wilt pathogen Raffaelea sp. *Journal of Microbiology and Biotechnology, 19*(8), 760–764.
22. Jung, J.-H., Kim, S.-W., Min, J.-S., Kim, Y.-J., Lamsal, K., Kim, K. S., et al. (2010). The effect of nano-silver liquid against the white rot of the green onion caused by Sclerotium cepivorum. *Mycobiology, 38*(1), 39–45.
23. Iravani, S., Korbekandi, H., Mirmohammadi, S. V., & Zolfaghari, B. (2014). Synthesis of silver nanoparticles: Chemical, physical and biological methods. *Research in Pharmaceutical Sciences, 9*(6), 385–406.
24. Guilger-Casagrande, M., & Lima, R. (2019). Synthesis of silver nanoparticles mediated by fungi: A review. *Frontiers in Bioengineering and Biotechnology, 7*, 287.
25. Masum, M. M. I., Siddiqa, M. M., Ali, K. A., Zhang, Y., Abdallah, Y., Ibrahim, E., et al. (2019). Biogenic synthesis of silver nanoparticles using Phyllanthus emblica fruit extract and its inhibitory action against the pathogen Acidovorax oryzae strain RS-2 of rice bacterial brown stripe. *Frontiers in Microbiology, 10*, 820.
26. Sahu, N., Soni, D., Chandrashekhar, B., Satpute, D. B., Saravanadevi, S., Sarangi, B. K., et al. (2016). Synthesis of silver nanoparticles using flavonoids: Hesperidin, naringin and diosmin, and their antibacterial effects and cytotoxicity. *International Nano Letters, 6*(3), 173–181.
27. Sap-Iam, N., Homklinchan, C., Larpudomlert, R., Warisnoicharoen, W., Sereemaspun, A., & Dubas, S. (2010). UV irradiation-induced silver nanoparticles as mosquito larvicides. *Journal of Applied Sciences, 10*(23), 3132–3136.
28. Musial, J., Krakowiak, R., Mlynarczyk, D. T., Goslinski, T., & Stanisz, B. J. (2020). Titanium dioxide nanoparticles in food and personal care products-what do we know about their safety? *Nanomaterials (Basel), 10*(6), 1110.
29. Agarwal, K., & Chibber, S. (2017). Titanium dioxide (Tio2) nanoparticles induced ROS generation and its effect on cellular antioxidant defense in WRL-68 cell. *Global Journal of Medical Research, 3*(3), 70–74.
30. Saranya, S., Selvi, A., Babujanarthanam, R., Rajasekar, A., & Madhavan, J. (2020). 12 Insecticidal activity of nanoparticles and mechanism of action. In *Model organisms to study biological activities and toxicity of nanoparticles* (p. 243). Springer.

31. Pho, Q. H., Losic, D., Ostrikov, K. K., Tran, N. N., & Hessel, V. (2020). Perspectives on plasma-assisted synthesis of N-doped nanoparticles as nanopesticides for pest control in crops. *Reaction Chemistry & Engineering, 5*(8), 1374–1396.
32. De Filpo, G., Palermo, A. M., Rachiele, F., & Nicoletta, F. P. (2013). Preventing fungal growth in wood by titanium dioxide nanoparticles. *International Biodeterioration and Biodegradation, 85*, 217–222.
33. Oluwaferanmi, O., & Ogunleti, D. (2008). Effects of titanium dioxide on the diseases, development and yield of edible cowpea. *Journal of Plant Protection Research, 48*(3), 329. https://doi.org/10.2478/v10045-008-0042-5
34. Mathew, S. S., Sunny, N. E., & Shanmugam, V. (2021). Green synthesis of anatase titanium dioxide nanoparticles using Cuminum cyminum seed extract; effect on Mung bean (Vigna radiata) seed germination. *Inorganic Chemistry Communications, 126*, 108485.
35. Santhoshkumar, T., Rahuman, A. A., Jayaseelan, C., Rajakumar, G., Marimuthu, S., Kirthi, A. V., et al. (2014). Green synthesis of titanium dioxide nanoparticles using Psidium guajava extract and its antibacterial and antioxidant properties. *Asian Pacific Journal of Tropical Medicine, 7*(12), 968–976.
36. Hou, J., Wang, L., Wang, C., Zhang, S., Liu, H., Li, S., et al. (2019). Toxicity and mechanisms of action of titanium dioxide nanoparticles in living organisms. *Journal of Environmental Sciences, 75*, 40–53.
37. Tripathy, B. C., & Oelmüller, R. (2012). Reactive oxygen species generation and signaling in plants. *Plant Signaling & Behavior, 7*(12), 1621–1633.
38. Itelima, J., Bang, W., Onyimba, I., Sila, M., & Egbere, O. (2018). Bio-fertilizers as key player in enhancing soil fertility and crop productivity: A review. *Direct Research Journal of Agriculture and Food Science, 6*, 73.
39. Butt, B. Z., & Naseer, I. (2020). Nanofertilizers. In *Nanoagronomy* (pp. 125–152). Springer.
40. Lee, Y.-C., & Moon, J.-Y. (2020). Bionanotechnology in agriculture, food, cosmetic and cosmeceutical. In *Introduction to bionanotechnology* (pp. 199–217). Springer.
41. Pirzadah, B., Pirzadah, T. B., Jan, A., & Hakeem, K. R. (2020). Nanofertilizers: A way forward for green economy. In *Nanobiotechnology in agriculture* (pp. 99–112). Springer.
42. Papa, E., Medri, V., Amari, S., Manaud, J., Benito, P., Vaccari, A., et al. (2018). Zeolite-geopolymer composite materials: Production and characterization. *Journal of Cleaner Production, 171*, 76–84.
43. Eroglu, N., Emekci, M., & Athanassiou, C. G. (2017). Applications of natural zeolites on agriculture and food production. *Journal of the Science of Food and Agriculture, 97*(11), 3487–3499.
44. Jakkula, V. S., & Wani, S. (2018). Zeolites: Potential soil amendments for improving nutrient and water use efficiency and agriculture productivity. *Scientific Reviews & Chemical Communications, 8*(1), 1–15.
45. Liu, R., & Lal, R. (2015). Potentials of engineered nanoparticles as fertilizers for increasing agronomic productions. *Science of the Total Environment, 514*, 131–139.
46. Kalia, A., Sharma, S. P., Kaur, H., & Kaur, H. (2020). Novel nanocomposite-based controlled-release fertilizer and pesticide formulations: Prospects and challenges. In *Multifunctional hybrid nanomaterials for sustainable agri-food and ecosystems* (pp. 99–134). Elsevier.
47. Okey-Onyesolu, C. F., Hassanisaadi, M., Bilal, M., Barani, M., Rahdar, A., Iqbal, J., et al. (2021). Nanomaterials as nanofertilizers and nanopesticides: An overview. *ChemistrySelect, 6*(33), 8645–8663.
48. Ashraf, S. A., Siddiqui, A. J., Abd Elmoneim, O. E., Khan, M. I., Patel, M., Alreshidi, M., et al. (2021). Innovations in nanoscience for the sustainable development of food and agriculture with implications on health and environment. *Science of the Total Environment, 768*, 144990.
49. Ramesh, K., Biswas, A. K., Somasundaram, J., & Rao, A. S. (2010). Nanoporous zeolites in farming: Current status and issues ahead. *Current Science, 99*, 760–764.

50. Preetha, P. S., & Balakrishnan, N. (2017). A review of nano fertilizers and their use and functions in soil. *International Journal of Current Microbiology and Applied Sciences, 6*(12), 3117–3133.
51. Rop, B. K. (2019). *Development of slow release nano composite fertilizer using biodegradable super absorbent polymer*. University of Nairobi.
52. Mani, P. K., & Mondal, S. (2016). Agri-nanotechniques for plant availability of nutrients. In *Plant nanotechnology* (pp. 263–303). Springer.
53. Guo, M., Liu, M., Zhan, F., & Wu, L. (2005). Preparation and properties of a slow-release membrane-encapsulated urea fertilizer with superabsorbent and moisture preservation. *Industrial and Engineering Chemistry Research, 44*(12), 4206–4211.
54. Khodakovskaya, M., Dervishi, E., Mahmood, M., Xu, Y., Li, Z., Watanabe, F., et al. (2009). Carbon nanotubes are able to penetrate plant seed coat and dramatically affect seed germination and plant growth. *ACS Nano, 3*(10), 3221–3227.
55. Srinivasan, C., & Saraswathi, R. (2010). Nano-agriculture-carbon nanotubes enhance tomato seed germination and plant growth. *Current Science, 99*(3), 274–275.
56. Siddaiah, C. N., Prasanth, K. V. H., Satyanarayana, N. R., Mudili, V., Gupta, V. K., Kalagatur, N. K., et al. (2018). Chitosan nanoparticles having higher degree of acetylation induce resistance against pearl millet downy mildew through nitric oxide generation. *Scientific Reports, 8*(1), 1–14.
57. Nadendla, S. R., Rani, T. S., Vaikuntapu, P. R., Maddu, R. R., & Podile, A. R. (2018). HarpinPss encapsulation in chitosan nanoparticles for improved bioavailability and disease resistance in tomato. *Carbohydrate Polymers, 199*, 11–19.
58. Nguyen, H. M., Hwang, I. C., Park, J. W., & Park, H. J. (2012). Photoprotection for deltamethrin using chitosan-coated beeswax solid lipid nanoparticles. *Pest Management Science, 68*(7), 1062–1068.
59. Pankaj, Shakil, N. A., Kumar, J., Singh, M., & Singh, K. (2012). Bioefficacy evaluation of controlled release formulations based on amphiphilic nano-polymer of carbofuran against Meloidogyne incognita infecting tomato. *Journal of Environmental Science and Health, Part B, Pesticides, Food Contaminants, and Agricultural Wastes, 47*(6), 520–528.
60. Adak, T., Kumar, J., Shakil, N., & Walia, S. (2012). Development of controlled release formulations of imidacloprid employing novel nano-ranged amphiphilic polymers. *Journal of Environmental Science and Health, Part B, Pesticides, Food Contaminants, and Agricultural Wastes, 47*(3), 217–225.
61. Sarkar, D. J., Kumar, J., Shakil, N., & Walia, S. (2012). Release kinetics of controlled release formulations of thiamethoxam employing nano-ranged amphiphilic PEG and diacid based block polymers in soil. *Journal of Environmental Science and Health, Part A, Toxic/Hazardous Substances & Environmental Engineering, 47*(11), 1701–1712.
62. Loha, K. M., Shakil, N. A., Kumar, J., Singh, M. K., Adak, T., & Jain, S. (2011). Release kinetics of β-Cyfluthrin from its encapsulated formulations in water. *Journal of Environmental Science and Health, Part B, Pesticides, Food Contaminants, and Agricultural Wastes, 46*(3), 201–206.
63. Loha, K. M., Shakil, N. A., Kumar, J., Singh, M. K., & Srivastava, C. (2012). Bio-efficacy evaluation of nanoformulations of β-cyfluthrin against Callosobruchus maculatus (Coleoptera: Bruchidae). *Journal of Environmental Science and Health, Part B, 47*(7), 687–691.
64. Abdel-Hafez, S. I., Nafady, N. A., Abdel-Rahim, I. R., Shaltout, A. M., Daròs, J.-A., & Mohamed, M. A. (2016). Assessment of protein silver nanoparticles toxicity against pathogenic Alternaria solani. *3 Biotech, 6*(2), 1–12.
65. Madbouly, A. K., Abdel-Aziz, M. S., & Abdel-Wahhab, M. A. (2017). Biosynthesis of nanosilver using Chaetomium globosum and its application to control Fusarium wilt of tomato in the greenhouse. *IET Nanobiotechnology, 11*(6), 702–708.
66. Elshahawy, I., Abouelnasr, H. M., Lashin, S. M., & Darwesh, O. M. (2018). First report of Pythium aphanidermatum infecting tomato in Egypt and its control using biogenic silver nanoparticles. *Journal of Plant Protection Research, 58*(2), 137.

67. Roseline, T. A., Murugan, M., Sudhakar, M., & Arunkumar, K. (2019). Nanopesticidal potential of silver nanocomposites synthesized from the aqueous extracts of red seaweeds. *Environmental Technology and Innovation, 13*, 82–93.
68. Farahat, G. (2018). Biosynthesis of nano zinc and using of some nanoparticles in reducing of Cercospora leaf spot disease of sugar beet in the field. *Environment, Biodiversity and Soil Security, 2018*(2), 103–117.
69. Chen, J., Mao, S., Xu, Z., & Ding, W. (2019). Various antibacterial mechanisms of biosynthesized copper oxide nanoparticles against soilborne Ralstonia solanacearum. *RSC Advances, 9*(7), 3788–3799.
70. Liu, F., Wen, L.-X., Li, Z.-Z., Yu, W., Sun, H.-Y., & Chen, J.-F. (2006). Porous hollow silica nanoparticles as controlled delivery system for water-soluble pesticide. *Materials Research Bulletin, 41*(12), 2268–2275.
71. Gao, Y., Li, D., Li, D., Xu, P., Mao, K., Zhang, Y., et al. (2020). Efficacy of an adhesive nanopesticide on insect pests of rice in field trials. *Journal of Asia-Pacific Entomology, 23*(4), 1222–1227.
72. Xu, C., Shan, Y., Bilal, M., Xu, B., Cao, L., & Huang, Q. (2020). Copper ions chelated mesoporous silica nanoparticles via dopamine chemistry for controlled pesticide release regulated by coordination bonding. *Chemical Engineering Journal, 395*, 125093.
73. Cao, L., Zhang, H., Zhou, Z., Xu, C., Shan, Y., Lin, Y., et al. (2018). Fluorophore-free luminescent double-shelled hollow mesoporous silica nanoparticles as pesticide delivery vehicles. *Nanoscale, 10*(43), 20354–20365.
74. Wei, Y., Wu, Y., Chang, Q., Xie, M., Wang, X., Mo, J., et al. (2017). Ultrasonic-assisted modification of a novel silkworm-excrement-based porous carbon with various Lewis acid metal ions for the sustained release of the pesticide thiamethoxam. *RSC Advances, 7*(48), 30020–30031.
75. Sarlak, N., Taherifar, A., & Salehi, F. (2014). Synthesis of nanopesticides by encapsulating pesticide nanoparticles using functionalized carbon nanotubes and application of new nanocomposite for plant disease treatment. *Journal of Agricultural and Food Chemistry, 62*(21), 4833–4838.
76. Garrido-Herrera, F., Gonzalez-Pradas, E., & Fernández-Pérez, M. (2006). Controlled release of isoproturon, imidacloprid, and cyromazine from alginate– bentonite-activated carbon formulations. *Journal of Agricultural and Food Chemistry, 54*(26), 10053–10060.
77. Subramanian, K. S., & Thirunavukkarasu, M. (2017). *Nano-fertilizers and nutrient transformations in soil. Nanoscience and plant–soil systems* (pp. 305–319). Springer.
78. Sharmila, R. (2010). *Nutrient release pattern of nano-fertilizer formulations*. Tamil Nadu Agricultural University, Coimbatore.
79. Subramanian, K., & Rahale, C. S. (2012). Ball milled nanosized zeolite loaded with zinc sulfate: A putative slow release Zn fertilizer. *International Journal of Innovative Horticulture, 1*(1), 33–40.
80. Pohshna, C., Mailapalli, D. R., & Laha, T. (2020). Synthesis of nanofertilizers by planetary ball milling. In *Sustainable agriculture reviews* (Vol. 40, pp. 75–112). Springer.
81. Kalia, A., & Kaur, H. (2018). Nanofertilizers: An innovation towards new generation fertilizers for improved nutrient-use efficacy and environmental sustainability. In *Nanoagroceuticals & nanophytochemicals* (pp. 45–61). CRC Press.
82. Hasaneen, M., Abdel-Aziz, H., El-Bialy, D., & Omer, A. M. (2014). Preparation of chitosan nanoparticles for loading with NPK fertilizer. *African Journal of Biotechnology, 13*(31), 3158.
83. Aziz, H. M. A., Hasaneen, M. N., & Omer, A. M. (2016). Nano chitosan-NPK fertilizer enhances the growth and productivity of wheat plants grown in sandy soil. *Spanish Journal of Agricultural Research, 14*(1), 17.
84. Abdel-Aziz, H., Hasaneen, M. N., & Omar, A. (2018). Effect of foliar application of nano chitosan NPK fertilizer on the chemical composition of wheat grains. *Egyptian Journal of Botany, 58*(1), 87–95.

85. Ha, N. M. C., Nguyen, T. H., Wang, S.-L., & Nguyen, A. D. (2019). Preparation of NPK nanofertilizer based on chitosan nanoparticles and its effect on biophysical characteristics and growth of coffee in green house. *Research on Chemical Intermediates, 45*(1), 51–63.
86. Lee, D., & Seo, J. (2010). Preparation of carbon nanotubes from graphite powder at room temperature. *arXiv*, 10071062.
87. Hasaneen, M., Abdel-aziz, H. M. M., & Omer, A. M. (2016). Effect of foliar application of engineered nanomaterials: Carbon nanotubes NPK and chitosan nanoparticles NPK fertilizer on the growth of French bean plant. *Biochemistry and Biotechnology Research, 4*(4), 68–76.
88. Marchiol, L., Filippi, A., Adamiano, A., Degli Esposti, L., Iafisco, M., Mattiello, A., et al. (2019). Influence of hydroxyapatite nanoparticles on germination and plant metabolism of tomato (Solanum lycopersicum L.): Preliminary evidence. *Agronomy, 9*(4), 161.
89. Yoon, H. Y., Lee, J. G., Esposti, L. D., Iafisco, M., Kim, P. J., Shin, S. G., et al. (2020). Synergistic release of crop nutrients and stimulants from hydroxyapatite nanoparticles functionalized with humic substances: Toward a multifunctional nanofertilizer. *ACS Omega, 5*(12), 6598–6610.
90. Manik, A., & Subramanian, K. (2014). Fabrication and characterisation of nanoporous zeolite based N fertilizer. *African Journal of Agricultural Research, 9*(2), 276–284.
91. Hidayat, R., Fadillah, G., Chasanah, U., Wahyuningsih, S., & Ramelan, A. H. (2015). Effectiveness of urea nanofertilizer based aminopropyltrimethoxysilane (APTMS)-zeolite as slow release fertilizer system. *African Journal of Agricultural Research, 10*(14), 1785–1788.
92. Soliman, A. S., El-feky, S. A., & Darwish, E. (2015). Alleviation of salt stress on Moringa peregrina using foliar application of nanofertilizers. *Journal of Horticulture and Forestry, 7*(2), 36–47.
93. Liu, R., Zhang, H., & Lal, R. (2016). Effects of stabilized nanoparticles of copper, zinc, manganese, and iron oxides in low concentrations on lettuce (Lactuca sativa) seed germination: Nanotoxicants or nanonutrients? *Water, Air, and Soil Pollution, 227*(1), 42.
94. Palchoudhury, S., Jungjohann, K. L., Weerasena, L., Arabshahi, A., Gharge, U., Albattah, A., et al. (2018). Enhanced legume root growth with pre-soaking in α-Fe2O3 nanoparticle fertilizer. *RSC Advances, 8*(43), 24075–24083.
95. Wang, S.-L., & Nguyen, A. D. (2018). Effects of Zn/B nanofertilizer on biophysical characteristics and growth of coffee seedlings in a greenhouse. *Research on Chemical Intermediates, 44*(8), 4889–4901.
96. Shah, M., Fawcett, D., Sharma, S., Tripathy, S. K., & Poinern, G. E. J. (2015). Green synthesis of metallic nanoparticles via biological entities. *Materials, 8*(11), 7278–7308.
97. Chaudhuri, S. K., & Malodia, L. (2017). Biosynthesis of zinc oxide nanoparticles using leaf extract of Calotropis gigantea: Characterization and its evaluation on tree seedling growth in nursery stage. *Applied Nanoscience, 7*(8), 501–512.
98. Chahar, V., Sharma, B., Shukla, G., Srivastava, A., & Bhatnagar, A. (2018). Study of antimicrobial activity of silver nanoparticles synthesized using green and chemical approach. *Colloids and Surfaces A: Physicochemical and Engineering Aspects, 554*, 149–155.
99. de França Bettencourt, G. M., Degenhardt, J., Torres, L. A. Z., de Andrade Tanobe, V. O., & Soccol, C. R. (2020). Green biosynthesis of single and bimetallic nanoparticles of iron and manganese using bacterial auxin complex to act as plant bio-fertilizer. *Biocatalysis and Agricultural Biotechnology, 30*, 101822.
100. Priyam, A., Das, R. K., Schultz, A., & Singh, P. P. (2019). A new method for biological synthesis of agriculturally relevant nanohydroxyapatite with elucidated effects on soil bacteria. *Scientific Reports, 9*(1), 1–14.
101. Kalimuthu, K., Babu, R. S., Venkataraman, D., Bilal, M., & Gurunathan, S. (2008). Biosynthesis of silver nanocrystals by Bacillus licheniformis. *Colloids and Surfaces. B, Biointerfaces, 65*(1), 150–153.
102. Pandian, S. R. K., Deepak, V., Kalishwaralal, K., Viswanathan, P., & Gurunathan, S. (2010). Mechanism of bactericidal activity of Silver Nitrate-a concentration dependent bi-functional molecule. *Brazilian Journal of Microbiology, 41*(3), 805–809.

103. Chhipa, H. (2019). Applications of nanotechnology in agriculture. *Methods in Microbiology, 46*, 115–142.
104. Bratovcic, A., Hikal, W. M., Said-Al Ahl, H. A., Tkachenko, K. G., Baeshen, R. S., Sabra, A. S., et al. (2021). Nanopesticides and nanofertilizers and agricultural development: Scopes, advances and applications. *Open Journal of Ecology, 11*(4), 301–316.
105. Rani, A., Rani, K., Tokas, J., Singh, A., Kumar, R., Punia, H., et al. (2020). Nanomaterials for agriculture input use efficiency. In *Resources use efficiency in agriculture* (pp. 137–175). Springer.
106. Peshin, R., Bandral, R. S., Zhang, W., Wilson, L., & Dhawan, A. K. (2009). Integrated pest management: A global overview of history, programs and adoption. In *Integrated pest management: Innovation-development process* (pp. 1–49). Springer.
107. Koul, O., Walia, S., & Dhaliwal, G. (2008). Essential oils as green pesticides: Potential and constraints. *Biopesticides International, 4*(1), 63–84.
108. Guilger-Casagrande, M., Germano-Costa, T., Pasquoto-Stigliani, T., Fraceto, L. F., & de Lima, R. (2019). Biosynthesis of silver nanoparticles employing Trichoderma harzianum with enzymatic stimulation for the control of Sclerotinia sclerotiorum. *Scientific Reports, 9*(1), 1–9.
109. de Lima, R., Seabra, A. B., & Durán, N. (2012). Silver nanoparticles: A brief review of cytotoxicity and genotoxicity of chemically and biogenically synthesized nanoparticles. *Journal of Applied Toxicology, 32*(11), 867–879.
110. Ouda, S. M. (2014). Antifungal activity of silver and copper nanoparticles on two plant pathogens, Alternaria alternata and Botrytis cinerea. *Research Journal of Microbiology, 9*(1), 34.
111. Lopez-Lima, D., Mtz-Enriquez, A. I., Carrión, G., Basurto-Cereceda, S., & Pariona, N. (2021). The bifunctional role of copper nanoparticles in tomato: Effective treatment for Fusarium wilt and plant growth promoter. *Scientia Horticulturae, 277*, 109810.
112. Jo, Y.-K., Kim, B. H., & Jung, G. (2009). Antifungal activity of silver ions and nanoparticles on phytopathogenic fungi. *Plant Disease, 93*(10), 1037–1043.
113. Peréz, C. D., De La Torre, R. R., Zuverza-Mena, N., Ma, C., Shen, Y., White, J. C., et al. (2019). Metalloid and metal oxide nanoparticles suppress sudden death syndrome of soybean. *Journal of Agricultural and Food Chemistry, 68*(1), 77–87.
114. de Jesus, R. A., de Assis, G. C., de Oliveira, R. J., Costa, J. A. S., da Silva, C. M. P., Bilal, M., et al. (2021). Environmental remediation potentialities of metal and metal oxide nanoparticles: Mechanistic biosynthesis, influencing factors, and application standpoint. *Environmental Technology and Innovation, 24*, 101851.
115. Elmer, W. H., & White, J. C. (2016). The use of metallic oxide nanoparticles to enhance growth of tomatoes and eggplants in disease infested soil or soilless medium. *Environmental Science: Nano, 3*(5), 1072–1079.
116. Athanassiou, C., Kavallieratos, N., Benelli, G., Losic, D., Rani, P. U., & Desneux, N. (2018). Nanoparticles for pest control: Current status and future perspectives. *Journal of Pesticide Science, 91*(1), 1–15.
117. Kumar, S., Nehra, M., Dilbaghi, N., Marrazza, G., Hassan, A. A., & Kim, K.-H. (2019). Nano-based smart pesticide formulations: Emerging opportunities for agriculture. *Journal of Controlled Release, 294*, 131–153.
118. Kah, M., Beulke, S., Tiede, K., & Hofmann, T. (2013). Nanopesticides: State of knowledge, environmental fate, and exposure modeling. *Critical Reviews in Environmental Science and Technology, 43*, 1823.
119. de Oliveira, J. L., Campos, E. V. R., Goncalves da Silva, C. M., Pasquoto, T., Lima, R., & Fraceto, L. F. (2015). Solid lipid nanoparticles co-loaded with simazine and atrazine: Preparation, characterization, and evaluation of herbicidal activity. *Journal of Agricultural and Food Chemistry, 63*(2), 422–432.

120. Grillo, R., Pereira, A. E., Nishisaka, C. S., De Lima, R., Oehlke, K., Greiner, R., et al. (2014). Chitosan/tripolyphosphate nanoparticles loaded with paraquat herbicide: An environmentally safer alternative for weed control. *Journal of Hazardous Materials, 278*, 163–171.
121. Saini, P., Gopal, M., Kumar, R., & Srivastava, C. (2014). Development of pyridalyl nanocapsule suspension for efficient management of tomato fruit and shoot borer (Helicoverpa armigera). *Journal of Environmental Science and Health, Part B, Pesticides, Food Contaminants, and Agricultural Wastes, 49*(5), 344–351.
122. Kumar, S., Bhanjana, G., Sharma, A., Sidhu, M., & Dilbaghi, N. (2014). Synthesis, characterization and on field evaluation of pesticide loaded sodium alginate nanoparticles. *Carbohydrate Polymers, 101*, 1061–1067.
123. Memarizadeh, N., Ghadamyari, M., Adeli, M., & Talebi, K. (2014). Preparation, characterization and efficiency of nanoencapsulated imidacloprid under laboratory conditions. *Ecotoxicology and Environmental Safety, 107*, 77–83.
124. Lövestam, G., Rauscher, H., Roebben, G., Klüttgen, B. S., Gibson, N., Putaud, J.-P., et al. (2010). *Considerations on a definition of nanomaterial for regulatory purposes. Joint Research Centre (JRC) reference reports* (Vol. 80, pp. 1–41). European Union.
125. Elmer, W., & White, J. C. (2018). The future of nanotechnology in plant pathology. *Annual Review of Phytopathology, 56*, 111–133.
126. Kookana, R. S., Boxall, A. B., Reeves, P. T., Ashauer, R., Beulke, S., Chaudhry, Q., et al. (2014). Nanopesticides: Guiding principles for regulatory evaluation of environmental risks. *Journal of Agricultural and Food Chemistry, 62*(19), 4227–4240.
127. Hayles, J., Johnson, L., Worthley, C., & Losic, D. (2017). Nanopesticides: A review of current research and perspectives. In *New pesticides and soil sensors* (pp. 193–225). Academic Press.
128. Rajeshkumar, S., Malarkodi, C., Paulkumar, K., Vanaja, M., Gnanajobitha, G., & Annadurai, G. (2014). Algae mediated green fabrication of silver nanoparticles and examination of its antifungal activity against clinical pathogens. *International Journal of Metals, 2014*, 692643.
129. Krishnaraj, C., Ramachandran, R., Mohan, K., & Kalaichelvan, P. (2012). Optimization for rapid synthesis of silver nanoparticles and its effect on phytopathogenic fungi. *Spectrochimica Acta Part A: Molecular and Biomolecular Spectroscopy, 93*, 95–99.
130. Díez-Pascual, A. M. (2018). *Antibacterial activity of nanomaterials* (p. 359). Multidisciplinary Digital Publishing Institute.
131. Xin, Q., Shah, H., Nawaz, A., Xie, W., Akram, M. Z., Batool, A., et al. (2019). Antibacterial carbon-based nanomaterials. *Advanced Materials, 31*(45), 1804838.
132. Duhan, J. S., Kumar, R., Kumar, N., Kaur, P., Nehra, K., & Duhan, S. (2017). Nanotechnology: The new perspective in precision agriculture. *Biotechnology Reports, 15*, 11–23.
133. Avendaño, R., Chaves, N., Fuentes, P., Sánchez, E., Jiménez, J. I., & Chavarría, M. (2016). Production of selenium nanoparticles in Pseudomonas putida KT2440. *Scientific Reports, 6*(1), 1–9.
134. Mishra, S., Singh, B. R., Singh, A., Keswani, C., Naqvi, A. H., & Singh, H. (2014). Biofabricated silver nanoparticles act as a strong fungicide against Bipolaris sorokiniana causing spot blotch disease in wheat. *PLoS One, 9*(5), E97881.
135. Ocsoy, I., Paret, M. L., Ocsoy, M. A., Kunwar, S., Chen, T., You, M., et al. (2013). Nanotechnology in plant disease management: DNA-directed silver nanoparticles on graphene oxide as an antibacterial against Xanthomonas perforans. *ACS Nano, 7*(10), 8972–8980.
136. Cromwell, W., Yang, J., Starr, J., & Jo, Y.-K. (2014). Nematicidal effects of silver nanoparticles on root-knot nematode in bermudagrass. *Journal of Nematology, 46*(3), 261.
137. Rai, M., & Ingle, A. (2012). Role of nanotechnology in agriculture with special reference to management of insect pests. *Applied Microbiology and Biotechnology, 94*(2), 287–293.
138. Chhipa, H. (2017). Nanofertilizers and nanopesticides for agriculture. *Environmental Chemistry Letters, 15*(1), 15–22.
139. Burketova, L., Trda, L., Ott, P. G., & Valentova, O. (2015). Bio-based resistance inducers for sustainable plant protection against pathogens. *Biotechnology Advances, 33*(6), 994–1004.

140. Paret, M. L., Vallad, G. E., Averett, D. R., Jones, J. B., & Olson, S. M. (2013). Photocatalysis: Effect of light-activated nanoscale formulations of TiO2 on Xanthomonas perforans and control of bacterial spot of tomato. *Phytopathology, 103*(3), 228–236.
141. Adisa, I. O., Reddy Pullagurala, V. L., Rawat, S., Hernandez-Viezcas, J. A., Dimkpa, C. O., Elmer, W. H., et al. (2018). Role of cerium compounds in Fusarium wilt suppression and growth enhancement in tomato (Solanum lycopersicum). *Journal of Agricultural and Food Chemistry, 66*(24), 5959–5970.
142. Wang, X., Liu, X., Chen, J., Han, H., & Yuan, Z. (2014). Evaluation and mechanism of antifungal effects of carbon nanomaterials in controlling plant fungal pathogen. *Carbon, 68*, 798–806.
143. Zheng, M., Wang, C., Wang, Y., Wei, W., Ma, S., Sun, X., et al. (2018). Green synthesis of carbon dots functionalized silver nanoparticles for the colorimetric detection of phoxim. *Talanta, 185*, 309–315.
144. Stampoulis, D., Sinha, S. K., & White, J. C. (2009). Assay-dependent phytotoxicity of nanoparticles to plants. *Environmental Science & Technology, 43*(24), 9473–9479.
145. Mikkelsen, R. (2018). Nanofertilizer and nanotechnology: A quick look. *Better Crops with Plant Food, 102*(3), 18–19.
146. Bruinsma, J. (2002). *World agriculture: Towards 2015/2030: Summary report*. Food and Agriculture Organization of the United Nations (FAO).
147. Mejías, J. H., Salazar, F. J., Pérez, L., Hube, S., Rodriguez, M., & Alfaro, M. A. (2021). Nanofertilizers: A cutting-edge approach to increase nitrogen use efficiency in grasslands. *Frontiers in Environmental Science, 9*, 52.
148. Tilman, D., Cassman, K. G., Matson, P. A., Naylor, R., & Polasky, S. (2002). Agricultural sustainability and intensive production practices. *Nature, 418*(6898), 671–677.
149. Feizi, H., Kamali, M., Jafari, L., & Moghaddam, P. R. (2013). Phytotoxicity and stimulatory impacts of nanosized and bulk titanium dioxide on fennel (Foeniculum vulgare Mill). *Chemosphere, 91*(4), 506–511.
150. Mukherjee, A., Majumdar, S., Servin, A. D., Pagano, L., Dhankher, O. P., & White, J. C. (2016). Carbon nanomaterials in agriculture: A critical review. *Frontiers in Plant Science, 7*, 172.
151. Srivastava, A. K., & Hu, C. (2019). *Fruit crops: Diagnosis and management of nutrient constraints*. Elsevier.
152. Raliya, R., Saharan, V., Dimkpa, C., & Biswas, P. (2017). Nanofertilizer for precision and sustainable agriculture: Current state and future perspectives. *Journal of Agricultural and Food Chemistry, 66*(26), 6487–6503.
153. Singh, M. D. (2017). Nano-fertilizers is a new way to increase nutrients use efficiency in crop production. *International Journal of Agriculture Sciences, 9*, 3831–3833.
154. Liu, R., & Lal, R. (2014). Synthetic apatite nanoparticles as a phosphorus fertilizer for soybean (Glycine max). *Scientific Reports, 4*(1), 1–6.
155. Changmei, L., Chaoying, Z., Junqiang, W., Guorong, W., & Mingxuan, T. (2002). Research of the effect of nanometer materials on germination and growth enhancement of Glycine max and its mechanism. *Soybean Science, 21*(3), 168–171.
156. Janmohammadi, M., Amanzadeh, T., Sabaghnia, N., & Ion, V. (2016). Effect of nano-silicon foliar application on safflower growth under organic and inorganic fertilizer regimes. *Botanica Lithuanica, 22*(1), 53.
157. Subramanian, K. S., Manikandan, A., Thirunavukkarasu, M., & Rahale, C. S. (2015). Nano-fertilizers for balanced crop nutrition. In *Nanotechnologies in food and agriculture* (pp. 69–80). Springer.
158. Abd El-Azeim, M., Sherif, M., Hussien, M., Tantawy, I., & Bashandy, S. (2020). Impacts of nano-and non-nanofertilizers on potato quality and productivity. *Acta Ecologica Sinica, 40*(5), 388–397.

159. Abdel-Aziz, H., Hasaneen, M., & Omer, A. (2019). Impact of engineered nanomaterials either alone or loaded with NPK on growth and productivity of French bean plants: Seed priming vs foliar application. *South African Journal of Botany, 125*, 102–108.
160. Solanki, P., Bhargava, A., Chhipa, H., Jain, N., & Panwar, J. (2015). Nano-fertilizers and their smart delivery system. In *Nanotechnologies in food and agriculture* (pp. 81–101). Springer.
161. Zulfiqar, F., Navarro, M., Ashraf, M., Akram, N. A., & Munné-Bosch, S. (2019). Nanofertilizer use for sustainable agriculture: Advantages and limitations. *Plant Science, 289*, 110270.
162. Jorfi, S., Rezaee, A., Moheb-Ali, G.-A., & Alah Jaafarzadeh, N. (2013). Pyrene removal from contaminated soils by modified Fenton oxidation using iron nano particles. *Journal of Environmental Health Science and Engineering, 11*(1), 1–8.
163. Zhang, Z., Li, M., Chen, W., Zhu, S., Liu, N., & Zhu, L. (2010). Immobilization of lead and cadmium from aqueous solution and contaminated sediment using nano-hydroxyapatite. *Environmental Pollution, 158*(2), 514–519.
164. Dong, A., Ye, X., Li, H., Zhang, Y., & Wang, G. (2016). Micro/nanostructured hydroxyapatite structurally enhances the immobilization for Cu and Cd in contaminated soil. *Journal of Soils and Sediments, 16*(8), 2030–2040.
165. Lazaratou, C., Vayenas, D., & Papoulis, D. (2020). The role of clays, clay minerals and clay-based materials for nitrate removal from water systems: A review. *Applied Clay Science, 185*, 105377.
166. Iqbal, M., & Umar, S. (2019). Nano-fertilization to enhance nutrient use efficiency and productivity of crop plants. In *Nanomaterials and plant potential* (pp. 473–505). Springer.
167. Kah, M., Kookana, R. S., Gogos, A., & Bucheli, T. D. (2018). A critical evaluation of nanopesticides and nanofertilizers against their conventional analogues. *Nature Nanotechnology, 13*(8), 677–684.
168. Achari, G. A., & Kowshik, M. (2018). Recent developments on nanotechnology in agriculture: Plant mineral nutrition, health, and interactions with soil microflora. *Journal of Agricultural and Food Chemistry, 66*(33), 8647–8661.

The Janus Face of Nanomaterials: Physiological Responses as Inducers of Stress or Promoters of Plant Growth?

Harleen Kaur, Jashanpreet Kaur, Anu Kalia, and Kamil Kuca

Abstract The agri-applications of nanomaterials have been rising in the recent decade, particularly the use of nano-scale nutrient fertilizers and targeted release of nano-pesticides besides the utilization of nano-based sensor modules both for pre- and post-harvest rapid and sensitive identification of nutrient deficiencies, pests, and pathogen attack. Researchers have identified improved germination, seedling vigor, proficient vegetative growth, enhancement in yield, and yield attributing traits due to altered physiological profile of diverse crop plants. The predominant mechanism (s) of action of these engineered nanomaterials affecting the plant metabolic processes have to be further probed to identify variability in the physiological pathways followed in response to the application of nanomaterials at low, moderately high, and high concentrations leading to genotoxicity and cytotoxicity manifestations. It is, therefore, imperative to evaluate the impact of nanomaterials on plant physiology and stress response besides demonstrating the positive effects. Improved know-how has to be generated to predict the nuances of plant-nanomaterial interactions that may probably enhance the ecological safety of their widespread use. This manuscript compiles the details in the published literature on the influence of manufactured nanoparticles on plant physiology and vegetative characteristics to provide an improved viewpoint on the current status of nanoparticles as plant growth promoters.

Keywords Engineered nanomaterials · Metal toxicity · Phytohormone profile · Reactive oxygen species

H. Kaur · J. Kaur
Department of Microbiology, College of Basic Sciences and Humanities, Punjab Agricultural University, Ludhiana, Punjab, India

A. Kalia
Electron Microscopy and Nanoscience Laboratory, Department of Soil Science, College of Agriculture, Punjab Agricultural University, Ludhiana, Punjab, India
e-mail: kaliaanu@pau.edu

K. Kuca (✉)
Department of Chemistry, Faculty of Science, University of Hradec Kralove, Hradec Kralove, Czech Republic
e-mail: kamil.kuca@uhk.cz

J.-T. Chen (ed.), *Plant and Nanoparticles*,
https://doi.org/10.1007/978-981-19-2503-0_16

1 Introduction

Agriculture forms the backbone of the economy in majority of the developing nations. However, intensive and conventional agricultural practices tend to lay stress on land resources and can lead to long-term irreversible deterioration of the soil and water ecosystems. The technological and analytical advancements have led to the development of nanotechnology-enabled devices and products which are providing great promise for the improvement in the crop production and productivity standards. Nanotechnology is the science related to designing, fabrication, characterization, modification, and application of the nanomaterials. A nanomaterial is a substance having one of its dimensions ranging from 1 to 100 nm. Such a small size imparts certain characteristic properties to these materials such as strength, optical property, catalytic activity, magnetic property, bioavailability, and chemical stability [1]. Therefore, nanoparticles find an inclusive range of applicability in varied fields ranging from electronics, biosensors, drug delivery, diagnosis, cosmetics and personal care, crop production, and post-harvest food processing. In agriculture, nanotechnological research is expected to assist and shape the future stages of the utilization of genetically customized crops, agro-chemical delivery, and precision cultivation.

2 Nanomaterials in Plant Science

The published research studies indicate that nanotechnology holds promise for meeting future agricultural and food security demands by providing novel nanoscale tools and strategies [2]. Nanomaterials functionalized with biomolecules exhibit unique physicochemical properties. Furthermore, a range of nanomaterials including the metal (gold), metal oxide (magnetic), polymeric, and hybrid nanomaterials exhibit controlled release of macromolecules both spatially and temporally in response to external stimuli. The applications of nanotechnology in plant science are gaining growing interest lately, especially in light of the remarkable potential to enhance crop yield and the use of nanomaterials for transportation or as delivery systems for agrochemicals and biomolecules [3]. However, the nano-agri-applications have not yet been unleashed to their full potential primarily because of the growing concerns about absorption, translocation, bioavailability, and toxicity of the applied nanomaterials, as well as the unavailability of information on long-term nanomaterial exposure studies which can provide the quantitative statistics of probable aftermaths of continued use. Furthermore, the skepticism in the masses and the stringent regulatory guidelines discourage the embracement of the nano-enabled agri-products by the growers [4].

The initial studies including nanomaterials have been focused on determining the negative impacts of nanoparticles particularly their toxicity when applied at higher concentrations [3]. Therefore, only a few studies provide details about the plant

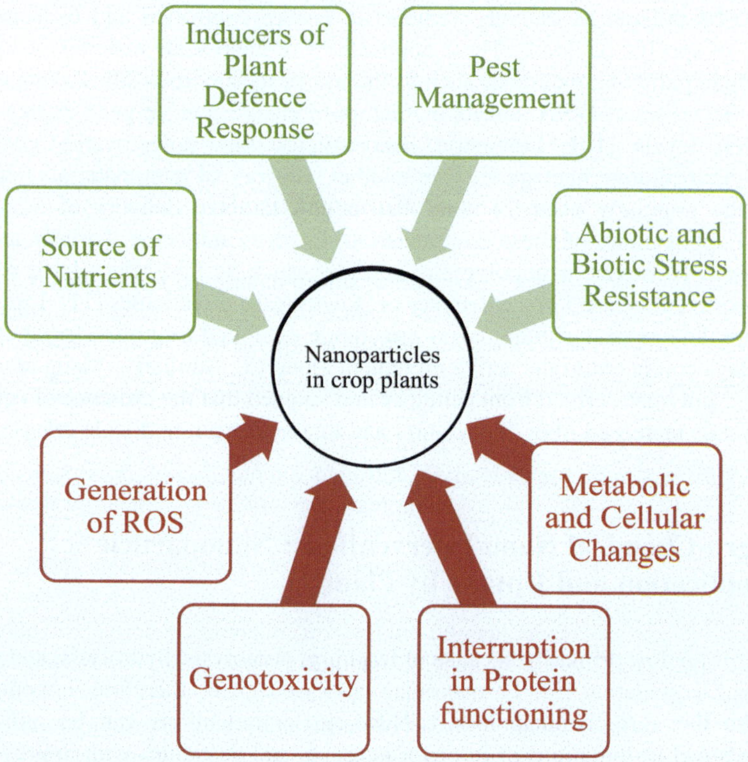

Fig. 1 The yin and yang of applications of nanomaterials in crops

beneficial effects of the nanomaterials. Plant scientists and researchers in other fields could benefit from nanotechnology by developing advanced tools for incorporating nanoparticles into plants, to enhance their functions [5] including designing miniaturized and effective tools to enhance seed germination, plant growth, and security against both environmental and biological stresses [3]. Depending on their size as well as interactions with the plants or microorganisms, the nanomaterials exert their discrete positive and negative effects. In a hydroponic study, when cut stems of *Lolium*, *Allium cepa*, and *Chrysanthemum* plants were placed in aqueous solutions of CdSe/ZnS quantum dots (QDs), these nanoparticles were visibly transported in the vascular system of the whole plants through the roots [5]. Kole et al. [6] have described the positive effects of multiwalled carbon nano-tubes (MWCNT) on tomato plant growth and germination. Lin and Xing [7] have also reported the affirmative effects of MWCNTs on several vegetable crops and *Zea mays*. Furthermore, the hydrophobic nano-silica particles have been demonstrated to kill insect pests of agricultural importance [8]. Thus, the direct use of nanomaterials has a substantial effect on the plant developmental processes (Fig. 1).

Thus, the use of nanoparticles is emerging as an effective way to suppress pathogenic infections, resulting in augmented crop yield [9]. The nanomaterials can also be utilized as delivery vehicles to ensure controlled and even targeted delivery of specific payloads. These applications of nano-scale materials in agriculture include particular examples such as the use of mesoporous silica nanoparticles (MSNs) to transport DNA and its regulators to ensure precise gene expression at single-cell levels [10]. Polymeric nanoparticles like polyethylene glycol or polyvinylpyrrolidone nanospheres are another category of nanomaterials that have been quite popularly used for controlled and/or targeted delivery of molecules. Medical applications of these nanomaterials for drug molecule delivery are well established. However, nanopolymeric particles, such as liposomes, have recently been reported to be used in the delivery of slow-release insecticides [11]. Liposomal nanoformulations of pyrifluquinazon insecticide exhibited improved lethal efficacy for 14 days compared to the active ingredient alone [5]. Similarly, Xiang et al. [12] examined polymer derived from cellulose and showed that the presence of cellulose nanocrystals increased fiber degradation and thiamethoxam herbicide release.

3 Agro-Chemical Nano-Interventions: Nanoparticle Application and Uptake by Plant

Chemicals are lost primarily because of leaching, photolysis, hydrolysis, and microbiological degradation [5]. To maximize crop yield, it is, therefore, imperative to minimize the agro-chemical losses. Nano-agri interventions can be utilized to develop novel formulations of fertilizer nutrients and pesticides with improved use efficiency and target efficacy. The nano-encapsulated or nano-scaled nutrients fertilizers might possess properties useful for crops, such as the controlled release of chemical fertilizers, controlled release of nutrients as per requirement, and enhanced plant growth [13]. However, determination of an effective concentration (along with high efficiency, solubility, and stability), a temporally controlled release as a result of stimuli, enhanced targeted activity, and higher levels of safety are required for nanoformulated agrochemicals.

Fertilizer management is considered an important prerequisite for sustainable agricultural development. The use of quality seeds, effective management of fertilizers, and irrigation water can increase agricultural production by 35–40%. Nanoformulated fertilizers have proven to be highly effective in increasing crop productivity [9]. By adding carbon nanoparticles as fertilizer to *Oryza sativa*, *Zea mays*, *Glycine max*, *Triticum*, and vegetables (11.34–19.76%), grain yields were reported to get increased by 10.29%, 10.93%, 16.74%, and 28.81%, respectively [9]. Due to their high efficacy and environmental friendliness, nanocomposites have been extensively researched and used in plant protection research in the recent years (Fig. 2). Zeolite chip fertilized with urea are being used to slow-release nitrogen fertilizers, according to a study by Millán et al. [14]. Zeolite charged with ammonia has been proven to improve P uptake and crop yield by increasing the solubilization of P-related minerals. Kottegoda et al. [15] synthesized encapsulated urea-modified

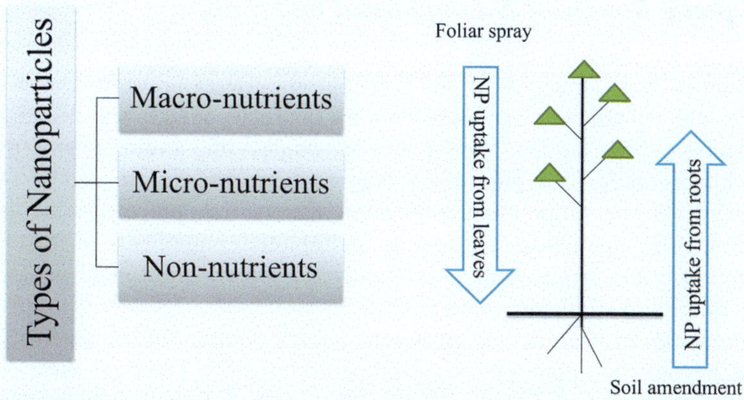

Fig. 2 The role of nanomaterials as nano-scale nutrient fertilizers to enhance crop productivity

hydroxyapatite nanoparticles under compression in the soft wood of *Gliricidia sepium* to deliver sustained nitrogen release to the soil on a long-term basis.

Nano-encapsulated formulation of pesticide Avermectin helped in ensuring slow release of the compound till 30 days vis-a-vis the non-nanoformulated version which got UV-degraded within 6 h of application. Likewise, Loha et al. [16] found that α-cyfluthrin was released throughout 1–20 days from a nanoformulation, while its commercially available counterpart was released within 4–5 days. Piola et al. [17] found that nanoformulations of glyphosate increased the bioavailability of the herbicide when compared to current formulations, which contain several adjuvants that may be toxic to nontarget organisms. Apart from encapsulation, the nano-clay materials including halloysite and montmorillonite minerals can be utilized to develop cost-effective formulations of fertilizers and pesticides [18]. These natural nanomaterials may ensure a better contact with a minimum impact on the environment, as well as an extended-release rate of active ingredients. Hydrophobic nano-silica can be easily absorbed through the cuticle layer of the insect skin and ensure a slow release of the adsorbed insecticide causing the death of the insect [9]. Alternatively, other nanomaterials such as nanofibers and nanoshells can also be useful candidates. Xiang et al. [12] reported that nanofiber networks loaded with thiamethoxam have shown efficacy against whitefly as compared to the 50% of the thiamethoxan dose recommended by the manufacturer. Alternatively, Pendimethalin loaded into hollow-shell particles made of manganese carbonate cores can be possibly utilized for field applications [5]. The development of nanocomposites of pesticides is another nanoformulation strategy. Compared to Ag+ chitosan nanocomposites (Ag@CS) and antracol alone, Ag+ chitosan nanocomposites with fungicide exhibited higher anti-fungal activity [19]. Herbicide-loaded pectin (polysaccharide) nanoparticles exhibited stronger cytotoxic activity towards *Chenopodium album* plants compared to the commercial herbicide under laboratory and field conditions as the nanoherbicides prevented the regrowth of the weeds [20].

3.1 Uptake Routes of Nanomaterial by Plants

Application of nano-strategies in plants require the use of preventive assessments including understanding the processes of nanoparticle uptake, translocation, and accumulation. These studies should also encompass the evaluation of possible contrary effects on the plant physiology. Nanoparticle uptake is affected by a variety of factors which govern the process including nanoparticle properties, the mode of application, their interactions with biotic and abiotic components of the environment, the constraint imposed by a cell wall, and the variation in physiology and anatomy of individual plant species [3].

The dynamics of nanoparticle–plant interactions should also be considered from the plant anatomy and plant physiological perspective [3]. There are three possible ways for uptake of nanoparticles by the plants. Firstly, uptake via roots on the addition of nanoparticles to the medium/substrate and soil. Secondly, uptake through the cuticle and other surfaces on foliar application, and lastly by coating or priming the seeds. Small sized nanoparticles can pass through protein channels or porins in the plant cell barrier membranes and can be absorbed passively.

3.1.1 Nanoparticle Uptake from Roots

A comparison of plant aerial and soil dynamics of nanoparticle uptake indicated that soil dynamics appear to be more complex [3]. The first step in the uptake of nanoparticles by roots is called adsorption which leads to the enhanced presence of the applied nanoparticles in the rhizosphere soil environment, the bioaccumulation. The root adsorption of nanoparticles depends on their size. Thus, forces such as osmotic pressure and capillary forces allow penetration of smaller nanoparticles through root epidermal cells. The root epidermal cells being semi-permeable in nature restrict the entry of larger nanoparticles as pores are much smaller. Although, some nanoparticles are known to produce new pores in the epidermal cells to facilitate their entry. As NPs enter the root, there are two possible transport pathways. First, is the apoplastic pathway/transportation in which nanoparticles after crossing the cell wall are transported via extra-cellular spaces to the vascular bundle and then to the xylem for unidirectional transport [5]. The other pathway is symplast transport, for this Casparian strip barrier is crossed to gain entry to the vascular bundle. Nanoparticles bind to carrier proteins occurring in the plasma membrane of the endodermal cells followed by endocytosis. This transportation occurs through the cytoplasm (internal transport) from one cell to another. Although, nanoparticles of larger size which are unable to pass through the Casparian strip, aggregate at the junction. However, the smaller nanoparticles reaching the xylem can be transported acropetally as well as basipetal towards the shoot and back to the roots through the phloem. Nanoparticles absorbed through the root are transported to various parts and may be found within the cortical cell, cytoplasm, epidermal cell wall as well as nuclei. Moreover, nanoparticles that are not taken up by the roots show aggregation on the root surface and these can alter the adsorption of other nutrients [5]. Mostly the root tips are easily accessible for nanomaterials, however, the presence of suberin makes the upper parts impermeable.

Some factors influence the availability of nanoparticles in the soil such as symbiotic organisms, root and microbial exudates, and soil organic matter. The symbiotic microorganisms present in the soil play a crucial role in the nanoparticle availability and uptake by plants. For example, these microbes may increase the accumulation of various heavy metal nanoparticles in grasses but reduce their occurrence in legumes [3]. The root exudates and mucilage discharged in the rhizosphere region play a dual role, i.e. it promotes adhesion of nanoparticles on the root surface, which in turn accelerates the rate of nanoparticle internalization, and/or may also trigger trapping and agglomeration of the nanoparticles. Recent detection techniques have been used to evaluate the root border cells and associated mucilage which led to trapping of the gold nanoparticles irrespective of particle charge, whereas negatively charged nanoparticles may translocate directly into the root tissues without being sequestered by the mucilages [3].

3.1.2 Nanoparticle Uptake from Leaves

The leaves of the plant can be a preferred site for the application of NPs which ensures entry of nanoparticles through stomata or cuticles. However, the cuticle of the leaf acts as a barrier gradient which restricts the entry of larger nanoparticles (more than 5 nm). The nanoparticles of size >10 nm (10–50 nm) can enter through natural openings (stomata) and are transported through a symplastic route into the vascular system of the plant. For larger nanoparticles (between 50 and 200 nm) apoplastic pathway is favored, which leads to nanoparticle translocation between cells [5]. After entering the cells, the nanoparticles are transported through phloem leading to bidirectional transport such that the nanoparticles accumulate in different plant tissues to varying degree [9].

Several factors affect the adsorption, transportation, or accumulation of nanoparticles in plants on foliar application including the nanoparticle attributes (size, shape, and concentration), and plant leaf characteristics such as leaf surface morphological features (presence of trichome, cuticle thickness, and nature), and chemical composition (cuticular wax types, leaf exudates). Therefore, these essential factors should be considered before the elucidation of the uptake mechanism of nanoparticles in plants [9]. According to El-Feky et al. [21], the application mode of nanomaterials affects the plant performance. They showed that nano Fe_3O_4 foliar spray significantly enhanced the plant vegetative parameters, compared to soil application, due to enhancement of the total chlorophyll, total carbohydrates, essential oils, and the iron content in basil plants.

4 Effect of Application of Nanomaterials on Plant Physiology

4.1 Photosynthetic Apparatus

The process of photosynthesis comprises about 50 redox reactions [22] clustered as photo- and biochemical reactions. The primary light-dependent phase of reactions takes place in the thylakoid membranes [23] and the following phase occurs in the stroma of the chloroplasts [24]. The structural configuration of the sub-stomatal cavity regulating gaseous concentrations within the chloroplast is responsible for the circulation of carbon dioxide (CO_2) to the carboxylation sites to carry out photosynthesis [25]. The factors influencing the photosynthesis rate include the organizational robustness of chloroplasts and mesophyll cells and grana, CO_2 aggregation, RuBisCo activity, regulatory proteins, and the presence of chlorophyll pigments [26, 27]. Nanoparticles can both positively and negatively affect photosynthesis by influencing any of the above factors. Nanoparticles can prevent the metabolic pathways associated with photosynthesis by affecting the light-harvesting photosystems, pigment complexes, electron transport system, and activity of certain enzymes such as RuBisCo, carbonic anhydrase, and phosphoenolpyruvate carboxylase [28, 29].

Different types of nanomaterials affect photosynthesis via different mechanisms. Silver nanoparticles tend to alter the optical behavior of chlorophyll depending on the nanoparticle size and dose [30]. Biochemical and physiological changes may also be initiated by nanoparticles and include reduction in the content of photosynthetic pigments, changes in the development of grana, structural disintegration of chloroplasts, decreased chlorophyll fluorescence, and lower stomatal conductance. The fluorescence suppression occurs due to reduction in the quantum yield of the photosystem and cessation of the electron transport chain [31]. On the contrary, silver nanoparticles, if applied at well-calculated doses, may leave a progressive effect on the photosynthesis process by increasing light absorption [32]. Zinc oxide nanoparticles tend to improve gaseous exchange and chlorophyll content leading to a higher photosynthesis rate [33]. Several studies have reported positive effects of zinc as it induces chlorophyll synthesis, and chloroplast development resulting in enhancing the photosynthetic process [34, 35].

The nitrogen content of leaves is positively correlated with plant photosynthesis [36]. Studies have revealed that TiO_2 (Titanium dioxide) nanoparticles can enhance nitrate reductase activity in spinach and speed up the conversion of inorganic forms of nitrogen to organic forms which can be assimilated as protein and chlorophyll, and increase photosynthesis rate [37]. However, Gao et al. [38] reported a reduction in the net photosynthetic rate of *Ulmus elongata* seedlings on nano-TiO_2 treatment. Zsiros et al. [39] have shown that selenite nanoparticles induced no significant structural and functional changes in the photosynthetic machinery as the nanoparticle penetration was limited in the leaf tissue.

4.2 Antioxidant Mechanism

Nanoparticles can perform a variety of functions in crops, and their long-term effects need to be understood [40]. According to several reports, nanoparticles could negatively impact plants through the release of reactive oxygen species (ROS) [13]. Plants get severely damaged by the formation of superoxide (O_2^-), hydrogen peroxide (H_2O_2), and hydroxyl (OH^-) radicals which can harm the tissues, proteins, lipids, and DNA at the cellular level [40]. Besides, ROS may play a key role as signaling molecules or may induce generation of signal molecules that regulate cellular stress responses [41]. As nanoparticles enter the root and leaf tissues and interact with the plant system, reactive oxygen species (ROS) are generated which activate plants' antioxidant mechanism [40]. Therefore, plants exposed to nanoparticles scavenge or detoxify ROS-generated substances through enzymes and low-molecular-weight antioxidants [42]. Increasing ROS values indicate an increase in antioxidant enzymes. The nanomaterials also upsurge the number of non-enzymatic antioxidants such as flavonoids and phenols.

Plant cell–nanoparticle interactions get initiated at the cell surface. These interactions cause a sequence of cellular responses, such as the generation of stress indicating biomolecules including the non-enzymatic antioxidant molecules and antioxidant enzymes that induce defense responses based on their physical characteristics. The rise in the activity of the antioxidant system reduces the oxidative damage of the cell membranes and may also help impart other benefits such as improved salt stress tolerance trait [42]. Ismail et al. [41] have studied the antioxidant defense responses of plants to nanoparticles as a biomarker for ecotoxicological analysis. The ROS generation and activation of the antioxidant machinery are established methods of determining whether a plant is experiencing stress caused by a specific compound or not. In plants, the presence of nanoparticles results in the initiation of processes that impart protection of cellular components from impairment caused by oxidative stress [42]. The nanoparticles are known to enhance the activity of a variety of antioxidant enzymes which mitigate different types of reactive oxygen species and therefore protect cells against oxidative damage [40].

The salinity tolerance was enhanced in *S. lycopersicum* by the application of Si (silica) nanoparticles through enhancement in the catalase and ascorbate peroxidase activities [43]. According to Ioannou et al. [44] nanoparticles can reduce the harmful effects of salinity in *Dracocephalum moldavica* L, probably by enhancing antioxidant enzyme activity, causing more systematic ROS detoxification. According to a study on tomatoes improved antioxidant enzyme activity imparted protection to cell membranes from ROS attack and also prevention of the lipid peroxidation events [40]. Enhanced ROS scavenging and plant enzymes in response to engineered nanoparticles were observed in plant systems [42]. Using copper and zinc nano-oxides in sand-grown wheat plants, Dimkpa et al. [45] examined the degree of phytotoxicity endured. Using either of the nanoparticles as additives to autoclaved sand, analyses were performed on root samples for catalase and peroxidase enzyme activities. The results obtained indicated no increase in the catalase or peroxidase activities by zinc nanoparticle treatment though the copper nanoparticle-treated plants showed improved activity concerning the control [40]. In *Oryza sativa*, silver nanoparticles resulted in an escalation in the ROS production. Moreover, carotenoids activated the antioxidant machinery in the plant to decrease ROS damage due to nanoparticles. Dimkpa et al. [46] showed that the presence of silver (Ag) nanoparticles caused the oxidized glutathione levels to be elevated, as well as activation of the metallothionein gene involved in detoxification in *Triticum aestivum*.

4.3 Plant Protein–Nanomaterial Interactions: Alterations in the Plant Proteome Composition

Nanoparticles may interact with biomolecules including proteins, nucleic acids, and lipids. The adsorption of protein on nanoparticles' surfaces results in the formation of a complex known as the nanoparticle-protein corona. The adsorption is aided by hydrogen (H) bonds and Van der Waals interactions and the longevity of the formed nanoparticle-protein corona is governed by the association–dissociation rates for each protein [47]. The interaction of proteins with nanoparticles is of two types, i.e. irreversible or long-term binding leading to the creation of a hard corona while the reversible or quick binding of proteins results in a soft corona [48].

The protein corona generation can influence the biological reactivity of the nanoparticle as well as the functioning of the proteins. For example, nanoparticles with curved surfaces can cause irreversible changes in the secondary structures of proteins [49]. Changes induced in the protein conformation by nanoparticles affect the related downstream processes such as protein–protein interactions, cellular signaling, and DNA transcription. Other factors controlling the adsorption of protein on the surface of nanoparticles include its composition, presence of precise functional groups, and properties of the biological medium such as pH and temperature [50–53]. Wang et al. [54] through their study showed that nanoparticles altered with ionic ligands bind higher levels of proteins as compared to those possessing neutral surface charge. Moreover, nanoparticles with smaller sizes showed a higher level of protein association, thus elucidating the role of surface area and curvature in protein binding.

Several researchers have reported the effects of different nanoparticles on the protein composition at the cellular scale (proteome analysis). In gel-based proteome analysis, Mirzajani et al. [55] have revealed that on exposure of *Oryza sativa* cells to Ag NPs, specific elicitation of diverse proteins occurred including the proteins involved in defense-related processes and antioxidant pathways. The Ag NPs caused a decrease in cellular metabolic activities through the involvement of receptors and channels based on Ca^{2+}, and Ca^{2+}/Na^{+}–ATPases. Likewise, a change in the number of proteins related to the regulation of redox activities of the cell, vacuolar, and endoplasmic reticulum specific proteins, and those involved in the sulfur metabolic pathway has been reported in *Eruca sativa* [56]. In the case of soybean grown under flooding conditions, silver nanoparticles mitigated the stress by regulating the wax formation and protein metabolism-related proteins [57]. Elemental copper forms an integral component of metabolic process-related proteins. The application of copper nanoparticles in wheat seedlings increased the relative amount of proteins associated with glucose catabolism and decreased the proteins associated with photosynthesis as well as tetrapyrrole synthesis [58]. Cerium nanoparticle application increased the amount of membrane lipids degradation thereby resulting in electrolyte leakage and oxidative stress [59].

4.4 Disease Resistance

The nano-applications in agriculture include the development of innovative nanomaterials capable of controlling phytopathogenic diseases. Nanoparticles can be applied directly to crops via soil spiking or foliar spray to suppress the pathogens and help in increasing plant growth and crop yields. Thus, the nanoparticles can impart a dual role of crop protection (as nano-pesticides) as well as plant nutrition (as nanofertilizers).

On nanoparticle application, the general plant response includes an increase in reactive oxygen species (ROS) levels. The generated ROS restricts the pathogen ingression and distribution besides inducing systemic and local defense responses in plants [60]. Plant resistance against diseases operate through an elaborate defense pathway triggered primarily by ethylene, jasmonic acid, and salicylic concentrations. Other plant growth hormones including gibberellin, auxin, cytokinin, brassinosteroids, abscisic acid, and strigolactone can alter the defense-related responses. Nanoparticle stress is known to up or down-regulate pathways associated with the synthesis of growth hormones thereby stimulating plant defense mechanisms [61].

Nanoparticles exhibit remarkable anti-viral, anti-bacterial, and anti-fungal properties [62] as given in Table 1. The nanoparticles can also directly act upon the pathogens. Further, the nanoparticle can interact with virus capsid proteins. Cai et al. [60] have described the direct inactivation of tobacco mosaic virus (TMV) by nanoparticles (zinc oxide and silica nanoparticles) via interaction with capsid glycoproteins, causing the direct injury of virus shell proteins resulting in viral aggregation. The mechanism via which silver nanoparticles act against microbes is initiated via the binding of Ag^+ to membrane proteins, thereby compromising membrane integrity [67]. Chitosan nanoparticles can act as inducers of the antioxidant and defense systems in plants [82]. Transcriptome analysis of plants treated with chitosan nanoparticles resulted in elevated levels of defense-related genes thereby augmenting the innate immunity of plants against different stimuli [82].

Table 1 Nanoparticles providing resistance against various pathogens in different crops

Nanoparticle	Host plant	Concentration	Mode of application	Pathogen	Reference
Zinc (Zn) nanoparticles	Eggplant	0.2%	Foliar spray	Cucumber Mosaic Virus	[63]
Carbon nano-tubes (CNT)	*Nicotiana benthamiana*	100, 200 and 500 mg/L	Foliar application	TMV	[64]
Silver (Ag) nanoparticles	Tomato	50, 60 and 70 mg/L	Foliar application	Tomato Mosaic Virus and Potato Virus Y	[65]
Iron oxide (Fe_2O_3), TiO_2, Multiwalled CNT, fullerenes	*Nicotiana benthamiana*	50 and 200 mg/L	Foliar application	Turnip Mosaic Virus	[66]
DNA-directed Ag NPs grown on grapheme oxide (GO)	*Solanum lycopersicum*	16, 50 and 100 mg/L	Foliar application	*Xanthomonas performance*	[67]
Ag NPs	*Solanum lycopersicum*	5 μg/mL	Foliar spray	*Alternaria solani*	[68]
ZnO NPs	*Solanum lycopersicum*	100 μg/mL	Foliar spray	TMV	[69]
Ag NPs	Wheat	2, 4 and 10 μg/mL	Foliar spray	*Bipolaris sorokiniana*	[70]
Boron (B), CuO, MnO, SiO, TiO, and ZnO NPs	Watermelon	500–1000 μg/mL	Foliar spray	*Fusarium oxysporum* f. sp. niveum	[71]
Thiamine loaded chitosan NPs	Chickpea	0.1% (w/v)	Foliar spray	*Fusarium oxysporum* f. sp. ciceri	[72]
Cu-chitosan nanoparticles	*Eleusine coracana* Gaertn.	0.01, 0.05, 0.1, 0.15%	Foliar application and seed coating	*Pyricularia grisea*	[73]
Sulfur (S) nanoparticles	*Solanum lycopersicum*	30–100 mg/L	Foliar spray and seed treatment	*Fusarium oxysporum* f. sp. lycopersici	[74]
Titanium dioxide (TiO_2) nanostructures	*Vicia faba* L. Fabaceae	150 μM	Foliar spray and soil drench	Broad Bean Stain Virus	[75]
Hairpin pss loaded chitosan nanoparticles	*Solanum lycopersicum*	20 μg	Leaf infiltration	*Rhizoctonia solani*	[76]

NPs of oxides of Aluminum (Al), Manganese (Mn), Nickel (Ni), Cu, Fe and Zn	*Solanum lycopersicum* and Eggplant	–	Root priming	*Fusarium oxysporum* f. sp. lycopersici	[77]
Selenium (Se) NPs	*Solanum lycopersicum*	100 mg/L	Seed coating	*Phytophthora infestans*	[78]
Curcumin enhanced copper (Cu) nanoparticles	*Cicer arietinum* L.	0.01%	Seed treatment	*Fusarium oxysporum* f. sp. Cicero	[79]
Chitosan NPs	Pearl millet	0.5, 1.0, 2.5 and 5.0 mg/mL	Seed treatment	*Sclerospora graminicola*	[80]
Magnesium oxide (MgO) nanoparticles	*Solanum lycopersicum*	0.1%, 0.5%, 0.7% and 1.0%	Soil drench	*Ralstonia solanacearum*	[81]

NP nanoparticle

4.5 *Drought Resistance*

Climate change has strongly affected crop morphological, biochemical, and physiological behaviors globally. Water shortage has increased because of global warming and has laid stress on the crop plants. Plants undergo morphological adaptations such as a decrease in leaf area as adaptive mechanism against water stress. The nanoparticles can be utilized to mitigate drought stress. The colloidal suspension of copper and zinc nanoparticles can result in a decrease in the TBARS accumulation, an increase in enzymes activities related to ROS scavenging, a changed ratio of leaf chlorophyll pigments (Chl a:b), and an increase in carotenoid content which implies the increase of plant antioxidant status in response to nanoparticles under drought conditions [83]. Application of chitosan nanoparticles increased leaf parameters such as leaf area, relative water, chlorophyll content, photosynthesis rate, antioxidant enzyme activities, and yield parameters as compared to the control in wheat plants under drought conditions [84]. The mechanisms via which nanoparticles alleviate the effect of water stress include an increase in the endogenous level of cytokinins, which stimulated chlorophyll synthesis and promoted plant growth [85]. Selenium nanoparticles can increase the level of chlorophyll pigments, osmolytes, antioxidant enzymes, and abscisic acid resulting in better physical and chemical parameters [86].

5 Nanoparticles Affect Plant Vegetative Parameters

5.1 *Plant Regeneration*

The plant tissue growth eliciting aspects of nanoparticles have paved towards their use in plant tissue culture techniques. Application of nanoparticles for improving germination index and plant development, and alteration of plant genetic machinery for escalated generation of secondary metabolites form the basis of nanoparticle use in plant tissue culture. Nanomaterials can improve tissue multiplication (both root and shoot tissues), callus culture formation, and somatic embryogenesis. Various studies showing positive effects of nanoparticles in plant regeneration via tissue culture techniques have been published [87–92].

A variety of nanoparticle and plant factors decide the impact of nanoparticles on plant regeneration. The action of nanoparticles can be rightly related to the developmental potentials of plant genotypes. Nalci et al. [93] have shown a higher rate of embryogenic callus formation as compared to control on the treatment of ferric oxide nanoparticles. The concentration of nanoparticles applied can also affect the growth rate of the plant cell. Nanocarbon can promote regeneration by improving the water and/or nutrient transport, increasing aquaporin levels besides regulating genes participating in cell division and cell wall extension [87]. Citrate-coated gold nanoparticles can lead to regeneration in barley plants to a certain extent

[94]. Application of Cu nanoparticles on *Ocimum basilicum* explants resulted in increased explant production, somatic embryogenesis, and mean explant regeneration concerning control [88]. Silver nanoparticles have been reported to reduce auxin accumulation and down-regulate the auxin receptor-related genes in plants [95]. Other studies have shown that the application of silver nanoparticles results in elevated total cytokinin levels along with decreased ethylene levels in regeneration experiments [96, 97].

5.2 Seedling Vigor

As discussed in Sect. 3, the application of nanoparticles in agriculture has extensively increased in the past two decades. Nanoparticles can be applied as nanofertilizers to increase plant growth and boost crop production (Table 2). Nanofertilizers can be categorized as macro-nutrient nanofertilizers, micro-nutrient nanofertilizers, and non-nutrient nanofertilizers [2]. Nano-scale nutrient particles act as sources of essential nutrients thereby increasing germination, shoot and root growth, and plant biomass development. For example, nanoparticles of important mineral elements (such as copper, zinc, manganese, calcium, and sodium) play an important role in cellular metabolic processes. The availability of such minerals and nutrients as nanoparticles is anticipated to enhance germination and plant growth. Moreover, the small size of nanoparticles facilitates their uptake by seeds and plants thereby enhancing their role in growth promotion [123].

Nanomaterials can penetrate the seed coat and improve water adsorption and utilization, which thereby can boost the enzymatic system and results in better germination and plant growth. The published reports have demonstrated the positive impacts of nanoparticles including improved seedling vigor in different crop plants such as chilli [124, 125], pigeonpea [126], wheat [127–129], maize [130, 131], green gram [132, 133], rice [134–136], and lentil [137]. In a study, nanomaterials were found to have the potential to improve crops and quality products [138]. Nanomaterials arguably have the potential to help plant roots for absorption of a higher amount of nutrients and water, which increases root system vigor and enzymatic activity, though the exact mechanism of promoting plant growth and quality is unclear [139].

5.3 Yield Attributes

Nanomaterials when applied at appropriate concentrations positively affect the crop yield and yield attributing traits. Many varieties of crops, including *Arachis hypogaea*, *Glycine max*, *Vigna radiata*, *T. aestivum*, *A. cepa*, *Spinacia oleracea*, *S. lycopersicum*, *S. tuberosum* and *B. nigra*, have been benefited by nanomaterials that increased the crop growth and development, as well as increased the produce

Table 2 Effect of nanoparticles on plant growth, physiology, and yield attributes of different crops

Crop	Nanoparticle	Concentration	Mode of application	Effect on crop	Reference
Cereal crops					
Wheat	Cu NPs	0.2–50 mg/L	Soil amendment	Significant increase in leaf area, chlorophyll content, plant weight, grain quantity and weight	[98]
	Ag NPs	20, 200 and 2000 mg/kg	Soil amendment	Inhibition in plant growth, a significant decrease in seed number and 100-grain weight	[99]
	K NPs	20, 40 and 60 mg/L	Foliar application	Significant increase in the number of spikes and spikelets/spike, photosynthetic pigments, and grain yield	[100]
	ZnO NPs	1.5 and 10 ppm	Seed priming	significant enhancement in heat stress tolerance, improved SOD and GPX activity	[101]
	TiO_2 NPs	1.5 and 10 ppm	Seed priming	Higher FRAP, DPPH and OH scavenging activities	[101]
	Si NPs	0, 300, 600, 900, 1200 mg/L	Seed priming	Significant enhancement in shoot and spike length, weight of root/shoot/spike, and grain weight (dry weight basis)	[102]
Maize	ZnO NPs/ZnONPs-Starch	0, 20, 40 mg/L	Seed priming	Enhance plant biomass and height, stover yield, hemicellulose and acid detergent fiber, and shoot zinc content	[103]
	Zn-chitosan NPs	0.01%, 0.04%, 0.08%, 0.12% and 0.16% w/v	Seed priming	Proliferation in shoot and root length, stem diameter, 100 seed weight, grain yield	[104]
	CNPs	0, 50, 100, 200, 400 and 800 mg/kg	Soil amendment	Increase in plant height, root and shoot biomass and nutrient uptake	[105]
Rice	ZnO NPs	20, 40 and 60 mg/L	Foliar spray	Significant increase in chlorophyll content, leaf area index, dry matter, yield attributing traits and grain yield	[106]
	ZnO NPs	5, 10, 25, 50, 100 and 200 mg/L	Hydroponic application and foliar spray	Increase in germination rate, root and shoot length, and leaf length. Non-significant effect on yield parameters	[107]

	ZnO NPs	0, 0.5, 1.0 and 5.0 g/L	Foliar spray	Reversion of Zn-deficiency symptoms by ZnONPs treatments, enhancement of plant Zn contents, improvement of plant growth and yield parameters (5.0 g/L concentration)	[108]
	Ag NPs	5%, 15% and 30%	Foliar spray	Significant increase in leaf length, plant height and grain yield	[109]
	Poly-acrylic coated CeO_2 NPs	1.0 μM (hydroponic) and 10 μM (greenhouse)	Hydroponic application and foliar spray	Increase in shoot height, chlorophyll content, fresh weight, grain weight	[110]
Soybean	ZnO NPs	0, 40, 80, 160 and 400 mg Zn/kg soil	Soil amendment	Significant increase in seed yield	[111]
	Fe, Co and Cu NPS	0.080 g/ha	Seed treatment	Increase in % germination and seedling vigor, shoot height, root length, and seed yield	[112]
Other crops					
Fenugreek	Ag NPs	0, 20, 40 and 60 mg/L	Soil amendment	Significant increases in all photosynthetic pigment contents plant growth, and yield quantity and quality	[113]
Chili	Ag NPs	10.4 ± 2.1 mg Ag/kg soil	Soil amendment	No effect on fruiting time and yield	[114]
Lettuce	Ag NPs	10.4 ± 2.1 mg Ag/kg soil	Soil amendment	Significant reduction in plant biomass	[114]
	CNPs	0.3% w/v	Seed priming	Significant enhancement in seed germination	[115]
Radish	Ag NPs	10.4 ± 2.1 mg Ag/kg soil	Soil amendment	Significant reduction in plant biomass	[114]
Phaseolus vulgaris L.	Si NPs	2.5 and 5.0 mmol/L	Foliar spray	Significant rise in plant growth, photosynthetic pigments, transpiration and net photosynthetic rates	[116]
Phaseolus vulgaris L.	CeO_2 NPs	250–2000 mg/L	Foliar spray	Increase in yield attributing traits including pods formed per plant, seed number per pod, seed per plant and seed weight	[117]
Pomegranate	Se NPs	10 and 50 mg/L	Foliar spray	Significant increase in photosynthetic pigments, nutrient status, fruit yield and fruit diameter	[86]

(continued)

Table 2 (continued)

Crop	Nanoparticle	Concentration	Mode of application	Effect on crop	Reference
Tomato	ZnO, CuO Al_2O_3 and TiO_2 NPs	20–2000 mg/kg	Soil amendment	Increase in seed germination	[118]
	Se and Cu NPs	Se NPs: 1, 10 and 20 mg/L	Through irrigation system	Interaction of both NPs increased the average weight of tomato. Se NPs (10 mg/L) increased the yield	[119]
		Cu NPs: 10, 50 and 250 mg/L			
Potato	Nano-chitosan-urea	0%, 50%, 75% and 100% recommended dose of N fertilizer	Soil application	Improved content of photosynthetic pigments, yield highest in 75% RDN nano-chito-urea formulation equivalent to 100% RDN conventional fertilizer	[120]
	Nano-zeolite; Zn, B and Si NPs	Nano-zeolite: 1.3 L/ha	Nano-zeolite: Through irrigation system	Significant increase in plant vegetative parameters, relative water content, number of tubers per plant and tuber yield	[121]
		Zn, B and Si NPs: 20 mg/L, 12 mg/L, and 15 mg/L, respectively	Zn, B and Si NPs: Soil amendment		
	Nano-chitosan NPK	10%, 50% and 100%	Foliar application	Significantly improved root/shoot length and weight, yield parameters	[122]

NP nanoparticle

quality [139]. The effect of soil amendment with cerium oxide (CeO) nanoparticles (125–500 mg/kg) was examined in the wheat (*T. aestivum* L.) crop. The NP amendment increased vegetative traits such as shoot growth and biomass besides the grain yield. As the concentration of CeO increased, Ce accumulation predominantly increased in root tissues indicating lower or no Ce-mobility to above-ground tissues. Yasmeen et al. [140] investigated the effect of Cu and iron NP supplementation on physiological and overall protein composition alterations in wheat seeds. With the use of 25 mg/L, both Cu and Fe NPs improved the yield attributing traits in wheat. A consequent increase in the yield was also reported. Proteomic studies revealed enhanced levels of glycolysis and starch degradation enzymes. The Cu nanoparticles may increase the stress tolerance towards temperature in wheat [140]. Likewise, Arora et al. [141] evaluated the stimulation of *Brassica juncea* (L.) Coss. growth and yield by NP application under field conditions. Several parameters related to plant growth and yield were positively impacted by foliar spray application of Au NPs (0–100 mg/L), including stem thickness, number of branches and pods, and seed yield. Treatment with Au nanoparticles (10 mg/L) enhanced the seed yield optimally. Therefore, the application of Au nanoparticles can be a useful alternative to ensure food security as proven under field conditions [141].

Research by Bradfield et al. [142] recorded a decrease in yield in sweet potatoes on exposure to high concentrations of ZnO, CuO, and CeO_2 nanoparticles (1000 mg/kg). Using ZnO nanoparticles, Subbaiah et al. [143] examined their effect on maize growth, productivity, and zinc biofortification. They observed improved germination rate and seedling vigor index by application of ZnO nanoparticles (1500 mg/L). Plants treated with $ZnSO_4$ at 2000 mg/L produced a yield that was 42% higher as compared to untreated plants. As a result of the experiment, ZnO nanoparticles significantly affected the yield, growth, and zinc levels of maize grains [142].

Several comparative studies evaluated the effect of nanofertilizers versus conventional fertilizers on different cereal crops. The results of such studies showed that compared to control, nanofertilizer application improved the vegetative and yield traits. For example, increased plant length, photosynthetic pigment content, several panicles, and spikelets in rice by 3.6%, 2.72%, 9.10%, 9.10%, and 15.42%, respectively, were recorded on nanofertilizer application [144]. Significant increase in root and shoot lengths, as well as biomass changes, have been observed on exposure to Zn nanoparticles (25–150 mg/L). Shoot and leaf dry matter indices were increased by 63.8% and 69.7%, respectively, on application of ZnO nanoparticles [145]. Corn yield and dry weight varied significantly depending on the concentration and size of TiO_2 nanoparticles applied to the plant [146]. With a silver nanoparticle concentration of 25 parts per million (ppm), maximum leaf area and grain yield significantly improved in wheat, however, at a 75 mg/L concentration grain yield decreased [9].

6 Detection of Nanoparticles in Plants: Tracing the Fate of the Applied Nanomaterials in Plant System

The nanomaterials applied to plants must be traced to identify their uptake dynamics, and translocation followed by relative distribution rates in various parts of the plant [147]. A variety of techniques can be applied to serve this purpose [148]. Certain common techniques employed for nanoparticle identification in plants have been discussed here.

6.1 Electron Microscopy Techniques

In Scanning EM, the interaction between surface atoms and electrons produces various signals providing data about the composition as well as the topography of the sample surface [149]. This technique is useful to elucidate the aggregate morphology of the nanomaterials. However, SEM is a useful technique to identify the surface adsorption of the nanomaterials on tissue or cell surface. In a foliar application study of ZnO nanoparticles in rice, the SEM analysis revealed that the applied nanoparticles got adsorbed on the surface of the leaves and exhibited internalization through the stomatal apertures [108]. Furthermore, it can be used to capture images of sections as well as intact samples to achieve resolution up to 1 nm [150].

The TEM is considered as the most popular technique used to capture the presence of the nanoparticles in tissue or cells of plants. This technique is also one of the basic techniques for morphological characterization including the particle size, shape, and dimensions of nanomaterials. It has the potential to generate images at significantly higher resolution (up to sub-nanometer) [150]. The TEM analysis of the plant tissues can be carried out by generating the serial ultra-thin sectioning of the sample tissue(s). The presence of nanomaterial is generally identified as electron-dense structures in the ultra-sectioned tissue on viewing in a transmission electron microscope. The TEM analysis of the ultra-cut sections of isolated protoplasts of *Petunia hybrida* showed internalization of electron-dense PEG-coated Ag nanoparticles across the cell membrane possibly through endocytosis [151]. Likewise, Deng et al. [152] have reported the occurrence of condensed dark spots in the TEM micrographs representing the movement of the TiO_2 nanoparticles across the cell wall and membrane of the rice roots exposed to TiO_2 nanoparticles in a hydroponic system. Another hydroponic study on nanoTiO_2-*Rhizobium*-pea interaction used the TEM micrographs of the pea noduled treated with nTiO_2 particles and revealed the occurrence of a small number of unbranched bacteroid in infected host cells compared to more, larger and Y-shaped branched bacteroides in control cells [153]. However, the nTiO_2 particles could not be identified and established in these TEM micrographs.

Transmission EM analysis of the plant tissues can provide valuable information on tissue internalization of the nanomaterials but it can only be used for small sections of the sample tissue. Therefore, a variant hybrid TEM technique called scanning transmission EM (STEM) can be utilized to study larger volumes of the biological tissues and thus can provide valuable information on the occurrence of engineered nanomaterials particularly the electron-dense metal/metalloid nanoparticles in the sample [154]. Though the initial studies have been performed for visualization of nanoparticles in microbial cells such as Ag nanoparticles in bacteria [155], and animal tissues such as peripheral mononuclear blood cells [156] but the reports on plant-nanoparticles are also emerging.

Coupling TEM with an energy-dispersive X-ray spectroscopy module can help in the direct identification of the nanomaterials through the semi-quantitative elemental composition of the nanomaterials. The EDS measures the amount of the X-rays radiated from a sample to represent the sample's elemental composition as each element generates unique electromagnetic emission peaks due to its atomic structure [150]. A nano-TiO_2 pea seed treatment study revealed the concentration-dependent presence of the Ti element on the surface of treated seeds [153].

6.2 *Advanced Conjugate Microscopy-Spectroscopy Techniques*

Obtaining direct and unambiguous evidence of NP uptake is one of the biggest challenges in studying NP–plant interactions. Metal analyses of roots and shoots, as well as imaging tools, have been used to assess the NP association and translocation in roots. The localization of the nanoparticles in plant tissue through a single microscopy or spectroscopy technique would not yield reliable results specifically at low concentrations. Also, a few visible nanoparticles are usually barely distinguishable from natural nanoparticles or background interference. In addition, the sample processing protocols which include labels, stains, sputter coating, and ultrathin sectioning steps may introduce artifacts besides these approaches are destructive. These limitations of the in planta NP visualization techniques can be addressed through the use of multiple techniques, e.g., the elemental analysis combined with nanoparticle identification and mapping [157].

6.2.1 X-ray Tomography

Computed tomography at micro-scale with the use of X-rays (μCT) involves the generation of virtual cross-sections of the sample through the use of X-rays to obtain digital 3-D images of the internal tissues of the plants at a micron level spatial resolution in a non-invasive manner. Monochromatic X-rays are passed through the sample which leads to the conversion of transmitting X-rays into visible light which is later captured by the use of a photodetector to obtain the 2D images. Later, a 3D image is reconstructed by combining two dimensional serial images, thus allowing the identification of the element distribution in three dimensions within the sample [150].

6.2.2 Scanning Transmission Microscopy

This is a technique that enables the in situ measurement of elements within a specimen with extremely high lateral resolution. The technique involves the use of a Fresnel zone plate for focusing a small spot of the X-ray beam followed by detection of the transmitted X-ray. Coupling STXM with X-ray Absorption Spectroscopy (XAS) provides a unique in situ chemical speciation technique that allows sub-nanometer spatial resolution [150]. An XAS study validated the occurrence of nano-ceria in lyophilized and macerated root tissues of alfalfa, corn, cucumber, and tomato on the treatment of the seeds which provided the evidence that plants roots exhibited both uptake and storage of the nano-ceria particles as indicated from the Ce^{4+} oxidation state [158]. A similar observation of non-biotransformation of nano-TiO_2 particles in cucumber has been reported on micro-XANES spectroscopy analysis of the root tissues [159].

6.2.3 Other Techniques

The enhanced resolution dark-field microscopy along with hyperspectral imaging are emerging techniques that have a high potential for incorporation into the integrative approach. By use of this 2D visualization tool, surfaces at a nano-scale can be detected and mapped in comparatively shorter time in complex environments, in a narrow focus plane. This technique can be utilized to study interactions of NPs with organisms, such as in protozoa, bacteria, and green algae; as well as in vivo interactions in fish or worms. However, the technique has not been used to check the presence of nanoparticles in terrestrial plants, despite its demonstrated value in locating nanoparticles in cells and small organisms [157].

For non-invasive in planta imaging, X-ray computed tomography (CT) complements DF-HSI as well. This technique can help yield 3D images by using X-ray attenuation phenomena and also help in reducing the sample preparation artifacts as samples are neither cut nor tagged [157]. A micro-CT and nano-CT systems possess a resolution limit of 1 mm and 50 nm, respectively. By using micro-CT, microscopic plant features can be obtained with a resolution of a few millimeters. Benchtop nano-CT recently provided 3D information on roots-nanoparticle interactions (adsorption vs. internalization) with a good resolution. Furthermore, it permits scanning of a large volume of the sample which eases the analysis of low-nanoparticle concentrations. A cross-validation technique specific to nanoparticles needs to be used to identify nanoparticles [157]. A 3D cross-validated analysis of the scattering of Au nanoparticles (up to 12 nm in diameter) in *Arabidopsis thaliana* model plant throughout and on the roots was performed using DF-HSI together with nano-CT [157]. Exposure of plants with positively and negatively charged gold nanoparticles (−/+ Au nanoparticles) resulted in translocation of negatively charged nanoparticles by apoplast while positively charged nanoparticles were prevented from uptake as revealed via imaging techniques [157].

7 Conclusion and Future Roadmap

In recent years, nanotechnology has taken a great leap in research and development foci for identification of potential agricultural applications. Despite huge developments in analytical techniques, there is still a paucity of the specific studies relating to the effect of nanoparticles on plant physiology. The interaction of nanoparticles with plant system is dependent on a multitude of factors based on plant and nanoparticle properties. The environmental abiotic and biotic factors also determine the extent of the nanoparticle effect. The nanoparticles depending on dose and time of duration levy positive and negative effects on plant physiology and yield. Despite being inducers of plant growth, the engineered nanomaterials also generate ROS which can induce stress response(s) in plants. It is necessary to standardize the synthesis, functionalization, and dosage concentrations of nanoparticles for plant-based applications for achieving sustainability of agro-ecosystems. It can be reasonably argued that the applications of nanoparticles in agriculture are yet to be fully exploited and suffer some potential bottlenecks. The need for a synthesis of safer nanoparticles, studies related to the mechanism of nanoparticle action, and sustainable application of nanoparticles still need a long road to map.

References

1. Singh, S., Sangwan, S., Sharma, P., Devi, P., & Moond, M. (2021). Nanotechnology for sustainable agriculture: An emerging perspective. *Journal of Nanoscience and Nanotechnology, 21*, 3453.
2. Kalia, A., Sharma, S. P., & Kaur, H. (2019). Nanoscale fertilizers: Harnessing boons for enhanced nutrient use efficiency and crop productivity. In K. A. Abd-Elsalam & R. Prasad (Eds.), *Nanobiotechnology applications in plant protection*. Springer Nature.
3. Sanzari, I., Leone, A., & Ambrosone, A. (2019). Nanotechnology in plant science: To make a long story short. *Frontiers in Bioengineering and Biotechnology, 7*, 120.
4. Ali, S., Mehmood, A., & Khan, N. (2021). Uptake, translocation, and consequences of nanomaterials on plant growth and stress adaptation. *Journal of Nanomaterials, 2021*, 6677616.
5. Thul, S. T., & Bijaya, K. S. (2015). Implications of nanotechnology on plant productivity and its rhizospheric environment. In *Nanotechnology and plant sciences*. Springer.
6. Kole, C., Kumar, D. S., & Khodakovskaya, M. V. (2016). *Plant nanotechnology: Principles and practices*. Springer.
7. Lin, D., & Xing, B. (2007). Phytotoxicity of nanoparticles: Inhibition of seed germination and root growth. *Environmental Pollution, 150*, 243.

8. Thabet, A. F., Boraei, H. A., Galal, O. A., El-Samahy, M. F. M., Mousa, K. M., Zhang, Y. Z., Tuda, M., Helmy, E. A., Wen, J., & Nozaki, T. (2021). Silica nanoparticles as pesticide against insects of different feeding types and their non-target attraction of predators. *Scientific Reports, 11*, 14484.
9. Shang, Y., Hasan, K., Ahammed, G. J., Li, M., Yin, H., & Zhou, J. (2019). Applications of nanotechnology in plant growth and crop protection: A review. *Molecules, 24*, 2558.
10. Torney, F., Trewyn, B. G., Lin, V. S. Y., & Wang, K. (2007). Mesoporous silica nanoparticles deliver DNA and chemicals into plants. *Nanoscience and Nanotechnology Letters, 2*, 295.
11. Zhang, Z., Yang, J., Yang, Q., Tian, G., & Cui, Z. K. (2021). Fabrication of non-phospholipid liposomal nanocarrier for sustained-release of the fungicide cymoxanil. *Frontiers in Molecular Biosciences, 8*, 627817.
12. Xiang, C., Taylor, A. G., Hinestroza, J. P., & Frey, M. W. (2013). Controlled release of nonionic compounds from poly(lactic acid)/cellulose nanocrystal nanocomposite fibers. *Journal of Applied Polymer Science, 127*, 79.
13. Siddiqui, M. H., Al-Whaibi, M. H., & Mohammad, F. (2015). *Nanotechnology and plant sciences: Nanoparticles and their impact on plants*. Springer.
14. Millán, G., Agosto, F., Vázquez, M., Botto, L., Lombardi, L., & Juan, L. (2008). Use of clinoptilolite as a carrier for nitrogen fertilizers in soils of the Pampean regions of Argentina. *Ciencia e Investigacion Agraria, 35*, 245.
15. Kottegoda, N., Khalil, M., Madusanka, N., & Karunaratne, V. (2011). A green slow-release fertilizer composition based on urea-modified hydroxyapatite nanoparticles encapsulated wood. *Current Science, 101*, 73.
16. Loha, K. M., Shakil, N. A., Kumar, J., Singh, M. K., Adak, T., & Jain, S. (2011). Release kinetics of β-cyfluthrin from its encapsulated formulations in water. *Journal of Environmental Science and Health, Part B, Pesticides, Food Contaminants, and Agricultural Wastes, 46*, 201.
17. Piola, L., Fuchs, J., Oneto, M. L., Basack, S., Kesten, E., & Casabé, N. (2013). Comparative toxicity of two glyphosate-based formulations to Eisenia andrei under laboratory conditions. *Chemosphere, 91*, 545.
18. Gayathari, U., & Ramasamy, V. (2019). Utilization of liquid fertilizers for agro-industrial waste management and reducing challenges through nano-encapsulation-a review. *Indian Journal of Agricultural Research, 53*, R-1898.
19. Le, V. T., Bach, L. G., Pham, T. T., Le, N. T. T., Ngoc, U. T. P., Tran, D.-H. N., & Nguyen, D. H. (2019). Synthesis and antifungal activity of chitosan-silver nanocomposite synergize fungicide against *Phytophthora capsici*. *Journal of Macromolecular Science, Part A: Pure and Applied Chemistry, 56*, 522.
20. Kumar, S., Bhanjana, G., Sharma, A., Dilbaghi, N., Sidhu, M. C., & Kim, K. H. (2017). Development of nanoformulation approaches for the control of weeds. *Science of the Total Environment, 586*, 1272.
21. El-Feky, S. A., Mohammed, M. A., Khater, M. S., Osman, Y. A., & Elsherbini, E. (2013). Effect of magnetite nano-fertilizer on growth and yield of *Ocimum basilicum* L. *International Journal of Indigenous and Medicinal Plants, 46*, 1286.
22. Taiz, L., & Zeiger, E. (2002). *Plant physiology* (3rd ed.). Sinauer.
23. Sekar, N., & Ramasamy, R. P. (2015). Recent advances in photosynthetic energy conversion. *Journal of Photochemistry and Photobiology C: Photochemistry Reviews, 22*, 19.
24. Sharkey, T. D., & Weise, S. E. (2016). The glucose 6-phosphate shunt around the Calvin–Benson cycle. *Journal of Experimental Botany, 67*, 4067.
25. Mediavilla, S., Escudero, A., & Heilmeier, H. (2001). Internal leaf anatomy and photosynthetic resource use efficiency: Interspecific and intraspecific comparisons. *Tree Physiology, 21*, 251.
26. Wang, L., Czedik-Eysenberg, A., Mertz, R. A., Si, Y., Tohge, T., Nunes-Nesi, A., Arrivault, S., Dedow, L. K., Bryant, D. W., Zhou, W., & Xu, J. (2014). Comparative analyses of C4 and C3 photosynthesis in developing leaves of maize and rice. *Nature Biotechnology, 32*, 1158.

27. Sáez, P. L., Bravo, L. A., Cavieres, L. A., Vallejos, V., Sanhueza, C., Font-Carrascosa, M., Gil-Pelegrín, E., Peguero-Pina, J., & Galmés, J. (2017). Photosynthetic limitations in two Antarctic vascular plants: Importance of leaf anatomical traits and Rubisco kinetic parameters. *Journal of Experimental Botany, 68*, 2871.
28. Da Costa, M. V. J., & Sharma, P. K. (2016). Effect of copper oxide nanoparticles on growth, morphology, photosynthesis, and antioxidant response in *Oryza sativa*. *Photosynthetica, 54*, 110.
29. Kataria, S., Jain, M., Rastogi, A., Živčák, M., Brestic, M., Liu, S., & Tripathi, D. K. (2019). Role of nanoparticles on photosynthesis: Avenues and applications. In *Nanomaterials in plants, algae and microorganisms: Concepts and controversies* (Vol. 2). Elsevier.
30. Queiroz, A. M., Mezacasa, A. V., Graciano, D. E., Falco, W. F., M'Peko, J. C., Guimarães, F. E. G., Colbeck, I., Oliveira, S. L., & Caires, A. R. L. (2016). Quenching of chlorophyll fluorescence induced by silver nanoparticles. *Spectrochimica Acta Part A: Molecular and Biomolecular Spectroscopy, 168*, 73.
31. Matorin, D. N., Todorenko, D. A., Seifullina, N. K., Zayadan, B. K., & Rubin, A. B. (2013). Effect of silver nanoparticles on the parameters of chlorophyll fluorescence and P700 reaction in the green alga *Chlamydomonas reinhardtii*. *Microbiology, 82*, 809.
32. Sarmast, M. K., & Salehi, H. (2016). Silver nanoparticles: An influential element in plant nanobiotechnology. *Molecular Biotechnology, 58*, 441.
33. Salam, A., Khan, A. R., Liu, L., Yang, S., Azhar, W., Ulhassan, Z., et al. (2022). Seed priming with zinc oxide nanoparticles downplayed ultrastructural damage and improved photosynthetic apparatus in maize under cobalt stress. *Journal of Hazardous Materials, 423*, 127021.
34. Sadak, M. S., & Bakry, B. A. (2020). Zinc-oxide and nano ZnO oxide effects on growth, some biochemical aspects, yield quantity, and quality of flax (*Linum uitatissimum* L.) in absence and presence of compost under sandy soil. *Bulletin of National Research Centre, 44*, 98.
35. Del Buono, D., Di Michele, A., Costantino, F., Trevisan, M., & Lucini, L. (2021). Biogenic ZnO nanoparticles synthesized using a novel plant extract: Application to enhance physiological and biochemical traits in maize. *Nanomaterials, 11*, 1270.
36. Zhang, Y., Niu, S., Xu, W., & Han, Y. (2008). Species-specific response of photosynthesis to burning and nitrogen fertilization. *Journal of Integrative Plant Biology, 50*, 565.
37. Yang, F., Hong, F., You, W., Liu, C., Gao, F., Wu, C., & Yang, P. (2006). Influences of nanoanatase TiO_2 on the nitrogen metabolism of growing spinach. *Biological Trace Element Research, 110*, 179.
38. Gao, J., Xu, G., Qian, H., Liu, P., Zhao, P., & Hu, Y. (2013). Effects of nano-TiO_2 on photosynthetic characteristics of *Ulmus elongata* seedlings. *Environmental Pollution, 176*, 63.
39. Zsiros, O., Nagy, V., Párducz, Á., Nagy, G., Ünnep, R., El-Ramady, H., et al. (2018). Effects of selenate and red Se-nanoparticles on the photosynthetic apparatus of *Nicotiana tabacum*. *Photosynthesis Research, 139*, 449.
40. Vera-Reyes, I., Edgar, V. N., Lira-Saldivar, R. H., & Mende-Arguello, B. (2018). Effects of nanoparticles on germination, growth, and plant crop development. In *Agricultural nanobiotechnology*. Springer.
41. Ismail, A., Takeda, S., & Nick, P. (2014). Life and death under salt stress: Same players, different timing? *Journal of Experimental Botany, 65*, 2963.
42. Kumar, V., Sharma, M., Khare, T., & Wani, S. H. (2018). *Impact of nanoparticles on oxidative stress and responsive antioxidative defense in plants*. Elsevier.
43. Haghighi, M., Afifipour, Z., & Mozafarian, M. (2012). The effect of N-Si on tomato seed germination under salinity levels. *Journal of Biological and Environmental Sciences, 6*, 87.
44. Ioannou, A., Gohari, G., Papaphilippou, P., Panahirad, S., Akbari, A., Dadpour, M. R., Krasia-Christoforou, T., & Fotopoulos, V. (2020). Advanced nanomaterials in agriculture under a changing climate: The way to the future? *Environmental and Experimental Botany, 176*, 104048.

45. Dimkpa, C., Bindraban, P., McLean, J. E., Gatere, L., Singh, U., & Hellums, D. (2017). Methods for rapid testing of plant and soil nutrients. In *Sustainable agricultural reviews*. Springer.
46. Dimkpa, C. O., McLean, J. E., Martineau, N., Britt, D. W., Haverkamp, R., & Anderson, A. J. (2013). Silver nanoparticles disrupt wheat (*Triticum aestivum* L.) growth in a sand matrix. *Environmental Science and Technology, 47*, 1082.
47. Saptarshi, S. R., Duschl, A., & Lopata, A. L. (2013). Interaction of nanoparticles with proteins: Relation to bio-activity of the nanoparticle. *Journal of Nanobiotechnology, 11*, 26.
48. Cedervall, T., Lynch, I., Foy, M., Berggad, T., Donnelly, S., Cagney, G., Linse, S., & Dawson, K. (2007). Detailed identification of plasma proteins adsorbed on copolymer nanoparticles. *Angewandte Chemie International Edition, 46*, 5754.
49. Worrall, J. W. E., Verma, A., Yan, H. H., & Rotello, V. M. (2006). "Cleaning" of nanoparticle inhibitors via proteolysis of adsorbed proteins. *Chemical Communications, 22*, 2338.
50. Kopac, T., Bozgeyik, K., & Yener, J. (2008). Effect of pH and temperature on the adsorption of bovine serum albumin onto titanium dioxide. *Colloids and Surfaces A: Physicochemical and Engineering Aspects, 322*, 19.
51. Turci, F., Ghibaudi, E., Colonna, M., Boscolo, B., Fenoglio, I., & Fubini, B. (2010). An integrated approach to the study of the interaction between proteins and nanoparticles. *Langmuir, 26*, 8336.
52. Gojova, A., Guo, B., Kota, R. S., Rutledge, J. C., Kennedy, I. M., & Barakat, A. I. (2007). Induction of inflammation in vascular endothelial cells by metal oxide nanoparticles: Effect of particle composition. *Environmental Health Perspectives, 115*, 403.
53. Sohaebuddin, S. K., Thevenot, P. T., Baker, D., Eaton, J. W., & Tang, L. P. (2010). Nanomaterial cytotoxicity is composition, size, and cell type dependent. *Particle and Fibre Toxicology, 7*, 22.
54. Wang, M., Gustafsson, O. J. R., Pilkington, E. H., Kakinen, A., Javed, I., Faridi, A., et al. (2018). Nanoparticle–proteome in vitro and in vivo. *Journal of Materials Chemistry B, 6*, 6026.
55. Mirzajani, F., Askari, H., Hamzelou, S., Schober, Y., Rompp, A., Ghassempour, A., & Spengler, B. (2014). Proteomics study of silver nanoparticles toxicity on *Oryza sativa* L. *Ecotoxicology and Environmental Safety, 108*, 335.
56. Vannini, C., Domingo, G., Onelli, E., Prinsi, B., Marsoni, M., Espen, L., & Bracale, M. (2002). Morphological and proteomic responses of *Eruca sativa* exposed to Ag nanoparticles or Ag nitrate. *PLoS One, 8*, e68752.
57. Mustafa, G., Sakata, K., & Komatsu, S. (2016). Proteomic analysis of soybean root exposed to varying sizes of silver nanoparticles under flooding stress. *Journal of Proteomics, 148*, 113.
58. Yasmeen, F., Raja, N. I., Ilyas, N., & Komatsu, S. (2018). Quantitative proteomic analysis of shoot in stress tolerant wheat varieties on copper nanoparticle exposure. *Plant Molecular Biology Reports, 36*, 326.
59. Salehi, H., Chehregani, A., Lucini, L., Majd, A., & Gholami, M. (2018). Morphological, proteomic and metabolomic insight into the effect of cerium dioxide nanoparticles to *Phaseolus vulgaris* L. under soil or foliar application. *Science of the Total Environment, 616–617*, 1540.
60. Cai, L., Liu, C., Fan, G., Liu, C., & Sun, X. (2019). Preventing viral disease by ZnONPs through directly deactivating TMV and activating the plant immunity in *Nicotiana benthamiana*. *Environmental Science: Nano, 6*, 3653.
61. Rastogi, A., Zivcak, M., Sytar, O., Kalaji, H., Xiaolan, H., Mbarki, S., & Brestic, M. (2017). Impact of metal and metal oxide nanoparticles on plant: A critical review. *Frontiers in Chemistry, 5*, 78.
62. Kalia, A., Sharma, S. P., Kaur, H., & Kaur, H. (2020). Novel nanocomposite-based controlled-release fertilizer and pesticide formulations: Prospects and challenges. In *Multifunctional hybrid nanomaterials for sustainable agri-food and ecosystems*. Elsevier.

63. El-Sawy, M. M., Elsharkawy, M. M., Abass, J. M., & Kasem, M. H. (2017). Antiviral activity of 2-nitromethyl phenol, zinc nanoparticles and seaweed extract against Cucumber mosaic virus (CMV) in eggplant. *Journal of Virology & Antiviral Research, 6*, 2.
64. Adeel, M., Farooq, T., White, J. C., Hao, Y., He, Z., & Rui, Y. (2021). Carbon-based nanomaterials suppress tobacco mosaic virus (TMV) infection and induce resistance in *Nicotiana benthamiana*. *Journal of Hazardous Materials, 404*, 124167.
65. El-Dougdoug, N. K., Bondok, A. M., & El-Dougdoug, K. A. (2018). Evaluation of silver nanoparticles as antiviral agent against ToMV and PVY in tomato plants. *Middle East Journal of Applied Science, 8*, 100.
66. Hao, Y., Yuan, W., Ma, C., White, J. C., Zhang, Z., Adeel, M., Zhou, T., Rui, Y., & Xing, B. (2018). Engineered nanomaterials suppress turnip mosaic virus infection in tobacco (*Nicotiana benthamiana*). *Environmental Science: Nano, 5*, 1685.
67. Ocsoy, I., Paret, M. L., Ocsoy, M. A., Kunwar, S., Chen, T., You, M., & Tan, W. (2013). Nanotechnology in plant disease management: DNA-directed silver nanoparticles on graphene oxide as an antibacterial against *Xanthomonas perforans*. *ACS Nano, 7*, 8972.
68. Kumari, M., Pandey, S., Bhattacharya, A., Mishra, A., & Nautiyal, C. S. (2017). Protective role of biosynthesized silver nanoparticles against early blight disease in *Solanum lycopersicum*. *Plant Physiology and Biochemistry, 121*, 216.
69. Abdelkhalek, A., & Al-Askar, A. A. (2020). Green synthesized ZnO nanoparticles mediated by *Mentha Spicata* extract induce plant systemic resistance against Tobacco Mosaic Virus. *Applied Science, 10*, 5054.
70. Mishra, S., Singh, B. R., Singh, A., Keswani, C., Naqvi, A. H., & Singh, H. B. (2014). Biofabricated silver nanoparticles act as a strong fungicide against Bipolaris sorokiniana causing spot blotch disease in wheat. *PLoS One, 9*, e97881.
71. Elmer, W., De La Torre-Roche, R., Pagano, L., Majumdar, S., Zuverza-Mena, N., Dimkpa, C., Gardea-Torresdey, J., & White, J. C. (2018). Effect of metalloid and metal oxide nanoparticles on Fusarium wilt of watermelon. *Plant Disease, 102*, 1394.
72. Sathiyabama, M., Indhumathi, M., & Muthukumar, S. (2019). Chitosan nanoparticles loaded with thiamine stimulate growth and enhances protection against wilt disease in chickpea. *Carbohydrate Polymers, 212*, 169.
73. Sathiyabama, M., & Manikandan, A. (2018). Application of Copper-chitosan nanoparticles stimulate growth and induce resistance in finger millet (*Eleusine coracana* Gaertn.) plants against blast disease. *Journal of Agricultural and Food Chemistry, 66*, 1784.
74. Cao, X., Wang, C., Luo, X., Yue, L., White, J. C., Elmer, W., Dhankher, O. P., Wang, Z., & Xing, B. (2021). Elemental sulfur nanoparticles enhance disease resistance in tomatoes. *ACS Nano, 15*, 11817.
75. Elsharkawy, M. M., & Derbalah, A. (2019). Antiviral activity of titanium dioxide nanostructures as a control strategy for broad bean strain virus in faba bean. *Pest Management Science, 75*, 828.
76. Nadendla, S. R., Rani, T. S., Vaikuntapu, P. R., Maddu, R. R., & Podile, A. R. (2018). Hairpinpss encapsulation in chitosan nanoparticles for improved bioavailability and disease resistance in tomato. *Carbohydrate Polymers, 199*, 11.
77. Elmer, W. H., & White, J. C. (2016). The use of metallic oxide nanoparticles to enhance growth of tomatoes and eggplants in disease infested soil or soilless medium. *Environmental Science: Nano, 3*, 1072.
78. Joshi, S. M., De Britto, S., & Jogaiah, S. (2021). Myco-engineered selenium nanoparticles elicit resistance against tomato late blight disease by regulating differential expression of cellular, biochemical and defense responsive genes. *Journal of Biotechnology, 325*, 196.
79. Sathiyabama, M., Indhumathi, M., & Amutha, T. (2020). Preparation and characterization of curcumin functionalized copper nanoparticles and their application enhances disease resistance in chickpea against wilt pathogen. *Biocatalysis and Agricultural Biotechnology, 29*, 101823.

80. Siddaiah, C. N., Prasanth, K. V. H., Satyanarayana, N. R., et al. (2018). Chitosan nanoparticles having higher degree of acetylation induce resistance against pearl millet downy mildew through nitric oxide generation. *Scientific Reports, 8*, 2485.
81. Imada, K., Sakai, S., Kajihara, H., Tanaka, S., & Ito, S. (2016). Magnesium oxide nanoparticles induce systemic resistance in tomato against bacterial wilt disease. *Plant Pathology, 65*, 551.
82. Chandra, S., Chakraborty, N., Dasgupta, A., Sarkar, J., Panda, K., & Acharya, K. (2015). Chitosan nanoparticles: A positive modulator of innate immune responses in plants. *Scientific Reports, 5*, 15195.
83. Taran, N., Storozhenko, V., Svietlova, N., Batsmanova, N., Shvartau, V., & Kovalenko, M. (2017). Effect of zinc and copper nanoparticles on drought resistance of wheat seedlings. *Nanoscale Research Letters, 12*, 60.
84. Behboudi, F., Tahmasebi-Sarvestani, Z., Kassaee, M. Z., Modarres-Sanavy, S. A. M., Sorooshzadeh, A., & Mokhtassi-Bidgoli, A. (2019). Evaluation of chitosan nanoparticles effects with two application methods on wheat under drought stress. *Journal of Plant Nutrition, 42*, 1439.
85. Chibu, H., & Shibayama, H. (2001). Effects of chitosan applications on the growth of several crops. In *Chitin and chitosan in life science*. Kodansha Scientific.
86. Zahedi, S. M., Hosseini, M. S., Meybodi, N. D. H., & Peijnenburg, W. (2021). Mitigation of the effect of drought on growth and yield of pomegranates by foliar spraying of different sizes of selenium nanoparticles. *Journal of the Science of Food and Agriculture, 101*, 5202.
87. Chutipaijit, S., & Sutjaritvorakul, T. (2017). Application of activated charcoal and nanocarbon to callus induction and plant regeneration in aromatic rice. *Chemical Speciation & Biolavailability, 30*, 1.
88. Ibrahim, A. S., Fahmy, A. H., & Ahmed, S. S. (2019). Copper nanoparticles elevate regeneration capacity of (*Ocimum basilicum* L.) plant via somatic embryogenesis. *Plant Cell, Tissue and Organ Culture, 136*, 41.
89. Malik, W. A., Mahmood, I., Razzaq, A., Afzal, M., Shah, G. A., Iqbal, A., Zain, M., Ditta, A., Asad, S. A., Ahmad, I., Mangi, N., & Ye, W. (2021). Exploring potential of copper and silver nano particles to establish efficient callogenesis and regeneration system for wheat (*Triticum aestivum* L.). *GM Crops & Food, 12*, 564.
90. Al-Mayahi, A. M. W. (2021). The effect of humic acid (HA) and zinc oxide nanoparticles (ZnO-NPS) on in vitro regeneration of date palm (*Phoenix dactylifera* L.) cv. Quntar. *Plant Cell, Tissue and Organ Culture, 145*, 445.
91. Shafique, S., Jabeen, N., Ahmad, K. S., Irum, S., Anwaar, S., Ahmad, N., et al. (2020). Green fabricated zinc oxide nanoformulated media enhanced callus induction and regeneration dynamics of *Panicum virgatum* L. *PLoS One, 15*, e0230464.
92. Tripathi, D., Rai, K. K., & Pandey-Rai, S. (2021). Impact of green synthesized WcAgNPs on in-vitro plant regeneration and withanolides production by inducing key biosynthetic genes in *Withania coagulans*. *Plant Cell Reports, 40*, 283.
93. Nalci, O. B., Nadaroglu, H., Pour, A. H., Gungor, A. A., & Haliloglu, K. (2018). Effects of ZnO, CuO and γ-Fe_3O_4 nanoparticles on mature embryo culture of wheat (*Triticum aestivum* L.). *Plant Cell, Tissue and Organ Culture, 136*, 269.
94. Feichtmeier, N. S., Walther, P., & Leopold, K. (2015). Uptake, effects, and regeneration of barley plants exposed to gold nanoparticles. *Environmental Science and Pollution Research, 22*, 8549.
95. Sun, J., Wang, L., Li, S., Yin, L., Huang, J., & Chen, C. (2017). Toxicity of silver nanoparticles to Arabidopsis: Inhibition of root gravitropism by interfering with auxin pathway. *Environmental Toxicology and Chemistry, 36*, 2773.
96. Vinkovic, T., Novák, O., Strnad, M., Goessler, W., Jurasin, D. D., Paradikovic, N., & Vrcek, I. V. (2017). Cytokinin response in pepper plants (*Capsicum annuum* L.) exposed to silver nanoparticles. *Environmental Research, 156*, 10.

97. Aghdaei, M., Salehi, H., & Sarmast, M. K. (2012). Effects of silver nanoparticles on *Tecomella undulate* (Roxb.) seem, micropropagation. *Advances in Horticultural Science, 26*, 21.
98. Hafeez, A., Razzaq, A., Mahmood, T., & Jhanzab, H. M. (2015). Potential of copper nanoparticles to increase growth and yield of wheat. *Journal of Nanoscience with Advanced Technology, 1*, 6.
99. Yang, J., Jiang, F., Ma, C., Rui, Y., Rui, M., Muhammad, A., Cao, W., & Xing, B. (2018). Alteration of crop yield and quality of wheat upon exposure to silver nanoparticles in a life cycle study. *Journal of Agricultural and Food Chemistry, 66*, 2589.
100. Sheoran, P., Goel, S., Boora, R., Kumari, S., Yashveer, S., & Grewal, S. (2021). Biogenic synthesis of potassium nanoparticles and their evaluation as a growth promoter in wheat. *Plant Gene, 27*, 100310.
101. Thakur, S., Asthir, B., Kaur, G., Kalia, A., & Sharma, A. (2021). Zinc oxide and titanium dioxide nanoparticles influence heat stress tolerance mediated by antioxidant defense system in wheat. *Cereal Research Communications.*
102. Hussain, A., Rizwan, M., Ali, Q., et al. (2019). Seed priming with silicon nanoparticles improved the biomass and yield while reduced the oxidative stress and cadmium concentration in wheat grains. *Environmental Science and Pollution Research, 26*, 7579.
103. Tondey, M., Kalia, A., Singh, A., Dheri, G. S., Taggar, M. S., Nepovimova, E., Krejcar, O., & Kuca, K. (2021). Seed priming and coating by nano-scale zinc oxide particles improved vegetative growth, yield and quality of fodder maize (*Zea mays*). *Agronomy, 11*, 729.
104. Choudhary, R. C., Kumaraswamy, R. V., Kumari, S., Sharma, S. S., Pal, A., Raliya, R., Biswas, P., & Saharan, V. (2019). Zinc encapsulated chitosan nanoparticle to promote maize crop yield. *International Journal of Biological Macromolecules, 127*, 126.
105. Zhao, F., Xin, X., Cao, Y., Su, D., Ji, P., Zhu, Z., & He, Z. (2021). Use of carbon nanoparticles to improve soil fertility, crop growth and nutrient uptake by corn (*Zea mays* L.). *Nanomaterials, 11*, 2717.
106. Elshayb, O. M., Farroh, K. Y., Amin, H. E., & Atta, A. M. (2021). Green synthesis of zinc oxide nanoparticles: Fortification for rice grain yield and nutrients uptake enhancement. *Molecules, 26*, 584.
107. Itroutwar, P. D., Govindaraju, K., Tamilselvan, S., Kannan, M., Raja, K., & Subramanian, K. S. (2020). Seaweed-based biogenic ZnO nanoparticles for improving agro-morphological characteristics of rice (*Oryza sativa* L.). *Journal of Plant Growth Regulation, 39*, 717.
108. Bala, R., Kalia, A., & Dhaliwal, S. S. (2019). Evaluation of efficacy of ZnO nanoparticles as remedial zinc nanofertilizer for rice. *Journal of Soil Science and Plant Nutrition, 19*, 379.
109. Ikhajiagbe, B., Igiebor, F. A., & Ogwu, M. C. (2021). Growth and yield performances of rice (*Oryza sativa* var. nerica) after exposure to biosynthesized nanoparticles. *Bulletin of the National Research Centre, 45*, 62.
110. Zhou, H., Wu, H., Zhang, F., Su, Y., Guan, W., Xie, Y., Giraldo, J. P., & Shen, W. (2021). Molecular basis of cerium oxide nanoparticle enhancement of rice salt tolerance and yield. *Environmental Science: Nano, 8*, 3294.
111. Yusefi-Tanha, E., Fallah, S., Rostamnejadi, A., & Pokhrel, L. R. (2020). Zinc oxide nanoparticles (ZnONPs) as a novel nanofertilizer: Influence on seed yield and antioxidant defense system in soil grown soybean (*Glycine max* cv. Kowsar). *Science of the Total Environment, 738*, 140240.
112. Ngo, Q. B., Dao, T. H., Nguyen, H. C., Tran, X. T., Nguyen, T. V., Khuu, T. D., & Huynh, T. H. (2014). Effects of nanocrystalline powders (Fe, Co and Cu) on the germination, growth, crop yield and product quality of soybean (Vietnamese species DT-51). *Advances in Natural Sciences: Nanoscience and Nanotechnology, 5*, 015016.
113. Sadak, M. S. (2019). Impact of silver nanoparticles on plant growth, some biochemical aspects, and yield of fenugreek plant (*Trigonella foenum-graecum*). *Bulletin of the National Research Centre, 43*, 38.

114. Li, M., Liu, H., Dang, F., Hintelmann, H., Yin, B., & Zhou, D. (2020). Alteration of crop yield and quality of three vegetables upon exposure to silver nanoparticles in sludge-amended soil. *ACS Sustainable Chemistry & Engineering, 8*, 2472.
115. Baz, H., Creech, M., Chen, J., Gong, H., Bradford, K., & Huo, H. (2020). Water-soluble carbon nanoparticles improve seed germination and post-germination growth of lettuce under salinity stress. *Agronomy, 10*, 1192.
116. El-Saadony, M. T., Desoky, E. S. M., Saad, A. M., Eid, R. S. M., Selem, E., & Elrys, A. S. (2021). Biological silicon nanoparticles improve Phaseolus vulgaris L. yield and minimize its contaminant contents on a heavy metals-contaminated saline soil. *Journal of Environmental Sciences, 106*, 1.
117. Salehi, H., Chehregani, R. A., Raza, A., & Chen, J. T. (2021). Foliar application of CeO2 nanoparticles alters generative components fitness and seed productivity in bean crop (*Phaseolus vulgaris* L.). *Nanomaterials, 11*, 862.
118. Ahmed, B., Syed, A., Rizvi, A., Shahid, M., Bahkali, A. H., Khan, M. S., & Musarrat, J. (2020). Impact of metal-oxide nanoparticles on growth, physiology and yield of tomato (Solanum lycopersicum L.) modulated by *Azotobacter salinestris* strain ASM. *Environmental Pollution, 269*, 116218.
119. Hernández-Hernández, H., Quiterio-Gutiérrez, T., Cadenas-Pliego, G., Ortega-Ortiz, H., Hernández-Fuentes, A. D., Cabrera de la Fuente, M., Valdés-Reyna, J., & Juárez-Maldonado, A. (2019). Impact of selenium and copper nanoparticles on yield, antioxidant system, and fruit quality of tomato plants. *Plants, 8*, 355.
120. Kondal, R., Kalia, A., Krejcar, O., Kuca, K., Sharma, S. P., Luthra, K., Dheri, G. S., Vikal, Y., Taggar, M. S., Abd-Elsalam, K. A., & Gomes, C. L. (2021). Chitosan-urea nanocomposite for improved fertilizer applications: The effect on the soil enzymatic activities and microflora dynamics in N cycle of potatoes (*Solanum tuberosum* L.). *Polymers (Basel), 13*, 2887.
121. Mahmoud, A. W. M., Abdeldaym, E. A., Abdelaziz, S. M., El-Sawy, M. B. I., & Mottaleb, S. A. (2020). Synergetic effects of zinc, boron, silicon, and zeolite nanoparticles on confer tolerance in potato plants subjected to salinity. *Agronomy, 10*, 19.
122. Elshamy, M. T., El Khallal, S. M., Husseiny, S. M., & Farroh, K. Y. (2019). Application of nano-chitosan NPK fertilizer on growth and productivity of potato plant. *Journal of Scientific Research in Science, 36*, 424.
123. Kalia, A., & Kaur, H. (2019). Nanofertilizers: An innovation towards new generation fertilizers for improved nutrient-use efficacy and environmental sustainability. In *Nanoagroceuticals & nanophytochemicals*. CRC Press, Taylor & Francis.
124. Afrayeem, S. M., & Chaurasia, A. K. (2017). Effect of zinc oxide nanoparticles on seed germination and seed vigour in chilli (*Capsicum annuum* L.). *Journal of Pharmacognosy and Phytochemistry, 8*, 215.
125. Kumar, G. D., Raja, K., Natarajan, N., Govindaraju, K., & Subramanian, K. S. (2020). Invigouration treatment of metal and metal oxide nanoparticles for improving the seed quality of aged chilli seeds (*Capsicum annum* L.). *Materials Chemistry and Physics, 242*, 122492.
126. Raju, B. B., & Rai, P. K. (2017). Studies on effect of polymer seed coating, nanoparticles and hydro priming on seedlings characters of Pigeonpea (*Cajanus cajan* L.) seed. *Journal of Pharmacognosy and Phytochemistry, 6*, 140.
127. Rawat, P. S., Kumar, R., Ram, P., & Pandey, P. (2018). Effect of nanoparticles on wheat seed germination and seedling growth. *International Journal of Agricultural and Biosystems Engineering, 12*, 13.
128. Rai-Kalal, P., & Jajoo, A. (2021). Priming with zinc oxide nanoparticles improve germination and photosynthetic performance in wheat. *Plant Physiology and Biochemistry, 160*, 341.
129. Solanki, P., & Laura, J. S. (2018). Effect of ZnO nanoparticles on seed germination and seedling growth in wheat (*Triticum aestivum*). *Journal of Pharmacognosy and Phytochemistry, 7*, 2048.
130. Saharan, V., Kumaraswamy, R. V., Choudhary, R. C., Kumari, S., Pal, A., Raliya, R., & Biswas, P. (2016). Cu-chitosan nanoparticle mediated sustainable approach to enhance

seedling growth in maize by mobilizing reserved food. *Journal of Agricultural and Food Chemistry, 64*, 6148.

131. Shah, T., Latif, S., Saeed, F., Ali, I., Ullah, S., Alsahli, A. A., Jan, S., & Ahmad, P. (2021). Seed priming with titanium dioxide nanoparticles enhances seed vigor, leaf water status, and antioxidant enzyme activities in maize (*Zea mays* L.) under salinity stress. *Journal of King Saud University-Science, 33*, 101207.
132. Anand, K. V., Anugraga, A. R., Kannan, M., Singaravelu, G., & Govindaraju, K. (2020). Bio-engineered magnesium oxide nanoparticles as nano-priming agent for enhancing seed germination and seedling vigour of green gram (*Vigna radiata* L.). *Materials Letters, 271*, 127792.
133. Chakraborty, A., & Bordolui, S. K. (2021). Impact of seed priming with Ag-nanoparticle and GA3 on germination and vigour in green gram. *International Journal of Current Microbiology and Applied Sciences, 10*, 941.
134. Afzal, S., Sharma, D., & Singh, N. K. (2021). Eco-friendly synthesis of phytochemical-capped iron oxide nanoparticles as nano-priming agent for boosting seed germination in rice (*Oryza sativa* L.). *Environmental Science and Pollution Research, 28*, 40275.
135. Divya, K., Vijayan, S., Nair, S. J., & Jisha, M. S. (2019). Optimization of chitosan nanoparticle synthesis and its potential application as germination elicitor of *Oryza sativa* L. *International Journal of Biological Macromolecules, 24*, 1053.
136. Mahakham, W., Sarmah, A. K., Maensiri, S., & Theerakulpisut, P. (2017). Nanopriming technology for enhancing germination and starch metabolism of aged rice seeds using phytosynthesized silver nanoparticles. *Scientific Reports, 7*, 8263.
137. Khan, Z., & Ansari, M. Y. K. (2018). Impact of engineered Si nanoparticles on seed germination, vigour index and genotoxicity assessment via DNA damage of root tip cells in *Lens culinaris*. *Journal of Plant Biochemistry and Physiology, 6*, 3.
138. Hu, J., Wu, X., Wu, F., Chen, W., White, J. C., Yang, Y., Wang, B., Xing, B., Tao, S., & Wang, X. (2020). Potential application of titanium dioxide nanoparticles to improve nutritional quality of coriander (*Coriandrum sativum* L.). *Journal of Hazardous Materials, 389*, 121837.
139. Shang, Y., et al. (2019). Applications of nanotechnology in plant growth and crop protection: A review. *Molecules, 24*, 2558.
140. Yasmeen, F., Raja, N. I., Razzaq, A., & Komatsu, S. (2017). Proteomic and physiological analyses of wheat seeds exposed to copper and iron nanoparticles. *Biochimica et Biophysica Acta, Proteins and Proteomics, 1865*, 28.
141. Arora, S., Sharma, P., Kumar, S., Nayan, R., Khanna, P. K., & Zaidi, M. G. H. (2012). Gold-nanoparticle induced enhancement in growth and seed yield of *Brassica juncea*. *Plant Growth Regulation, 66*, 303.
142. Bradfield, S. J., Kumar, P., White, J. C., & Ebbs, S. D. (2017). Zinc, copper, or cerium accumulation from metal oxide nanoparticles or ions in sweet potato: Yield effects and projected dietary intake from consumption. *Plant Physiology and Biochemistry, 110*, 128.
143. Subbaiah, L. V., Prasad, T. N. V. K. V., Krishna, T. G., Sudhakar, P., Reddy, B. R., & Pradeep, T. (2016). Novel effects of nanoparticulate delivery of zinc on growth, productivity, and zinc biofortification in maize (Zea mays L.). *Journal of Agriculture and Food Chemistry, 64*, 3778.
144. Jyothi, T. V., & Hebsur, N. S. (2017). Effect of nanofertilizers on growth and yield of selected cereals-a review. *Agricultural Reviews, 38*, R-1663.
145. Taheri, M., Qarache, H. A., Qarache, A. A., & Yoosefi, M. (2015). The effects of zinc-oxide nanoparticles on growth parameters of corn (SC704). *STEM Fellowship Journal, 1*, 17.
146. Karvar, M., Azari, A., Rahimi, A., Maddah-Hosseini, S., & Ahamdi-Lahijani, M. J. (2022). Titanium dioxide nanoparticles (TiO2-NPs) enhance drought tolerance and grain yield of sweet corn (*Zea mays* L.) under deficit irrigation regimes. *Acta Physiologiae Plantarum, 44*, 14.

147. Ostrowski, A., Nordmeyer, D., Boreham, A., Holzhausen, C., Mundhenk, L., Graf, C., Meinke, M. C., Vogt, A., Hadam, S., Lademann, J., Ruhl, E., Alexiev, U., & Gruber, A. D. (2015). Overview about the localization of nanoparticles in tissue and cellular context by different imaging techniques. *Beilstein Journal of Nanotechnology, 6*, 263.
148. Lombi, E., Scheckel, K. G., & Kempson, I. M. (2011). In situ analysis of metal(loid)s in plants: State of the art and artefacts. *Environmental and Experimental Botany, 72*, 3.
149. Yan, A., & Chen, Z. (2018). Detection methods of nanoparticles in plant tissues. In *New visions in plant science*. IntechOpen.
150. Ivan, P., & Cristina, B. (2018). Nanoparticle uptake by plants: Beneficial or detrimental? In *Phytotoxicity of nanoparticles*. Springer.
151. Naderi, R., Rad, S. J., Yaraghi, A., Farhoodi, M., & Nemati, M. (2013). Detection of silver nanoparticles internalization into petunia (*Petunia hybrida*) isolated protoplasts. *Advances in Materials Research, 622–623*, 878.
152. Deng, Y., Petersen, E. J., Challis, K. E., Rabb, S. A., Holbrook, R. D., Ranville, J. F., Nelson, B. C., & Xing, B. (2017). Multiple method analysis of TiO2 nanoparticle uptake in rice (*Oryza sativa* L.) plants. *Environmental Science and Technology, 51*, 10615.
153. Fan, R., Huang, Y. C., Grusak, M. A., Huang, C. P., & Sherrier, D. J. (2014). Effects of nano-TiO2 on the agronomically-relevant Rhizobium-legume symbiosis. *Science of the Total Environment, 466–467*, 503.
154. Kempen, P. J., Thakor, A. S., Zavaleta, C., Gambhir, S. S., & Sinclair, R. (2013). A scanning transmission electron microscopy approach to analyzing large volumes of tissue to detect nanoparticles. *Microscopy and Microanalysis, 19*, 1290.
155. Ponce, A., Mejía-Rosales, S., & José-Yacamán, M. (2012). Scanning transmission electron microscopy methods for the analysis of nanoparticles. In M. Soloviev (Ed.), *Nanoparticles in biology and medicine: Methods and protocols. Methods in molecular biology*. Springer.
156. Klein, N. D., Hurley, K. R., Feng, Z. V., & Haynes, C. L. (2015). Dark field transmission electron microscopy as a tool for identifying inorganic nanoparticles in biological matrices. *Analytical Chemistry, 87*, 4356.
157. Avellan, A., Schwab, F., Masion, A., Chaurand, P., Borschneck, D., Vidal, V., Rose, J., Santaella, C., & Levard, C. (2017). Nanoparticle uptake in plants: Gold nanomaterial localized in roots of Arabidopsis thaliana by X-ray computed nanotomography and hyperspectral imaging. *Environmental Science & Technology, 51*, 8682.
158. López-Moreno, M. L., de la Rosa, G., Hernández-Viezcas, J. A., Peralta-Videa, J. R., & Gardea-Torresdey, J. L. (2010). X-ray absorption spectroscopy (XAS) corroboration of the uptake and storage of CeO2 nanoparticles and assessment of their differential toxicity in four edible plant species. *Journal of Agriculture and Food Chemistry, 58*, 3689.
159. Servin, A. D., Castillo-Michel, H., Hernandez-Viezcas, J. A., Diaz, B. C., Peralta-Videa, J. R., & Gardea-Torresdey, J. L. (2012). Synchrotron micro-XRF and micro-XANES confirmation of the uptake and translocation of TiO_2 nanoparticles in cucumber (*Cucumis sativus*) plants. *Environmental Science & Technology, 46*, 7637.

The manufacturer's authorised representative in the EU is Springer Nature Customer Service Centre GmbH, Europaplatz 3, 69115 Heidelberg, Germany. If you have any concerns regarding our products, please contact ProductSafety@springernature.com

Printed and bound by CPI Group (UK) Ltd, Croydon, CR0 4YY
30/09/2026
02225299-0007